Univariate and Multivariate General Linear Models

Theory and Applications with SAS

Second Edition

STATISTICS: Textbooks and Monographs

Recent Titles

Handbook of Stochastic Analysis and Applications, *edited by D. Kannan and V. Lakshmikantham*

Testing for Normality, *Henry C. Thode, Jr.*

Handbook of Applied Econometrics and Statistical Inference, *edited by Aman Ullah, Alan T. K. Wan, and Anoop Chaturvedi*

Visualizing Statistical Models and Concepts, *R. W. Farebrother and Michaël Schyns*

Financial and Actuarial Statistics: An Introduction, *Dale S. Borowiak*

Nonparametric Statistical Inference, Fourth Edition, Revised and Expanded, *Jean Dickinson Gibbons and Subhabrata Chakraborti*

Computer-Aided Econometrics, *edited by David E.A. Giles*

The EM Algorithm and Related Statistical Models, *edited by Michiko Watanabe and Kazunori Yamaguchi*

Multivariate Statistical Analysis, Second Edition, Revised and Expanded, *Narayan C. Giri*

Computational Methods in Statistics and Econometrics, *Hisashi Tanizaki*

Applied Sequential Methodologies: Real-World Examples with Data Analysis, *edited by Nitis Mukhopadhyay, Sujay Datta, and Saibal Chattopadhyay*

Handbook of Beta Distribution and Its Applications, *edited by Arjun K. Gupta and Saralees Nadarajah*

Item Response Theory: Parameter Estimation Techniques, Second Edition, *edited by Frank B. Baker and Seock-Ho Kim*

Statistical Methods in Computer Security, *edited by William W. S. Chen*

Elementary Statistical Quality Control, Second Edition, *John T. Burr*

Data Analysis of Asymmetric Structures, *Takayuki Saito and Hiroshi Yadohisa*

Mathematical Statistics with Applications, *Asha Seth Kapadia, Wenyaw Chan, and Lemuel Moyé*

Advances on Models, Characterizations and Applications, *N. Balakrishnan, I. G. Bairamov, and O. L. Gebizlioglu*

Survey Sampling: Theory and Methods, Second Edition, *Arijit Chaudhuri and Horst Stenger*

Statistical Design of Experiments with Engineering Applications, *Kamel Rekab and Muzaffar Shaikh*

Quality by Experimental Design, Third Edition, *Thomas B. Barker*

Handbook of Parallel Computing and Statistics, *Erricos John Kontoghiorghes*

Statistical Inference Based on Divergence Measures, *Leandro Pardo*

A Kalman Filter Primer, *Randy Eubank*

Introductory Statistical Inference, *Nitis Mukhopadhyay*

Handbook of Statistical Distributions with Applications, *K. Krishnamoorthy*

A Course on Queueing Models, *Joti Lal Jain, Sri Gopal Mohanty, and Walter Böhm*

Univariate and Multivariate General Linear Models: Theory and Applications with SAS, Second Edition, *Kevin Kim and Neil Timm*

Univariate and Multivariate General Linear Models

Theory and Applications with SAS

Second Edition

Kevin Kim
University of Pittsburgh
Pennsylvania, U.S.A.

Neil Timm
University of Pittsburgh
Pennsylvania, U.S.A.

Chapman & Hall/CRC
Taylor & Francis Group
Boca Raton   London   New York

Chapman & Hall/CRC is an imprint of the
Taylor & Francis Group, an informa business

Chapman & Hall/CRC
Taylor & Francis Group
6000 Broken Sound Parkway NW, Suite 300
Boca Raton, FL 33487-2742

© 2007 by Taylor & Francis Group, LLC
Chapman & Hall/CRC is an imprint of Taylor & Francis Group, an Informa business

No claim to original U.S. Government works
Printed in the United States of America on acid-free paper
10 9 8 7 6 5 4 3 2 1

International Standard Book Number-10: 1-58488-634-X (Hardcover)
International Standard Book Number-13: 978-1-58488-634-1 (Hardcover)

Library of Congress Cataloging-in-Publication Data

Kim, Kevin.
      Univariate and multivariate general linear models : theory and applications with
SAS. -- 2nd ed. / Kevin Kim and Neil Timm.
            p. cm. -- (Statistics : textbooks and monographs)
      Rev. ed. of: Univariate & multivariate general linear models / Neil H. Timm, Tammy
A. Mieczkowski. c1997.
      Includes bibliographical references and index.
      ISBN-13: 978-1-58488-634-1 (acid-free paper)
      ISBN-10: 1-58488-634-X (acid-free paper)
      1. Linear models (Statistics)--Textbooks. 2. Linear models (Statistics)--Data
processing--Textbooks. 3. SAS (Computer file) I. Timm, Neil H. II. Timm, Neil H.
Univariate & multivariate general linear models. III. Title. IV. Series.

QA279.T56 2007
519.5'35--dc22   2006026561

Visit the Taylor & Francis Web site at
http://www.taylorandfrancis.com

and the CRC Press Web site at
http://www.crcpress.com

Contents

List of Tables

Preface

The general linear model is often first introduced to graduate students during a course on multiple linear regression, analysis of variance, or experimental design; however, most students do not fully understand the generality of the model until they have taken several courses in applied statistics. Even students in graduate statistics programs do not fully appreciate the generality of the model until well into their program of study. This is due in part to the fact that theory and applications of the general linear model are discussed in discrete segments throughout the course of study rather than within a more general framework. In this book, we have tried to solve this problem by reviewing the theory of the general linear model using a general framework. Additionally, we use this general framework to present analyses of simple and complex models, both univariate and multivariate, using data sets from the social and behavioral sciences and other disciplines.

AUDIENCE

The book is written for advanced graduate students in the social and behavioral sciences and in applied statistics who are interested in statistical analysis using the general linear model. The book may be used to introduce students to the general linear model; at the University of Pittsburgh it is used in a one-semester course on linear models. The book may also be used as a supplement to courses in applied statistical methods covering the essentials of estimation theory and hypothesis testing, simple linear regression, and analysis of variance. They should also have some familiarity with matrix algebra and with running SAS procedures.

OVERVIEW

Each chapter of this book is divided into two sections: theory and applications. Standard SAS procedures are used to perform most of the analyses. When standard SAS procedures are not available, PROC IML code to perform the analysis is discussed. Because SAS is not widely used in the social and behavioral sciences, SAS code for analyzing general linear model applications is discussed in detail. The code can be used as a template.

Chapter 1 provides an overview of the general linear model using matrix algebra and an introduction to the multivariate normal distribution as well as to the general theory of hypothesis testing. Applications include the use of graphical methods to evaluate univariate and multivariate normality and the use of transformations to normality. In Chapter 2 the general linear model without restrictions is introduced and used to analyze multiple regression and ANOVA designs. In Chapter 3 the general linear model with restrictions is discussed and used to analyze ANCOVA designs and repeated measurement designs.

Chapter 4 extends the concepts of the first three chapters to general linear models with heteroscedastic errors and illustrates how the model may be used to perform weighted least squares regression and to analyze categorical data. Chapter 5 extends the theory of Chapter 2 to the multivariate case; applications include multivariate regression analysis, MANOVA, MANCOVA, and analyses of repeated measurement data. This chapter also extends "standard" hypothesis testing to extended linear hypotheses. In Chapter 6, the double multivariate linear model is discussed.

Chapter 7 extends the multivariate linear model to include restrictions and considers the growth curve model. In Chapter 8, the seeming unrelated regression (SUR) and the restricted GMANOVA models are analyzed. Many of the applications in this chapter involve PROC IML code. Finally, Chapter 9 includes analyses of hierarchical linear models, and Chapter 10 treats the analysis of incomplete repeated measurement data.

While the coverage given the general linear model is extensive, it is not exhaustive. Excluded from the book are Bayesian methods, nonparametric procedures, nonlinear models, and generalized linear models, among others.

ACKNOWLEDGMENTS

We would like to thank the reviewers at SAS Institute Inc. and the technical reviewer Vernon M. Chinchilli for their helpful comments and suggestions on the book. We thank Hanna Hicks Schoenrock and Caroline Brickley for making the process of producing this book run so smoothly. We also appreciate the helpful suggestions made on an early draft by doctoral students in the linear models course.

We would especially like to extend our gratitude to Roberta S. Allen. Ms. Allen expertly typed every draft of the book from inception through every revision, including all equations. Thank you for your excellent work and patience with us, Roberta. The authors also want to thank the authors and publishers of copyrighted material for permission to reproduce tables and data sets used in the book.

This book was completed while Neil H. Timm was on sabbatical leave during the 1996-1997 academic year. He would like to thank his colleagues for their support and the School of Education for this opportunity. He would also like to thank his wife, Verena, for her support and encouragement.

Tammy A. Mieczkowski was a full time doctoral student in the Research Methodology program in the School of Education when this book was completed. She also works as a Graduate Student Researcher at the University of Pittsburgh School of Medicine, Department of Family Medicine and Clinical Epidemiology. She would like to thank her family, especially her sister and friend, Karen, for their support, encouragement, understanding, and love.

Neil H. Timm
Tammy A. Mieczkowski

SECOND EDITION

For the second edition, we note that the authorship has changed. The former second author has removed herself from the second edition and a new author, Kevin H. Kim, appears. The order of the authors is alphabetical.

The revision includes corrections to the first edition, expanded material, additional examples, and new material. The theory in Chapters 2, 5, and 8 has been expanded to include recent developments; Chapters 9 and 10 have been rewritten and expanded to include recent developments in structural equation modeling (SEM), Growth Mixture Modeling, Longitudinal Data Analysis, and Hierarchical Linear Models (HLM), Chapters 11, 12, and 13. The material in Chapter 12 on missing data has been expanded to include multiple imputation and the EM algorithm. Also included in the second edition is the addition of Chapter 9 and 10 on Finite Intersection Tests and Power Analysis which illustrates the experimental GLMPOWER procedure. All examples have been revised to include options available using SAS version 9.0.

The second addition also includes applications of the MI, MIANALYZE, TRANSREG, and CALIS procedures. In addition, use of ODS capabilities are illustrated in the examples and functions not known to all users of SAS such as the PROBMC distribution are illustrated. While we still include some output from the SAS procedures in the book, the output has been abbreviated. SAS code and data files have been excluded from the book to save space and are available at the Pitt site www.pitt.edu/~timm.

Kevin H. Kim
Neil H. Timm

CHAPTER 1

Overview of General Linear Model

1.1 INTRODUCTION

In this chapter, we introduce the structure of the general linear model (GLM) and use the structure to classify the linear models discussed in this book. The multivariate normal distribution which forms the basis for most of the hypothesis testing theory of the linear model is reviewed, along with a general approach to hypothesis testing. Graphical methods and tests for assessing univariate and multivariate normality are also reviewed. The generation of multivariate normal data, the construction of Quantile-Quantile (Q-Q) plots, chi-square plots, scatter plots, and data transformation procedures are reviewed and illustrated to evaluate normality.

1.2 GENERAL LINEAR MODEL

Data analysis in the social and behavioral sciences and numerous other disciplines is associated with a model known as the GLM. Employing matrix notation, univariate and multivariate linear models may be represented using the general form

$$\Omega_0 : y = X\beta + e \qquad (1.1)$$

where $y_{n \times 1}$ is a vector of n observations, $X_{n \times k}$ is a known design matrix of full column rank k, $\beta_{k \times 1}$ is a vector of k fixed parameters, $e_{n \times 1}$ is a random vector of errors with mean zero, $\mathcal{E}(e) = 0$, and covariance matrix $\Omega = \text{cov}(e)$. If the design matrix is not of full rank, one may reparameterize the model to create an equivalent model of full rank. In this book, we systematically discuss the GLM specified by (1.1) with various structures for X and Ω.

Depending on the structure of X and Ω, the model in (1.1) has many names in the literature. To illustrate, if $\Omega = \sigma^2 I_n$ in (1.1), the model is called the classical linear regression model or the standard linear regression model. If we partition X to have the form $X = (X_1, X_2)$ where X_1 is associated with fixed effects and X_2 is associated with random effects, and if covariance matrix Ω has the form

$$\Omega = X_2 V X_2' + \Psi \qquad (1.2)$$

1

where V and Ψ are covariance matrices, then (1.1) becomes the general linear mixed model (GLMM). If we let X and Ω take the general form

$$X = \begin{pmatrix} X_1 & 0 & \cdots & 1 \\ 0 & X_2 & \cdots & 0 \\ \vdots & \vdots & \ddots & \vdots \\ 0 & 0 & \cdots & X_p \end{pmatrix} = \oplus_{i=1}^p X_i \tag{1.3}$$

$$\Omega = \Sigma \otimes I_n \tag{1.4}$$

where $\Sigma_{p \times p}$ is a covariance matrix, $A \otimes B$ denotes the Kronecker product of two matrices A and B ($A\otimes = a_{ij}B$), and $\oplus_{i=1}^p$ represents the direct sum of the matrices X_i, then (1.1) is Zellner's seemingly unrelated regression (SUR) model or a multiple design multivariate (MDM) model. The SUR model may also be formulated as p separate linear regression models that are not independent

$$y_i = X_i \beta_{ii} + e_i \tag{1.5}$$

$$\text{cov}(y_i, y_j) = \sigma_{ij} I_n \tag{1.6}$$

for $i, j = 1, 2, \ldots, p$ where y, β and e in (1.1) are partitioned

$$y' = \begin{pmatrix} y'_1 & y'_2 & \cdots & y'_p \end{pmatrix} \qquad \text{where } y_i : n^* \times 1 \tag{1.7}$$

$$\beta' = \begin{pmatrix} \beta'_1 & \beta'_2 & \cdots & \beta'_p \end{pmatrix} \qquad \text{where } \beta_{ii} : k_i \times 1 \tag{1.8}$$

$$e' = \begin{pmatrix} e'_1 & e'_2 & \cdots & e'_p \end{pmatrix} \qquad \text{where } e_i : n^* \times 1 \tag{1.9}$$

and $\Sigma_{p \times p} = (\sigma_{ij})$. Alternatively, we may express the SUR model as a restricted multivariate regression model. To do this, we write

$$\Omega_0 : Y_{n^* \times p} = X_{n^* \times k} \tilde{\beta}_{k \times p} + U_{n^* \times p} \tag{1.10}$$

where $Y = (y_1, y_2, \ldots, y_p)$, $X = (X_1, X_2, \ldots, X_p)$, $U = (e_1, e_2, \ldots, e_p)$ and

$$\tilde{\beta} = \begin{pmatrix} \beta_{11} & 0 & \cdots & 0 \\ 0 & \beta_{22} & \cdots & 0 \\ \vdots & \vdots & \ddots & \vdots \\ 0 & 0 & \cdots & \beta_{pp} \end{pmatrix}. \tag{1.11}$$

Letting $X_1 = X_2 = \cdots = X_p = \tilde{X}_{n^* \times k}$ and $\tilde{\beta} = (\beta_{11}, \beta_{22}, \ldots, \beta_{pp})$ in the SUR model, (1.1) becomes the classical multivariate regression model or the multivariate analysis of variance (MANOVA) model. Finally, letting

$$X = X_1 \otimes X'_2 \tag{1.12}$$

$$\Omega = I_n \otimes \Sigma \tag{1.13}$$

model (1.1) becomes the generalized MANOVA (GMANOVA) or the generalized growth curve model. All these models with some further extensions are special forms of the GLM discussed in this book.

The model in (1.1) is termed the "classical" model since its orientation is subjects or observations by variables where the number of variables is one. An alternative orientation for the model is to assume $y' = (y_1, y_2, \ldots, y_n)$ is a $(1 \times n)$ vector of observations where the number of variables is one. For each observation y_i, we may assume that there are $x_i' = (x_1, x_2, \ldots, x_k)$ or k independent (possible dummy) variables. With this orientation (1.1) becomes

$$y = X'\beta + e \tag{1.14}$$

where $X = (x_1, x_2, \ldots, x_k)$, $e' = (e_1, e_2, \ldots, e_n)$ and each x_i contains k independent variables for the ith observation. Model (1.14) is often called the "response-wise" form. Model (1.1) is clearly equivalent to (1.14) since the design matrix has the same order for either representation; however, in (1.14) X is of order $k \times n$. Thus, $X'X$ using (1.14) becomes XX' for the responsewise form of the classical model.

The simplest example of the GLM is the simple linear regression model

$$y = \beta_0 + \beta_1 x + e \tag{1.15}$$

where x represents the independent variable, y the dependent variable and e a random error. Model (1.15) states that the observed dependent variable for each subject is hypothesized to be a function of a common parameter β_0 for all subjects and an independent variable x for each subject that is related to the dependent variable by a weighting (i.e., regression) coefficient β_1 plus a random error e. For $k = p + 1$ with p variables, (1.15) becomes (1.16)

$$y = \beta_0 + \beta_1 x_1 + \beta_2 x_2 + \cdots + \beta_p x_p + e \tag{1.16}$$

or using matrix notation, (1.16) is written as

$$y = x'\beta + e \tag{1.17}$$

where $x' = (x_0, x_1, \ldots, x_p)$ denotes k independent variables, and x_0 is a dummy variable in the vector x'. Then for a sample of n observations, (1.17) has the general form (1.14) where $y' = (y_1, y_2, \ldots, y_n)$, $e' = (e_1, e_2, \ldots, e_n)$ and $X = (x_1, x_2, \ldots, x_n)$ of order $k \times n$ since each column vector x_i in X contains k variables. When using the classical form (1.1), $X \equiv X'$, a matrix of order $n \times k$. In discussions of the GLM, many authors will use either the classical or the response-wise version of the GLM, while we will in general prefer (1.1). In some applications (e.g., repeated measurement designs) form (1.14) is preferred.

1.3 RESTRICTED GENERAL LINEAR MODEL

In specifying the GLM using (1.1) or (1.14), we have not restricted the k-variate parameter vector β. A linear restriction on the parameter vector β will affect the characterization of the model. Sometimes it is necessary to add restrictions to the GLM of the general form

$$R\beta = \theta \tag{1.18}$$

where $R_{s \times k}$ is a known matrix with full row rank, $\text{rank}(R) = s$, and θ is a known parameter vector, often assumed to be zero. With (1.18) associated with the GLM, the model is commonly called the restricted general linear model (RGLM). Returning to (1.15), we offer an example of this in the simple linear regression model with a restriction

$$y = \beta_0 + \beta_1 x + e \qquad \beta_0 = 0 \qquad (1.19)$$

so that the regression of y on x is through the origin. For this situation, $R \equiv (1, 0)$ and $\theta = (0, 0)$. Clearly, the estimate of β_1 using (1.19) will differ from that obtained using the general linear model (1.15) without the restriction.

Assuming (1.1) or (1.16), one first wants to estimate β with $\hat{\beta}$ where the estimator $\hat{\beta}$ has some optimal properties like unbiasness and minimal variance. Adding (1.18) to the GLM, one obtains a restricted estimator of β, $\hat{\beta}_r$, which in general is not equal to the unrestricted estimator. Having estimated β, one may next want to test hypotheses regarding the parameter vector β and the structure of Ω. The general form of the null hypothesis regarding β is

$$H : C\beta = \xi \qquad (1.20)$$

where $C_{g \times k}$ is a matrix of full row rank g, $\text{rank}(C) = g$ and $\xi_{g \times 1}$ is a vector of known parameters, usually equal to zero. The hypothesis in (1.20) may be tested using the GLM with or without the restriction given in (1.18). Hypotheses in the form (1.20) are in general testable, provided β is estimable; however, testing (1.20) assuming (1.18) is more complicated since the matrix C may not contain a row identical, inconsistent or dependent on the rows of R and the rows of C must remain independent. Thus, the rank of the augmented matrix must be greater than s, $\text{rank}\binom{R}{C} = s + g > s$.

Returning to (1.15), we may test the null hypotheses

$$H : \beta = \binom{\beta_0}{\beta_1} = \xi = \binom{\xi_0}{\xi_1} \qquad (1.21)$$

where ξ is a known parameter vector. The hypothesis in (1.21) may not be inconsistent with the restriction $\beta_0 = 0$. Thus, given the restriction, we may test

$$H : \beta_1 = \xi_1 \qquad (1.22)$$

so that (1.22) is not inconsistent or dependent on the restriction.

1.4 MULTIVARIATE NORMAL DISTRIBUTION

To test hypotheses of the form given in (1.20), one usually makes distributional assumptions regarding the observation vector y or e, namely the assumption of multivariate normality. To define the multivariate normal distribution, recall that the definition of a standard normal random variable y is defined by the density

$$f(y) = (2\pi)^{-1/2} \exp\left(-\frac{1}{2}y^2\right) \qquad (1.23)$$

denoted by $y \sim N(0, 1)$. A random variable y has a normal distribution with mean μ and variance $\sigma^2 > 0$ if y has the same distribution as the random variable

$$\mu + \sigma e \tag{1.24}$$

where $e \sim N(0, 1)$. The density for y is given by

$$f(y) = \frac{1}{\sigma\sqrt{2\pi}} \exp\left[-\frac{1}{2}\frac{(y-\mu)^2}{\sigma^2}\right]$$
$$= (2\pi\sigma^2)^{-1/2} \exp\left[-\frac{1}{2}\frac{(y-\mu)^2}{\sigma^2}\right] \tag{1.25}$$

with this as motivation, the definition for a multivariate normal distribution is as follows.

Definition 1.1. A p-dimensional random vector y is said to have a multivariate normal distribution with mean μ and covariance matrix Σ [$y \sim N_p(\mu, \Sigma)$] if y has the same distribution as $\mu + Fe$ where $F_{p \times p}$ is a matrix of rank p, $\Sigma = FF'$ and each element of e is distributed: $e_i \sim N(0, 1)$. The density of y is given by

$$f(y) = (2\pi)^{-p/2}|\Sigma|^{-1/2} \exp\left[-\frac{1}{2}(y-\mu)'\Sigma^{-1}(y-\mu)\right]. \tag{1.26}$$

Letting $(\mathrm{diag}\Sigma)^{-1/2}$ represent the diagonal matrix with diagonal elements equal to the square root of the diagonal elements of Σ, the population correlation matrix for the elements of the vector y is

$$P = (\mathrm{diag}\Sigma)^{-1/2}\Sigma(\mathrm{diag}\Sigma)^{-1/2} = \left(\frac{\sigma_{ij}}{\sigma_{ii}^{1/2}\sigma_{jj}^{1/2}}\right) = (\rho_{ij}). \tag{1.27}$$

If $y \sim N_p(\mu, \Sigma)$ and $w = F^{-1}(y-\mu)$, then the quadratic form $(y-\mu)'\Sigma^{-1}(y-\mu)$ has a chi-square distribution with p degrees of freedom, written as

$$w'w = (y-\mu)'\Sigma^{-1}(y-\mu) \sim \chi_p^2. \tag{1.28}$$

The quantity $\left[(y-\mu)'\Sigma^{-1}(y-\mu)\right]^{1/2}$ is called the Mahalanobis distance between y and μ.

For a random sample of n independent p-vectors $(y_1, y_2, \ldots, y_n)$ from a multivariate normal distribution, $y_i \sim IN_p(\mu, \Sigma)$, we shall in general write the data matrix $Y_{n \times p}$ in the classical form

$$Y_{n \times p} = \begin{pmatrix} y_1' \\ y_2' \\ \vdots \\ y_n' \end{pmatrix} = \begin{pmatrix} y_{11} & y_{12} & \cdots & y_{1p} \\ y_{21} & y_{22} & \cdots & y_{2p} \\ \vdots & \vdots & \ddots & \vdots \\ y_{n1} & y_{n2} & \cdots & y_{np} \end{pmatrix}. \tag{1.29}$$

The corresponding responsewise representation for Y is

$$Y_{p \times n} = \begin{pmatrix} y_1 & y_2 & \cdots & y_n \end{pmatrix} = \begin{pmatrix} y_{11} & y_{12} & \cdots & y_{1n} \\ y_{21} & y_{22} & \cdots & y_{2n} \\ \vdots & \vdots & \ddots & \vdots \\ y_{p1} & y_{p2} & \cdots & y_{pn} \end{pmatrix}. \tag{1.30}$$

The joint probability density function (pdf) for $(y_1, y_2, \ldots, y_n)$ or the likelihood function is

$$L = L(\mu, \Sigma | y) = \prod_{i=1}^{n} f(y_i). \tag{1.31}$$

Substituting $f(y)$ in (1.26) for each $f(y_i)$, the pdf for the multivariate normal distribution is

$$[(2\pi)^p |\Sigma|]^{-n/2} \exp\left[-\frac{1}{2} \sum_{i=1}^{n} (y_i - \mu)' \Sigma^{-1} (y_i - \mu) \right]. \tag{1.32}$$

Using the property of the trace of a matrix, $\operatorname{tr}(x'Ax) = \operatorname{tr}(Axx')$, (1.32) may be written as

$$[(2\pi)^p |\Sigma|]^{-n/2} \operatorname{etr} \left\{ -\frac{1}{2} \Sigma^{-1} \left[\sum_{i=1}^{n} (y_i - \mu)(y_i - \mu)' \right] \right\} \tag{1.33}$$

where etr stands for the exponential of a trace of a matrix.
   If we let the sample mean be represented by

$$\bar{y} = n^{-1} \sum_{i=1}^{n} y_i \tag{1.34}$$

and the sum of squares and products (SSP) matrix, using the classical form (1.29), is

$$E = \sum_{i=1}^{n} (y_i - \bar{y})(y_i - \bar{y})' = Y'Y - n\bar{y}\bar{y}' \tag{1.35}$$

or using the responsewise form (1.30), the SSP matrix is

$$E = YY' - n\bar{y}\bar{y}'. \tag{1.36}$$

In either case, we may write (1.33) as

$$[(2\pi)^p |\Sigma|]^{-n/2} \operatorname{etr} \left\{ -\frac{1}{2} \Sigma^{-1} \left[E + n(y_i - \mu)(y_i - \mu)' \right] \right\} \tag{1.37}$$

so that by Neyman's factorization criterion $(E, \bar{y})$ are sufficient statistics for estimating (Σ, μ), (Lehmann, 1991, p. 16).

Theorem 1.1. *Let $y_i \sim IN_p(\mu, \Sigma)$ be a sample of size n, then $\bar{y}$ and E are sufficient statistics for μ and Σ.*

It can also be shown that $\bar{y}$ and E are independently distributed. The distribution of E is known as the Wishart distribution, a multivariate generalization of the chi-square distribution, with $\nu = n - 1$ degrees of freedom. The density of the Wishart distribution is

$$c|\Sigma|^{-\nu/2}|E|^{(\nu-p-1)/2}\mathrm{etr}\left(-\frac{1}{2}\Sigma^{-1}E\right) \tag{1.38}$$

where c is an appropriately chosen constant so that the total probability is equal to one. We write that $E \sim W_P(\nu, \Sigma)$. The expectation of E is $\nu\Sigma$.

Given a random sample of observations from a multivariate normal distribution, we usually estimate the parameters μ and Σ.

Theorem 1.2. *Let $y_i \sim IN_p(\mu, \Sigma)$, then the maximum likelihood estimators (MLEs) of μ and Σ are $\bar{y}$ and $E/n = \widehat{\Sigma}$.*

Furthermore, $\bar{y}$ and

$$S = \left(\frac{n}{n-1}\right)\widehat{\Sigma} = \frac{E}{n-1} \tag{1.39}$$

are unbiased estimators of μ and Σ, so that $\mathcal{E}(\bar{y}) = \mu$ and $\mathcal{E}(S) = \Sigma$. Hence, the sample distributions of S is Wishart, $(n-1)S \sim W_p(\nu = n-1, \Sigma)$ or $S = (s_{ij}) \sim W_p[1, \Sigma/(n-1)]$. Since S is proportional to the MLE $\widehat{\Sigma}$ of Σ, the MLE of the population correlation coefficient matrix is

$$R = (\mathrm{diag}S)^{-1/2}S(\mathrm{diag}S)^{-1/2} = \left(\frac{s_{ij}}{s_{ii}^{1/2}s_{jj}^{1/2}}\right) = (r_{ij}), \tag{1.40}$$

where r_{ij} is the sample correlation coefficient.

For a random sample of n independent identically distributed (iid) p-vectors $(y_1, y_2, \ldots, y_n)$ from any distribution with mean μ and covariance matrix Σ, by the central limit theorem (CLT), the pdf for the random variable $z = \sqrt{n}(\bar{y} - \mu)$ converges in distribution to a multivariate normal distribution with mean 0 and covariance matrix Σ,

$$z = \sqrt{n}(\bar{y} - \mu) \xrightarrow{d} N_p(0, \Sigma). \tag{1.41}$$

And, the quadratic form,

$$n(\bar{y} - \mu)'\Sigma^{-1}(\bar{y} - \mu) = z'z \xrightarrow{d} \chi_p^2, \tag{1.42}$$

converges in distribution to a chi-square distribution with p degrees of freedom. The quantity $\left[(\bar{y} - \mu)'\Sigma^{-1}(\bar{y} - \mu)\right]^{1/2}$ is the Mahalanobis distance from $\bar{y}$ to μ.

The mean μ and covariance matrix Σ are the first two moments of a random vector y. We now extend the classical measures of skewness and kurtosis, $\mathcal{E}(y - \mu)^3/\sigma^3$ and the $\mathcal{E}(y - \mu)^4/\sigma^4$, to the multivariate case following Mardia (1970). Letting $y_i \sim (\mu, \Sigma)$, Mardia's sample measures of multivariate skewness $(\beta_{1,p})$ and kurtosis $(\beta_{2,p})$ are based upon the scaled random variables, $z_i = S^{-1/2}(y_i - \bar{y})$, $i = 1, \ldots, n$. Mardia's measures of the population values are respectively,

$$b_{1,p} = \frac{1}{n^2} \sum_{i,j=1}^{n} \left[(y_i - \bar{y})' S^{-1}(y_i - \bar{y}) \right]^3 \qquad (1.43)$$

$$b_{2,p} = \frac{1}{n} \sum_{i,j=1}^{n} \left[(y_i - \bar{y})' S^{-1}(y_i - \bar{y}) \right]^2. \qquad (1.44)$$

When $y_i \sim IN_p(\mu, \Sigma)$, the population values of the moments are $\beta_{1,p} = 0$ and $\beta_{2,p} = p(p + 2)$. Under normality, Mardia showed that the statistic $X^2 = nb_{1,p}/6$ converges to a chi-square distribution with $\nu = p(p + 1)(p + 2)/6$ degrees of freedom. And, that the multivariate kurtosis statistic converges to a normal distribution with mean $\mu = p(p + 2)$ and variance $\sigma^2 = 8p(p + 2)/n$. When $n > 50$, one may use the test statistics to evaluate multivariate normality. Rejection of normality indicates either the presence of outliers or that the distribution is significantly different from a multivariate normal distribution. Romeu and Ozturk (1993) performed a comparative study of goodness-of-fit tests for multivariate normality and showed that Mardia's tests are most stable and reliable. They also calculated small sample empirical critical values for the tests.

When one finds that a distribution is not multivariate normal, one usually replaces the original observations with some linear combination of variables which may be more nearly normal. Alternatively, one may transform each variable in the vector using a Box-Cox power transformation, as outlined for example by Bilodeau and Brenner (1999, p. 95). However, because marginal normality does not ensure ensure multivariate normality a joint transformation may be desired. Shapiro and Wilk (1965) W statistic or Royston (1982, 1992) approximation may be used to test for univariate normality one variable at a time. Since marginal normality does not ensure multivariate normality, a multivariate test must be used to evaluate joint normality for a set of p variables.

1.5   ELEMENTARY PROPERTIES OF NORMAL RANDOM VARIABLES

Theorem 1.3. *Let* $y \sim IN_p(\mu, \Sigma)$ *and* $w = A_{m \times p} y$, *then* $w \sim IN_m(A\mu, A\Sigma A')$.

Thus, linear combinations of multivariate normal random variables are again normally distributed. If one assumes that the random variable y is multivariate normal, $y \sim N_n(X\beta = \mu, \Omega = \sigma^2 I)$, and $\hat{\beta}$ is an unbiased estimate of β such that $\hat{\beta} = (X'X)^{-1}X'y$ then by Theorem 1.3,

$$\hat{\beta} \sim N_k \left[\beta, \text{cov} \left(\hat{\beta} \right) \right]. \qquad (1.45)$$

Given that a random vector y is multivariate normal, one may partition the p elements of a random vector y into two subvectors y_1 and y_2 where the number of elements $p = p_1 + p_2$. The joint distribution of the partitioned vector is multivariate normal and written as

$$y = \begin{pmatrix} y_1 \\ y_2 \end{pmatrix} \sim N_p \left[\begin{pmatrix} \mu_1 \\ \mu_2 \end{pmatrix}, \begin{pmatrix} \Sigma_{11} & \Sigma_{12} \\ \Sigma_{21} & \Sigma_{22} \end{pmatrix} \right]. \tag{1.46}$$

Given the partition of the vector y, one is often interested in the conditional distribution of y_1 given the subset y_2, $y_1|y_2$. Provided Σ_{22} is nonsingular, we have the following general result.

Theorem 1.4. $y_1|y_2 = z \sim N_{p_1}(\mu_{1 \cdot 2}, \Sigma_{11 \cdot 2})$, where $\mu_{1 \cdot 2} = \mu_1 + \Sigma_{12}\Sigma_{22}^{-1}(y_2 - \mu_2)$ and $\Sigma_{11 \cdot 2} = \Sigma_{11} - \Sigma_{12}\Sigma_{22}^{-1}\Sigma_{21}$.

Letting the subset y_1 contain a single element, then

$$y = \begin{pmatrix} y_1 \\ y_2 \end{pmatrix} \sim N_p \left[\begin{pmatrix} \mu_1 \\ \mu_2 \end{pmatrix}, \begin{pmatrix} \sigma_{11} & \sigma'_{12} \\ \sigma_{21} & \Sigma_{22} \end{pmatrix} \right]. \tag{1.47}$$

The multiple correlation coefficient R is the maximum correlation possible between y_1 and the linear combination of the random vector y_2, $a'y_2$. Using the Cauchy-Schwarz inequality, one can show that the multiple correlation coefficient $R = [\sigma'_{12}\Sigma_{22}^{-1}\sigma_{21}/\sigma_{11}]^{1/2} \geq 0$, which is seen to be the correlation between y_1 and $z = \mu_1 + \Sigma_{12}\Sigma_{22}^{-1}(y_2 - \mu_2)$. The sample, biased overestimate of R^2, the squared multiple correlation coefficient, is $\hat{R}^2 = s'_{12}S_{22}^{-1}s_{21}/s_{11}$. Even if $R^2 = 0$, $\mathcal{E}(\hat{R}^2) = (p - 1)/(n - 1)$. Thus, if the sample size n is small relative to p, the bias can be large.

The partial correlation coefficient between variables y_i and y_j is the ordinary simple correlation ρ between y_i and y_j with the variables in the subset y_2 held fixed and represented by $\rho_{ij|y_2}$. Letting $\Sigma_{11 \cdot 2} = (\sigma_{ij|y_2})$, the matrix of partial variances and covariances, the partial correlation between the (i, j) element is

$$\rho_{ij|y_2} = \frac{\sigma_{ij|y_2}}{\sigma_{ii|y_2}^{1/2}\sigma_{jj|y_2}^{1/2}}. \tag{1.48}$$

Replacing $\Sigma_{11 \cdot 2}$ with the sample estimate $S_{11 \cdot 2} = (s_{ij|y_2})$, the MLE of $\rho_{ij|y_2}$ is

$$r_{ij|y_2} = \frac{s_{ij|y_2}}{s_{ii|y_2}^{1/2}s_{jj|y_2}^{1/2}}. \tag{1.49}$$

1.6 HYPOTHESIS TESTING

Having assumed a linear model for a random sample of observations, used the observations to obtain an estimate of the population parameters, and decided upon the structure of the restriction R (if any) and the hypothesis test matrix C, one next test hypotheses. Two commonly used procedures for testing hypotheses are the likelihood ratio (LR) and union-intersection (UI) test. To construct a LR test, two likelihood functions are compared for a random sample of observations, $L(\hat{\omega})$, the likelihood function maximized under the hypothesis H in (1.20), and the likelihood $L(\hat{\Omega}_0)$, the likelihood function maximized over the entire parameter space Ω_0 unconstrained by the hypothesis. Defining λ as the ratio

$$\lambda = \frac{L(\hat{\omega})}{L(\hat{\Omega}_0)} \tag{1.50}$$

the hypothesis is rejected for small values of λ since $L(\hat{\omega}) < L(\hat{\Omega}_0)$ does not favor the hypothesis. The test is said to be of size α if for a constant λ_0, the

$$P(\lambda < \lambda_0 | H) = \alpha \tag{1.51}$$

where α is the size of the Type I error rate, the probability of rejecting H given H is true. For large sample sizes and under very general conditions, Wald (1943) showed that $-2 \ln \lambda$ converges in distribution to a chi-square distribution as $n \to \infty$, where the degrees of freedom ν is equal to the number of independent parameters estimated under Ω_0 minus the number of independent parameters estimated under ω.

To construct a UI test according to Roy (1953), we write the null hypothesis H as an intersection of an infinite number of elementary tests

$$H : \bigcap_i H_i \tag{1.52}$$

and each H_i is associated with an alternative A_i such that

$$A : \bigcup_i A_i. \tag{1.53}$$

The null hypothesis H is rejected if any elementary test of size α is rejected. The overall rejection region being the union of all the rejection regions of the elementary tests of H_i vs. A_i. Similarly, the region of acceptance for H is the intersection of the acceptance regions. If T_i is a test statistic for testing H_i vs. A_i, the null hypothesis H is accepted or rejected if the $T_i \lessgtr c_\alpha$ where the

$$P\left(\sup_i T_i \leq c_\alpha | H\right) = 1 - \alpha \tag{1.54}$$

and c_α is chosen such that the Type I error is α.

1.7   GENERATING MULTIVARIATE NORMAL DATA

In hypothesis testing of both univariate and multivariate linear models, the assumption of multivariate normality is made. The multivariate normal distribution of a random vector y with p variables has the density function given in (1.26), written as $y \sim N_p(\mu, \Sigma)$. For $p = 1$, the density function reduces to the univariate normal distribution. Some important properties of normally distributed random variables were reviewed in Section 1.5. To generate data having a multivariate normal distribution with mean $\mu' = (\mu_1, \mu_2, \ldots, \mu_p)$ and covariance matrix $\Sigma = (\sigma_{ij})$, we use Definition 1.1. Program 1_7.sas uses the IML procedure to generate 50 observations from a multivariate normal distribution with structure

$$\mu = \begin{pmatrix} 10 \\ 20 \\ 30 \\ 40 \end{pmatrix} \text{ and } \Sigma = \begin{pmatrix} 3 & 1 & 0 & 0 \\ 1 & 4 & 0 & 0 \\ 0 & 0 & 1 & 4 \\ 0 & 0 & 4 & 20 \end{pmatrix}.$$

In Program 1_7.sas, PROC IML is used to produce a matrix Z that contains $n = 50$ observation vectors with $p = 4$ variables. Each vector is generated by using the standard normal distribution, $N(0, 1)$. Using Theorem 1.3, new variables $y_i = z_i A + \mu$ are created where the matrix A is such that the $\text{cov}(y) = A'\text{cov}(z)A = A'IA = A'A = \Sigma$ and $E(y) = \mu$. The Cholesky factorization procedure is used to obtain A from Σ: the ROOT function in PROC IML performs the Cholesky decomposition and stores the result in the matrix named a in the program. Next, the matrix u is created by repeating the mean row vector u 50 times to produce a 50×4 data matrix. The multivariate normal random variables are created using the statement: y=(z*a)+uu. The observations are printed and output to the file named 1_7.dat. The seed in the program allows one to always create the same data set.

1.8 ASSESSING UNIVARIATE NORMALITY

Before one tests hypotheses, it is important that one examines the distributional assumptions for the sample data under review. While the level of the test (Type I error) for means is reasonably robust to nonnormality, this is not the case when investigating the covariance structure. However, very skewed data and extreme outliers may result in errors in statistical inference of the population means. Thus, one usually wants to verify normality and investigate data for outliers.

If a random vector y is distributed multivariate normally, then its components y_i are distributed univariate normal. Thus, one step in evaluating multivariate normality of a random vector is to evaluate the univariate normality of it components. One can construct and examine histograms, stem-and-leaf plots, box plots, and Quantile-Quantile (Q-Q) probability plots for the components of a random p-vector.

Q-Q plots are plots of the observed, ordered quantile versus the quantile values expected if the observed data are normally distributed. Departures from a straight line are evidence against the assumption that the population from which the observations are drawn is normally distributed. Outliers may be detected from these plots as points well separated from the other observations. The behavior at the ends of the plots can provide information about the length of the tails of the distribution and the symmetry or asymmetry of the distribution (Singh, 1993).

One may also evaluate normality by performing the Shapiro and Wilk (1965) W test when sample sizes are less than or equal to 50. The test is known to show a reasonable sensitivity to nonnormality (Shapiro, Wilk, & Chen, 1968). For $50 \leq n \leq$

2000, Royston (1982, 1992) approximation is recommended and is implemented in the SAS procedure UNIVARIATE.

When individual variables are found to be nonnormal, one can often find a Box and Cox (1964) power transformation that may be applied to the data to achieve normality. The Box-Cox power transformation has the general structure for $y > 0$:

$$
x = \begin{cases} \frac{(y^\lambda - 1)}{\lambda} & : \lambda \neq 0 \\ \log(y) & : \lambda = 0 \end{cases}.
$$

Note that the random dependent random variable y must be positive for all values. Thus, one may have to add a constant to the sample data before applying the transformation. After applying a Box-Cox type transformation, one should again check whether transformed data are more nearly normal.

1.8.1 Normally and Nonnormally Distributed Data

Program 1_8_1.sas produces Q-Q plots for the univariate normally distributed random variables generated by Program 1_7.sas. The Q-Q plots for the normal data show that the observations lie close to a line, but not exactly; the tail, especially, falls off from the line (Graph 1.8.1). Recall that we know that these data are normally distributed. Thus, when using Q-Q plots for diagnostic purposes, we cannot expect that even normally distributed data will lie exactly on a straight line. Note also that some of the Shapiro-Wilk test statistics are significant at the nominal $\alpha = 0.05$ level even for normal data. When performing tests for sample sizes less than or equal to 50, it is best to reduce the level of the normality test to the nominal $\alpha = 0.01$ level.

In Program 1_8_1.sas we transform the normal data using the transformations: $ty1 = 1/y^2$, $ty2 = e^y$, $ty3 = \log(y)$, and $ty4 = y^2$ for the four normal variables and generating Q-Q plots for the transformed nonnormal data (Graph 1.8.2). Inspection of the plots clearly show marked curvilinear patterns. The plots are not linear. Next we illustrate how one may find a Box-Cox power transformation using the transformed variable $x = ty4$. To achieve normality the value of λ should be near one-half, the back transformation for the variable. Using the macro %adxgen and %adxtran in SAS, the value $\lambda = -0.4$ (found in the log file) is obtained for the data; the transformed data are stored in the data set named result. The λ value is near the correct value of -0.5 (or a square root transformation) for the variable. The λ plot in the output indicates that the value of λ should be within the interval: $-0.2 \leq \lambda \leq -0.6$. While the macro uses the minimal value, one often trys other values within the interval to attain near normality.

Finally, we introduce an outlier into the normally distributed data, and again generate Q-Q plots (Graph 1.8.3). Inspection of the Q-Q plot clearly shows the extreme observation. The data point is far from the line.

Graph 1.8.1: Q-Q plot of y1-y4

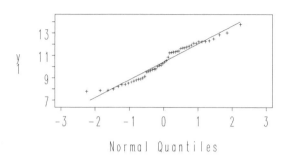

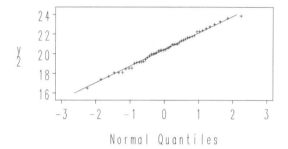

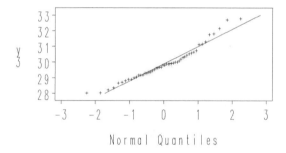

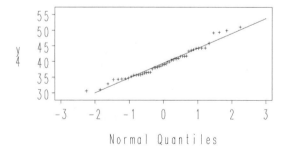

Graph 1.8.2: Q-Q plot of ty1-ty4

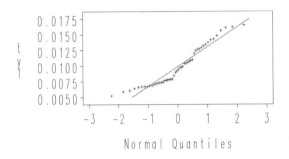

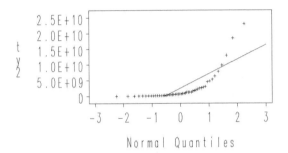

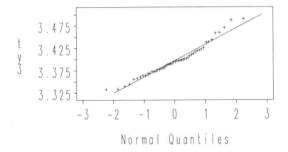

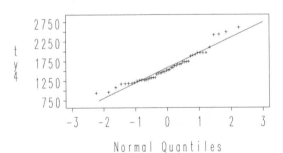

Graph 1.8.3: Q-Q plot of y1 with an outlier

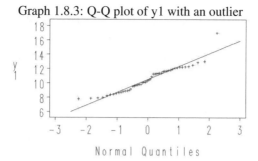

1.8.2   Real Data Example

To illustrate the application of plots and tests to evaluate normality, data from Rohwer given in (Timm, 2002, p. 213) are used. The data are in the data set Rohwer.dat.

The data are for 32 selected school children in an upper-class, white residential school and contain three standardized tests: Peabody Picture Vocabulary (y1), Student Achievement (y2), and the Raven Progressive Matrices test (y3) and five paired-associate, learning-proficiency tasks: Named (x1), Still (x2), Named Still (x3), Named Action (x4), and sentence still (x5). While we will use the data to evaluate multivariate prediction in Chapter 5, we use the raw data to investigate the normality of the three standardized test variables using univariate Q-Q plots and tests for univariate normality. Also illustrated is the use of the Box-Cox transformation. The code for the analysis is contained in Program 1_8_2.sas.

The program produces Q-Q plots for each dependent variable (Graph 1.8.4). Review of the plots indicates that the variables y1 and y2 appear normal. However, this is not the case for the variable y3. Using the Box-Cox power transformation, a value of $\lambda = 0.4$ is used to transform the data to near normality, yt3. The plot is more nearly linear and the Shapiro-Wilk test appears to marginally support normality at the nominal level $\alpha = 0.01$ for the transformed data.

1.9   ASSESSING MULTIVARIATE NORMALITY WITH CHI-SQUARE PLOTS

Even though each variable in a vector of variables is normally distributed, marginally normality does not ensure multivariate normality. However, multivariate normality does ensure marginal normality. Thus, one often wants to evaluate whether or not a vector of random variables follows a multivariate normal distribution. To evaluate multivariate normality, one may compute the Mahalanobis distance for the ith observation:

$$D_i^2 = (y_i - \bar{y})' S^{-1} (y_i - \bar{y}) \tag{1.55}$$

Graph 1.8.4: Q-Q plot of y1-y3 and yt3

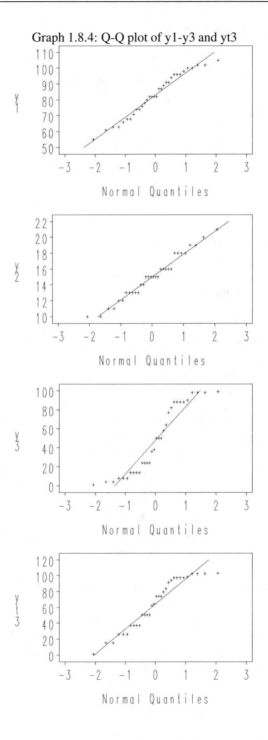

and plot these distances against the ordered chi-square percentile values $q_i = \chi^2_p \cdot$ $[(i - \frac{1}{2})/n]$ where q_i $(i = 1, 2, \ldots, n)$ is the $100(i - \frac{1}{2})/n$ sample quantile of the chi-square distribution.

Singh (1993) constructed probability plots resembling Shewart-type control charts, where warning points were placed at the $\alpha100\%$ critical value of the distribution of Mahalanobis distances, and a maximum point limit was also defined. Thus, any observation falling beyond the maximum limit was considered an outlier, and any point between the warning limit and the maximum limit required further investigation.

Singh (1993) constructs multivariate probability plots with the ordered Mahalanobis distances versus quantiles from a beta distribution, rather than a chi-square distribution. The exact distribution of $b_i = nD_i^2/(n-1)^2$ follows a beta $[a = p/2, b = (n-p-1)/2]$ distribution (Gnanadesikan & Kettenring, 1972). Small (1978) found that as p gets large ($p > 5\%$ of n) relative to n that the chi-square approximation may not be adequate unless $n \geq 25$ and in these cases recommends a beta plot.

When evaluating multivariate normality, one should also compute measures of multivariate skewness and kurtosis. If data follow a multivariate normal distribution, these measures should be near zero. If the distribution is leptokurtic (has heavy tails), the measure of kurtosis will be large. If the distribution is platykurtic (has light tails) the kurtosis coefficient will be small.

Mardia (1970) defined the measures of multivariate skewness and kurtosis:

$$\beta_{1,p} = \mathcal{E}[(x - \mu)'\Sigma^{-1}(y - \mu)]^3 \tag{1.56}$$

$$\beta_{2,p} = \mathcal{E}[(y - \mu)'\Sigma^{-1}(y - \mu)]^2 \tag{1.57}$$

where x and y are identically and independently distributed. Sample estimates of these quantities are:

$$\hat{\beta}_{1,p} = \frac{1}{n^2} \sum_{i=1}^{n} \sum_{j=1}^{n} \left[(y_i - \bar{y})'S^{-1}(y_j - \bar{y})\right]^3 \tag{1.58}$$

$$\hat{\beta}_{2,p} = \frac{1}{n} \sum_{i=1}^{n} \sum_{j=1}^{n} \left[(y_i - \bar{y})'S^{-1}(y_j - \bar{y})\right]^2. \tag{1.59}$$

If $y \sim N_p(\mu, \Sigma)$, then $\beta_{1,p} = 0$ and $\beta_{2,p} = p(p + 2)$. Mardia showed that the sample estimate of multivariate kurtosis $X^2 = n\hat{\beta}_{1,p}/6$ has an asymptotic chi-square distribution with $\nu = p(p + 1)(p + 2)/6$ degrees of freedom. And that $Z = [\hat{\beta}_{2,p} - p(p+2)]/[8p(p+2)/n]^{1/2}$ converges in distribution to a standard normal distribution. Provided the sample size $n \geq 50$ one may develop tests of multivariate normality. Mardia (1974) developed tables of approximate percentiles for $p = 2$ and $n \geq 10$ and alternative large sample approximations. Romeu and Ozturk (1993) investigated ten tests of goodness-of-fit for multivariate normality. They show that the multivariate tests of Mardia are most stable and reliable for assessing multivariate normality. In general, tests of hypotheses regarding means are sensitive to high values of skewness and kurtosis for multivariate data.

Graph 1.9.1: Chi-square Q-Q plot

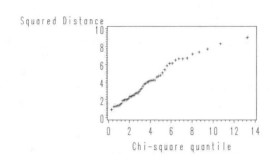

Output 1.9.1: MULTNORM Macro Univariate and Multivariate Normality Tests for
y1-y4.

Variable	n	Test	Multivariate Skewness & Kurtosis	Test Statistic Value	p-value
y1	50	Shapiro-Wilk	.	0.95519	0.0560
y2	50	Shapiro-Wilk	.	0.99170	0.9775
y3	50	Shapiro-Wilk	.	0.95347	0.0475
y4	50	Shapiro-Wilk	.	0.96571	0.1540
	50	Mardia Skewness	1.0846	9.81142	0.9715
	50	Mardia Kurtosis	20.5357	-1.76789	0.0771

While Andrews, Gnanadesikan, and Warner (1971) have developed a multivariate
extension of the Box-Cox power transformation for multivariate data, determination
of the appropriate transformation is complicated (see, Chambers, 1977; Velilla &
Barrio, 1994). In general, one applies the Box-Cox transformation a variable at a
time or uses some linear combination of the variables in the analysis when multivariate normality is not satisfied.

1.9.1   Multivariate Normal Data

To illustrate the construction of a chi-square plot, the data in the multivariate data
set 1_7.dat are used. Program 1_9_1.sas contains the code for the chi-square plots.
The program uses the SAS macro %multnorm which calculates Mardia's test statistics for multivariate skewness and kurtosis and also the Shapiro-Wilk W statistics for
each variable. Inspection of the plot and the multivariate statistics indicate that the
data are clearly multivariate normal (Graph 1.9.1, Output 1.9.1).

1.9.2   Real Data Example

Using the Rohwer data set described in Section 1.8.2, we developed chi-square plots for the raw data and the transformed data. Program 1_9_2.sas contains the code for the example. The p-values for Mardia Skewness and Kurtosis for the raw data are: 0.93807 and 0.05000, respectively. Upon transformation of the third variable, the corresponding values become: 0.96408 and 0.05286. While the data are nearly normal, the transformation does not show a significant improvement in joint normality.

1.10   USING SAS INSIGHT

Outliers in univariate data only occur in the tail of the Q-Q plot since the plots are based upon ordered variables. However, for multivariate data this is not the case since multivariate vector observations cannot be ordered. Instead, ordered squared distances are used so that the location of an outlier within the distribution is uncertain. It may involve any distance in the multivariate chi-square Q-Q plot. To evaluate the data for potential outliers, one may use the tool SAS INSIGHT interactively.

When SAS is executed, it creates temporary data sets in the Library WORK. To access the Library interactively, click on Solution → Analysis → Interactive Data Analysis. This executes the SAS INSIGHT software. Using SAS INSIGHT, click on the data set called WORK. The data sets used and created by the SAS program are displayed. For the multivariate Q-Q plot, select the data set _CHIPLOT. Displayed will be the coordinates of the multivariate Q-Q plot. From the tool bar select Analyze → Fit(Y X). This will invoke a Fit(Y X) software window; next, move the variables MAHDIST to the window labeled Y and the variable CHISQ to the window labeled X. Then, select Apply from the menu. This will produce the multivariate Q-Q plot generated by macro %multnorm (Graph 1.10.1). The observation number will appear by clicking on a data point. By double clicking on a value, the window Examine Observations appears which display the residual and predicted squared distances (Figure 1.10.1). Also contained in the output is a plot of these values. By clicking on data points, extreme observations are easily located. To illustrate the use of SAS INSIGHT two data sets using real data are investigated.

1.10.1   Ramus Bone Data

To illustrate the use of SAS INSIGHT for the location of outliers Ramus bone data from Elston and Grizzle (1962) are used. The data are in the data set Ramus.dat. Using Program 1_10_1.sas, the data are investigated for normality. One observes that while all variables are univariate normal, the test of multivariate normality is rejected (Output 1.10.1). This is due in part to the small sample size. Following the procedure discussed above, we observe that observation 9 appears extreme (Graph 1.10.1). Removing this observation from the data set using Program 1_10_1a.sas, the data become more normal, but remain skewed (Output 1.10.2). For a multivariate

Graph 1.10.1: Chi-square Q-Q plot generated by SAS INSIGHT

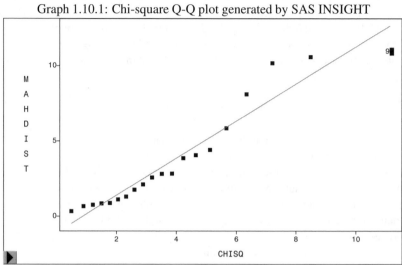

Figure 1.10.1: Examine Observations

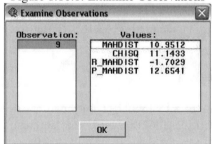

Output 1.10.1: MULTNORM Macro Univariate and Multivariate Normality Tests for Ramus Bone Data.

Variable	n	Test	Multivariate Skewness & Kurtosis	Test Statistic Value	p-value
y1	20	Shapiro-Wilk	.	0.9479	0.3360
y2	20	Shapiro-Wilk	.	0.9628	0.6020
y3	20	Shapiro-Wilk	.	0.9578	0.5016
y4	20	Shapiro-Wilk	.	0.9180	0.0905
	20	Mardia Skewness	11.3431	46.1170	0.0008
	20	Mardia Kurtosis	28.9174	1.5871	0.1125

Output 1.10.2: MULTNORM Macro Univariate and Multivariate Normality Tests for Ramus Bone Data without Observation 9.

Variable	n	Test	Multivariate Skewness & Kurtosis	Test Statistic Value	p-value
y1	19	Shapiro-Wilk	.	0.9436	0.3064
y2	19	Shapiro-Wilk	.	0.9519	0.4249
y3	19	Shapiro-Wilk	.	0.9533	0.4490
y4	19	Shapiro-Wilk	.	0.9210	0.1180
	19	Mardia Skewness	11.0359	43.0477	0.0020
	19	Mardia Kurtosis	29.0259	1.5810	0.1139

analysis of this data set, one should consider linear combination of the Ramus data over the years of growth since the skewness is not easily removed from the data.

1.10.2   Risk-Taking Behavior Data

For our second example, data from a large study by Dr. Stanley Jacobs and Mr. Ronald Hritz at the University of Pittsburgh are used. Students were assigned to three experimental conditions and administered two parallel forms of a test given under high and low penalty. The data set is in the file Stan_Hz.dat. Using Program 1_10_2.sas, the data are investigated for multivariate normality. The test of multivariate normality is clearly rejected (Output 1.10.3). Using SAS INSIGHT, observation number 82 is clearly an outlier (Graph 1.10.2). Removing the observation (Program 1_10_2a.sas), the data are restored to multivariate normality (Output 1.10.4). These examples clearly indicate the importance of removing outliers from multivariate data.

Graph 1.10.2: Chi-square Q-Q plot of risk-taking behavior data

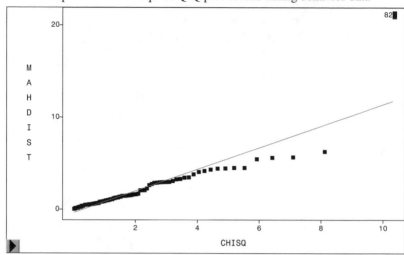

Output 1.10.3: MULTNORM Macro Univariate and Multivariate Normality Tests of Risk-taking Behavior Data.

Variable	n	Test	Multivariate Skewness & Kurtosis	Test Statistic Value	p-value
resL	87	Shapiro-Wilk	.	0.9888	0.6674
resH	87	Shapiro-Wilk	.	0.9520	0.0027
	87	Mardia Skewness	0.7450	11.4348	0.0221
	87	Mardia Kurtosis	10.7652	3.2240	0.0013

Output 1.10.4: MULTNORM Macro Univariate and Multivariate Normality Tests of Risk-taking Behavior Data with Observation 82.

Variable	n	Test	Multivariate Skewness & Kurtosis	Test Statistic Value	p-value
resL	86	Shapiro-Wilk	.	0.98969	0.7354
resH	86	Shapiro-Wilk	.	0.97449	0.0863
	86	Mardia Skewness	0.13474	2.04568	0.7274
	86	Mardia Kurtosis	7.04198	-1.11054	0.2668

Graph 1.11.1: Bivariate Normal Distribution with u=(0, 0), var(y1)=3, var(y2)=4, cov(y1,y2)=1, r=.289

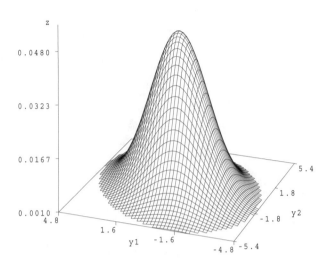

1.11   THREE-DIMENSIONAL PLOTS

Three dimensional scatter plots of multivariate data often help with the visualization of data. They are generated using the G3D procedure in SAS. The first part of Program 1_11.sas (adapted from Khattree & Naik, 1995, p. 65) produces a plot of a bivariate normal distribution with mean and covariance matrix:

$$\mu = \begin{pmatrix} 0 \\ 0 \end{pmatrix} \text{ and } \Sigma = \begin{pmatrix} 3 & 1 \\ 1 & 4 \end{pmatrix}.$$

This is the covariance matrix of variables y1 and y2 from the simulated multivariate normal data generated by Program 1_7.sas. The three-dimensional plot is given in Graph 1.11.1.

To see how plots vary, a second plot is generated in Program 1_11.sas using variables y1 and y4 from the simulated data in data set 1_7.dat. The covariance matrix for population parameters for the plot are:

$$\mu = \begin{pmatrix} 0 \\ 0 \end{pmatrix} \text{ and } \Sigma = \begin{pmatrix} 3 & 0 \\ 0 & 20 \end{pmatrix}.$$

The plot is displayed in Graph 1.11.2.

For the first plot, a cross-wise plot would result in an oval shape, whereas in the second plot, a circular shape results. This is due to the structure of the covariance matrix. Using SAS INSIGHT for the data set, one may generate contour plots for

Graph 1.11.2: Bivariate Normal Distribution with u=(0, 0), var(y1)=3, var(y2)=20, cov(y1,y2)=0, r=0

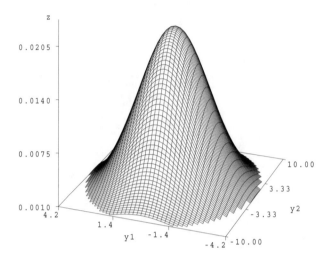

the data. See Khattree and Naik (1995) for more graphical displays of multivariate data using SAS.

CHAPTER 2

Unrestricted General Linear Models

2.1 INTRODUCTION

Unrestricted (univariate) linear models are linear models that specify a relationship between a set of random, independent, identically distributed (iid) dependent variables $y' = (y_1, y_2, \ldots, y_n)$ and a matrix of fixed, nonrandom, independent variables $X = (x_{ik})$ such that

$$\mathcal{E}(y_i) = x_{i1}\beta_1 + x_{i2}\beta_2 + \cdots + x_{ik}\beta_k \qquad i = 1, 2, \ldots, n. \qquad (2.1)$$

The variance of each y_i is constant (σ^2) or homogeneous, and the relationship is linear in the unknown, nonrandom parameters $\beta' = (\beta_1, \beta_2, \ldots, \beta_k)$. Special classes of such models are called multiple linear regression models, analysis of variance (ANOVA) models, and intraclass covariance models. In this chapter, we review both ordinary least squares (OLS) and maximum likelihood (ML) estimation of the model parameters, hypothesis testing, model selection and prediction in multiple linear regression model, and the general linear mixed model (GLMM) is introduced. Estimation theory and hypothesis testing for the GLMM are not discussed until Chapter 11. Applications discussed include multiple linear regression analyses and the analysis of variance for several experimental designs.

2.2 LINEAR MODELS WITHOUT RESTRICTIONS

For multiple regression, ANOVA, and intraclass covariance models, we assume that the covariance matrix for the vector y has the structure

$$\Omega = \sigma^2 I_n \qquad (2.2)$$

where I_n is an $n \times n$ identity matrix. The error structure for the observation is said to be homogeneous or spherical. Models of the form (1.1) with covariance structure (2.2) are called unrestricted (univariate) linear models.

To estimate β, the vector of unknown, nonrandom, fixed effects regression coefficients, the method of OLS is commonly utilized. The least squares criterion requires minimizing the error sum of squares, $\sum_{i=1}^{n} e_i^2 = \text{tr}(ee')$, where $\text{tr}(\cdot)$ is the trace operator. Minimizing the error sum of squares leads to the normal equations

$$(X'X)\hat{\beta} = X'y. \tag{2.3}$$

Because X has full column rank, $\text{rank}(X) = k$, the ordinary least squares estimator (OLSE) of β is the unique solution to the normal equation (2.3),

$$\hat{\beta} = (X'X)^{-1}X'y. \tag{2.4}$$

This is also called the best linear unbiased estimator (BLUE) of β since among all parametric functions $\psi = c'\beta$, $\hat{\psi} = c'\hat{\beta}$ is unbiased for ψ and has smallest variance. The mean and variance of the parametric functions are

$$\mathcal{E}(\hat{\psi}) = \psi \tag{2.5}$$

$$\text{var}(\hat{\psi}) = \sigma^2 c'(X'X)^{-1}c. \tag{2.6}$$

If the matrix X is not of full rank k, one may either reparameterize the model to full rank or use a generalized inverse of $X'X$ in (2.4) to solve the normal equations.

Definition 2.1. A generalized inverse of a real matrix A is any matrix G that satisfies the condition $AGA = A$. The generalized inverse of A is written as $G = A^-$.

Because A^- is not unique, (2.3) has no unique solution if $X'X$ is not full rank k; however, linear combinations of β, $\psi = c'\beta$, may be found that are unique even though a unique estimate of β is not available. Several SAS procedures use a full rank design matrix while others do not; more will be said about this when the applications are discussed. For a thorough discussion of the analysis of univariate linear models see Searle (1971) and Milliken and Johnson (1984). Following Searle (1987), and Timm and Carlson (1975), we will usually assume in our discussion in this chapter that the design matrix X is of full rank.

2.3 HYPOTHESIS TESTING

Once the parameter β has been estimated, the next step is usually to test hypotheses about β. For hypothesis testing, we assume that the vector e follows a spherical multivariate normal distribution,

$$y \sim N_n(\mu = X\beta, \Omega = \sigma^2 I_n). \tag{2.7}$$

The MLE of the population parameters β and σ^2 assuming normality are

$$\hat{\beta}_{MLE} = (X'X)^{-1}X'y \tag{2.8}$$

$$\hat{\sigma}^2_{MLE} = \frac{(y - X\hat{\beta})'(y - X\hat{\beta})}{n}$$

$$= \frac{(y'y - n\bar{y}^2)}{n} \tag{2.9}$$

where the likelihood function using (1.31) has the form

$$L(\sigma^2, \beta | y) = (2\pi\sigma^2)^{-n/2} \exp\left[-\frac{(y - X\beta)'(y - X\beta)}{2\sigma^2}\right]. \qquad (2.10)$$

We see that the OLSE (2.4) and the MLE (2.8) estimate of β are identical.

To test the hypothesis $H : C\beta = \xi$ (1.20), we may create a likelihood ratio test which requires maximizing (2.10) with respect to β and σ^2 under the hypothesis $L(\hat{\omega})$ and over the entire parameter space $L(\hat{\Omega}_0)$. Over the entire parameter space, Ω_0, the MLE of β and σ^2 are given in (2.8). The corresponding estimates under the hypothesis are

$$\hat{\beta}_\omega = \hat{\beta} - (X'X)^{-1}C'[C(X'X)^{-1}C']^{-1}(C\hat{\beta} - \xi) \qquad (2.11)$$

$$\hat{\sigma}_\omega^2 = \frac{(y - X\hat{\beta}_\omega)'(y - X\hat{\beta}_\omega)}{n} \qquad (2.12)$$

(see Timm, 1975, p. 178). Substituting the estimates under ω and Ω_0 into the likelihood function (2.10), the likelihood ratio defined in (1.50) becomes

$$\lambda = \frac{L(\hat{\omega})}{L(\hat{\Omega}_0)} = \frac{(2\pi\hat{\sigma}_\omega^2)^{-n/2}}{(2\pi\hat{\sigma}^2)^{-n/2}}$$

$$= \left[\frac{(y - X\hat{\beta})'(y - X\hat{\beta})}{(y - X\hat{\beta}_\omega)'(y - X\hat{\beta}_\omega)}\right]^{n/2} \qquad (2.13)$$

so that

$$\Lambda = \lambda^{2/n} = \frac{E}{E + H} \qquad (2.14)$$

where

$$E = y'y - n\bar{y}^2 = y'(I - X(X'X)^{-1}X')y \qquad (2.15)$$

$$H = (C\hat{\beta} - \xi)'[C(X'X)^{-1}C']^{-1}(C\hat{\beta} - \xi). \qquad (2.16)$$

Details are provided in Timm (1975, 1993b, Chapter 3) and Searle (1987, Chapter 8).

The likelihood ratio test is to reject

$$H : C\beta = \xi \qquad \text{if } \Lambda < c \qquad (2.17)$$

where c is determined such that the $P(\Lambda < c | H) = \alpha$. The statistic Λ is related to a beta distribution represented generally as U_{p,ν_h,ν_e} where p is the number of variables, ν_h is the degrees of freedom for the hypothesis, $\nu_h = \text{rank}(C) = g$, and ν_e is the degrees of freedom for error, $\nu_e = n - \text{rank}(X)$.

Theorem 2.1. *When $p = 1$,*

$$\left[\frac{\nu_e(1 - U_{1,\nu_h,\nu_e})}{\nu_h U_{1,\nu_h,\nu_e}}\right] = F_{\nu_h,\nu_e}. \qquad (2.18)$$

From Theorem 2.1, we see that rejecting H for small values of U is equivalent to rejecting H for large values F where

$$F = \frac{H/\nu_h}{E/\nu_e} = \frac{MS_h}{MS_e} \tag{2.19}$$

is the F statistic, and MS_h refers to the mean square for hypothesis and MS_e refers to the mean square for error.

2.4   SIMULTANEOUS INFERENCE

While the parametric function that led to the rejection of H may not be of interest to the researcher, one can easily find the combination of the parameters β that led to rejection. To find the function $\psi_{\max} = c'\beta$, observe that by (2.19) and (2.15)-(2.16) with $\hat{\psi} = C\hat{\beta}$ that

$$(\hat{\psi} - \psi)'(C(X'X)^{-1}C')^{-1}(\hat{\psi} - \psi) > gMS_eF^{1-\alpha} \tag{2.20}$$

where $F^{1-\alpha}$ is the upper $1 - \alpha$ percentage value of the F distribution. using the Cauchy-Schwarz (C-S) inequality, $(X'y)^2 \leq (X'x)(y'y)$, with $x = Fa$ and $y = F^{-1}b$ and $G = F'F$, we have that $(a'b)^2 \leq (a'Ga)(b'G^{-1}b)$. Letting $b = (\hat{\psi} - \psi)$ and $G = C(X'X)^{-1}C'$, the

$$\sup_a \frac{[a'(\hat{\psi} - \psi)]^2}{a'C(X'X)^{-1}C'a} \leq (\hat{\psi} - \psi)'(C(X'X)^{-1}C')^{-1}(\hat{\psi} - \psi) \tag{2.21}$$

or for (2.20), the

$$\sup_a \frac{[a'(\hat{\psi} - \psi)]^2}{a'C(X'X)^{-1}C'a} \geq (gMS_eF^{1-\alpha})^{1/2}. \tag{2.22}$$

By again applying the C-S inequality

$$(a'a)^2 \leq (a'a)(a'a) \text{ or } (a'a) \leq [(a'a)(a'a)]^{1/2},$$

we have that

$$[a'(\hat{\psi} - \psi)]^2 \leq a'(\hat{\psi} - \psi)(\hat{\psi} - \psi)'a. \tag{2.23}$$

Hence, for (2.22) the

$$\sup_a \frac{a'G_1a}{a'G_2a} \geq (gMS_eF^{1-\alpha}) \tag{2.24}$$

where $G_1 = (\hat{\psi} - \psi)(\hat{\psi} - \psi)'$ and $G_2 = C(X'X)^{-1}C'$. Recall, however, that the supremum of the ratio of two quadratic forms is the largest characteristic root of the determinantal equation $|G_1 - \lambda G_2| = 0$ with associated eigenvector a_*. Solving $|G_2^{-1}G_1 - \lambda I| = 0$, we find that there exists a matrix P say such that $G_2^{-1}G_1P = $

ΛP where Λ are the roots and P is the matrix of associated eigenvectors of the determinantal equation. For (2.24), we have that

$$
\begin{aligned}
G_2^{-1}G_1|G_2^{-1}(\hat{\psi} - \psi)| &= G_2^{-1}(\hat{\psi} - \psi)(\hat{\psi} - \psi)'G_2^{-}(\hat{\psi} - \psi) \\
&= [(\hat{\psi} - \psi)'G_2^{-1}(\hat{\psi} - \psi)]G_2^{-1}(\hat{\psi} - \psi) \\
&= \lambda G_2^{-1}(\hat{\psi} - \psi)
\end{aligned}
\tag{2.25}
$$

so that $a_* = G_2^{-1}(\hat{\psi} - \psi)$ is the eigenvector of $G_2^{-1}G_1$ for the maximum root of $|G_2^{-1}G_1 - \lambda I| = 0$. Thus, the eigenvector a_* may be used to find the linear parametric function of β that is most significantly different from ξ. The function is

$$
\psi_{\max} = (a_*'C)\beta = c'\beta.
\tag{2.26}
$$

The Scheffé-type simultaneous confidence interval for $\psi = c'\beta$ for all nonnull vectors c' such that the $\sum_i c_i = 0$ and the c_i are elements of the vector c is given by

$$
\hat{\psi} - c_0\hat{\sigma}_{\hat{\psi}} \leq \psi \leq \hat{\psi} + c_0\hat{\psi}_{\hat{\psi}}
\tag{2.27}
$$

where $\psi = c'\beta$, $\hat{\psi} = c'\hat{\beta}$, $\hat{\sigma}_{\hat{\psi}}$ is an estimate of the standard error of $\hat{\psi}$ given by (2.6) and

$$
c_0^2 = g F_{g,\nu_e}^{1-\alpha}
\tag{2.28}
$$

where $g = \nu_h$ (see Scheffé, 1959, p. 69).

With the rejection of the test of size α for the null overall hypothesis $H : C\beta = 0$, one may invoke Scheffé's S_2-method to investigate the infinite, nonhierarchical family of contrasts orthogonal to the significant contrast c found using the S-method. Any contrast $\hat{\psi}$ is significantly different from zero if

$$
|\hat{\psi}| > [(\nu_h - 1)F_{\nu_h-1,\nu_e}^{1-\alpha}]^{1/2} = S_2
\tag{2.29}
$$

where ν_h is the degrees of freedom of the null hypothesis, and $F_{\nu_h-1,\nu_e}^{1-\alpha}$ is the upper $(1 - \alpha)$ 100% critical value of the F distribution with degrees of freedom $\nu_h - 1$ and ν_e. Scheffé (1970) showed that the experimentwise Type I error rate for the procedure is controlled at the nominal level α, the level of the overall test. However, the Per-Family Error Rate (PFE), the expected number of Type I errors within the family will increase, Klockars, Hancock, and Krishnaiah (2000). Because these tests are guaranteed to control the overall experimentwise error rate at the level α, they are superior to Fisher's protected t-tests which only weakly control the experimentwise Type I error rate, due to the protection of the significant overall F test, Rencher and Scott (1990).

To construct an UI test of $H : C\beta = \xi$ one writes the null hypothesis as the intersection hypothesis

$$
H = \bigcap_a H_a
\tag{2.30}
$$

where a is a nonnull g-dimensional vector. Hence, H is the intersection of a set of elementary tests H_a. We would reject H if we can reject H_a for any a. By the UI principle, it follows that if we could reject for any a, we could reject for the $a = a_*$ that maximizes (2.22); thus, the UI test for this situation is equivalent to the F test or a likelihood ratio test. For additional details, see for example Casella and Berger (1994, Section 11.2.2).

2.5 MULTIPLE LINEAR REGRESSION

Multiple linear regression procedures are widely applied in the social and physical sciences, in business and industry, and in the health sciences to explain variation in a dependent (criterion) variable by employing a set of independent (predictor) variables using observational (nonexperimental) data. In these studies, the researcher's objective is to establish an optimal model by selecting a subset of available predictors that accounts for the variation in the dependent variable. In such studies, the primary goal is to discover the relationship between the dependent variable and the "best" subset of predictor variables. Multiple linear regression analysis is also used with experimental data. In these situations, the regression coefficients are employed to evaluate the marginal or partial effect of a predictor on the dependent variable given the other predictor variables in the model. In both of these cases, one is usually concerned with estimating model parameters, model specification and variable selection. The primary objective is to develop an "optimal" model using a sampling plan with fixed or random predictors, based upon an established theory. Generally speaking, one is concerned with model calibration using sample data that employs either fixed or random predictors. A second distinct phase of the study may involve model validation. For this phase of the study, one needs to define a measure of predictive precision.

Regression models are also developed to predict some random continuous outcome variable. In these situations, predictor variables are selected to maximize the predictive power of the linear regression model. Studies in this class are not concerned with model calibration, but predictive precision. As a result, regression coefficients are not interpreted as indices of the effects of a predictor variable on the criterion. And, variable selection methods that maximize prediction accuracy and/or minimize the mean squared error of prediction are of primary interest. While the predictors may be fixed, they are usually considered to be random when investigation prediction.

Even though both paradigms are widely used in practice and appear to be interrelated, they are distinct and depend on a set of model assumptions. And, the corresponding model assumptions affect measures of model fit, prediction, model validation, and variable selection which are not always clearly understood when put into practice. Model calibration studies are primarily concerned with understanding the relationship among the predictors to account for variation in the criterion variable and prediction studies are primarily concerned with selecting variables that maximize predictive precision.

Regression models may be applied using a random dependent variable and several fixed predictors making specific assumptions regarding the random errors: (i) the classical multiple linear regression (CR) model. Models are also developed with fixed predictors, assuming that the random errors have normal structure: (ii) the classical normal multiple linear regression (CNR) model. The CR model and CNR model are called the general linear model and the normal general linear model, respectively. For these models the set of predictors is not random; it remains fixed under repeated stratified sampling of the dependent variable. For the CR or CNR models, the model calibration phase of the study and the model validation phase of a study are usually separated into two distinct phases called model calibration and model validation. The validation phase of the study may require a second data set to initiate the cross-validation process. The second data set may be obtained from a single sample by splitting the sample or by obtaining an independent set of observations.

An alternative framework for model development occurs when the dependent variable and the set of predictors are obtained from a multivariate population as a random sample of independent and identically distributed observations. If one develops linear models of the joint variation of all variables in the study, we have what is called the (iii) random, classical (distribution-free) multiple linear regression (RCR) model. For the RCR model, the joint distribution of the dependent and independent random variables is unknown. While one may employ robust regression procedures to develop a RCR model, in most applications of multiple linear regression, one assumes a structural linear model where the dependent and independent variables follow a multivariate normal distribution: this is the (iv) jointly normal multiple linear regression (JNR) model. Another model closely related to the JNR model is the (iv) random, classical normal multiple linear regression (RCN) model. For the RCN model, the conditional distribution of the dependent variable is assumed to be normal and the marginal distribution of the independent variables is unknown. For both the JNR and RCN models, model calibration and model validation need not be separated. They may be addressed in a single study without cross-validation.

In the following sections, model assumptions, sampling plans, model calibration and model prediction, goodness of model fit criteria, model selection criteria, predictive precision, mean squared error of prediction in multiple linear regression models, and model validation when the independent variables are consider fixed or random are reviewed. While the concepts of variation free-ness and weak exogeneity may also be of interest, the topic of exogeneity in linear regression models is discussed in detail by Ericsson (1994) and will not be reviewed here.

2.5.1 Classical and Normal Regression Models

Estimation

The CR model is most commonly used in experimental and observational studies to "discover" the relationship between a random dependent variable y and p fixed independent variables x_i. The goal of the study is model specification or model calibration. Assuming the relationship between y and the independent variable is linear in the elements of the parameter vector β the linear regression model is represented as $y = \beta_0 + \beta_1 x_1 + \cdots + \beta_p x_p + e$ where the unknown parameter vector $\beta' = (\beta_0, \beta'_\ell)$, and the vector of regression coefficients $\beta'_\ell = (\beta_1, \ldots, \beta_p)$ is associated with the p independent variables and β_0 is the model's intercept. The random unknown error e is assumed to have mean zero, $\mathcal{E}(e) = 0$, and common unknown error variance σ^2, $V(e) = \sigma^2$. Organizing the n observations into a vector, the observation vector is y and the n fixed row vectors x'_i in the matrix $X = (1, X_\ell)$ is the $n \times k$ (design) matrix of fixed variables with full rank $k = p+1$ for a model with p parameters. The vector 1 is a column vector of n 1's and the matrix X_ℓ contains the independent variables. The linear model $y = X\beta + e$ has mean $\mathcal{E}(y) = X\beta$ and covariance matrix $\Omega = \sigma^2 I_n$. The primary goal of an analysis is to estimate the unknown parameter vector β and σ^2 using the collected data, the calibration sample.

To estimate the unknown parameter vector β, the n row vectors x'_i of the matrix X are fixed for some set of (optimally) defined values that define the strata for the ith subpopulation. For each subpopulation or strata, a single observation y_i is selected and the n observations are organized to form the elements of the observation vector y. Using the stratified sampling scheme, the elements in y are independent, but not identically distributed since they are obtained from distinct subpopulations. Given this sampling process (for an elementary discussion of this sampling scheme, one may consult Graybill, 1976, p. 154-158), it is not meaningful to estimate the population mean or variance of the row vectors in the matrix X since each row is not sampled from the joint distribution of independent variables. Furthermore, σ^2 is the variance of the random variable y given x'_i and the means $\mathcal{E}(y_i | x'_i) = x'_i \beta = \mu_i$ are the conditional means of y_i given x'_i with population mean vector $\mu = X\beta$. For the CR model, we are not directly concerned with estimating the marginal or unconditional mean of y, $\mathcal{E}(y_i) = \mu_y$, or the marginal or unconditional variance of y, $V(y_i) = \sigma_y^2$ for all i. In the CR model, the variance of the unknown random error e is the same as variance of y, namely σ^2. Thus, as correctly noted by Goldberger (1991, p. 179), the sample estimator $\sum_i (y_i - \bar{y})^2 / (n-1) = SST/(n-1) = \tilde{\sigma}_y^2$ is not an unbiased estimator of the population variance of y. Using properties of quadratic forms, the expected value of the sample estimator is $\mathcal{E}|\tilde{\sigma}_y^2| = \sigma^2 + \beta'_\ell (X'_\ell P_1 X_\ell)\beta_\ell / (n-1)$ where $P_1 = (I - 1(1'1)^{-1}1')$ is the projection (symmetric and idempotent) matrix for the CR model. Furthermore, because the matrix X is fixed in the CR model the sample covariance matrix associated with the independent variables may not be used as an estimate of the unknown population covariance matrix Σ_{xx} since X is not a random matrix. The row vectors x'_i are not selected from the joint multivariate distribution of the independent variables. Given the CR model, the OLSE for the parameter vector is given in (2.4). Adding the assumption of normality to the CR model yields the CNR model. Then the MLE of the parameters is given in (2.8). The mean squared error "risk" of the estimate $\hat{\beta}$, assuming either a CR or CNR model, is

$$\mathcal{E}[(\beta - \hat{\beta})'(\beta - \hat{\beta})] = \sigma^2 \text{tr}[(X'X)^{-1}]. \tag{2.31}$$

While the OLS estimator is the best linear unbiased estimator, shrinkage estimators due to Stein may have uniformly smaller risk (Dempster, Laird, & Rubin, 1977; Srivastava & Bilodeau, 1989).

Having found the estimate of β, the vector $\hat{y} = X\hat{\beta}$ characterizes the empirical relationship or fit between the random variable y and the vector of fixed independent variables $x' = (x_0, x_1, \ldots, x_p)$ where $x_0 = 1$, based upon the sampling plan. The vector $e = y - \hat{y}$ is the vector of estimated errors or the vector of residuals, where the residuals $e_i = y_i - \hat{y}_i$ have mean zero. Letting $SSE = \sum_i^n (y_i - \hat{y}_i)^2 = \|e\|^2$ represent the error sum of squares, the minimum variance unbiased estimator and maximum likelihood estimators of σ^2 are, respectively,

$$s^2 = \frac{SSE}{(n-k)} = MSE \qquad (2.32)$$

and

$$\hat{\sigma}^2 = \frac{SSE}{n}. \qquad (2.33)$$

These variance estimates are estimates of the conditional variance of the random variable y given the fixed vector of observations.

Model Fit

Having established the relationship between y and x it is customary to report a measure of the proportion of the variation about the sample mean y, $\bar{y}$, that can be accounted for by the regression function. The measure of sample fit is Fisher's correlation-like ratio, $\tilde{\eta}^2$, defined as

$$\begin{aligned}
\tilde{\eta}^2 &= \frac{\|\hat{y} - 1\bar{y}\|^2}{\|y - 1\bar{y}\|^2} = \frac{\sum_i^n (\hat{y}_i - \bar{y})^2}{\sum_i^n (y_i - \bar{y})^2} = \frac{SSB}{SST} \\
&= 1 - \frac{SSE}{SST} = 1 - \frac{\|e\|^2}{(y'y - n\bar{y}^2)} \\
&= 1 - \frac{\sum_i^n e_i^2}{\sum_i^n (y_i - \bar{y})^2} \qquad (2.34)
\end{aligned}$$

where the vector 1 represents a vector of n 1's and from an analysis of the variation about the mean, the total sum of squares (SST) is equal to the sum of the squares deviations between the fitted values and the mean (SSB) plus the sum of the squares error (SSE). The correlation-like ratio lies between zero and one (provided an intercept is included in the model). If the relationship between y and x is linear, then Fisher's correlation-like ratio becomes the coefficient of multiple determination, $\tilde{R}^2$. And, expression (2.34) becomes

$$\tilde{\eta}^2 = \tilde{R}^2 = \frac{(\hat{\beta}'Xy - n\bar{y}^2)}{(y'y - n\bar{y}^2)} = \frac{SSR}{SST} = 1 - \frac{SSE}{SST} \qquad (2.35)$$

where the deviations between the fitted values and the mean are replaced by deviations due to the linear relationship, represented by (SSR), and is called the sum of square regression. The coefficient of determination $\tilde{R}^2 = 1$ if $e = 0$, and the vector $y = X\beta$ so that y is an exact linear function of the variables in x. The sample fit is exact in that all the variation in the elements of the vector y is accounted for (linearly) or explained by the variation among the elements in the vector x. When the coefficient of determination $\tilde{R}^2 = 0$, each element of the vector $\hat{y}$ is identical to the sample mean of the observations, $\hat{y}_i = \bar{y}$ for all i observations. Then, the best regression equation is the sample mean vector $\bar{y} = 1\bar{y}$ so that none of the variation in y is accounted for (linearly) by the variation in the independent variables. Thus the quantity $\tilde{R}^2$ is often used as a measure of goodness-of-fit for the estimated regression model. However, since $\tilde{R}^2$ tends to increase as additional independent variables are included in the CR or CNR model, many authors suggest the adjusted coefficient of determination as a measure of goodness of model fit which takes into account the size of k relative to n. While one may account for the size of k relative to n in any number of ways, the most popular correction is the adjusted coefficient of determination defined as

$$\tilde{R}_a^2 = 1 - \left(\frac{n-1}{n-k}\right)\left(\frac{SSE}{SST}\right) = 1 - \left(\frac{SSE/(n-k)}{SST/(n-1)}\right)$$

$$= 1 - \frac{MSE}{\tilde{\sigma}_y^2}$$

$$= 1 - \frac{(n-1)(1-\tilde{R}^2)}{n-k}. \tag{2.36}$$

While the numerator MSE is an unbiased estimate of σ^2, the denominator of the ratio is not an unbiased estimate of the variance of y for either the CR or CNR models. Although $\tilde{R}_a^2$ may be expressed using a "sample variance-like" formula

$$\tilde{R}_a^2 = 1 - \frac{s_{y \cdot x}^2}{\tilde{\sigma}_y^2}. \tag{2.37}$$

Neither $\tilde{R}^2$ or $\tilde{R}_a^2$ is an estimator of the population coefficient of determination R^2 since we have selected the rows of the matrix X selectively and not at random. Furthermore, with a fixed matrix X, we can always find a design matrix X that makes $\tilde{R}^2 = \tilde{R}_a^2 = 1$. Since the matrix X is not randomly created, but fixed, one can always obtain a set of n-linearly independent $n \times 1$ column vectors to create a basis for the matrix X. Then, y may be represented by $y = X\beta$ exactly making the coefficient of determination or adjusted coefficient of determination unity. Given these limitations, Goldberger (1991, p. 177) concluded that the most important thing about $\tilde{R}^2$ and $\tilde{R}_a^2$ for the CR and CNR models is that they are not very useful.

In summary, for the CR or CNR models: (i) the predictor variables are nonrandom and fixed, (ii) the sampling plan for the model employs a stratified sampling process where the criterion variable is obtained for fixed values of the predictor so that the dependent variables are independent, but are not identically distributed, (iii) because the predictors are fixed, the values of the predictors may be chosen to create an optimal design to minimize the mean square error of the estimate $\hat{\beta}$, and (iv) neither $\tilde{R}^2$ or $\tilde{R}_a^2$ are necessarily very useful in the evaluation of model fit.

Model Selection

Because X is fixed and not random in the CR and CRN models, the matrix X may represent an over fitted model. Letting X^t represent the true model and X^u an under fitted model, the relationship among the variables are often assumed to be nested in that $X^u \subseteq X^t \subseteq X$ with corresponding parameter vectors β^u, β^t, and β, respectively. Next, suppose a sample of n observations from an over fitted model is used to estimate the unknown parameter vector β so that the estimator has the expression given in (2.4). Then, if we obtain n new observations y where the observations have the linear identical form $y = X\beta + e$, the predicted value of y is $\hat{y} = X\hat{\beta}$. Since the matrix X is fixed, the average mean squared error of prediction for the n new observations is

$$
\delta_f^2 = \frac{\mathcal{E}[(y - \hat{y})'(y - \hat{y})]}{n} = \frac{\mathcal{E}[(X\beta - X\hat{\beta})'(X\beta - X\hat{\beta})]}{n} = \frac{\mathcal{E}[e'e]}{n}
$$

$$
= \frac{\operatorname{tr}[(X'X)\sigma^2(X'X)^{-1}]}{n} + \frac{n\sigma^2}{n}
$$

$$
= \sigma^2 \left(1 + \frac{k}{n} \right). \tag{2.38}
$$

The quantity δ_f^2 is called the final prediction error (FPE).

An unbiased estimator of δ_f^2 is obtained by substituting for σ^2 in (2.38), the unbiased estimator given in (2.32). An unbiased estimate of the FPE and its associated variance follow

$$
\hat{\delta}_u^2 = s^2 \left(1 + \frac{k}{n} \right), \tag{2.39}
$$

$$
\operatorname{var}(\hat{\delta}_u^2) = \left(\frac{2\sigma^4}{n-k} \right) \left(1 + \frac{2k}{n} + \frac{k^2}{n} \right), \tag{2.40}
$$

Picard and Berk (1990). For the CNR model, the errors follow a multivariate normal distribution so the MLE of the FPE is

$$
\hat{\delta}_{MLE}^2 = \frac{SSE}{(n-k)} \left(\frac{n+k}{n} \right) = \frac{SSE}{n} \left(\frac{n+k}{n-k} \right)
$$

$$
= \hat{\sigma}^2 \left(\frac{n+k}{n-k} \right)
$$

$$
= \hat{\sigma}^2 \left(1 + \frac{2k}{n-k} \right) \tag{2.41}
$$

where $\hat{\sigma}^2$ is given in (2.33). As the number of parameters vary, the final prediction error balances the variance between the best linear predictor of y and the variance of $X\hat{\beta}$. Models with small final prediction error are examined to select the "best" candidate model. The effect of data splitting on the estimate of FPE is discussed by Picard and Berk (1990).

Mallows (1973) took an alternative approach in developing a model selection criterion, again for fixed X. There are $2^p - 1$ possible submodels. Let subscript j represent different models where $j = 1, \ldots, (2^p - 1)$ and k_j represents number of parameters in the jth model. He considered the relative error of estimating y for a submodel $\hat{y}_j = X_j \hat{\beta}_j$ defined by

$$
\begin{aligned}
J_j &= \frac{\mathcal{E}[(y - \hat{y}_j)'(y - \hat{y}_j)]}{\sigma^2} \\
&= \frac{\sum_i^n \operatorname{var}(\hat{y}_i)^2 + \sum_i^n (\text{bias in } \hat{y}_i)^2}{\sigma^2} \\
&= k_j + \frac{(\hat{\beta}_j - \beta)' X' X (\hat{\beta}_j - \beta)}{2\sigma^2} = k_j + \hat{\lambda}_j
\end{aligned}
\tag{2.42}
$$

where $\lambda_j = \mathcal{E}(\hat{\lambda}_j)$ is the noncentrality parameter in the CR and CNR models. Letting $SSE_j = \|e_j\|^2$ for k_j parameters, Mallows proposed an estimator $\hat{J}_j = SSE_j / s^2 - n + 2k_j$ of J_j by obtaining unbiased estimators of the numerator and denominator of the relative error ratio. His familiar C_p criterion for model selection is

$$
C_p = \left(\frac{SSE_j}{s^2} - n + 2k_j \right).
\tag{2.43}
$$

Mallows (1995) suggests that any model in which $C_p < k_j$ may be a potential candidate model. Both the FPE and C_p criteria may be used with either the CR and CNR models.

In 1973, Hirotugu Akaike derived an estimator of the (relative) Kullback-Leibler distance based on Fisher's maximized log-likelihood for the CNR model. His measure for model selection is called Akaike's information criterion (AIC). Akaike (1973) criterion is defined as

$$
AIC = -2 \log(\text{likelihood}) + 2(\text{number of parameters estimated}).
\tag{2.44}
$$

For the CNR model,

$$
-2 \log(\text{likelihood}) = n \log(2\pi) + n \log(\sigma^2) + \frac{(y - X\beta)'(y - X\beta)}{\sigma^2}
\tag{2.45}
$$

and since the number of parameters to be estimated is k_j for β and 1 for σ_j^2, the AIC criterion is

$$
AIC_j = n \log(2\pi) + n \log(\sigma_j^2) + n + 2(k_j + 1).
\tag{2.46}
$$

Ignoring the constants, the AIC criterion becomes $AIC_j = n \log(\sigma^2) + 2d$, for $d = k_j + 1$. Substituting a MLE and an unbiased estimator for the unknown parameter σ_j^2, the AIC criteria follow

$$
AIC_{MLE} = n \log(\hat{\sigma}_j^2) + 2d
\tag{2.47}
$$

$$
AIC_u = n \log(s_j^2) + 2d.
\tag{2.48}
$$

The model with the smallest AIC value is said to fit best. McQuarrie and Tsai scale the AIC criterion by dividing it by the sample size n.

When using Akaike's AIC fit criterion, one selects a model with too many variables (an overfit model) when there are too many parameters relative to the sample size n, $[n/d < 40]$. For this situation, Sugiura (1978) proposed a corrected AIC (CAIC) criterion defined as

$$CAIC_j = AIC_j + \frac{2d(d+1)}{n-d-1}. \tag{2.49}$$

A word of caution, most statistical packages do not calculate $CAIC_j$, but calculate AIC_{MLE}. Users must make their own adjustments. One may simply, for example, substitute AIC_{MLE} for the AIC_j. Monte Carlo studies performed by McQuarrie and Tsai (1998) use both the statistic AIC_u and the criterion $CAIC$, where the latter includes the biased MLE for the variance estimate and Suguira's penalty correction. An estimate of AIC that includes an unbiased estimate of the variance and Suguira's penalty correction is represented by $CAIC_u$, a "doubly" corrected criterion.

Schwarz (1978) and Akaike (1978) developed model selection criteria using a Bayesian approach which incorporates a large penalty factor for over fitting. The criteria select models based upon the largest posterior probability of being correct. In large samples, their posterior probabilities are approximated using a Taylor series expansion. Scaling the first two terms in the series by n their criterion is labeled BIC for Bayesian information criterion (or also SIC for Schwarz information criterion or SBC for Schwarz-Bayesian criterion). Hannan and Quinn (1979) developed another criterion when analyzing autoregressive time series models. Applying their criterion to CNR models, the criterion is represented by HQ. Formula for the two criteria are

$$BIC_j = n \log(\sigma_j^2) + d \log(n), \tag{2.50}$$

$$HQ_j = n \log(\sigma_j^2) + 2d \log[\log(n)]. \tag{2.51}$$

One may again substitute either a MLE for σ_j^2, $\hat{\sigma}_j^2$, or the minimum variance unbiased estimator, s_j^2. When an unbiased estimate is substituted, the criteria are represented by BIC_u, and HQ_u, respectively. In either case, one investigates potential candidate models for a subset of variables that have the smallest BIC and HQ values. For very large samples, the HQ criterion behaves very much like the AIC criterion. Using the scaling factor $n/(n-d-1)$ for the HQ criteria, the scaled corrected criteria (CHQ) proposed by McQuarrie and Tsai (1998, p. 35) is defined

$$CHQ_j = \log(\hat{\sigma}_j^2) + \frac{2k_j \log[\log(n)]}{n-d-1}. \tag{2.52}$$

Many authors suggest the investigation of models that minimize s_j^2 or equivalently the sum of squares error criterion SSE_j since $s_j^2 = SSE_j/(n-k_j)$. However, since $1 - \tilde{R}_a^2 = s_j^2/(SST/n)$, the denominator is constant as k_j varies, so that minimizing s_j^2 is equivalent to selecting a model that maximizes that statistic $\tilde{R}_a^2$.

One may also relate Mallow's criterion to $\tilde{R}^2$. Letting $\tilde{R}_j^2$ denote the coefficient of determination for the submodel,

$$C_p = \left(\frac{1 - \tilde{R}_j^2}{1 - \tilde{R}^2} \right) (n - k) - n + 2k_j. \tag{2.53}$$

A common practice in the CR or CNR models with fixed X is to assume that the "true" model is within the class of models under study, a nested set of models. Thus, the best model is defined by a parameter space defined by a subset of the collected variables. Model selection criteria that select the true nested model asymptotically with probability one are said to be consistent. The criteria BIC and HQ are consistent. This is not the case for the selection criteria AIC, C_p, and $\tilde{R}_a^2$. When using these criteria, one tends to select a model with too many independent variables, McQuarrie and Tsai (1998, p. 42 & 370); however, the AIC_u and CAIC criteria tend to overfit least. If a researcher is not sure that the "true" model is among the nested variables, the model is nonnested; however, one may still want to locate a model that is an approximation to the true model. In this case, the approximation is usually evaluated by comparing the average minimum mean square error of prediction for any two models.

In large samples, a model selection criterion that chooses the model with minimum mean squared error is said to be asymptotically efficient. The FPE criterion δ_f^2, and the criteria AIC, C_p, and $\tilde{R}_a^2$ are all asymptotically efficient criteria. But, in small samples, they may lead to overfitting. No selection criterion is both consistent and asymptotically efficient, so there is not a single criterion that is best for all situations. However, based on extensive Monte Carlo studies conducted by McQuarrie and Tsai (1998) using random normal errors, they found that the asymptotically efficient criterion CAIC and the consistent criterion CHQ performed best. The criteria were most likely to find the correct or closest candidate model. They are least likely to under fit and minimize over fitting. For weakly identified models no criterion is best; however, criteria with weak penalty functions tend to overfit excessively.

Finally, the C_p and FPE criteria may only be used to select a submodel from within the class of nested models. The information criteria allow one to rank order potential candidate models whether the models are nested or nonnested. For a comprehensive review of model selection criteria for nonnested models, one may consult McAleer (1995).

Model Selection, Likelihood Ratio Tests

For the CNR model, the likelihood ratio test statistic for testing the null hypothesis that a subset of the regression coefficients β_i associated with any $h = p - m$ variables (excluding the intercept-even thought it is included in the regression model) is zero versus the alternative hypothesis that the coefficients are not zero, one may employ the F statistic

$$F = \frac{(n - k)}{n} \cdot \frac{(\tilde{R}^2 - \tilde{R}_m^2)}{(1 - \tilde{R}^2)} \sim F(\nu_h, \nu_e) \tag{2.54}$$

which has a central F distribution with degrees of freedom $\nu_h = h$ and $\nu_e = (n - k)$ when the null hypothesis is true, Neter, Kutner, Nachtsheim, and Wasserman (1996, p. 271). To evaluate the marginal effect of a single variable in the linear regression model, given that the other variables are in the model, one sets $h = 1$ and the F statistic reduces to the partial F test where $t = \sqrt{F}$ is the familiar Student t statistic.

To evaluate the contribution of the variables simultaneously, one is interested in testing the simultaneous hypotheses $H_i : \beta_i = 0$ against the alternative hypotheses $H_{a_i} : \beta_i \neq 0$. To test these hypotheses, we have p simultaneous tests, a simultaneous test procedure (STP). To control the overall familywise error rate (FWE) at a nominal level α, one may test the p hypotheses using either a step-down, step-up, or single-step procedure, Hochberg and Tamhane (1987), Troendle (1995), and Shaffer (2002). Using Holm's (1979) step-down sequential rejective Bonferroni method, one first (1) selects the α level for the model selection process and ranks the calculated P-values $P_{(i)}$ from largest to smallest: $P_{(1)} \geq P_{(2)} \geq \cdots \geq P_{(p)}$ associated with the p null hypotheses: $H_{(p)} : \beta_{(p)} = 0$. The second step (2) is to set $i = 1$. Then, for step (3) if $P_{(p-i+1)} > \alpha/(p - i + 1)$, one accepts all the remaining hypotheses $H_{(p-i+1)}, \ldots, H_{(1)}$ and STOPS. All variables are removed from the model. If, however, step (4) is $P_{(p-i+1)} \leq \alpha/(p - i + 1)$, then the hypothesis $H_{(p-i+1)}$ is rejected (the corresponding variable remains in the model) and one increments the index i by one, and returns to step (3). For Holm's method, we are sequentially rejecting hypotheses. Alternatively, by starting with the largest P-value and sequentially accepting hypotheses, one obtains Hochberg's (1988) step-up sequential Bonferroni method. For Hochberg's procedure, steps(1) and (2) above are identical to that of Holm's sequential procedure; however steps(3) and (4) differ. Step (3) becomes step (3*): if $P_{(i)} \leq \alpha/i$, one rejects all the remaining hypotheses $H_{(i)}, \ldots, H_{(p)}$ and STOPS, all variables are retained in the model. While step (4) becomes step (4*), if $P_{(i)} > \alpha/i$, one accepts the hypothesis $H_{(i)}$ (the variable is removed), and one again increments the index i by one and returns to step (3*). Again the process continues iteratively, repeating instead steps (3*) and (4*) until the process stops. Both these simultaneous test procedures control the FWE rate at the nominal level α and are both uniformly more powerful than the modified single-step Bonferroni (Dunn-Sidák) method which uses $\alpha_p = 1 - (1 - \alpha)^{1/p}$ for each test. Even though Hochberg's method is uniformly more powerful than Holm's method, many texts do not discuss the procedure. For example, Neter et al. (1996, p. 739-742) only illustrate Holm's procedure and N. R. Draper and Smith (1998) discuss neither. Even though the STP procedure may be used to evaluate the significance of a regression coefficient, it must be used with caution to screen variables since they are marginal tests and as such are only investigating the effect of a variable given the others are included in the linear model.

Exact single-step procedures are in general more complicated to apply in practice than step-down or step-up procedures since they require the calculation of the percentage points of the multivariate F distribution, Schmidhammer (1982). One may also use resampling based methods as suggested by Westfall and Young (1993). These procedures are again evaluating the marginal effect of a variable.

Pseudo single-step procedures have been proposed by many authors. For example, Takeuchi, Yanai, and Mukherjee (1982, p. 118) suggest keeping any variable in the model if the partial F statistic is larger than 2, while N. R. Draper and Smith (1998, p. 62) suggest the value 4. For a single model with p variables, the actual α level may be nearer to $1 - (1 - \alpha)^p$. If one were to evaluate the $q = 2^p - 1$ all possible models, the "actual" α value is much larger. As suggested by Akaike (1974) and Burnham and Anderson (1998, p. 62), the size of the test must be adjusted to reflect differences in the number of parameters and degrees of freedom. However, how to make the appropriate adjustments to ensure a FWE rate at some nominal level is not clear; therefore, such pseudo single-step methods are to be avoided.

When using a simultaneous test procedure to select variables in multiple linear regression model, no overall test of significance is performed. However, if one sets $m = 0$ in (2.54), the F test may be used to test that all the regression coefficients are simultaneously zero. Rejecting the overall test at some nominal level α, Rencher and Scott (1990) show that if the p partial F tests are performed at the same nominal α level that the FWE rate remains near the nominal level since the individual tests are protected by the rejected overall test. These tests are called protected F tests after Fisher's famous least significant difference (LSD) procedure. Using this approach, one would only retain those variables for which the calculated P value of the protected test is less than α. Again these tests are marginal tests and are not recommended for variable selection.

The likelihood ratio test defined in (2.54) may only be used to compare two nested models. Timm and Al-Subaihi (2001) developed a test statistic to compare any two nonnested candidate models.

Prediction

Having established a linear regression model using a calibration sample, one may be interested in knowing how well the estimated regression model predicts the criterion in the population from which it was drawn. The population parameter for population validity or predictive precision is ρ. This is usually evaluated using a sample estimate of the population multiple correlation coefficient, R. However, since the predictors are fixed and not random an estimate of R is not available. Thus, one must address the problem indirectly and ask how well does the estimated regression model predict the criterion in the population using another sample from the same population. To obtain the second (validation) sample one usually splits the original sample into two parts at random. This results in the calibration and validation samples. The validation sample is always less than $\frac{1}{2}$ of the original sample and usually contains $\frac{1}{4}$ to $\frac{1}{3}$ of the original sample observations. Now one is interested in estimating the population cross-validity coefficient, ρ_c. The square of this correlation is represented by ρ_c^2 to emphasize the calibration-validation nature of the parameter and to distinguish it from the population squared multiple correlation coefficient R^2. Lord (1950) and Nicholson (1960) were among the first to develop formula-based estimators for ρ_c^2 using fixed predictors. Burket (1963) developed a formula that does not require data splitting. For a comprehensive review of formula-based estimators using a cross-validation sample with fixed predictors, one may consult Raju, Bilgic, Edwards, and Fleer (1997).

Instead of measuring predictive precision using a correlation coefficient, one may instead use the mean squared error of prediction, δ_{cf}^2. Again, the subscript is modified to indicate the cross-validation nature of the statistic. For a discussion of this approach, one may consult the papers by Mallows (1973), Picard and Cook (1984), and Picard and Berk (1990). Using either criterion as a measure of predictive precision, data splitting results in less precise predictions, Horst (1966, p. 379), Picard and Cook (1984), and Picard and Berk (1990, p. 142). This is the case because data splitting severely affects the stability of the regression coefficients estimated from the calibration sample because the estimated standard errors of the estimated coefficients may be large due to the smaller sample size. Thus, the usefulness of data splitting in models with fixed predictors has questionable value.

Finally, one may obtain an independent sample of size n^* observations to validate a model with fixed predictor variables. Then, an unbiased estimate of the final FPE is obtained by evaluating the expression

$$\hat{\delta}_u^{2*} = \frac{\sum_{i=1}^n (y_i^* - \hat{\beta}_0 - \hat{\beta}_1' x_i^*)^2}{n^*}$$

$$= s^{2*} \left(1 + \frac{k}{n^*} \right). \tag{2.55}$$

Here the asterisk is used to indicate values in the independent validation sample. As a measure of predictive prediction one may simply compare $\hat{\delta}_u^{2*}$ with $\hat{\delta}_u^2$. The ratio of the estimates should be fairly close to one. If $\hat{\delta}_u^{2*} \gg \hat{\delta}_u^2$, the predictive value of the original model may not be reasonable.

As an alternative measure of predictive precision using the new n^* observations in the independent validation sample, one may calculate the Pearson product moment correlation between y_i^* and $\hat{y}_i^* | \hat{\beta}$ where $\hat{y}_i^* = (X_*'X_*)^{-1}X_*\hat{\beta}$ represents an estimate of the observation vector using the regression coefficients $\hat{\beta}$ from the calibration sample. We represent this criterion as $\hat{\rho}_c^2$. Model reliability is evaluated by the shrinkage in the fit index difference: $\tilde{R}_a^2 - \hat{\rho}_c^2$. If the fit shrinkage is small, less than about 10%, one usually concludes that the calibration estimates of the regression ceofficients are reliable.

Additional Comments

The procedures we have included in the discussion of CR and CNR models are a subset of a wider class of techniques known as all-possible regression procedures, since the researcher investigates all subsets of the variables to locate candidate models. There are of course many other procedures in this general category such as the $PRESS_j$ (prediction sum of squares) criterion which creates a cross-validation sample by withholding a single observation from the sample and numerous bootstrap techniques. Because Monte Carlo studies conducted by McQuarrie and Tsai (1998) show that the CAIC and CHQ criteria usually outperform these procedures, we have not included them in this text. One may consult the review by Krishnaiah (1982), the text by A. J. Miller (2002), and the review by Kandane and Lazar (2004). Model selection criteria that are based upon automatic statistical search algorithms such as forward, backward and stepwise methods have been excluded, by design, since the identification of a single model as "best" often eliminates the exploration of a wider class of potential models. In general, stepwise procedures are exploratory, do not control the experimentwise error rate at a nominal α level, and should be avoided except to perhaps find an appropriate subset of a very large set of variables for further evaluation and study by a procedure that evaluates all-possible regressions of the subset, Mantel (1970). Except for the FPE criterion, normality of the errors has been required for the techniques reviewed with fixed matrix X. When a continuous data set is essentially normal except for a few outliers, robust regression techniques are used to locate candidate models. If the dependent variable is categorical, quasi-likelihood techniques are employed. And if the errors are nonnormal, nonparametric regression procedures and wavelets may be used. For a discussion of these procedures, one may consult McQuarrie and Tsai (1998).

2.5.2   Random Classical and Jointly Normal Regression Models

Estimation and Model Fit

To give $\tilde{R}^2$ and $\tilde{\eta}^2$ meaning, the CR model must hold conditionally for every value of x. The sampling scheme may be performed in such a manner that the observations (y_i, x_i') are independently and identically distributed to yield the random pairs, the JNR model. The observations x_i are stochastic or random. By Theorem 1.4, the conditional joint distribution is multivariate normal with mean $\mu_{y \cdot x} = \mathcal{E}(y|x) \equiv \mu_{1 \cdot 2}$ and variance $\sigma_{y \cdot x}^2 \equiv \sigma_{1 \cdot 2}^2$. Then, the correlation ratio is defined as

$$\eta^2 = 1 - \frac{\sigma_{y \cdot x}^2}{\sigma_y^2}. \tag{2.56}$$

The correlation ratio is the maximum correlation between y and any function of a random vector x of independent variables. If the relationship between y and x is linear in x, then the correlation ratio is equal to the square of the population coefficient of determination or the population squared multiple correlation, R^2. And,

$$\eta^2 = R^2 = 1 - \frac{\sigma_{y \cdot x}^2}{\sigma_y^2}. \tag{2.57}$$

Although (2.57) looks very much like (2.37), as noted earlier, (2.37) is not an estimate of R^2. When $R^2 = 1$, the population coefficient of determination also measures the proportional reduction in the expected squared prediction error that is associated with the best linear unbiased prediction of $\mu_{y \cdot x}$ rather than the marginal expectation $\mathcal{E}(y)$ for predicting y given x in the population. Adding the (sufficient) condition that the sample of observations (y_i, x_i') is sampled from a multivariate normal distribution with mean vector $\mu_{y \cdot x} \equiv \mu_{1 \cdot 2} = z$ and nonsingular covariance matrix

$$\Sigma = \begin{pmatrix} \sigma_y^2 & \sigma_{yx}' \\ \sigma_{xy} & \Sigma_{xx} \end{pmatrix} \equiv \begin{pmatrix} \sigma_1^2 & \sigma_{12}' \\ \sigma_{21} & \Sigma_{22} \end{pmatrix}, \tag{2.58}$$

by Theorem 1.4 the conditional normal distribution of $y|x$ has mean

$$\begin{aligned} \mu_{y \cdot x} &= \mu_y + \sigma_{yx}' \Sigma_{xx}^{-1} (x - \mu_x) \\ &= (\mu_y - \sigma_{yx}' \Sigma_{xx}^{-1} \mu_x) + \sigma_{yx}' \Sigma_{xx}^{-1} x \\ &= \beta_0 + \beta_\ell' x \end{aligned} \tag{2.59}$$

with unknown parameter vector $\beta' = (\beta_0, \beta_\ell')$ and conditional variance

$$\sigma_{y \cdot x}^2 = \sigma_y^2 - \sigma_{yx}' \Sigma_{xx}^{-1} \sigma_{xy}. \tag{2.60}$$

To stress that the conditional variance is independent of x, it is often simply denoted by σ^2. The square of the correlation between y and $y|x$ is

$$\begin{aligned} R^2 &= [\mathrm{corr}(y, y|x)]^2 = \frac{[\mathrm{cov}(y, y|x)]^2}{\sigma_y^2 \sigma_x^2} \\ &= \frac{\sigma_{yx}' \Sigma_{xx}^{-1} \sigma_{xy}}{\sigma_y^2} \\ &= 1 - \frac{\sigma_{y \cdot x}^2}{\sigma_y^2}, \end{aligned} \tag{2.61}$$

since $\sigma_z^2 = \sigma_{yx}' \Sigma_{xx}^{-1} \sigma_{xy} = \mathrm{cov}(y, y|x)$, the population coefficient of determination. R^2 measures the "linear relation" between y and x. If $R^2 = 1$, then y is an exact linear function of $\mu_{y \cdot x}$ for the JNR or RCN models.

To estimate $\mu_{y \cdot x}$, one substitutes maximum likelihood estimates for the unknown parameter vector β. The maximum likelihood estimates for the elements of the parameter vector are

$$\hat{\beta}_\ell = S_{xx}^{-1} s_{xy}, \tag{2.62}$$

$$\hat{\beta}_0 = \bar{y} - s_{yx}' S_{xx}^{-1} s_{xy}, \tag{2.63}$$

and are easily shown to be identical to the estimates obtained using the CR or CRN models, Graybill (1976, p. 380). To estimate R^2, Fisher (1924) suggested replacing σ_y^2 and $\sigma_{y \cdot x}^2$ with unbiased estimators. And, given that the independent variables are random, we may estimate the population squared multiple correlation coefficient using (2.35), and represented as $\widehat{R}^2$ since x is random. For random x, the coefficient of determination R^2 is estimated by

$$\widehat{R}^2 = 1 - \frac{s_{y \cdot x}^2}{s_y^2} \tag{2.64}$$

where $s_y^2 = \sum_i (y_i - \bar{y})^2/(n-1) = SST/(n-1)$ and $\sigma_{y \cdot x}^2 = MSE$. And, the adjusted coefficient of determination for random x is

$$R_a^2 = 1 - \frac{(n-1)(1-\widehat{R}^2)}{(n-k)} = \widehat{R}^2 - \frac{p(1-\widehat{R}^2)}{(n-p-1)}. \qquad (2.65)$$

Since R_a^2 is linear in $\widehat{R}^2$, one can show (assuming joint multivariate normality of y and x) that the $\mathcal{E}(R_a^2) - R^2 \simeq -2R^2(1-R^2)/n$ so that the bias in R_a^2 is independent of the number of random independent variables p and never positive, M. W. Browne (1975a). When $R^2 \neq 0$ in the population, then R_a^2 may lead to negative estimates. For a review of other formula used to estimate R^2, one may consult Raju et al. (1997).

Prediction

Having fit a model using the jointly normal (JNR) model or a random conditionally normal calibration sample, RCN model, one may evaluate the predictive precision of the model by evaluating the correlation between $\hat{y} = X\hat{\beta}$ in the calibration sample with an independent set of observation y in another independent validation sample. The square of the population correlation between the fitted values estimated in the calibration sample and the observations in the new validation sample is represented as ρ_{rc}^2. The square of the correlation between y and $\hat{y}$ is

$$\rho_{rc}^2 = [\text{corr}(y, \hat{y})]^2 = \frac{[\text{cov}(y, \hat{y})]^2}{\sigma_y^2 \sigma_{\hat{y}}^2}$$

$$= \frac{(\hat{\beta}_\ell' \sigma_{xy})^2}{\sigma_y^2 (\hat{\beta}_\ell' \Sigma_{xx} \hat{\beta}_\ell)}. \qquad (2.66)$$

The quantity ρ_{rc}^2 is a measure of prediction between the observation vector in the validation sample that is predicted using the parameter vector estimated in the calibration sample. Although (2.61) and (2.66) are equal; in (2.61) R^2 is being used to summarize a linear model that describes the linear relationship between y and a set of p random predictors in the calibration sample and in (2.66) ρ_{rc}^2 is being used to determine predictive validity or predictive precision using a validation sample, M. W. Browne (1975a). If $\widehat{R}^2$ is used to estimate R^2 with random x then it is well known that $\widehat{R}^2$ will never decrease as more independent variables are added to the linear model. However, if $\hat{\beta}_\ell'$ in (2.66) is estimated from the calibration sample, the estimate for ρ_{rc}^2 may decrease when p is increased.

To estimate ρ_{rc}^2 is complicated since it is a function of both the population parameters and the calibration sample, thus the subscript rc. The parameter ρ_{rc}^2 is a random variable while R^2 is a population parameter. Using the fact that the distribution of the estimated parameter vector $\hat{\beta}_\ell$ with random x has a multivariate t distribution (see for example, Wegge, 1971, Corollary 1), assuming (y_i, x_i') are sampled from a multivariate normal distribution, M. W. Browne (1975a) obtains a linear biased estimator of the expected value of ρ_{rc}^2 defined as

$$\hat{\rho}_{rc}^2 = \mathcal{E}(\rho_{rc}^2) = \frac{(n-p-3)(R_a^2)^2 + R_a^2}{(n-2p+2)R_a^2 + p} \tag{2.67}$$

where R_a^2 is given in (2.65), provided $R_a^2 \geq 0$. If $R_a^2 < 0$, Browne recommends using $(R_a^2)^2 = (\widehat{R})^2 - 2p(1 - \widehat{R}^2)/(n-1)(n-p-1)$ where $R_a^2 = \sqrt{(R_a^2)^2}$ since $\widehat{R}^2$ is always greater than or equal to zero. The estimator $\hat{\rho}_{rc}^2$ is sometimes called the shrunken estimate of the coefficient of determination since $\hat{\rho}_{rc}^2 < \widehat{R}^2$. Precision is lost by using $\hat{\beta}$ to estimate β in the population. In developing the estimate for $\mathcal{E}(\rho_{rc}^2)$, Browne makes the assumption that the random $(1 \times p)$ vector x_i' is not chosen from a larger set of n predictors. Thus, model selection criteria are not applied to the independent random variables; the researcher must specify relevant random predictors. And, secondly, the unknown parameter vector β must be the same for any other sample of $(1 \times p)$ predictors in the population.

For the JNR and RCN models, the estimator $\hat{\rho}_{rc}^2$ was developed by M. W. Browne (1975a) as a measure of predictive precision. Another method for estimating precision in prediction is to use the mean squared error of prediction, Feldstein (1971). This is an alternative, but not equivalent method for evaluating predictive precision. This problem was also addressed by M. W. Browne (1975b). To use the mean squared error of prediction as a measure of precision, Browne includes the same assumptions as employed in the development $\hat{\rho}_{rc}^2$: the random x_i contains all relevant variables and, secondly, that the estimator of the unknown parameter vector β may be applied to any other sample in the population. Then, the mean squared error of prediction, called random predictive error (RPE), is defined as

$$d_r^2 = \mathcal{E}[y - \hat{y}(x|\hat{\beta})]^2 \tag{2.68}$$

where the expectation operation is taken over all random vectors x and y for fixed values of $\hat{\beta}$. Then assuming that the pairs (y_i, x_i') are sampled from a multivariate normal distribution, the maximum likelihood estimates of the parameters β_0 and β_ℓ' are

$$\hat{\beta}_\ell = \widehat{\Sigma}_{xx}^{-1} \hat{\sigma}_{xy} = S_{xx}^{-1} x_{xy}' \tag{2.69}$$

$$\hat{\beta}_0 = \bar{y} - \hat{\beta}_\ell' \bar{x} = \bar{y} - s_{xy}' S_{xx}^{-1} \bar{x} \tag{2.70}$$

where the covariance matrix is $S_{xx} = n\widehat{\Sigma}_{xx}/(n-1)$. Under joint multivariate normality or conditional normality, Stein (1960) shows that the mean squared error of prediction in multiple regression is

$$d_r^2 = \sigma_{y \cdot x}^2 + [(\hat{\beta}_0 - \beta_0) + (\hat{\beta}_\ell - \beta_\ell)' \mu_x)]^2 + (\hat{\beta}_\ell - \beta_\ell)' \Sigma_{xx}^{-1} (\hat{\beta}_\ell - \beta_\ell) \tag{2.71}$$

where $\sigma_{y \cdot x}^2$ is the conditional variance of y given x. The expected value of RPE, $\mathcal{E}(d_r^2)$, is (Helland & Almoy, 1994, Theorem 1)

$$\delta^2 = \mathcal{E}(d_r^2) = \sigma_{y \cdot x}^2 \left(1 + \frac{1}{n}\right) \left(\frac{n-1}{n-p-2}\right). \tag{2.72}$$

The value is defined provided $n > p + 2$ and the value of δ^2 (the expected predicted error) is seen to depend on the sample size n and the number of predictors p. Its value becomes very large as p gets closer to n. The MLE of δ^2 is obtained by estimating the conditional variance with its MLE. Then,

$$\hat{\delta}^2_{MLE} = \mathcal{E}(d_r^2) = \hat{\sigma}^2 \left(\frac{n+1}{n}\right) \left(\frac{n-2}{n-p-2}\right). \tag{2.73}$$

To remove the bias, one may replace $\sigma^2_{y \cdot x}$ with its minimum variance unbiased estimator s^2. The unbiased estimator of δ^2 and its associated variance are, respectively,

$$\hat{\delta}^2_{ru} = \frac{s^2(n+1)(n-2)}{(n-k)(n-p-2)}$$

$$= \frac{s_p^2(n+1)(n-2)}{n}, \tag{2.74}$$

$$\text{var}\left(\hat{\delta}^2_{ru}\right) = \frac{2\sigma^4(n+1)^2(n-2)^2}{n^2(n-p-2)^2(n-p-1)}, \tag{2.75}$$

where $s_p^2 = MSE/(n-p-2)$. However, $\hat{\delta}^2_{ru}$ is still a biased estimator of δ^2. Roecker (1991) showed that models selected using the criterion tend to under estimate prediction error, even though over all calibration samples, we expect that the $\mathcal{E}(\hat{\delta}^2_{ru} - d_r^2) = 0$. M. W. Browne (1975b) showed that the mean squared error of estimation is

$$\mathcal{E}([\hat{\delta}^2_{ru} - d_r^2])^2 = \left[\frac{2\sigma^2(n-2)}{n^2(n-p-2)^2}\right] \left[\frac{(n-2)(n^2+3n-p)}{n-p-1} + \frac{p(n^2+2n+3)}{n-p-4}\right]. \tag{2.76}$$

To obtain an unbiased estimator of δ^2 involves a second independent validation sample of size n^* of new observations. M. W. Browne (1975b) showed that the cross-validation estimate of δ^2, based upon n^* new observations, is

$$\hat{d}_r^{2*} = \sum_{i=1}^{n} \frac{(y_i^* - \hat{\beta}_0 - \hat{\beta}_\ell' x_i^*)^2}{n^*} \tag{2.77}$$

a conditional unbiased estimator of $\mathcal{E}(d_r^2)$. The conditional mean squared error of estimation for the independent cross-validation procedure is also provided by Browne as

$$\mathcal{E}([\hat{d}_r^{2*} - d_r^2])^2 = \mathcal{E}([\hat{d}_r^{2*} - d_r^2]|\hat{\beta})^2 = \left[\frac{2\sigma^4(n+2)(n-2)(n-4)}{n^*n(n-p-2)(n-p-4)}\right]. \tag{2.78}$$

Comparing the two estimates of mean squared error, one observes that if the calibration sample and the validation sample are of equal size $(n^* = n)$, then $\hat{d}_r^{2*}$ is a more precise estimator of δ^2 than $\hat{\delta}^2_{ru}$. This is not the case if the two samples are pooled. However, the variance of $\hat{d}_r^{2*}$ will be less than the variance of $\hat{\delta}^2_{ru}$ if one splits the calibration sample to obtain the validation sample, since $n^* < n$. The variance of the RPE is

$$\text{var}(d_r^2) = \left[\frac{2\sigma^2(n-2)}{n^2(n-p-2)^2}\right] \left[n - 2 + \frac{p(n^2+2n+3)}{(n-p-4)}\right]. \tag{2.79}$$

Roecker (1991) showed that data splitting for model validation should be avoided since it neither improves predictive precision or the precision of the estimate.

Model Selection, Likelihood Ratio Tests

Random predictive error (RPE) assumes that the observations (y_i, x_i') are identically distributed and sampled from a multivariate normal distribution where $\mathcal{E}(y) = X\beta$ represents the true model or that $(y_i | X_i = x_i')$ is conditionally normal, the RCN model. The value of p is fixed and d_r^2 is a random variable with expected value $\delta^2 = \mathcal{E}(d_r^2)$. While one may expect $\hat{\delta}_{u,m}^2 \le \hat{\delta}_u^2$ for some subset of $m < p$ random predictors, this does not necessarily ensure that the population values satisfy the inequality $\delta_m^2 \le \delta^2$ since d_r^2 is a random variable and $\mathcal{E}(\hat{\delta}_u^2) \ne d_r^2$. Thus, instead of calculating values for RPE it is more appropriate to test the hypothesis that some subset of predictors $x_1, x_2, \ldots, x_m$ will result in a lower expected value. The null hypothesis to be tested is $H : \delta_m^2 \le \delta^2$ versus the alternative hypothesis $H_A : \delta_m^2 > \delta^2$ where δ_m^2 and δ^2 denote the expected prediction errors based upon m and p variables, respectively. To test the equivalent null hypothesis $H : \rho_{h|m}^2 \le {}^h/_{(n-m-2)}$, Browne showed that one may use the sampling distribution of the square of the partial multiple coefficient of determination of y with the $h = p - m$ variables $x_{m+1}, \ldots, x_{m+h}$ given the subset of variables $x_1, x_2, \ldots, x_m$. While one may use percentage tables developed by S. Y. Lee (1972) based upon S. Y. Lee (1971) or newer methods developed by Marchand (1997), to calculate the exact percentage points of the distribution of the square of the partial multiple coefficient of determination, M. W. Browne (1975b) developed a simple approximate F-like likelihood ratio statistic for the test. Given the wider availability of percentage points for Fisher's F distribution, the statistic is

$$F^* = \left[\frac{(n-p-1)(n-p-2)}{h(2n-p-m-3)} \right] \left[\frac{(\widehat{R}^2 - \widehat{R}_m^2)}{(1 - \widehat{R}^2)} \right]$$

$$= \frac{(n-p-2)F}{(2n-p-m-3)} \tag{2.80}$$

where $\widehat{R}^2$ and $\widehat{R}_m^2$ are the coefficient of determination for p and m independent variables included in the model, respectively. Browne recommends approximating the distribution of the statistic F^* with Fisher's F distribution, $F(\nu_1, \nu_2)$, the degrees of freedom $\nu_1 = h(2n-p-m-3)^2/\{(n-m-1)92n-p-m-4) + (n-p-2)^2\}$ and $\nu_2 = (n-p-1)$.

Setting $m = p - 1$, $h = 1$ in (2.80) the statistic $F^* \approx F/2$, one may also use the test statistic to evaluate the contribution of a single variable that has one reducing the expected mean squared error in prediction. To evaluate whether or not a single variable should be removed from the model, one may again use Hochber's step-up sequential Bonferroni procedure. However, as in variable selection, the tests are only evaluating the marginal effect of a variable to the overall predictive precision of the model.

Prediction, Reduction Methods

In the evaluation of the expected value of predicted error defined in (2.72), the predictor variables were considered as a set of predefined variables or a subset of variables was chosen for deletion to improve prediction precision, since the expected prediction error increases as the number of variables approaches the sample size. An alternative strategy to improve predictive precision is to reduce the number of predictor variables by considering, for example, linear combinations of variables. Such methods include principal component regression (PCR), latent root regression (LRR), partial least squares regression (PLS), and reduced rank regression (RRR), among others. For example, PCR creates linear combinations of the independent variables that account for maximum variation in the independent (predictor) variables, ignoring the criterion variable y; RRR selects linear combinations of the predictor variables that have maximum covariance with the response variable; and PLS seeks to explain both response and predictor variation. Ridge regression and the "lasso" method of Tibshirani (1996) are often used to reduce or eliminate collinearity among the predictors. These methods are not included in this text; instead, one may consult Helland and Almoy (1994) for additional references.

Additional Comments

While the unbiased estimator $\hat{\delta}^2_{ru}$ of $\delta^2 = \mathcal{E}(d_r^2)$ may be used as a measure of model fit, it may not always reliably choose a best subset of predictors since it is not an unbiased estimator of RPE, d_r^2. However, the C_p and information criteria may be used to select an appropriate submodel since the CRN model may be obtained from the JNR model by conditioning on the design matrix, Graybill (1976, p. 382). While the test statistic in (2.54) may also be used to evaluate the marginal effect of a single variable whether or not the matrix X is fixed or random, the power of the F test is not the same for fixed and random X. Cohen (1988, p. 414) related a quantity f^2 which he calls effect size to the noncentrality parameter λ of the F statistic given in (2.54) assuming fixed X and then relates η^2 defined in (2.56) for random X using the relation that $\eta^2 = f^2/(1 + f^2)$. As pointed out by Gatsonis and Sampson (1989), Cohen's approach is only an approximation to exact power with random X, although quite good in many applications. They recommend increasing sample size calculations by about 5 observations if the number of independent variables p is less than 10, when using Cohen's approximation if the independent variables are random, but are being treated as fixed. Maxwell (2000) also discusses sample size calculations when one is only interested in a single regression coefficient; however, he also employs Cohen's approximation. Mathematical details for exact power and sample size calculations are provided by Sampson (1974). Taylor and Muller (1995) showed how to construct exact bounds for the noncentrality parameter λ, for fixed X. And, Glueck and Muller (2003) developed a procedure for estimating the power of the F test for linear models when the design matrix X is random, fixed or contains both fixed and random predictors. Jiroutek, Muller, Kupper, and Stewart (2003) developed new methods for sample size estimation using confidence interval-based methods instead of power. Power analysis for regression analysis and ANOVA are discussed in Chapter 10.

2.6 LINEAR MIXED MODELS

The general fixed effects univariate linear model was given in (1.1) as

$$y_{n \times 1} = X_{n \times k} \beta_{k \times 1} + e_{n \times 1}. \tag{2.81}$$

To establish a general form for the GLMM, we add to the model a vector of random effects b and a corresponding known design matrix Z. The model is

$$\Omega_0 : y_{n \times 1} = X_{n \times k} \beta_{k \times 1} + Z_{n \times h} b_{h \times 1} + e_{n \times 1}. \tag{2.82}$$

The vector b is usually partitioned into a set of t subvectors to correspond to the model parameters as

$$b' = \begin{pmatrix} b'_1 & b'_2 & \dots & b'_t \end{pmatrix}. \tag{2.83}$$

For example, for a two-way random model, $t = 3$ and $\mu_1 = \alpha$, $\mu_2 = \beta$ and $\mu_3 = \gamma$ if $y_{ijk} = \mu + \alpha_i + \beta_j + \gamma_{ij} + e_{ijk}$ and μ is a fixed effect. For the linear model in (2.82), we assume that

$$\mathcal{E}(y) = X\beta \qquad \mathcal{E}(y|u) = X\beta + Zb \tag{2.84}$$

so that the random error vector e is defined as

$$e = y - \mathcal{E}(y|u) \tag{2.85}$$

where

$$\mathcal{E}(y) = X\beta \qquad \mathcal{E}(e) = 0 \qquad \mathcal{E}(b) = 0. \tag{2.86}$$

The vector e has the usual structure, $\text{cov}(e) = \sigma^2 I_n$. However, because the b_i are random, we assume

$$\text{cov}(b_i) = \sigma_i^2 I_{q_i} \qquad \qquad \text{for all } i \tag{2.87}$$
$$\text{cov}(b_i, b_j) = 0 \qquad \qquad \text{for all } i \neq j \tag{2.88}$$
$$\text{cov}(b, e) = 0 \tag{2.89}$$

where q_i represents the number of elements in b_i. Thus the covariance matrix for b is

$$G = \text{cov}(b) = \begin{pmatrix} \sigma_1^2 I_{q_1} & 0 & \dots & 0 \\ 0 & \sigma_w^2 I_{q_2} & \dots & 0 \\ \vdots & \vdots & \ddots & \vdots \\ 0 & 0 & \dots & \sigma_t^2 I_{q_t} \end{pmatrix}. \tag{2.90}$$

Partition Z conformable with b, so that $Z = (z_1, z_2, \ldots, z_t)$ yields

$$
\begin{aligned}
y &= X\beta + Zb + e \\
&= X\beta + \sum_{i=1}^{t} Z_i b_i + e
\end{aligned}
\tag{2.91}
$$

so that the covariance matrix for y becomes

$$
\begin{aligned}
\Omega &= ZGZ' + \sigma^2 I_n \\
&= \sum_{i=1}^{t} \sigma_i^2 Z_i Z_i' + \sigma^2 I_n.
\end{aligned}
\tag{2.92}
$$

Since e is a random vector just like b_i, we may write (2.82) as

$$
y = X\beta + Zb + b_0
\tag{2.93}
$$

and (2.91) becomes

$$
y = X\beta + \sum_{i=0}^{t} Z_i b_i
\tag{2.94}
$$

where $Z_o \equiv I_n$ and $\sigma_0^2 = \sigma^2$. Thus, the covariance structure for Ω may be written as

$$
\Omega = \sum_{i=0}^{t} \sigma_i^2 Z_i Z_i'.
\tag{2.95}
$$

As an example of the structure given in (2.92) assume b has one random effect where

$$
y_{ij} = \mu + \alpha_i + e_{ij} \qquad i = 1,2,3,4 \text{ and } j = 1,2,3,
\tag{2.96}
$$

the α_i are random effects with mean 0 and variance σ_α^2, and the covariances among the α_i, and the e_{ij} and α_i are zero; then the

$$
\operatorname{cov}(y_{ij}, y_{i'j'}) = \begin{cases} \sigma_\alpha^2 + \sigma^2 & \text{for } i = i' \text{ and } j = j' \\ \sigma^2 & \text{for } i = i' \text{ and } j \neq j' \\ 0 & \text{otherwise} \end{cases}
\tag{2.97}
$$

or

$$
\begin{aligned}
\operatorname{cov}\begin{pmatrix} y_{11} \\ y_{12} \\ y_{13} \end{pmatrix} &= \begin{pmatrix} \sigma_\alpha^2 + \sigma^2 & \sigma_\alpha^2 & \sigma_\alpha^2 \\ \sigma_\alpha^2 & \sigma_\alpha^2 + \sigma^2 & \sigma_\alpha^2 \\ \sigma_\alpha^2 & \sigma_\alpha^2 & \sigma_\alpha^2 + \sigma^2 \end{pmatrix} \\
&= \sigma_\alpha^2 11' + \sigma^2 I \\
&= \sigma_\alpha^2 J + \sigma^2 I \\
&= \Sigma_1
\end{aligned}
\tag{2.98}
$$

where $J = 11'$. Hence, $\Omega = \text{diag}(\Sigma_i)$,

$$\Omega = \begin{pmatrix} \Sigma_1 & 0 & 0 & 0 \\ 0 & \Sigma_2 & 0 & 0 \\ 0 & 0 & \Sigma_3 & 0 \\ 0 & 0 & 0 & \Sigma_4 \end{pmatrix} \tag{2.99}$$

where $\Sigma_i = \sigma_\alpha^2 J + \sigma^2 I$ or using direct sum notation,

$$\Omega = \oplus_{i=1}^4 \Sigma_i = \oplus_{i=1}^4 (\sigma_\alpha^2 J + \sigma^2 I). \tag{2.100}$$

For our simple example, we let $\Sigma = \Sigma_i = \sigma_\alpha^2 11' + \sigma^2 I$. The covariance matrix Σ is said to have compound symmetry structure.

Experimental designs that permit one to partition the random vector b as in (2.82) where $b_i \sim IN(0, \sigma_i^2 I)$, $e \sim IN(0, \sigma^2 I)$, and the $\text{cov}(b_i, e) = 0$ for all i such that the covariance matrix Ω has the simple linear structure defined in (2.92) permits one to analyze mixed linear designs called variance component models. Because the components σ_i^2 are unknown, estimation of β and b_i, and tests of hypotheses are more complex. We address estimation of the parameter vectors β and b in Chapter 4 and discuss an alternative analysis of the model using the hierarchical (multilevel) modeling approach in Chapter 11. Because estimation of the variance components is of critical interest in variance component models, we review the estimation of the variance components and the construction of confidence intervals for balanced GLMM designs using the ANOVA model.

For variance component models, expression (2.82) and (2.92) state that both the mean and the variance of the observation vector have linear structure. While the matrices $X, Z_1, \ldots, Z_t$ are known, the mean of y and the covariance of y contain the unknown parameter β and the unknown variance components $\theta = (\sigma_1^2, \sigma_2^2, \ldots, \sigma_t^2)$. To estimate the variance components using the ANOVA method, only the mean square terms of the random effects are used to estimate the variance components. Letting SS_t represent the sum of squares for the tth effect and ν_t the associated degrees of freedom, one equates the mean square (SS_t / ν_t) to their expected values and solves the associated equations to estimate the unknown variance components. Letting ε represent vector of expected mean squares, the set of equations is represented as

$$\varepsilon = L\theta \tag{2.101}$$

for a suitable nonsingular matrix L. By ordering the unknown elements of θ appropriately, the matrix L may be represented as an upper triangular matrix. The unique estimate of the variance components is then given by

$$\hat{\theta} = L^{-1}\hat{\varepsilon} \tag{2.102}$$

where $\hat{\varepsilon}$ is the vector of mean squares. While the elements of the expected mean square vector are nonnegative, some of the elements $\hat{\sigma}_i^2$ in the variance component vector may be negative. When this occurs, the associated variance component parameter is set to zero.

Letting

$$\phi = \sum_t c_t \varepsilon_t \tag{2.103}$$

be a linear function of the expected mean squares, an unbiased estimate of ϕ is

$$\hat{\phi} = \sum_t c_t \left(\frac{SS_t}{\nu_t} \right). \tag{2.104}$$

Following Burdick and Graybill (1992), one may estimate an individual component, a linear function of the components, or sums and differences of the components. The three situations are represented as

Case 1 : Individual Component

$$\phi = \varepsilon_t \tag{2.105}$$

for some $t > 0$.

Case 2 : Linear Combination of Components

$$\phi = \sum_t c_t \varepsilon_t \tag{2.106}$$

for some $c_t > 0$ and $t > 0$.

Case 3 : Sum and Differences of Components

$$\phi = \sum_t c_t \varepsilon_t - \sum_t d_t \varepsilon_t \tag{2.107}$$

for some $c_t, d_t > 0$ and $t > 0$.

To obtain estimates for the variance components, we consider three cases.

Case 1 : to obtain a confidence interval for an individual component involving one mean square, recall that SS_t / ε_t is distributed as a chi-square distribution with ν_t degrees of freedom. Thus a $1 - 2\alpha$ confidence interval for the component ϕ is

$$\hat{\phi} - \frac{SS_t}{\chi^2_{\alpha, \nu_t}} \leq \phi \leq \hat{\phi} + \frac{SS_t}{\chi^2_{1-\alpha, \nu_t}} \tag{2.108}$$

where $\chi^2_{1-\alpha, \nu_t}$ is the $(1 - \alpha)$ critical value for a chi-square random variable with ν_t degrees of freedom.

Case 2 : for this situation there are several approximate methods for calculating a confidence set, Burdick and Graybill (1992). The SAS procedure PROC MIXED uses Sattherthwaite's method by using the CL option. However, the modified large sample (MLS) method, a modification of Welch's method proposed by Graybill and Wang (1980), is superior to Sattherthwaite's approximation. The MLS method yields exact intervals when all but one of the ε_t are zero or when all the degrees of freedom ν_t tend to infinity, except one. Graybill and Wang's confidence set is defined as

$$\hat{\phi} - \sqrt{\sum_t G_t^2 c_t^2 \left(\frac{SS_t}{\nu_t}\right)^2} \leq \sum_t c_t^2 \varepsilon_t \leq \hat{\phi} + \sqrt{\sum_t H_t^2 c_t \left(\frac{SS_t}{\nu_t}\right)^2} \quad (2.109)$$

where $G_i = 1 - \nu_t / \chi^2_{\alpha,\nu_t}$ and $H_i = \nu_t / \chi^2_{1-\alpha,\nu_t} - 1$.

Case 3 : for this case, one may use a method proposed by Howe (1974) since ϕ is the difference of two independent random variables or use a procedure proposed by Ting, Burdick, Graybill, Jeyaratnam, and Lu (1990). Their procedure is based on the Cornish-Fisher approximation and except for the extreme tails of the chi-square distribution, the procedure yields confidences set near the nominal level. While one may perform the necessary calculations to create MLS confidence sets, the calculations for Ting's et al. method are more complicated and currently neither is available in SAS. However, Hess and Iyer (2002) have developed a SAS macro available at the web site: www.stat.colostate.edu/~hess/MixedModels.htm, to perform the calculations for any mixed saturated ANOVA model involving five or fewer factors (not including the error term). A GLMM is saturated if it includes all possible main effect and interactions of all orders. We illustrate the macro in the analysis of mixed model designs in Chapter 11.

2.7 ONE-WAY ANALYSIS OF VARIANCE

For ANOVA designs without restrictions, model (1.1) is again assumed. ANOVA designs can be analyzed using a full rank model or an overparameterized less than full rank model. The full rank model is easy to understand, whereas the overparameterized model requires a knowledge of estimable functions, generalized inverses or side conditions. For the one-way ANOVA design, the full rank model is

$$y_{ij} = \beta_i + e_{ij} \quad i = 1, \ldots, I; j = 1, \ldots, J$$
$$e_{ij} \sim N(0, \sigma^2). \quad (2.110)$$

In this chapter we will discuss only the full rank model.

The vector of observations $y_{n \times 1}$ contains $n = IJ$ elements, the design matrix $X_{n \times k}$ contains 1's and 0's representing cell membership and has rank $k = I \leq n$, $\beta_{k \times 1}$ is a parameter vector of cell means, and $e_{n \times 1}$ is a vector of random errors. The parameter vector is estimated using (2.4), which is equivalent to a vector of observed cell means, $\hat{\beta} = \bar{y}$.

All linear parametric functions $\psi = c'\beta$ are estimable and estimated by $\hat{\psi} = c'\hat{\beta}$, with the estimated variance $\hat{\sigma}^2_{\hat{\psi}} = \text{var}(\hat{\psi})$ given in (2.6). The unbiased estimator for the unknown variance is

$$s^2 = \frac{(y - X\hat{\beta})'(y - X\hat{\beta})}{(n - k)} \tag{2.111}$$

where $k = I$. The null hypothesis may be written in the form $H_0 : C\beta = \xi$, where the hypothesis test matrix C has full row rank; thus, the number of linearly independent rows in the matrix C is $\nu_h = (I - 1)$, where I is the numbers of cells in the design or the column rank of the design matrix. The hypothesis test matrix C is not unique. One constructs the matrix based upon the specific contrasts of interest to the researcher. For a design with an equal number of observations n in each cell, the matrix C is an orthogonal contrast matrix if $C'C = D$, where D is a diagonal matrix. For unequal cell sizes n_i the matrix C is an orthogonal contrast matrix if $C(X'X)^{-1}C' = CD^{-1}C'$, where D is a diagonal matrix with diagonal entries n_i. For designs with unequal sample sizes n_i, nonorthogonal designs, the matrix C of contrasts is data dependent; it depends on the sample sizes in a study. While one may construct an orthogonal hypothesis test matrix that is not data dependent, it may not be used to test hypotheses since the rows of the matrix are no longer linearly independent.

2.7.1   Unrestricted Full Rank One-Way Design

As a numerical example of the analysis of an unrestricted one-way design with equal sample sizes n, the data from Winer (1971, p. 213) are analyzed using SAS. The data were also analyzed by Timm and Carlson (1975, p. 51). The data for the example are in the file 2_7_1.dat and represent measurements from four treatment conditions.

The hypothesis of interest is whether the means for the four normal populations are equal: $H_0 : \beta_1 = \beta_2 = \beta_3 = \beta_4$. The null hypothesis is equivalently written as:

$$\begin{aligned}
\beta_1 - \beta_4 &= 0 \\
\beta_2 - \beta_4 &= 0 \\
\beta_3 - \beta_4 &= 0.
\end{aligned} \tag{2.112}$$

Thus, using the representation $H_0 : C\beta = \xi$, we have

$$\begin{pmatrix} 1 & 0 & 0 & -1 \\ 0 & 1 & 0 & -1 \\ 0 & 0 & 1 & -1 \end{pmatrix} \begin{pmatrix} \beta_1 \\ \beta_2 \\ \beta_3 \\ \beta_4 \end{pmatrix} = \begin{pmatrix} 0 \\ 0 \\ 0 \end{pmatrix}.$$

In Program 2_7_1.sas the data set is read and printed. The analysis is performed using the PROC GLM. The variable TREAT represents a nominal variable and is thus specified in the CLASS statement. The option /E XPX tells SAS to print the vector of estimable functions (cell means for this problem) and the matrix $X'X$ matrix. The MEANS statement tells the program to print the sample means and the option /HOVTEST=BF invokes the M. B. Browne and Forsythe (1974) test (modified

Levine's test based on medians) for homogeneity of population variances. The hypothesis test matrix C is specified in the CONTRAST statement, and the ESTIMATE statements are used to obtain least squares estimates of the associated contrasts and their standard errors. The option /E on the contrast statements tells SAS to display the coefficients used in the contrast. We also include among the contrast statements that the contrast $\psi = \beta_1 - \beta_2 - \beta_3 - 3\beta_4 = 0$. Using the representation $H_0 : C\beta = \xi$, we have that

$$
\begin{pmatrix} -1 & -1 & -1 & 3 \end{pmatrix} \begin{pmatrix} \beta_1 \\ \beta_2 \\ \beta_3 \\ \beta_4 \end{pmatrix} = 0.
$$

Included in the SAS code are scatter plots of residuals versus predicted values which help with the evaluation of homogeneity of variance. If the scatter plot of the residuals has a fan structure, the assumption of homogeneity of population variances is usually not tenable. Then one usually uses some sort of weighted analysis for testing hypotheses of equal population means (Chapter 4). Finally we construct Q-Q plots for the example using studentized residuals (see Section 2.8) and tests of normality using the PROC UNIVARIATE.

The ANOVA summary table for testing the hypothesis of equal population means, taken from the Output 2.7.1, is shown below. Given a Type I level at $\alpha = 0.01$, we would reject the hypothesis that the four treatment means are equal in the population. The estimates of the individual contrasts $\hat{\psi} = C\hat{\beta}$ from the Output 2.7.1 are presented in Output 2.7.2. Finally, for the contrast $\psi = \beta_1 - \beta_2 - \beta_3 - 3\beta_4 = 0$ (labeled treats123 vs 4), the least square estimate is $\hat{\psi} = 16.04$ with a standard error $\hat{\sigma}_{\hat{\psi}}^2 = 2.43$. These results agree with those in Timm and Carlson (1975) using their FULRNK program.

Output 2.7.1: ANOVA Summary Table.

Source	DF	Sum of Squares	Mean Square	F Value	Pr > F
Model	3	239.9010989	79.9670330	24.03	<.0001
Error	22	73.2142857	3.3279221		
Corrected Total	25	313.1153846			

Reviewing the SAS output for this example and the SAS documentation, observe that SAS only generates confidence sets for all pairwise comparisons in the population means. This is done by using options on the LSMEANS statement. SAS does not generate confidence sets for complex contrasts nor does it find the contrast that led to the significant F test. Instead, SAS performs multiple t tests. To obtain confidence sets using the Scheffé method for arbitrary contrasts and to find the contrast that led to significant F tests the researcher must perform the calculations using the IML procedure. This is addressed in the next example.

Output 2.7.2: ANOVA Contrast Estimates Table.

Parameter	Estimate	Standard Error	t Value	Pr > |t|
treat1-treat4	-6.4285714	1.01492419	-6.33	<.0001
treat2-treat4	-2.4285714	1.06817688	-2.27	0.0331
treat3-treat4	-7.1785714	0.94414390	-7.60	<.0001
treats123 vs 4	16.0357143	2.43207428	6.59	<.0001

2.7.2   Simultaneous Inference for the One-Way Design

In exploratory data analysis an overall F test is performed and the S-method is used to locate significant differences in the population means. While one knows that a significant contrast exists when the overall test is significant, locating the significant contrast may be a problem. We now illustrate how to locate the contrast that led to a significant F test for a fixed effects one-way ANOVA design using the methodology in Section 2.4. For this example we are again testing the hypothesis that the means of four normal populations are equal: $H_0 : \beta_1 = \beta_2 = \beta_3 = \beta_4$. However, for this example the data in the data set 2_7_2.dat are used to investigate whether there are differences in four teaching methods. The four methods were taught by four experienced teachers: Groups two and three were taught by teachers with less than five years of experience while groups one and four were taught by teachers with more than five years of experience. To explore differences among the four teaching methods, an overall F test was performed using $\alpha = 0.05$. The program for the analysis uses the overparmeterized less than full rank model, the default option in SAS. The code for the example is provided in the Program 2_7_2.sas. The summary ANOVA table for the experiment is shown in Output 2.7.3.

Output 2.7.3: ANOVA Summary Table.

Source	DF	Sum of Squares	Mean Square	F Value	Pr > F
Model	3	205.363636	68.454545	2.99	0.0422
Error	40	915.818182	22.895455		
Corrected Total	43	1121.181818			

Using the nominal level $\alpha = 0.05$, the overall hypothesis is rejected. With the rejection of the hypothesis, one is interested in exploring differences among group means that may have led to the rejection of the overall test. Natural comparisons might include, for example, all pairwise comparisons or because groups two and three were taught by teachers with less experience than those who taught methods one and four, the complex contrast contrast $\psi = (\beta_1 + \beta_4)/2 - (\beta_2 + \beta_3)/2$ might be of interest. It involves the most extreme sample means. Using the LSMEANS statement, we observe that all six pairwise comparisons include zero (Output 2.7.4). This is also the case for the complex contrast. Some obvious comparisons are displayed in Table 2.1. None of the obvious comparisons are found to be significant. Indeed for this example it is not obvious which comparison led to the rejection of the hypothesis, try it!

Output 2.7.4: Pairwise Comparisons Using Scheffé Adjustment.

Least Squares Means for Effect method				
i	j	Difference Between Means	Simultaneous 95% Confidence Limits for LSMean(i)-LSMean(j)	
1	2	3.818182	-2.135932	9.772296
1	3	2.727273	-3.226841	8.681387
1	4	-1.636364	-7.590478	4.317750
2	3	-1.090909	-7.045023	4.863205
2	4	-5.454545	-11.408659	0.499568
3	4	-4.363636	-10.317750	1.590478

Table 2.1: Some Contrasts Using the S-Method.

Contrast	Confidence Set
$\psi_1 = \beta_1 - \beta_4$	$(-7.59, 4.32)$
$\psi_2 = \left(\frac{\beta_1+\beta_4}{2}\right) - \left(\frac{\beta_2+\beta_3}{2}\right)$	$(-0.12, 8.30)$
$\psi_3 = \left(\frac{\beta_1+\beta_2}{2}\right) - \left(\frac{\beta_3+\beta_4}{2}\right)$	$(-5.57, 2.84)$

To try to locate the maximal contrast we use (2.25) and (2.26). For the hypothesis test matrix

$$C = \begin{pmatrix} 1 & 0 & 0 & -1 \\ 0 & 1 & 0 & -1 \\ 0 & 0 & 1 & -1 \end{pmatrix} \quad \hat{\psi} = C\hat{\beta} = \begin{pmatrix} -1.6364 \\ -5.4546 \\ -4.3636 \end{pmatrix}.$$

Letting $\psi = 0$, the most significant contrast vector is

$$c'_* = (G_2^{-1}\hat{\psi})'C = a'_*C = \begin{pmatrix} 13.5 & -28.5 & -16.5 & 31.5 \end{pmatrix}.$$

The bounds for this contrast are: $(5.26, 405.47)$, and the value of the most significant contrast $\psi_{\max} = c'_*\beta$ is $\hat{\psi}_{\max} = c'_*\hat{\beta} = 205.36$, the hypothesis sum of squares for the F test. As expected, this contrast is not very meaningful for the study. Replacing the coefficients of the most significant contrast with perhaps the more meaningful coefficients in the vector: $c'_\psi = (.5, -1, -.5, 1)$, also results in a significant contrast. The confidence set for the contrast is: $(0.16, 13.08)$.

This example illustrates that the contrast that led to a significant F test may be difficult to find by just using simple linear combinations of cell means. However, we have illustrated how one may always find the exact contrast that let to the rejection of the overall test. The contrast defined by c_* may serve as a basis vector for the S_1-subspace in that all nonzero scalar multiples of the vector will result in a significant contrast. For example, suppose we divide the vector by the number of observations in each group, 11. Then, $c' = (1.277, -2.591, -1.500, 2.864)$. Thus the supremum identified by Hochberg and Tamhane (1987, p. 76). The ith element of the vector c' is defined: $c_i = \hat{\mu}_i - \hat{\mu}$ which is an estimated cell mean minus the overall mean and the $\sum_i c_i = 0$. This was considered by Scheffé (1947) to see if any group mean was significantly different from the overall mean in the analysis of means (ANOM) analysis (Nelson, 1993).

2.7.3 Multiple Testing

With the rejection of the test of size α for the overall null hypothesis $H_0 : C\beta = 0$, one may invoke Scheffé's S_2-method to investigate the infinite, nonhierarchical family of contrasts orthogonal to the significant contrast found using the S-method. Any contrast $\hat{\psi}$ is significantly different from zero if:

$$|\hat{\psi}| > \left[(\nu_h - 1)F_{(\nu_h, \nu_e)}^{(1-\alpha)}\right]^{1/2} = S_2 \qquad (2.113)$$

where ν_h is the degrees of freedom of the null overall hypothesis, and $F_{(\nu_h, \nu_e)}^{(1-\alpha)}$ is the critical value of the F distribution. Scheffé (1970) showed that the experimentwise Type I error rate for the procedure is controlled at the nominal level α, the level of the overall test.

To apply the S_2-method for the example in Section 2.7.2, the critical value for all contrast comparisons is

$$S_2 = 2.796 < S = 2.918.$$

The absolute value of any contrast larger than 2.796 is significantly different from zero. For example, consider the contrast ψ_2 in Table 2.1 that compares the mean difference in the extreme means. The absolute value for the contrast is: $|\hat{\psi}_2| = {}^{4.09}/_{1.44} = 2.840$ with P-value 0.0365 < 0.05. This contrast, while not significant using the S-method, is significant using the S_2-method. One may readily locate other significant contrasts.

2.8 MULTIPLE LINEAR REGRESSION: CALIBRATION

The multiple linear regression model can be expressed in the form of (1.1) where $y_{n \times 1}$ is a vector of observations for a dependent variable and $X_{n \times k}$ is a matrix fixed or random p independent variables. Here $k = p + 1$, and the first column of the design matrix X is a column vector of 1's and is used to estimate the intercept. The parameter vector has the form $\beta' = (\beta_0, \beta_1, \ldots, \beta_p)$ and the vector $e_{n \times 1}$ is a vector of random errors, or residuals. Whether the design matrix is fixed or random depends on the sampling plan.

Many applications of multiple linear regression are concerned with model development, discovering the relationship between a dependent variable and a set of fixed nonrandom independent variables, the CR and CRN models. The goal of the study is find a "best" model; the SAS procedure PROC REG is geared to this goal. In general, it is better to overfit a model by one or two variables than to underfit a model.

The Effects of Underfitting

Suppose the true CR or CNR model for the $n \times 1$ vector of observations is of the form:

$$y = X_1\beta_1 + X_2\beta_2 + e \tag{2.114}$$

where the rank of X_1 is p and the rank of X_2 is q. However, suppose the model fit to the data takes the reduced form:

$$y = X_1\beta_1 + e. \tag{2.115}$$

Then for the reduced model, the estimates for β_1 and σ^2 are:

$$\hat{\beta}_{1R} = (X_1'X_1)^{-1}X_1'y \tag{2.116}$$

$$\hat{\sigma}_R^2 = \frac{y'(I - X_1'(X_1'X_1)^{-1}X_1')y}{(n-p)}. \tag{2.117}$$

The expected value of the estimates are:

$$\mathcal{E}(\hat{\beta}_{1R}) = (X_1'X_1)^{-1}X_1'E(y) = \beta_1 + (X_1'X_1)^{-1}X_2'\beta_2 \neq \beta_1 \tag{2.118}$$

$$\mathcal{E}(\hat{\sigma}_R^2) = \sigma^2 + \frac{\beta_2'X_2'(I - X_1'(X_1'X_1)^{-1}X_1')X_2'\beta_2}{(n-p)}. \tag{2.119}$$

Thus, when the incorrect reduced model is fit to the data, the parameters estimated are unbiased if and only if either β_2 is zero in the population or the deleted variables are orthogonal to the retained variables. In general, underfitting leads to biased estimators. The matrix $A = (X_1'X_1)^{-1}X_1'X_2$ is called the alias or bias matrix. Finally, we observe that the mean squared error for the estimator $\hat{\beta}_{1R}$ is:

$$MSE(\hat{\beta}_{1R}) = \text{var}(\hat{\beta}_{1R}) + (\text{bias})^2 = \sigma^2(X_1'X_1)^{-1} + A\beta_2\beta_2'A'. \tag{2.120}$$

For the full model, the $MSE(\hat{\beta}_{1F}) = \text{var}(\hat{\beta}_{1F}) \geqslant MSE(\hat{\beta}_{1R})$. Hence, by deleting variables with small numerical values in the population, and which have large standard error, will result in estimates of the retained regression coefficients with higher prediction, smaller standard errors. While we may have gained precision, the estimates are no longer unbiased. We are always exactly wrong!

The Effects of Overfitting

Suppose the true CR or CNR model for the $n \times 1$ vector of observations is of the form:

$$y = X_1\beta_1 + e \tag{2.121}$$

where the rank of X_1 is p. However, suppose the model fit to the data takes the overfit form:

$$y = X_1\beta_1 + X_2\beta_2 + e \tag{2.122}$$

where the rank of X_2 is q. Thus, we have an overfitted model, a model with too many variables.

Letting $A = (X_1'X_1)^{-1}X_1'X_2$, $P_1 = X_1(X_1'X_1)^{-1}X_1'$, and $W = X_2'(I - P_1)X_2$, where $\hat{\beta}_{1F}$ and $\hat{\beta}_{2F}$ are the least squares estimates of the parameter vectors for the full model and $\hat{\beta}_{1R}$ is the estimator for the true reduced model, then the relationship between the estimates is:

$$\hat{\beta}_{1F} = \hat{\beta}_{1R} + A\hat{\beta}_{2R}$$
$$\hat{\beta}_{2F} = W^{-1}X_2'(I - P_1)y$$
$$\mathcal{E}(\hat{\beta}_{1F}) = \beta_1$$
$$\mathcal{E}(\hat{\beta}_{2F}) = 0 \quad \text{(for the reduced true model)}$$
$$\mathcal{E}(\hat{\sigma}_F^2) = \sigma^2$$
$$\text{cov}(\hat{\beta}_F) = \sigma^2 \begin{pmatrix} (X_1'X_1)^{-1} + AW^{-1}A' & -AW^{-1} \\ -W^{-1}A' & W^{-1} \end{pmatrix}.$$

Finally, the $\text{var}(\hat{\beta}_{1F}) \geqslant \text{var}(\hat{\beta}_{1FR})$.

Thus, from the standpoint of unbiasedness, the variance of fitted regression coefficients and predicted values, it is usually better to overfit by one or two variables, than to underfit.

Some guidelines for "t tests"

Case 1: β_i test is nonsignificant and other beta weights do not change substantially when variable is removed. IDEAL, remove variable.

Case 2: β_i test is significant and other beta weights do not change substantially when variable is removed. Retain variable, but investigate collinearity.

Case 3: β_i test is significant and other beta weights do not change substantially when variable is removed. IDEAL, retain variable.

Case 4: β_i test is nonsignificant and other beta weights are changed substantially when variable is removed. No decision, may have collinearity.

With these comments in mind, when developing a multiple linear regression model a primary goal is to find a set of independent variables that may be used to predict some dependent variable in an observational study, not an experiment. Thus, the independent variables are fixed and nonrandom. To find the best subset of variables, the researcher may be interested in testing hypotheses about the vector of regression coefficients β. Before performing a regression analysis and testing hypotheses, it is again advisable to evaluate normality of the dependent variable, conditioned on the independent variables, and to identify outliers. We again may use Box-Cox transformations to remove skewness and kurtosis, create Q-Q plots, and perform tests of normality.

Having obtained an estimate of the parameter vector, the predicted or fitted values of the observation vector y are represented as:

$$\hat{y} = X\hat{\beta} = X(X'X)^{-1}X'y \qquad (2.123)$$

where $H = X(X'X)^{-1}X'$ is the hat or projection matrix. The vector of residuals is defined as:

$$\hat{e} = (y - X\hat{\beta}) = (I - H)y. \qquad (2.124)$$

Using Theorem 1.3, for normal errors the distribution of the vector of residuals is: $\hat{e} \sim N_n[0, \sigma^2(I - H)]$. Hence, the residuals follow a singular normal distribution where the covariance matrix of order n with rank $n - k$, while the errors follow a nonsingular normal distribution since the dimension of the covariance matrix is equal to its rank, n. And, while the errors are independent, the residuals are not (except if $X = 0$). Provided the design matrix includes an intercept, then the residuals computed from a sample always have a mean of zero: $\sum_{i=1}^{n} \hat{e}_i = 0$. Without an intercept, errors in the sample never have a mean of zero. The distribution of an individual residual $\hat{e}_i \sim N(0, \sigma^2(1 - h_i))$ where h_i is the ith diagonal element of H. Potential outliers can be detected using the diagonal elements of the hat matrix, also called leverages. Stevens (1996) suggested that any observation with leverage $h_i \geq 3k/n$ should be examined.

Using (2.23), the unbiased estimate of σ^2 is the sample variance estimate $s^2 = SSE/(n-k) = SSE/(n-k-1)$. Dividing a residual by an estimate of its standard deviation creates a standardized residual:

$$r_i = \frac{\hat{e}_i}{\sqrt{s^2(1 - h_i)}}. \qquad (2.125)$$

Because the residuals are not independent and the distribution of s^2 is chi-square with $n - k$ degrees of freedom the standardized residuals do not follow a t distribution. To correct this situation, one may obtain a jackknifed residual: $y_i - \hat{y}_{(i)} = \hat{e}_{(i)}$ where the index (i) involves fitting a linear model n times by deleting the ith observation. Then the jackknifed residuals also called studentized residuals (or studentized deleted residuals) follow a t distribution (see for example, Chatterjee & Hadi, 1988,

p. 76-78):

$$r_{(i)} = \frac{\hat{e}_{(i)}}{\sqrt{s_{(i)}^2(1-h_i)}} = \frac{\hat{e}_i}{\sqrt{s_{(i)}^2(1-h_i)}} = r_i\sqrt{\frac{n-k-1}{n-k-r_i^2}} \equiv t_i \sim t(n-k-1).$$

$$(2.126)$$

Plots of studentized residuals versus predicted values are used to check for homogeneity of variance. If the assumptions of a linear relationship and of homoscedasticity are accurate, these plots should be linear (see Chapter 4 for an example of data with nonconstant variance), and plots of studentized residuals help with the identification of outliers. Since the studentized residuals are nearly normal for large sample sizes ($n > 30$), approximately 99% of the $r_{(i)}$ should be in the interval $-3 \le r_{(i)} \le 3$. Thus, any observation larger than three in absolute value should be examined as a possible outlier. For additional details regarding regression graphs, one may consult Cook and Weisberg (1994).

In regression analysis one often partitions the sum of squares:

$$\sum_{i=1}^{n}(y_i - \bar{y})^2 = \sum_{i=1}^{n}(\hat{y}_i - \bar{y})^2 + \sum_{i=1}^{n}(y_i - \hat{y}_i)^2$$

where the term on the left of the equality is the sum of squares total (SST), corrected for the mean, with $n-1$ degrees of freedom. The next term is the sum of squares due to regression with p degrees of freedom, and the last term is the sum of squares due to error (SSE) or the residual, with $n - p - 1$ degrees of freedom. The overall test of no linear regression, that all of the coefficients associated with the p independent variables are zero, is tested using the F statistic:

$$F = \frac{SSR/p}{SSE/(n-p-1)} \sim F(\nu_h, \nu_e) \qquad (2.127)$$

has a central F distribution with degrees of freedom $\nu_h = p$ and $\nu_e = n - p - 1$ when the null hypothesis is true. This F statistic reduces to the statistic defined in (2.54) for $h = p$ variables. The overall hypothesis is rejected if F is larger than the upper $F^{(1-\alpha)}$ critical value.

The independent variables should be investigated for multicollinearity; independent variables that are not linearly independent may cause the parameter estimates to be unstable, have inflated variance. A measure of multicollinearity may be obtained from the variance inflation factor (VIF). Recall that the var($\hat{\beta}$) $= \sigma^2(X'X)^{-1}$, the quantity $(X'X)^{-1}$ is called the VIF. Because the inverse of a matrix can be represented as:

$$(X'X)^{-1} = \frac{\text{adjoint}(X'X)}{|X'X|},$$

if the $|X'X| \simeq 0$ there is a dependency among the independent variables. The VIF increases as the independent variables become more dependent. Using (2.35), the

VIF for the kth variable is

$$VIF_k = (1 - \widetilde{R}_k^2)^{-1} \quad k = 1, 2, \ldots, p. \tag{2.128}$$

We usually would like the VIF to be near 1; however, any variable with a VIF of 10 or greater should be considered for exclusion or combined with other related variables.

Various methods are available for selecting independent variables to include in the prediction equation. These include forward, backward, and stepwise methods, methods based upon $\widetilde{R}^2$, adjusted $\widetilde{R}_a$, and Mallows C_p criterion, among others (see for example, Kleinbaum, Kupper, & Muller, 1998; Muller & Fetterman, 2002).

We saw in Chapter 1 how Mahalanobis' distance may be used to evaluate multivariate normality. It may also be used to detect outliers for independent variables in regression. Mahalanobis' distance for the ith independent observation is:

$$D_i^2 = (x_i - \bar{x})'S^{-1}(x_i - \bar{x}) \tag{2.129}$$

where x_i is the $p \times 1$ independent variable and $\bar{x}$ is the vector of means. Large values of D_i indicate that an observation vector of independent variables lies far from the other $n - 1$ observations (see, Stevens, 1996, for more details).

Influential data points are observations that have a large effect on the value of the estimated regression coefficients. Cook's distance is one measure for the detection of influential data points. Cook's distance for the ith observation is defined:

$$CD_i = \frac{\left[\hat{\beta} - \hat{\beta}_{(i)}\right]' X'X \left[\hat{\beta} - \hat{\beta}_{(i)}\right]}{ks^2} \tag{2.130}$$

where $\hat{\beta}_{(i)}$ is the vector of estimated parameters with the ith observation deleted, and s^2 is from the model with all variables included. It measures the standardized shift in predicted values and the shift in the least squares estimate for $\hat{\beta}$ due to deleting the ith observation. Values of $CD_i > 10$ are considered large (Cook & Weisberg, 1994, p. 118). Cook's distance can also be expressed as a function of the standardized residual:

$$CD_i = r_i^2 \left[\frac{h_i}{k(1 - h_i)}\right]. \tag{2.131}$$

Cook suggested comparing the statistic to the median of an F distribution with degrees of freedom equal to those of the test of all coefficients equal to zero $(k, N - k)$, provided one can assume normality.

To evaluate the influence of the ith observation y_i on the jth coefficient in the parameter vector β the $DFBETA$ statistics are used. The statistics are defined as:

$$DFBETA_{ij} = \frac{\hat{\beta}_j - \hat{\beta}_{j(i)}}{s_{(i)}\sqrt{(X'X)^{jj}}} \tag{2.132}$$

where $(X'X)^{jj}$ is the jj^{th} diagonal element of $(X'X)^{-1}$. An observation y_i is considered influential on the regression coefficient β_j if the $|DFBETA_{ij}| > 2/\sqrt{n}$ (Besley, Kuh, & Welch, 1980). They also derived a test statistic to evaluate the effect of the ith observation vector y_i, x_i on the ith fitted value $\hat{y}_i$. They estimated the closeness of $\hat{y}_i$ to $\hat{y}_{i(i)} = x_i'\hat{\beta}_{(i)}$. The statistic $DFFITS$ is defined as:

$$DFFITS_i = \frac{|\hat{y}_i - \hat{y}_{i(i)}|}{s_{(i)}\sqrt{h_i}} = \frac{|x_i'(\hat{\beta} - \hat{\beta}_{(i)})|}{s_{(i)}\sqrt{h_i}} = |r_{(i)}|\sqrt{\frac{h_i}{1 - h_i}}. \qquad (2.133)$$

The statistic $DFFITS_i \sim t\sqrt{k/(n-k)}$ and an observation is considered influential if $|DFFITS_i| > 2\sqrt{k/(n-k)}$. Finally, (Besley et al., 1980) used a covariance ratio to evaluate the influence of the ith observation on the $\mathrm{cov}(\hat{\beta})$. The covariance ratio (CVR) for the ith observation is defined:

$$CVR_i = \frac{s_{(i)}^2}{s^2(1 - h_i)}. \qquad (2.134)$$

An observation is considered influential if $|CVR_i - 1| > 3k/n$.

To illustrate the use of SAS using PROC REG to find the best regression model, the hypothetical data set given in (Kleinbaum et al., 1998, p. 117-118) is used. Here a sociologist is interested in developing a model to predict homicide rates per 100,000 city population (y) with the predictors of city population size (x1), the percent of families with yearly income less than \$5,000 (x2), and the rate of unemployment (x3). The general linear model (CRN) is assumed for this example. The SAS code for CVR the example is in Program 2_8_1.sas.

The first step in the regression analysis is include all variables in the model and use the diagnostic procedures in Chapter 1 to evaluate normality and homogeneity of the population variance. Because the linear model is a conditional regression model, the studentized residuals are analyzed by regressing the dependent variables on the independent variables. When analyzing a linear model, Q-Q plots are formed using the residuals and not the vector of observations. From the output of PROC UNIVARIATE and the Q-Q plots, there do not appear to be any violations in the normality. The Shapiro-Wilk test has P-value of 0.6110. To check for homogeneity of variance using PROC REG, one plots residuals versus fitted values. Currently these is no test for homogeneity using the regression procedure. To create a test of homogeneities, we form two groups of the residuals and use PROC GLM and Brown-Forsythe's test. Reviewing the residual plots and the output from PROC GLM, the P-Value for Brown-Forsythe test is 0.4043. Thus, for these, data do not appear to violate the assumptions of the normality or homogeneity of variance.

The model statement for the model with all the variables includes the options /VIF R COLLIN INFLUENCE. These provide information regarding the variance inflation factor for each variable, information about the residuals, collinearity measures, and influence measures. The VIF values of 1.06, 2.99, and 3.08 are all less than 10, thus multicollinearity is not a problem for the data.

Reviewing the studentized residuals, we would expect about 95% to lie between ± 2. From Output 2.8.1, the residual for observation 11 is 2.315 and for observation 20, the value is 1.986. The residual and diagnostic plots are created in SAS using ods graphics (Graph 2.8.1). Plots of the residuals versus the variables x1, x2, and x3 do not show any pattern. It does however indicate an influential case.

Output 2.8.1: Extreme Values of the Studentized Residuals.

Extreme Observations			
Lowest		Highest	
Value	Obs	Value	Obs
-2.314838	11	0.445152	3
-1.428798	19	0.567527	14
-1.294765	18	0.742739	5
-1.167020	2	1.986432	20
-0.931076	16	2.264353	8

Continuing with the analysis of the full model, the leverages appear in the column labeled Hat Diagonal H (Output 2.8.2). None of the observations have leverages greater than $3k/n = (3)(4)/20 = 0.6$. Reviewing Cook's distances, we observe that observation 11 again has a large value (also appear in Graph 2.8.1); this is also the case for the $DFBETA$ statistics for variable x1, its value $|-5.5353|$ exceeds $2/\sqrt{n} = 2/\sqrt{20} = 0.4472$. And, observation 20 appears influential. For $DFFITS$, an observation is influential if its absolute value exceeds $2\sqrt{k/(n-k)} = 2\sqrt{4/16} = 1$. Reviewing the Output 2.8.1, observation 11 again appears influential. Finally, for the CVR, the criterion is $|CVR_i - 1| > 3k/n = 0.6$. Thus, a value of $CVR_i \leq 0.4$ or ≥ 1.6 is suspect. This occurs for observation 3, 9, and 10. Our preliminary analysis of the full model suggests that observation 11 should be consider influential; thus, we refit the full model with the observation removed. Observe that with observation 11 included in the model that the coefficient associated with the variable x1 (population size) is not significant and the adjusted value for model fit is $\widetilde{R}_a^2 = 0.7843$.

Deleting observation 11 from the data set, tests of normality and homogeneity remain nonsignificant with P-values: 0.1807 and 0.4372, respectively. Removing the observation the adjusted fit index $\widetilde{R}_a^2 = 0.8519$ is increased And, with the deletion of the outlier, no observations appear to be influential.

We next check whether the number of variables in the model can be reduced using the C_p criterion (2.43) and the backward elimination criterion. In PROC REG, we use the option SELECTION=CP, and SELECTION=BACKWARD on the model statement. Reviewing Output 2.8.3, we see that all variables should remain in the model.

As a final note, if one does not remove the outlier from the data set one might be led to (incorrectly) exclude variable x1 from the model. With the variable x1 removed, the value for the adjusted fit index $\widetilde{R}_a^2$ is reduced: $\widetilde{R}_a^2 = 0.7787$ (Timm & Mieczkowski, 1997, p. 53).

Graph 2.8.1: Residual and diagnostic plots by SAS ODS Graphics

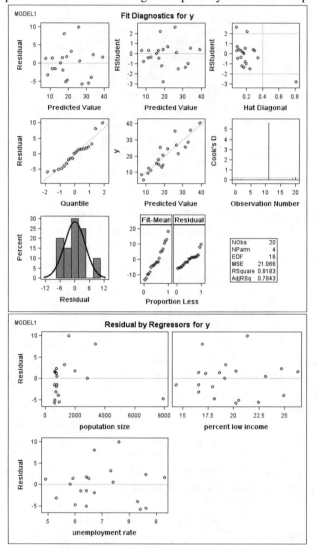

Output 2.8.2: List of Leverage, Cook's Distance, and $DFBETA$'s.

				Output Statistics			
					DFBETAS		
Obs	Hat Diag H	Cov Ratio	DFFITS	Intercept	x1	x2	x3
1	0.1283	1.4449	-0.1232	-0.0929	0.0487	0.0653	-0.0185
2	0.1383	1.0525	-0.4734	-0.1024	0.1848	-0.3120	0.3498
3	0.2814	1.7140	0.2714	-0.2117	-0.0055	0.0877	0.0598
4	0.1807	1.3651	-0.3510	-0.2827	0.1482	-0.0434	0.2014
5	0.0808	1.2239	0.2170	0.0243	-0.0010	-0.1229	0.1288
6	0.1252	1.4376	0.1247	0.1033	-0.0533	-0.0385	-0.0159
7	0.1027	1.3766	0.1423	0.0153	0.0176	0.0887	-0.0946
8	0.0681	0.2962	0.7188	-0.2635	0.1191	0.0359	0.1859
9	0.2972	1.7910	0.2116	0.1343	-0.0792	0.0955	-0.1732
10	0.3390	1.8843	-0.2729	-0.1589	0.0444	0.2410	-0.1605
11	0.8080	1.3192	-5.6375	0.6608	-5.3353	-0.0760	-0.0344
12	0.1484	1.5121	0.0590	-0.0185	-0.0131	0.0448	-0.0301
13	0.1644	1.5485	0.0193	0.0046	0.0060	0.0106	-0.0135
14	0.1775	1.4509	0.2578	-0.1779	-0.0169	0.1004	0.0273
15	0.0744	1.3568	0.0958	0.0539	-0.0460	-0.0050	-0.0188
16	0.1899	1.2788	-0.4488	0.2867	0.0315	-0.2769	0.0746
17	0.0880	1.3576	-0.1272	-0.0695	0.0440	0.0663	-0.0363
18	0.1642	0.9948	-0.5873	0.2911	-0.0154	0.1718	-0.4016
19	0.2399	0.9865	-0.8321	0.1288	0.0570	0.5633	-0.7145
20	0.2036	0.5237	1.1205	0.2169	0.6673	-0.7304	0.6152

Output 2.8.3: Model Selection Using Mallows C_p.

Number in Model	C(p)	R-Square	Adjusted R-Square	AIC	J(p)	MSE	SBC	Variables in Model
3	4.0000	0.8766	0.8519	54.8918	18.0913	14.94501	58.66954	x1 x2 x3
2	8.7346	0.8212	0.7988	59.9380	23.5069	20.30144	62.77131	x1 x3
2	11.5796	0.7978	0.7725	62.2752	26.5840	22.95887	65.10854	x2 x3
2	15.6567	0.7642	0.7347	65.1912	30.9936	26.76719	68.02452	x1 x2
1	16.2662	0.7427	0.7276	64.8470	30.3800	27.48669	66.73592	x3
1	21.3273	0.7011	0.6835	67.6977	35.2977	31.93601	69.58654	x2
1	104.1909	0.0193	-.0384	90.2726	115.8126	104.78287	92.16151	x1

2.8.1   Multiple Linear Regression: Prediction

While SAS allows one to evaluate the predicted value of a regression model as illustrated in Output 2.8.3 using the option SELECTION, for fixed independent variables, it does not allow one to select variables that optimize prediction using the JNM or the RCN model. To illustrate the prediction methods reviewed in Chapter 2, data provided by Dr. Keith E. Muller are utilized. The data are from Lewis and Taylor (1967) and represent the body weight, standing height, age, and gender for a sample of 237 subjects. The goal of the study is to develop a regression equation for predicting weight. Using a CR or CNR model where the predictor variables are fixed, Kleinbaum et al. (1998, Chapter 16, p. 403-309) provide an analysis of the data using data splitting with 118 subjects in the calibration sample and 119 in the validation sample. To predict body weight, Kleinbaum et al. (1998) considered thirteen predictor variables. The predictors included the linear effects of age, height, and gender; quadratic and cubic terms for both age and height were also included in the model since weight increases nonlinearly with age and height, and finally because the same equation should apply to both genders (male and female), interaction terms between gender and linear, quadratic, and cubic powers of age and height formed the initial set of variables for the analysis. Using the C_p criterion and the calibration sample to locate a "best" submodel for the analysis, they concluded that the "most appealing" candidate model should include five of the original thirteen variables. The final set contained the variables male (m), height (h), age (a), height-squared (h^2), and age-squared (a^2). The coefficient of determination for the model for the calibration sample was $\widetilde{R}^2 = 0.627$. To evaluate the predictive precision of the final equation, the square of the univariate Pearson product moment correlation between the predicted values in the validation sample estimated from the calibration sample and the observed values in the validation sample was used. The value for their study was: $\hat{\rho}_{rc}^2 = [\text{corr}(y, \hat{y}|\hat{\beta})]^2 = 0.621$. Given the small observed difference in the fit shrinkage index, one may pool the data to obtain a more reliable model for prediction.

Because the predictors are random and not fixed, an analysis with random predictors is more appropriate for the data. For the analysis, we assume a RCN model where the conditional distribution of the dependent variable is assumed to be normal and the marginal distribution of the independent variables is unknown. Following Kleinbaum et al. (1998), we began with 13 original (centered) prediction variables; however, the data are not divided into a calibration sample and a validation sample. Instead, the entire sample is used to obtain a subset of predictors (Program 2_8_2.sas). Because the ratio $n/(k+1) = n/d > 40$, and the sample size is large we may use the AIC_{MLE} information criterion defined in (2.47) to locate candidate models. This criterion is not designed to maximize predictive precision in the single sample. It is designed to locate the potential candidate models using the sample calibration data.

To select a model that minimizes the mean squared error of prediction, the expected value of the RPE defined in (2.72) and estimated in (2.74) is used to locate potential candidate models, even though it is not an unbiased estimate of the random parameter.

Using all thirteen variables, neither criterion was able to locate a reasonable model. Both criteria selected models that contained a third order variable (for example age) and excluded the associated second and first order terms. Furthermore, partial regression plots for the higher order terms did not support their inclusion in the model. Thus, the second step of the analysis was to evaluate models that included only first and second order variables. The models evaluated included the following nine variables: $a, a^2, h, h^2, m, m \times a, m \times a^2, m \times h,$ and $m \times h^2$. Table 2.2 presents two potential candidate models using two section criteria.

Table 2.2: Model Selection Summary.

Method	Variables Selected	R_a^2	R^2	$\hat{\rho}_{rc}^2 = \mathcal{E}(d_r^2)$
$AIC_{MLE} = 1172.28$	$a, a^2, h, m, m \times a$	0.674	0.678	0.6571
$\hat{\delta}_{ru} = 140.76$	$a, a^2, h, m, m \times a$			
$AIC_{MLE} = 1172.80$	a, a^2, h, m	0.672	0.678	0.6565
$\hat{\delta}_{ru} = 140.04$	a, a^2, h, m			

For a stochastic regression model, R_a^2 is an estimate of ρ^2 and $\hat{\rho}_{rc}^2$ is an estimate of $\rho_{rc}^2 = E(d_r^2)$ in the population. While we expect $\rho_{rc}^2 < \rho^2$ in the population, comparing the corresponding estimates for each model, the predictive precision difference between the two models is less than 3%. While the estimated predictive precision for the five variable model is higher, the "more appealing" was selected for the sample data. The final model selected follows:

$$w = 99.12 - 0.944m + 1.472(a - 13.69) + 1.221(a - 13.69)^2$$
$$+ 3.389(h - 61.36) \tag{2.135}$$

where the units of measurement for the variables are: age (a)-years, height (h)-inches, and weight (w)-pounds. A reasonable model was obtained without splitting the sample data which often results in loss of precision.

To compare the single sample approach with the cross-validation model used by Kleinbaum et al. (1998), we fit their model to the data. Their model included the five variables: $a, a^2, h, h^2,$ and m. For their model, $AIC_{MLE} = 1174.24$ and $\hat{\delta}_{ru} = 141.89$. Both estimates for their model are (marginally) larger while the estimates of $R_a^2 = 0.671$ and $R^2 = 0.678$ are (slightly) smaller.

Summary

Generally speaking multiple linear regression is used to determine the relationship between a dependent variable and several independent variables. The sample is used to discover the true linear relationship in the population between the dependent, criterion variable and a set of independent predictors. This is the calibration phase of the model development. In the calibration phase, one is trying to understand relationships among a set of variables in order to explain variation in the criterion. When the independent variables are fixed in the calibration phase, the final model developed may only apply to the sample of fixed variables. Thus, it has limited genaralizability. While model validation may occur with either fixed or random predictors, it is generally more difficult to validate a model developed using fixed independent variables than random independent variables since changes in the setting from the initial study and the replicated study are usually more difficult to control. Thus, while model calibration may be performed whether of not the design matrix is fixed or random, prediction is most useful with random predictors.

We have included two examples in this section to clarify some differences between model development and prediction when the design matrix is fixed or random.

For fixed models, neither $\tilde{R}^2$ or $\tilde{R}_a^2$ are estimates of ρ^2, however both may be used as a measure of model fit. While $\tilde{R}_a^2$ for fixed X and R_a^2 for random X are often used as a model selection criteria, they are both prone to excessive over fitting (McQuarrie & Tsai, 1998, p. 48 & 367). The best selection criteria are HQ and AIC since they allow one to rank order candidate models. The estimator R_a^2 is an estimate of ρ^2 for models with random X and may be viewed as an index of internal precision or the degree of the linear relationship between the dependent variable and a set of random independent variables. Under joint multivariate normality, the estimator $\hat{\rho}_{rc}^2 = E(\rho^2)$ is an estimate of the random population parameter ρ_{rc}^2 and may be viewed as a measure of predictive precision or predictive validity in the population. While the unbiased estimator $\hat{\delta}_{ru}$ of $\delta^2 = E(d_r^2)$ may be used as a measure of model fit, unless the sample is large, it may not always reliably choose a best subset of predictors since it is not an unbiased estimator of random predictive error (RPE), d_r^2. Finally, while protected t tests may be used to evaluate the marginal effect of a variable in a candidate model, they should not be used as a model selection method.

2.9   TWO-WAY NESTED DESIGNS

For our next example, we discuss the two-way nested design. In educational research, data often have a nested structure; students are nested in classes, classes are nested in schools, and schools are nested within school districts, and so on. If the factors are fixed and not random, the full rank general linear model without restrictions may be applied. We now discuss a simple two-way nested design where the fixed factor A is nested within the fixed factor B. Suppose there are $i = 1, 2, \ldots, I$ levels of A, and $j = 1, 2, \ldots, J_i$ levels of factor B within $A_{(i)}$, and $k = 1, 2, \ldots, n_{ij}$ observations within the cell $A_{(i)} B_j$. The full rank nested linear model can be written as:

$$y_{ijk} = \beta_{ij} + e_{ijk}, \tag{2.136}$$

where $e_{ijk} \sim IN(0, \sigma^2)$.

For this example, data analyzed by Searle (1971, p. 249) and by Timm and Carlson (1975, p. 84) are used. (Note there was a typographical error in the data set printed in Timm and Carson: Course 1, Section 1 should have read 5 and not 4.) For this data set, factor A represents course and factor B represents section. There are two sections nested with Course 1 and within Course 2 there are three sections. The dependent example is a test score. Using the unrestricted linear model, the model for the example follows:

$$y = X\beta + e$$

$$
\begin{pmatrix} 5 \\ 8 \\ 10 \\ 9 \\ 8 \\ 10 \\ 6 \\ 2 \\ 1 \\ 3 \\ 3 \\ 7 \end{pmatrix}
=
\begin{pmatrix}
1 & 0 & 0 & 0 & 0 \\
0 & 1 & 0 & 0 & 0 \\
0 & 1 & 0 & 0 & 0 \\
0 & 1 & 0 & 0 & 0 \\
0 & 0 & 1 & 0 & 0 \\
0 & 0 & 1 & 0 & 0 \\
0 & 0 & 0 & 1 & 0 \\
0 & 0 & 0 & 1 & 0 \\
0 & 0 & 0 & 1 & 0 \\
0 & 0 & 0 & 1 & 0 \\
0 & 0 & 0 & 0 & 1 \\
0 & 0 & 0 & 0 & 1
\end{pmatrix}
\begin{pmatrix} \beta_{11} \\ \beta_{12} \\ \beta_{21} \\ \beta_{22} \\ \beta_{23} \end{pmatrix}
+
\begin{pmatrix} e_{111} \\ e_{121} \\ e_{122} \\ e_{123} \\ e_{211} \\ e_{212} \\ e_{221} \\ e_{222} \\ e_{232} \\ e_{234} \\ e_{231} \\ e_{232} \end{pmatrix}.
$$

Program 2_9.sas contains the SAS code to analyze this data.

The MODEL statement is written as a y = a b(a), specifying that factor B is nested within factor A. The contrast statements are used for testing specific hypotheses (discussed below) and ESTIMATE statements are used to obtain contrast estimates and associated standard errors. The last analysis for the program uses PROC GLM to perform the less than full rank analysis for these data, the SAS default. Also included are residual plots, tests of normality, and tests of homogeneity.

There are several hypotheses which may be of interest with regard to these data. For example, one may be interested in testing that the mean of course 1 equals the mean of course 2. This hypothesis can be formulated as an unweighted test or as a weighted test. For the unweighted test the hypothesis is:

$$H_A : \frac{(\beta_{11} + \beta_{12})}{2} = \frac{(\beta_{21} + \beta_{22} + \beta_{23})}{3}$$

(a in the Type III SS table). This is tested in Program 2_9.sas for the full rank model using the CONTRAST statement contrast 'ha_1' a 1 -1;. For the less than full rank model, the unweighted test is obtained from the TYPE III Sum of Squares. Alternatively, one could perform the weighted test:

$$H_{A^*} : \frac{(1\beta + 3\beta_{12})}{4} = \frac{(2\beta_{21} + 4\beta_{22} + 2\beta_{23})}{8}.$$

For the less than full rank model, the weighted test is obtained from the TYPE I Sum of Squares.

Next suppose we wish to test the null hypothesis that the section population means are equal within course 1: $H_{\beta(1)} : \beta_{11} = \beta_{12}$ (hb_1). Or for course 2, the test becomes: $H_{\beta(2)} : \beta_{21} = \beta_{22} = \beta_{23}$ (hb_2). Or, the simultaneous test H_B : all β_{ij} are equal may be tested (b(a)). The ANOVA for these test is presented in Output 2.9.1. These results agree with those obtained by Timm and Carlson (1975, p. 85)

Output 2.9.1: Nested ANOVA Summary Table.

Source	DF	Type I SS	Mean Square	F Value	Pr > F
a	1	24.00000000	24.00000000	6.46	0.0386
b(a)	3	60.00000000	20.00000000	5.38	0.0309

Source	DF	Type III SS	Mean Square	F Value	Pr > F
a	1	3.76470588	3.76470588	1.01	0.3476
b(a)	3	60.00000000	20.00000000	5.38	0.0309

Contrast	DF	Contrast SS	Mean Square	F Value	Pr > F
ha_1	1	3.76470588	3.76470588	1.01	0.3476
hb_1	1	12.00000000	12.00000000	3.23	0.1153
hb_2	2	48.00000000	24.00000000	6.46	0.0257
hb_1and2	3	60.00000000	20.00000000	5.38	0.0309

using their FULRNK program.

Finally, tests of normality and homogeneity are performed for the example. The data appear normal and homogeneous. The P-values for the test are 0.6689 and 0.1075, respectively. Using the LSMEANS statement, one may obtain confidence sets for differences in unweighted cell means using the Scheffé procedure. The MEANS statement generates weighted means confidence sets for the example.

2.10   INTRACLASS COVARIANCE MODELS

In Section 2.7 we discussed the one-way fixed effect ANOVA design with four treatments. Suppose we have an additional variable in the experiment that is correlated with the dependent variable, but independent of the treatment variable. This type of variable may be included into the analysis of the ANOVA as a covariate. In analysis of covariance (ANCOVA), the regression slopes of the dependent variable on the covariables are assumed to be equal across the treatment groups (the assumption of parallelism, see Chapter 3). In this section we discuss a more general model that allows the slopes of the regression lines within the populations to differ, the general intraclass model, also called the slopes as variable model. To recover the ANCOVA model from the Intraclass model, one must place restrictions on the general model. The random ANCOVA model with varying slopes is discussed in Chapter 11.

In the general fixed effects intraclass model, it is assumed that each population has a distinct slope parameter. We write the one-way model with one covariate as:

$$y_{ij} = \alpha_i + \gamma_i z_{ij} + e_{ij} \quad i = 1, \ldots, I; j = 1, 2, \ldots, J_i \qquad (2.137)$$

where β_i and α_i represent the slope and intercept for the linear regression of y on z for the ith population. Using vector notation, the linear model matrices are:

$$X = \begin{pmatrix} 1_1 & 0_1 & \cdots & 0_1 & Z_1 & 0_1 & \cdots & 0_1 \\ 0_2 & 1_2 & \cdots & 0_2 & 0_2 & Z_2 & \cdots & 0_2 \\ \vdots & \vdots & \ddots & \vdots & \vdots & \vdots & \ddots & \vdots \\ 0_I & 0_I & \cdots & 1_I & 0_I & 0_I & \cdots & Z_I \end{pmatrix}$$

$$\beta' = \begin{pmatrix} \alpha_1 & \alpha_2 & \cdots & \alpha_I & \gamma_1 & \gamma_2 & \cdots & \gamma_I \end{pmatrix},$$

where Z_i is a vector of J_i covariates z in the ith population, and 1_i and 0_i are vectors of J_i ones and zeros, respectively. For the matrix $X'X$ to be of full rank, making the estimate for β unique, there must be at least two observations in each group in the sample having different values on the covariate.

One can test hypotheses about the slope and intercept parameters jointly or separately. The test that the slopes are equal is the test of parallelism:

$$H_P : \gamma_1 = \gamma_2 = \cdots = \gamma_I. \tag{2.138}$$

Representing the test in the form $H : C\beta = \xi$ for three groups (say), we have

$$C = \begin{pmatrix} 0 & 0 & 0 & 1 & 0 & -1 \\ 0 & 0 & 0 & 0 & 1 & -1 \end{pmatrix} \quad \text{and} \quad \beta' = \begin{pmatrix} \alpha_1 & \alpha_2 & \alpha_3 & \gamma_1 & \gamma_2 & \gamma_3 \end{pmatrix}.$$

Given parallelism (the standard assumption of the ANCOVA model), one may test that the intercepts are equal:

$$H_I : \alpha_1 = \alpha_2 = \cdots = \alpha_I. \tag{2.139}$$

However, to test this hypothesis using the intraclass model, one must add a restriction to the model, use the restricted linear discussed in Chapter 3. The usefulness of this test for the fixed effect model when the regression lines are not parallel is questionable since differences in intercepts are not constant for all values of the covariate. This is not the case if the intercepts and/or slopes are random, see Chapter 11. Or, one may test both simultaneously, the test of coincident regressions:

$$H_P : \begin{cases} \alpha_1 = \alpha_2 = \cdots = \alpha_I \\ \beta_1 = \beta_2 = \cdots = \beta_I \end{cases}. \tag{2.140}$$

To illustrate the intraclass model which reduces to the ANCOVA model when the slopes are equal, Program 2.10.sas is used; the data are in the file 2_10.dat and represent data from three treatment conditions with one covariate.

The data are first analyzed using the full rank model with the option NOINT in the model statement using PROC REG. Note how the variables are specified in the MODEL statement: three variables are needed to identify group membership, the variable y is the dependent variable, and the variables z1 z2 z3 are the covariates for each group. The TEST statement is used to specify the hypothesis test matrix C given above for three groups. Given the close association of the intraclass model with a regression model, the PROC REG is the natural procedure one should use in SAS to test for parallelism. The results for the test of H_P labeled `Test parallel results for dependent variable y` in the Output 2.10.1, the P-Value for the test is 0.8426. Hence, the test of parallelism is not rejected; the slopes are parallel. Next, we test that the regression lines are coincident (Output

Output 2.10.1: Intraclass Covariance Model — Parallelism Using PROC REG.

Test parallel Results for Dependent Variable y				
Source	DF	Mean Square	F Value	Pr > F
Numerator	2	0.09523	0.17	0.8462
Denominator	12	0.56245		

2.10.2). The p-value for the test of coincidence: coin is $p = 0.0012$. This indicates that the regression lines are not coincident. Thus, there must be a difference in intercepts. However, to test for differences in intercepts, we must use the restricted linear full rank model discussed in Chapter 3. The approach is illustrated in the program, with details provided in Section 3.6.1.

Output 2.10.2: Intraclass Covariance Model — Coincident Using PROC REG.

Test coin Results for Dependent Variable y				
Source	DF	Mean Square	F Value	Pr > F
Numerator	4	5.15440	9.16	0.0012
Denominator	12	0.56245		

Alternatively, using the less than rank model and PROC GLM, the test of parallelism is obtained by including the interaction term `z*a` in the `MODEL` statement (Output 2.10.3). The results are identical to the output found using PROC REG. And, the test of no difference in intercepts is directly tested since the less than full rank model assumes equal slopes or parallelism. To test for no differences in intercepts, given parallelism, one must use the Type III Sum of Squares; the F statistic is $F = 20.60$ with $p < 0.0001$ (Output 2.10.4). This is identical to the results obtained using the restricted linear full rank model. Hence, the test of no difference in intercepts is significant. The model is discussed more fully in Chapter 3.

Output 2.10.3: ANOVA Table with an Interaction Term z*a.

Source	DF	Type III SS	Mean Square	F Value	Pr > F
a	2	3.22575628	1.61287814	2.87	0.0960
z	1	17.74468236	17.74468236	31.55	0.0001
z*a	2	0.19045496	0.09522748	0.17	0.8462

Output 2.10.4: ANOVA Table for a Test of No Differences in Intercepts.

Source	DF	Type III SS	Mean Square	F Value	Pr > F
a	2	20.42714260	10.21357130	20.60	<.0001
z	1	21.51852314	21.51852314	43.41	<.0001

CHAPTER 3

Restricted General Linear Models

3.1 INTRODUCTION

In Chapter 2, we discussed the general linear model (GLM) without restrictions. There we assumed that the covariance structure of the GLM was $\Omega = \sigma^2 I_n$, but that no restrictions of the form (1.18) are associated with the model. For example, in the analysis of two-way factorial design without interaction one must place the restriction that the interaction is equal to zero. In this chapter, we consider the situation where restrictions of the form $R\beta = 0$ are associated with the model and the rank of the restriction matrix R is s, $\mathrm{rank}(R) = s > 0$. This model is called the restricted general linear model (RGLM). We discuss how to obtain ordinary least squares (OLS) and maximum likelihood (ML) estimates of β given the restrictions, and we also test hypotheses with restrictions and obtain simultaneous confidence sets for linear parametric functions of β.

We then discuss several examples of the RGLM using the full rank model. In the first example, we discuss the analysis of a two-way factorial design with unequal cell frequencies. We first discuss this model including the interaction (unrestricted model) and then excluding the interaction (restricted model). In the second example, we discuss the analysis of a Latin square design. In the third example, we discuss the split-plot repeated measures design. Later, we discuss the analysis of covariance model, with one covariate and with two covariates; we conclude with a discussion of a nested design.

3.2 ESTIMATION AND HYPOTHESIS TESTING

In the unrestricted linear model, the fixed parameter vector β was unconstrained in the parameter space. However, when one wants to perform a regression analysis and constrain the intercept to equal zero, or constrain independent regression lines to be parallel, or create additive analysis of variance models, one constrains the elements of β by imposing restrictions on the GLM. When the elements of β are constrained, we obtain the restricted OLS estimator of β by minimizing the error sum of squares with the estimable restrictions $R\beta = \theta$ as part of the model. For this situation, the restricted estimator of β is

$$\hat{\beta}_r = \hat{\beta} - (X'X)^{-1}R'[R(X'X)^{-1}R']^{-1}(R\hat{\beta} - \theta) \tag{3.1}$$

which is identical to the estimator given in (2.11) since the matrix R takes the place of C, with $\hat{\beta}$ defined by (2.4). The estimator in (3.1) is the restricted OLS estimator (ROLSE) which is again the best linear unbiased estimator (BLUE). Letting $G = (X'X)^{-1}$, we may rewrite (3.1) as

$$\begin{aligned}
\hat{\beta}_r &= (I - GR'(RGR')^{-1}R)\hat{\beta} + GR'(RGR')^{-1}\theta \\
&= F\hat{\beta} + GR'(RGR')^{-1}\theta \tag{3.2}
\end{aligned}$$

where F is idempotent, $F^2 = F = I - GR'(RGR')^{-1}R$ and symmetric so that F is a projection operator. When $\theta = 0$, the restrictions on the model project $\hat{\beta}$ into a smaller subspace of the parameter space.

To test hypotheses about β, we again assume spherical multivariate normality as given in (2.7), where the likelihood function in (2.10) is maximized with the added restriction that $R\beta = \theta$. For this restricted case, the ML estimator of β and σ^2 are

$$\hat{\beta}_{ML_r} = \hat{\beta}_r \tag{3.3}$$

$$\begin{aligned}
\hat{\sigma}^2_{ML_r} &= \frac{(y - X\hat{\beta}_r)'(y - X\hat{\beta} - r)}{n} \\
&= \frac{[E + (R\hat{\beta} - \theta)'(R(X'X)^{-1}R)^{-1}(R\hat{\beta} - \theta)]}{n} \tag{3.4}
\end{aligned}$$

where E is defined in (2.15). To obtain an unbiased estimator for σ^2, the value of n in (3.4) is replaced by $\nu_e = n - \text{rank}(X) + \text{rank}(R) = n - k + s$.

The likelihood ratio test of $H : C\beta = \xi$ in (1.20) again requires maximizing the likelihood function (2.10) under ω and Ω_0 which now includes the restrictions. If we let the matrix $Q = \binom{R}{C}$ and $\eta = \binom{\theta}{\xi}$, the ML estimator of β and σ^2 are

$$\hat{\beta}_{r,\omega} = \hat{\beta} - (X'X)^{-1}Q'[Q(X'X)^{-1}Q']^{-1}(Q\hat{\beta} - \eta) \tag{3.5}$$

$$\hat{\sigma}^2_{r,\omega} = \frac{(y - X\hat{\beta}_{r,\omega})'(y - X\hat{\beta}_{r,\omega})}{n}. \tag{3.6}$$

Forming the likelihood ratio,

$$\lambda = \frac{L(\hat{\omega})}{L(\hat{\Omega}_0)} = \left[\frac{(y - X\hat{\beta}_r)'(y - X\hat{\beta}_r)}{(y - X\hat{\beta}_{r,\omega})'(y - X\hat{\beta}_{r,\omega})} \right]^{n/2} \tag{3.7}$$

and

$$\Lambda = \lambda^{2/n} = \frac{E}{E_r + H_r} \tag{3.8}$$

where

$$E_r = E + (R\hat{\beta} - \theta)'(RGR')^{-1}(R\hat{\beta} - \theta) \tag{3.9}$$

$$H_r = (C\hat{\beta}_r - \xi)'[CGC' - CGR'(RGR')^{-1}RGC']^{-1}(C\hat{\beta}_r - \xi)$$
$$= (C\hat{\beta}_r - \xi)'[C(F'GF)C']^{-1}(C\hat{\beta}_r - \xi)$$
$$= (C\hat{\beta}_r - \xi)'(C'FGC)^{-1}(C\hat{\beta}_r - \xi) \tag{3.10}$$

on simplifying (3.7) using properties of partitioned matrices. If

$$A = \begin{pmatrix} A_{11} & A_{12} \\ A_{21} & A_{22} \end{pmatrix} \tag{3.11}$$

then

$$A^{-1} = \begin{pmatrix} A_{11}^{-1} + A_{11}^{-1}A_{12}T^{-1}A_{21}A_{11}^{-1} & -A_{11}^{-1}A_{12}^{-1}T^{-1} \\ -T^{-1}A_{21}A_{11} & T^{-1} \end{pmatrix} \tag{3.12}$$

where $T = A_{22} - A_{21}A_{11}^{-1}A_{12}$. For details, see Timm and Carlson (1975), or Searle (1987, p. 311).

Using the likelihood ratio statistic one can again test $H : C\beta = \xi$; however, the degrees of freedom for the hypothesis is the rank$(C) = g$, where no row of C is dependent on the rows of R, i.e., the intersection of R and C must be null. Having determined E_r and H_r, one may test hypotheses again using the F distribution (2.19) by replacing H with H_r, E with E_r, $\nu_h = \text{rank}(C) = g$, and $\nu_e = n - \text{rank}(X) + \text{rank}(R)$. The $\mathcal{E}(E_r) = (n - k + s)\sigma^2$. To obtain confidence intervals for the parametric functions $\psi = c'\beta$, we have that

$$\hat{\psi} = c'\hat{\beta}_r \tag{3.13}$$

$$\mathcal{E}(\hat{\psi}) = \psi$$

$$\text{var}(\hat{\psi}) = \sigma^2 c'(FGF')c = \sigma^2 c'(FG)c$$

where an unbiased estimate s^2 of σ^2 is obtained from (3.4) is $s^2 = \frac{(n\hat{\sigma}_{ML_r}^2)}{(n-k+s)}$.

3.3 TWO-WAY FACTORIAL DESIGN WITHOUT INTERACTION

In this section we discuss the two-way fixed effects factorial design without interaction, a restricted linear model. Because the restricted model is obtained by placing restrictions on an unrestricted model, we first discuss the unrestricted model. The unrestricted full rank two-way factorial design with interaction can be represented:

$$y_{ijk} = \mu_{ij} + e_{ijk} \quad i = 1, \ldots, I; j = 1, \ldots, J; k = 1, \ldots, n_{ij} \tag{3.14}$$

$$e_{ijk} \sim IN(0, \sigma^2)$$

where $n_{ij} \geq 1$ represents the number of observations in cell a_ib_j. Hence, we are assuming there are no empty cells in the design.

Table 3.1: Two-Way Design from Timm and Carlson.

		Factor B			
		b_1	b_2	b_3	b_4
Factor A	a_1	$3, 6, 3$	$3, 4, 5, 4, 3$	$7, 8, 7$	$6, 7, 8, 9, 8$
	a_2	$1, 2, 2, 2$	$2, 3, 4, 3$	$5, 6, 5, 6$	$10, 10, 9, 11$

Table 3.2: Population Cell Means.

		Factor B				
		b_1	b_2	b_3	b_4	
Factor A	a_1	μ_{11}	μ_{12}	μ_{13}	μ_{14}	$\mu_{1\cdot}$
	a_2	μ_{21}	μ_{22}	μ_{23}	μ_{24}	$\mu_{2\cdot}$
		$\mu_{\cdot 1}$	$\mu_{\cdot 2}$	$\mu_{\cdot 3}$	$\mu_{\cdot 4}$	$\mu_{\cdot\cdot}$

We consider the analysis of the data from Timm and Carlson (1975, p. 58). The data are displayed in Table 3.1. If we associate with each cell in Table 3.1 the population mean μ_{ij}, we may write the design as a full rank linear model. Table 3.2 is used to represent the cell and column means for the design. Because the cell frequencies may be unequal, the row and column means may be estimated using only the number of levels as weights (unweighted) or the cell sizes as weights (weighted). The full rank linear model for these data is:

$$y = X\beta + e$$

$$
\begin{pmatrix} 3 \\ 6 \\ 3 \\ 3 \\ 4 \\ 5 \\ \vdots \\ 10 \\ 10 \\ 9 \\ 11 \end{pmatrix}
=
\begin{pmatrix}
1 & 0 & 0 & 0 & 0 & 0 & 0 & 0 \\
1 & 0 & 0 & 0 & 0 & 0 & 0 & 0 \\
1 & 0 & 0 & 0 & 0 & 0 & 0 & 0 \\
0 & 1 & 0 & 0 & 0 & 0 & 0 & 0 \\
0 & 1 & 0 & 0 & 0 & 0 & 0 & 0 \\
0 & 1 & 0 & 0 & 0 & 0 & 0 & 0 \\
\vdots & \vdots & \vdots & \vdots & \vdots & \vdots & \vdots & \vdots \\
0 & 0 & 0 & 0 & 0 & 0 & 0 & 1 \\
0 & 0 & 0 & 0 & 0 & 0 & 0 & 1 \\
0 & 0 & 0 & 0 & 0 & 0 & 0 & 1 \\
0 & 0 & 0 & 0 & 0 & 0 & 0 & 1
\end{pmatrix}
\begin{pmatrix} \mu_{11} \\ \mu_{12} \\ \mu_{13} \\ \mu_{14} \\ \mu_{21} \\ \mu_{22} \\ \mu_{23} \\ \mu_{24} \end{pmatrix}
+
\begin{pmatrix} e_{111} \\ e_{112} \\ e_{113} \\ e_{121} \\ e_{122} \\ e_{123} \\ \vdots \\ e_{241} \\ e_{242} \\ e_{243} \\ e_{244} \end{pmatrix}.
$$

Now suppose we wish to test the hypothesis of no interaction. To test this hypothesis using the full rank model, we write the hypothesis as no differences in differences or that all tetrads are zero:

$$
\begin{aligned}
H_{AB} : (\mu_{ij} - \mu_{i'j}) &- (\mu_{ij'} - \mu_{i'j'}) \\
&= \mu_{ij} - \mu_{i'j} - \mu_{ij'} + \mu_{i'j'} = 0 \quad \text{for all } i, j, i', j'.
\end{aligned}
\tag{3.15}
$$

Output 3.3.1: Two-Way Unrestricted Factorial Design — AB using PROC REG.

Test AB Results for Dependent Variable y				
Source	DF	Mean Square	F Value	Pr > F
Numerator	3	9.05192	10.64	0.0001
Denominator	24	0.85069		

To construct the hypothesis test matrix C for H_{AB}, $(I-1)(J-1)$ independent tetrads are formed. One possible construction for the data in Table 3.1 is:

$$C\beta = 0$$

$$
\begin{pmatrix}
1 & -1 & 0 & 0 & -1 & 1 & 0 & 0 \\
0 & 1 & -1 & 0 & 0 & -1 & 1 & 0 \\
0 & 0 & 1 & -1 & 0 & 0 & -1 & 1
\end{pmatrix}
\begin{pmatrix}
\mu_{11} \\
\mu_{12} \\
\mu_{13} \\
\mu_{14} \\
\mu_{21} \\
\mu_{22} \\
\mu_{23} \\
\mu_{24}
\end{pmatrix}
=
\begin{pmatrix}
0 \\
0 \\
0
\end{pmatrix}
$$

which corresponds to the hypothesis

$$\mu_{11} - \mu_{21} - \mu_{12} + \mu_{22} = 0$$
$$\mu_{12} - \mu_{22} - \mu_{13} + \mu_{23} = 0$$
$$\mu_{13} - \mu_{23} - \mu_{14} + \mu_{24} = 0.$$

Program 3_3.sas contains the SAS code for testing the interaction hypothesis H_{AB} using the unrestricted full rank model. Also included in the program are tests of hypothesis using restricted full rank model and tests using the less than full rank model. In SAS, tests for the unrestricted and restricted full rank models are tested using PROC REG, and tests using the less than full rank model are tested using PROC GLM. Both approaches are illustrated and discussed.

In Program 3_3.sas, the data are first read and analyzed using PROC REG with the NOINT option specified to omit the intercept term from the model to obtain a full rank model. Notice that the independent variables specified in the MODEL statement are the columns of the design matrix X. The interaction hypothesis H_{AB} is tested using the TEST statement and labeled AB. The results of the test are found in Output 3.3.1 labeled Test AB Results for dependent variable y with the results summarized in Table 3.3.

To test hypotheses regarding the row effects or column effects using the unrestricted two-way design with interaction and unequal cell frequencies, a nonorthogonal design, we may test either unweighted or weighted tests. The unweighted test for row effects is:

$$H_A : \text{ all } \mu_{i\cdot} \text{ are equal} \tag{3.16}$$

Table 3.3: ANOVA Summary Table Unrestricted Two-Way Model.

Hypothesis	Reduction Notation	SS	df	MS	F	P-value
H_{AB}	$R(AB\|\mu)$	27.16	3	9.05	10.64	$< .0001$
H_A	$R(A\|\mu, B)$	2.98	1	2.98	3.51	.0733
H_B	$R(B\|\mu, A)$	188.98	3	62.99	74.05	$< .0001$
H_{A*}	$R(A\|\mu)$	3.13	1	3.13	3.67	.0673
H_{B*}	$R(B\|\mu)$	186.42	3	62.14	73.04	$< .0001$
$Error$		20.42	24	0.85		

while the weighted test is:

$$H_{A*} : \text{ all } \sum_{j=1}^{J} \left(\frac{n_{ij}}{n_{i.}} \right) \mu_{ij} \quad \text{are equal, for } i = 1, 2, \ldots, I \qquad (3.17)$$

where $n_{i.} = \sum_{j=1}^{J} n_{ij}$. Observe that the unweighted test does not depend on the cell frequencies, while the weighted test does depend on the frequencies. If upon replication of the study, one would expect the same unequal pattern to result, then one would test the weighted hypothesis. If on the other hand, the unequal cell frequencies are expected to be different upon replication, they are independent of the treatment; then the unweighted test is most appropriate.

The test of test H_A for the example data is: $H_A : \mu_{1.} = \mu_{2.}$ and the hypothesis test matrix is

$$C_A = \left(\frac{1}{4} \quad \frac{1}{4} \quad \frac{1}{4} \quad \frac{1}{4} \quad -\frac{1}{4} \quad -\frac{1}{4} \quad -\frac{1}{4} \quad -\frac{1}{4} \right).$$

For the test weighted test, $H_{A*} : \sum_{j=1}^{J} \left(\frac{n_{1j}}{n_{1.}} \right) \mu_{1j} = \sum_{j=1}^{J} \left(\frac{n_{2j}}{n_{2.}} \right) \mu_{2j}$, the hypothesis test matrix is

$$C_{A*} = \left(\frac{3}{16} \quad \frac{5}{16} \quad \frac{3}{16} \quad \frac{5}{16} \quad -\frac{4}{16} \quad -\frac{4}{16} \quad -\frac{4}{16} \quad -\frac{4}{16} \right)$$

where $n_1 = \sum_{j=1}^{J} n_{1j}$ and $n_2 = \sum_{j=1}^{J} n_{2j}$. For the weighted test, the weights are data dependent. Tests for factor B follow similarly.

Program 3_3.sas includes the SAS code to test the above hypotheses, both the weighted and unweighted test. The hypothesis test matrices are specified using the TEST statements and labeled: A, B, WTA, WTB. The results are provided in Output 3.3.2 and summarized in Table 3.3.

The analysis may also be performed using PROC GLM which uses the less than full rank model. The code is contained in the next section of Program 3_3.sas. Observe that one must execute the PROC two times, once with the model model y=a b a* b and then with the order: Model y=b a a* b. The different orders are used to obtain the reductions $R(A|\mu)$ and $R(B|\mu)$ for the weighed tests using the Type I sum of squares. The unweighted tests are obtained from the Type III sum of

Output 3.3.2: Two-Way Unrestricted Factorial Design Using PROC REG.

Test A Results for Dependent Variable y				
Source	DF	Mean Square	F Value	Pr > F
Numerator	1	2.98401	3.51	0.0733
Denominator	24	0.85069		

Test B Results for Dependent Variable y				
Source	DF	Mean Square	F Value	Pr > F
Numerator	3	62.99235	74.05	<.0001
Denominator	24	0.85069		

Test wtA Results for Dependent Variable y				
Source	DF	Mean Square	F Value	Pr > F
Numerator	1	3.12500	3.67	0.0673
Denominator	24	0.85069		

Test wtB Results for Dependent Variable y				
Source	DF	Mean Square	F Value	Pr > F
Numerator	3	62.13712	73.04	<.0001
Denominator	24	0.85069		

Output 3.3.3: Two-Way Unrestricted Factorial Design Using PROC GLM.

Source	DF	Type I SS	Mean Square	F Value	Pr > F
a	1	3.1250000	3.1250000	3.67	0.0673
b	3	184.8025794	61.6008598	72.41	<.0001
a*b	3	27.1557540	9.0519180	10.64	0.0001

Source	DF	Type III SS	Mean Square	F Value	Pr > F
a	1	2.9840054	2.9840054	3.51	0.0733
b	3	188.9770545	62.9923515	74.05	<.0001
a*b	3	27.1557540	9.0519180	10.64	0.0001

Source	DF	Type I SS	Mean Square	F Value	Pr > F
b	3	186.4206349	62.1402116	73.05	<.0001
a	1	1.5069444	1.5069444	1.77	0.1957
a*b	3	27.1557540	9.0519180	10.64	0.0001

Source	DF	Type III SS	Mean Square	F Value	Pr > F
b	3	188.9770545	62.9923515	74.05	<.0001
a	1	2.9840054	2.9840054	3.51	0.0733
a*b	3	27.1557540	9.0519180	10.64	0.0001

squares. Note also that the test of interaction does not depend on the order (Output 3.3.3).

This concludes the analysis of the unrestricted, full rank, two-way design. Now we turn to the same design, but assume no interaction. This results in the restricted full rank model since the interaction hypothesis becomes a model restrict. The restricted full rank model for the two-way factorial design follows:

$$y_{ijk} = \mu_{ij} + e_{ijk} \qquad i = 1, \ldots, I; j = 1, \ldots, J; k = 1, \ldots, n_{ij},$$

$\mu_{ij} - \mu_{i'j} - \mu_{ij'} + \mu_{i'j'} = 0$ for all linearly independent sets $(I-1)(J-1)$ and $e_{ijk} \sim IN(0, \sigma^2)$. To represent the restriction for this example, we have that $R\beta = \theta$ where

$$R = \begin{pmatrix} 1 & -1 & 0 & 0 & -1 & 1 & 0 & 0 \\ 0 & 1 & -1 & 0 & 0 & -1 & 1 & 0 \\ 0 & 0 & 1 & -1 & 0 & 0 & -1 & 1 \end{pmatrix},$$

$\beta = \mu$, and $\theta = 0$. Note that the matrix R is identical to the hypothesis test matrix C used to test for no interactions in the unrestricted linear model.

The SAS code for the analysis of the data in Table 3.1 using the restricted full

Table 3.4: ANOVA Summary Table Restricted Two-Way Model.

Hypothesis	Reduction Notation	SS	df	MS	F	P-value
H_A	$R(A\|\mu, B)$	1.51	1	1.51	0.86	.3633
H_B	$R(B\|\mu, A)$	184.80	3	61.60	34.96	$< .0001$
H_{A^*}	$R(A\|\mu)$	3.13	1	3.13	1.77	.1941
H_{B^*}	$R(B\|\mu)$	186.42	3	62.14	35.27	$< .0001$
$Error$		47.57	27	1.76		

rank model is also provided in Program 3_3.sas. Again we must use PROC REG; however, we must add the RESTRICTION statement which represent the matrix R.

To test hypotheses with the restricted model, we construct hypothesis test matrices C of rank g where the rows of C are independent and not dependent on the rows of the restriction matrix R. Again, because of the unequal cell frequencies, we can test either a weighted or an unweighted hypothesis for row and column means.

To test H_A : all $\mu_{i.}$ are equal, the unweighted test, the matrix $Q = \binom{R}{C}$ and η are:

$$Q = \begin{pmatrix} R \\ \cdots \\ C \end{pmatrix} = \begin{pmatrix} 1 & -1 & 0 & 0 & -1 & 1 & 0 & 0 \\ 0 & 1 & -1 & 0 & 0 & -1 & 1 & 0 \\ 0 & 0 & 1 & -1 & 0 & 0 & -1 & 1 \\ \cdots & \cdots & \cdots & \cdots & \cdots & \cdots & \cdots & \cdots \\ 1/4 & 1/4 & 1/4 & 1/4 & -1/4 & -1/4 & -1/4 & -1/4 \end{pmatrix}$$

$$\eta = \begin{pmatrix} \theta \\ \cdots \\ \xi \end{pmatrix} = \begin{pmatrix} 0 \\ 0 \\ 0 \\ \cdots \\ 0 \end{pmatrix}.$$

Similarly, one can test H_B : all $\mu_{.j}$ are equal.

Weighted test H_{A^*} and H_{B^*} may also be tested; again they depend on the cell frequencies. The results of the tests are in Output 3.3.4 and summarized in Table 3.4. Again, these hypotheses may be tested using PROC GLM which uses the unrestricted less than full rank model. For the less than full rank model, the interaction term is removed from the MODEL statement. The results are given in the Output 3.3.5 and are seen to agree with the full rank model.

Because the interaction term is significant for this example, we would not analyze these data by imposing the restriction on the model. This should only be done if the interaction hypothesis is nonsignificant. Formulating the model with the restrict forces the researcher to pool the interaction term with the error term. This is not the case when using the less than full rank model. One may use either a pool or no pool strategy.

Output 3.3.4: Two-Way Restricted Factorial Design Using PROC REG.

Test A Results for Dependent Variable y				
Source	DF	Mean Square	F Value	Pr > F
Numerator	1	1.50694	0.86	0.3633
Denominator	27	1.76194		

Test B Results for Dependent Variable y				
Source	DF	Mean Square	F Value	Pr > F
Numerator	3	61.60086	34.96	<.0001
Denominator	27	1.76194		

Test wtA Results for Dependent Variable y				
Source	DF	Mean Square	F Value	Pr > F
Numerator	1	3.12500	1.77	0.1941
Denominator	27	1.76194		

Test wtB Results for Dependent Variable y				
Source	DF	Mean Square	F Value	Pr > F
Numerator	3	62.14027	35.27	<.0001
Denominator	27	1.76194		

Output 3.3.5: Two-Way Restricted Factorial Design Using PROC GLM.

Source	DF	Type I SS	Mean Square	F Value	Pr > F
a	1	3.1250000	3.1250000	1.77	0.1941
b	3	184.8025794	61.6008598	34.96	<.0001

Source	DF	Type III SS	Mean Square	F Value	Pr > F
a	1	1.5069444	1.5069444	0.86	0.3633
b	3	184.8025794	61.6008598	34.96	<.0001

Source	DF	Type I SS	Mean Square	F Value	Pr > F
b	3	186.4206349	62.1402116	35.27	<.0001
a	1	1.5069444	1.5069444	0.86	0.3633

Source	DF	Type III SS	Mean Square	F Value	Pr > F
b	3	184.8025794	61.6008598	34.96	<.0001
a	1	1.5069444	1.5069444	0.86	0.3633

Table 3.5: 3×3 Latin Square Design I.

	b_1	b_2	b_3
a_1	C_2	C_3	C_1
a_2	C_3	C_1	C_2
a_3	C_1	C_2	C_3

3.4 LATIN SQUARE DESIGNS

In the previous example we discussed the two-way factorial design. In these designs both the row and the column effects are of interest. At times, it is the case that only one of the factors (say A) is of interest since the other factor may be a blocking factor where subjects are assigned to factor A within blocks. These designs are called completely randomized designs. And, the analysis of such designs for fixed treatments and blocks is identical to a factorial design. If blocks are random, we have a mixed model. These designs are discussed in Chapter 11.

When two fixed nonrandom blocking variables are included in a design (say A and B) each with m levels, and one is interested in a treatment factor (say C) then one may be interested in the analysis of an $m \times m$ Latin Square design. In a Latin Square design each row (A) and each column (B) receives each treatment (each level of C) only once (Box, Hunter, & Hunter, 1978). Levels of factors A and B occur together exactly once and thus the design is completely balanced. In most Latin Square designs there is only one observation in each cell, only one replication, as Table 3.5 illustrates. Thus, there is no interaction between the blocking factors and the treatment factor. There are 11 other ways of arranging three treatments in a 3×3 Square and the researcher randomly selects one of the 12 possible arrangements.

There are larger Latin Square designs, for example 4×4 and 5×5, but there is not 2×2 Square with one observation per cell. This is the case, since there would be no way of estimating the error variance in such a design. This can be seen from an examination of the degrees of freedom. With one experimental unit per cell in an $m \times m$ Latin Square design, the total number of degrees of freedom is $m^2 - 1$. With three factors with m levels each, the number of degrees of freedom for the main effects is $3(m - 1)$. Thus, the number of degrees of freedom remaining for the error term is $(m^2 - 1) - 3(m - 1) = (m - 1)(m - 2)$. Note that this is zero if $m = 2$. If there are more than one observation per cell, the within cell variation can be used to estimate the error variation and the $(m - 1)(m - 2)$ degrees of freedom can be used as a partial test of no interaction. With one observation per cell, we assume no interaction or an additive model and use the restrictions to specify the restricted full rank model.

The full rank restricted linear model for the $m \times m$ Latin Square design is:

$$y_{ijk} = \mu_{ijk} + e_{ijk} \tag{3.18}$$

for the m^2 treatments defined by the Design, with no interaction restriction, and

Table 3.6: 3×3 Latin Square Design II.

	b_1	b_2	b_3
a_1	$C_2(7)$	$C_3(5)$	$C_1(4)$
a_2	$C_3(5)$	$C_1(6)$	$C_2(11)$
a_3	$C_1(8)$	$C_2(8)$	$C_3(9)$

$e_{ijk} \sim IN(0, \sigma^2)$. We assume that the effects are additive and thus we place the restriction of no interaction on the model. The analysis of this design is identical to the analysis of a three-way factorial design with no interaction and no empty cells (see Timm, 2002, p. 289, for the analysis of a multivariate Latin Square design using the less than full rank model).

As an example of the analysis of a 3×3 Latin square design using SAS, and the restricted full rank model, we analyze the data in Table 3.6. Here, factors A and B are fixed blocking factors and factor C is the treatment factor. The numbers within the cells in parenthesis are the values of the dependent variable.

Using linear model notation, the design is written as:

$$y = X\beta + e$$

$$\begin{pmatrix} 7 \\ 5 \\ 4 \\ 5 \\ 6 \\ 11 \\ 8 \\ 8 \\ 9 \end{pmatrix} = \begin{pmatrix} 1 & 0 & 0 & 0 & 0 & 0 & 0 & 0 & 0 \\ 0 & 1 & 0 & 0 & 0 & 0 & 0 & 0 & 0 \\ 0 & 0 & 1 & 0 & 0 & 0 & 0 & 0 & 0 \\ 0 & 0 & 0 & 1 & 0 & 0 & 0 & 0 & 0 \\ 0 & 0 & 0 & 0 & 1 & 0 & 0 & 0 & 0 \\ 0 & 0 & 0 & 0 & 0 & 1 & 0 & 0 & 0 \\ 0 & 0 & 0 & 0 & 0 & 0 & 1 & 0 & 0 \\ 0 & 0 & 0 & 0 & 0 & 0 & 0 & 1 & 0 \\ 0 & 0 & 0 & 0 & 0 & 0 & 0 & 0 & 1 \end{pmatrix} \begin{pmatrix} \mu_{112} \\ \mu_{123} \\ \mu_{131} \\ \mu_{213} \\ \mu_{221} \\ \mu_{232} \\ \mu_{311} \\ \mu_{322} \\ \mu_{333} \end{pmatrix} + \begin{pmatrix} e_{112} \\ e_{123} \\ e_{131} \\ e_{213} \\ e_{221} \\ e_{232} \\ e_{311} \\ e_{322} \\ e_{333} \end{pmatrix}.$$

A matrix of restrictions R for this design may take the form

$$R = \begin{pmatrix} 1 & 0 & -1 & -1 & -1 & 0 & 0 & -1 & 1 \\ -1 & 1 & 0 & 0 & 0 & 1 & 1 & 0 & -1 \end{pmatrix}.$$

The number of rows in the restriction matrix is: $(m-1)(m-2) = (3-1)(3-2) = 2$, the rank of R or the degrees of freedom for error. There are three main effect hypotheses tested in the Latin Square design. The second hypothesis investigates the block effect and the third investigates the significance of the treatment factor:

$$H_A : \text{all } \mu_{i..} \text{ are equal}$$
$$H_B : \text{all } \mu_{.j.} \text{ are equal}$$
$$H_C : \text{all } \mu_{..k} \text{ are equal}$$

Table 3.7: ANOVA Summary for Latin Square Design.

Hypothesis	SS	df	MS	F	P-value
H_A	14.00	2	7.00	1.62	0.3824
H_B	4.67	2	2.33	0.54	0.6500
H_C	12.67	2	6.33	1.46	0.4063
$Error$	8.67	2	4.33		

where the population means of simple averages, for example $\mu_{1..} = \frac{(\mu_{112}+\mu_{123}+\mu_{131})}{3}$ for the design under study. Because the Latin Square design is balanced, the three hypotheses are orthogonal. To test these hypotheses, the hypothesis test matrices are:

$$C_A : \begin{pmatrix} \frac{1}{3} & \frac{1}{3} & \frac{1}{3} & 0 & 0 & 0 & -\frac{1}{3} & -\frac{1}{3} & -\frac{1}{3} \\ 0 & 0 & 0 & \frac{1}{3} & \frac{1}{3} & \frac{1}{3} & -\frac{1}{3} & -\frac{1}{3} & -\frac{1}{3} \end{pmatrix} \quad (3.19)$$

$$C_B : \begin{pmatrix} \frac{1}{3} & 0 & -\frac{1}{3} & \frac{1}{3} & 0 & -\frac{1}{3} & \frac{1}{3} & 0 & -\frac{1}{3} \\ 0 & \frac{1}{3} & -\frac{1}{3} & 0 & \frac{1}{3} & -\frac{1}{3} & 0 & \frac{1}{3} & -\frac{1}{3} \end{pmatrix} \quad (3.20)$$

$$C_C : \begin{pmatrix} 0 & -\frac{1}{3} & \frac{1}{3} & -\frac{1}{3} & \frac{1}{3} & 0 & \frac{1}{3} & 0 & -\frac{1}{3} \\ \frac{1}{3} & -\frac{1}{3} & 0 & -\frac{1}{3} & 0 & \frac{1}{3} & 0 & \frac{1}{3} & -\frac{1}{3} \end{pmatrix}. \quad (3.21)$$

Program 3_3.sas contains the SAS code to analyze these data using PROC REG. First, the data are read and PROC REG is used with the NOINT option to omit the intercept from the model so that the full rank model is defined. The columns of the design matrix X are included in the MODEL statement as the independent variables. The RESTRICT and TEST statements are used as in previous examples to specify the rows of the restriction matrix, R, and the hypothesis test, C, respectively. The results are given in Output 3.4.1; Table 3.7 summarizes the results.

3.5 REPEATED MEASURES DESIGNS

In social science and medical research, differences among subjects may be large relative to differences in treatment effects and hence the treatment effects are obscured by subject heterogeneity. One way to reduce this variability due to individual differences is to divide the subjects into blocks on the basis of some concomitant variable correlated with the dependent variable and then to randomly assign subjects to treatments within blocks; a randomized block design. This usually reduces the size of the error variance and makes it easier to detect treatment differences. To further separate true error variance from variability that is due to individual differences, each subject may serve as its own control if we take repeated measurements on a subject over several levels of an independent treatment variable where the order of presentation is randomized independently for each subject. Such designs are called one-sample repeated measures designs. Repeated measures designs have increased precision over completely randomized designs and randomized block designs and require fewer subjects. There are many experimental situations in the social sciences

Output 3.4.1: Latin Square using PROC REG.

Test A Results for Dependent Variable y1				
Source	DF	Mean Square	F Value	Pr > F
Numerator	2	7.00000	1.62	0.3824
Denominator	2	4.33333		

Test B Results for Dependent Variable y1				
Source	DF	Mean Square	F Value	Pr > F
Numerator	2	2.33333	0.54	0.6500
Denominator	2	4.33333		

Test C Results for Dependent Variable y1				
Source	DF	Mean Square	F Value	Pr > F
Numerator	2	6.33333	1.46	0.4062
Denominator	2	4.33333		

and in the analysis of medical trials research for which the design is appropriate.

3.5.1 Univariate Mixed ANOVA Model, Full Rank Representation for a Split Plot Design

Extending the notion of a repeated measures experiment having more than a single treatment leads one to a split plot repeated measures design. For our next example for a general linear model with restrictions, we discuss the univariate split plot design, also known as a two-sample profile analysis (Timm, 2002) or a mixed within-subjects design (Keppel & Wickens, 2004). The name split plot comes from agricultural research where each level of a treatment is applied to a whole plot of land, and all levels of the second treatments are applied to subplots within the larger plot (Winer, 1971).

A simple application of the design involves randomly assigning k subjects to I treatments, labeled factor A, and observing subjects repeatedly over J levels of a second factor B as illustrated in Table 3.8. In this section we analyze the split plot design as a restricted full rank univariate linear model. This approach is appropriate if we can assume equality of the covariance matrices and the sufficient condition of equal variances and covariances (compound symmetry) within each level of the repeated measures factor. The definition and test of compound symmetry or more generally circularity is discussed in Section 3.5.3. In Chapter 5, we analyze the design using the multivariate linear model where one requires equality, but not circularity for valid tests. Finally, in Chapter 11, the model is analyzed using the general linear mixed model.

Table 3.8: Simple Split Plot Design.

		b_1	b_2	b_3
		B		
A	a_1	y_{111}	y_{121}	y_{131}
		y_{112}	y_{122}	y_{132}
	a_2	y_{211}	y_{221}	y_{231}
		y_{212}	y_{222}	y_{232}

Table 3.9: Simple Split Plot Design Cell Means.

		b_1	b_2	b_3	
		B			
A	a_1	μ_{111}	μ_{121}	μ_{131}	$\mu_{1\cdot 1}$
		μ_{112}	μ_{122}	μ_{132}	$\mu_{1\cdot 2}$
		$\mu_{11\cdot}$	$\mu_{12\cdot}$	$\mu_{13\cdot}$	$\mu_{1\cdot\cdot}$
	a_2	μ_{211}	μ_{221}	μ_{231}	$\mu_{2\cdot 1}$
		μ_{212}	μ_{222}	μ_{232}	$\mu_{2\cdot 2}$
		$\mu_{21\cdot}$	$\mu_{22\cdot}$	$\mu_{23\cdot}$	$\mu_{2\cdot\cdot}$
		$\mu_{\cdot 1\cdot}$	$\mu_{\cdot 2\cdot}$	$\mu_{\cdot 3\cdot}$	$\mu_{\cdot\cdot\cdot}$

The full rank model with restrictions for the split plot design in Table 3.8 is:

$$y_{ijk} = \mu_{ijk} + s_{(i)k} + e_{ijk} \quad i = 1, \ldots, I; j = 1, \ldots, J; k = 1, \ldots, n_i \quad (3.22)$$

with a restriction (no interaction between S and B within each level of A), $s_{(i)k} \sim IN(0, \rho\sigma^2)$ and jointly independent of the errors, $e_{ijk} \sim IN(0, (1 - \rho)\sigma^2)$.

The population parameters for the design are displayed in Table 3.9.

For the split plot design, several hypotheses are of interest:

test for group differences

$$H_A : \mu_{1\cdot\cdot} = \mu_{2\cdot\cdot}$$

subject differences within treatment

$$H_{S(A)} : \mu_{1\cdot 1} = \mu_{1\cdot 2}$$
$$\mu_{2\cdot 1} = \mu_{2\cdot 2}$$

test for differences in conditions

$$H_B : \mu_{\cdot 1\cdot} = \mu_{\cdot 2\cdot} = \mu_{\cdot 3\cdot}.$$

Table 3.10: Expected Mean Squares for Split Plot Design.

Hypothesis	df	MS	F
H_A	$I-1$	$\sigma_e^2 + J\sigma_s^2 + \frac{\sigma_A}{(I-1)}$	$\frac{MS_A}{MS_{S(A)}}$
$H_{S(A)}$	$I(n_i-1)$	$\sigma_e^2 + J\sigma_s^2$	
H_B	$J-1$	$\sigma_e^2 + \frac{\sigma_B}{(J-1)}$	$\frac{MS_B}{MS_{error}}$
H_{AB}	$(I-1)(n_i-1)$	$\sigma_e^2 + \frac{\sigma_{AB}}{(I-1)(J-1)}$	$\frac{MS_{AB}}{MS_{error}}$
$Error$	$I(J-1)(n_i-1)$	σ_e^2	

test of parallelism

$$H_{AB} : \mu_{11\cdot} - \mu_{21\cdot} - \mu_{12\cdot} + \mu_{22\cdot} = 0$$
$$\mu_{12\cdot} - \mu_{22\cdot} - \mu_{13\cdot} + \mu_{23\cdot} = 0.$$

These hypotheses must be tested subject to the restriction that there is no interaction between subjects (S) and the levels of factor B within each level of A. The following equations represent the restrictions:

$$\mu_{111} - \mu_{112} - \mu_{121} + \mu_{122} = 0$$
$$\mu_{121} - \mu_{122} - \mu_{131} + \mu_{132} = 0$$
$$\mu_{211} - \mu_{212} - \mu_{221} + \mu_{222} = 0$$
$$\mu_{221} - \mu_{222} - \mu_{213} + \mu_{232} = 0.$$

To test all of the hypotheses for the split plot design, we again must define hypothesis test matrices C with each hypothesis. However, the subject factor is not fixed but random for the design. To obtain the correct F ratios for the tests, the sample means are expressed as a function of the population parameters and the random components. We then need to evaluate the expected mean squares for each hypothesis and error. Letting

$$y_{i\cdot\cdot} = \mu_{i\cdot} + s_{(i)\cdot} + e_{i\cdot\cdot}$$
$$y_{\cdot j} = \mu_{\cdot j} + s_{(\cdot)\cdot} + e_{\cdot j\cdot}$$
$$y_{\cdot\cdot k} = \mu_{\cdot\cdot} + s_{(\cdot)k} + e_{\cdot\cdot k}$$
$$y_{ij\cdot} = \mu_{\cdot j\cdot} + s_{(\cdot)\cdot} + e_{i\cdot j\cdot}$$
$$y_{i\cdot k} = \mu_{ij} + s_{(i)\cdot} + e_{i\cdot k}$$
$$y_{\cdot\cdot\cdot} = \mu_{\cdot\cdot} + s_{(\cdot)\cdot} + e_{\cdot\cdot}$$

and these into expressions for the hypothesis sums of squares for each hypothesis and error, the expected mean squares for each hypothesis result. The expected mean squares are presented in Table 3.10. From Table 3.10, we observe that the usual error term for the fixed effect model is only used to test H_B and H_{AB}. The hypothesis

Table 3.11: Split Plot Data Set.

		b_1	b_2	b_3
	a_1	3	4	3
		2	2	1
A	a_2	3	7	7
		5	4	6
	a_3	3	4	6
		2	3	5

mean square for $H_{S(A)}$ is the error term for testing H_A, while the test of H_A is valid independent of the covariance structure within subjects. The assumptions on the random components ensure that the structure of the covariance matrix is:

$$\Sigma = \rho\sigma^2 J_n + (1-\rho)\sigma^2 I_n = \sigma_s^2 J_n + \sigma_e^2 I_n \qquad (3.23)$$

where $\rho\sigma^2 = \rho(\sigma_s^2 + \sigma_e^2)$, the matrix $J_n = (1_{ij})$, a matrix of 1's and I_n is the identity matrix.

To illustrate the analysis of a split plot design using the restricted full rank linear model, the data in Table 3.11 is used. While it is more natural to analyze mixed models using PROC MIXED and to find expected mean squares using PROC GLM, we use the design to illustrate the flexibility of the restricted full rank model.

The parameter vector for the model follows: $\mu' = (\mu_{111}, \mu_{121}, \mu_{131}, \mu_{112}, \mu_{122}, \mu_{132}, \mu_{211}, \mu_{221}, \mu_{231}, \mu_{212}, \mu_{222}, \mu_{232}, \mu_{311}, \mu_{321}, \mu_{331}, \mu_{312}, \mu_{322}, \mu_{332})$. The hypothesis test matrices associated with the hypotheses in Table 3.10 follow:

$$C_A = \begin{pmatrix} 1 & 1 & 1 & 1 & 1 & 1 & -1 & -1 & -1 & -1 & -1 & -1 & 0 & 0 \\ 0 & 0 & 0 & 0 & 0 & 0 & 1 & 1 & 1 & 1 & 1 & 1 & -1 & -1 \end{pmatrix} \cdots$$
$$\begin{pmatrix} 0 & 0 & 0 & 0 \\ -1 & -1 & -1 & -1 \end{pmatrix}$$

$$C_B = \begin{pmatrix} 1 & 0 & -1 & 1 & 0 & -1 & 1 & 0 & -1 & 1 & 0 & -1 & 1 & 0 & -1 & 1 & 0 & -1 \\ 0 & 1 & -1 & 0 & 1 & -1 & 0 & 1 & -1 & 0 & 1 & -1 & 0 & 1 & -1 & 0 & 1 & -1 \end{pmatrix}$$

$$C_{S(A)} = \begin{pmatrix} 1 & 1 & 1 & -1 & -1 & -1 & 0 & 0 & 0 & 0 & 0 & 0 & 0 & 0 & 0 & 0 & 0 & 0 \\ 0 & 0 & 0 & 0 & 0 & 0 & 1 & 1 & 1 & -1 & -1 & -1 & 0 & 0 & 0 & 0 & 0 & 0 \\ 0 & 0 & 0 & 0 & 0 & 0 & 0 & 0 & 0 & 0 & 0 & 0 & 1 & 1 & 1 & -1 & -1 & -1 \end{pmatrix}$$

$$C_{AB} = \begin{pmatrix} 1 & -1 & 0 & 1 & -1 & 0 & -1 & 1 & 0 & -1 & 1 & 0 & 0 & 0 \\ 0 & 1 & -1 & 0 & 1 & -1 & 0 & -1 & 1 & 0 & -1 & 1 & 0 & 0 \\ 0 & 0 & 0 & 0 & 0 & 0 & 1 & -1 & 0 & 1 & -1 & 0 & -1 & 1 \\ 0 & 0 & 0 & 0 & 0 & 0 & 0 & 1 & -1 & 0 & 1 & -1 & 0 & -1 \end{pmatrix} \cdots$$
$$\begin{pmatrix} 0 & 0 & 0 & 0 \\ 0 & 0 & 0 & 0 \\ 0 & -1 & 1 & 0 \\ 1 & 0 & -1 & 1 \end{pmatrix}.$$

Finally, the restriction matrix for the example is:

$$R = \begin{pmatrix} 1 & -1 & 0 & -1 & -1 & 0 & 0 & 0 & 0 & 0 & 0 & 0 & 0 & 0 & 0 & 0 & 0 & 0 \\ 0 & 1 & -1 & 0 & -1 & 1 & 0 & 0 & 0 & 0 & 0 & 0 & 0 & 0 & 0 & 0 & 0 & 0 \\ 0 & 0 & 0 & 0 & 0 & 0 & 1 & -1 & 0 & -1 & 1 & 0 & 0 & 0 & 0 & 0 & 0 & 0 \\ 0 & 0 & 0 & 0 & 0 & 0 & 0 & 1 & -1 & 0 & -1 & 1 & 0 & 0 & 0 & 0 & 0 & 0 \\ 0 & 0 & 0 & 0 & 0 & 0 & 0 & 0 & 0 & 0 & 0 & 0 & 1 & -1 & 0 & -1 & 1 & 0 \\ 0 & 0 & 0 & 0 & 0 & 0 & 0 & 0 & 0 & 0 & 0 & 0 & 0 & 1 & -1 & 0 & -1 & 1 \end{pmatrix}.$$

The SAS code for the example is given in Program 3_5_1.sas. Again, the PROC REG is used with the NOINT option to specify a full rank model. To analyze these

Output 3.5.1: Split Plot Design Using PROC REG.

Test A Results for Dependent Variable y				
Source	DF	Mean Square	F Value	Pr > F
Numerator	2	12.05556	10.85	0.0102
Denominator	6	1.11111		

Test B Results for Dependent Variable y				
Source	DF	Mean Square	F Value	Pr > F
Numerator	2	4.22222	3.80	0.0859
Denominator	6	1.11111		

Test AB Results for Dependent Variable y				
Source	DF	Mean Square	F Value	Pr > F
Numerator	4	2.05556	1.85	0.2385
Denominator	6	1.11111		

Test SA Results for Dependent Variable y				
Source	DF	Mean Square	F Value	Pr > F
Numerator	3	2.11111	1.90	0.2307
Denominator	6	1.11111		

data using the restricted full rank model we must enter the columns of the design matrix as the independent variables in the model. For large designs, this becomes very cumbersome. The matrix R is specified using the RESTRICTION statement and the hypothesis test matrices are specified in the TEST statements similar to the previous examples in this chapter. The results are given in the Output 3.5.1 and summarized in the Table 3.12; the F value for the H_A is wrong in the output. TEST for A uses MS_{error} instead of $MS_{S(A)}$.

Also included in Program 3_5_1.sas is the analysis of the model using the PROC GLM using the REPEATED statement. This approach assumes the data are vector valued or multivariate in nature. Next, we use PROC GLM to define a univariate less than full rank model. This step evaluates expected mean squares for hypotheses, and F ratios are formed using the RANDOM statement. Finally, PROC MIXED, discussed in Chapter 11, is used to analyze the example. Using the multivariate approach and the REPEATED statement, tests occur in the output under the Title: Repeated Measures Analysis of Variance - Tests of Hypotheses for Between Subject and Repeated Measures Analysis of Variance-Univariate Tests of Hypotheses for Within Subjects (Output 3.5.2). Using PROC GLM and a univariate model, the tests ap-

Table 3.12: Test Result for the Split Plot Design.

Hypothesis	SS	df	MS	F	P-value
H_A	24.11	2	12.06	5.71	0.0949
$H_{S(A)}$	6.33	3	2.11		
H_B	8.44	2	4.22	3.80	0.0859
H_{AB}	8.22	4	2.06	1.85	0.2385
Error	6.67	6	1.11		

Output 3.5.2: Split Plot Design — PROC GLM Using repeated Statement.

Source	DF	Type III SS	Mean Square	F Value	Pr > F
a	2	24.11111111	12.05555556	5.71	0.0949
Error	3	6.33333333	2.11111111		

Source	DF	Type III SS	Mean Square	F Value	Pr > F	Adj Pr > F G - G	Adj Pr > F H - F
b	2	8.44444444	4.22222222	3.80	0.0859	0.1451	0.0859
b*a	4	8.22222222	2.05555556	1.85	0.2385	0.2985	0.2385
Error(b)	6	6.66666667	1.11111111				

pear in the output under Title: `Tests of Hypotheses for Mixed Model Analysis of Variance` (Output 3.5.3). Finally, using the PROC MIXED, the tests appear with the heading: `Type 3 Tests of Fixed Effects` (Output 3.5.4). More will be said regarding these options in SAS in Chapters 5 and 11.

3.5.2 Univariate Mixed Linear Model, Less Than Full Rank Representation

We next address the analysis of a split plot design using a real data set. The data are given in Timm (1975, p. 244) and were provided by Dr. Paul Ammon. The data represent Probe Word stimuli data for 10 subjects with low short-term memory capacity (Group 1), and for 10 subjects with high short-term memory capacity (Group 2). The two groups represent two levels for factor A. The dependent variable, seconds to recall, were obtained for 5 Probe-Word positions; factor B has 5 levels. The data are in the file 3_5_2.dat. To analyze these data using the restricted full rank model would involve a vector with over 100 cells. This approach would be very cumbersome. Hence, we use PROC GLM and a less than full rank model to analyze the data. The same data are analyzed in Chapter 5 using a multivariate model. The less than full rank unrestricted linear model for the design has the structure:

$$y_{ijk} = \mu + \alpha_i + \beta_j + (\alpha\beta)_{ij} + s_{(i)k} + e_{ijk}$$
$$i = 1, \ldots, I; j = 1, \ldots, J; k = 1, \ldots, n_i \quad (3.24)$$

Output 3.5.3: Split Plot Design — PROC GLM Using random Statement.

	Source	DF	Type III SS	Mean Square	F Value	Pr > F
*	a	2	24.111111	12.055556	5.71	0.0949
	Error: MS(subj(a))	3	6.333333	2.111111		
* This test assumes one or more other fixed effects are zero.						

	Source	DF	Type III SS	Mean Square	F Value	Pr > F
	subj(a)	3	6.333333	2.111111	1.90	0.2307
*	b	2	8.444444	4.222222	3.80	0.0859
	a*b	4	8.222222	2.055556	1.85	0.2385
	Error: MS(Error)	6	6.666667	1.111111		
* This test assumes one or more other fixed effects are zero.						

Output 3.5.4: Split Plot Design Using PROC Mixed.

Type 3 Tests of Fixed Effects				
Effect	Num DF	Den DF	F Value	Pr > F
a	2	3	5.71	0.0949
b	2	6	3.80	0.0859
a*b	4	6	1.85	0.2385

Output 3.5.5: Split Plot Design — Real Data Example.

Source	DF	Type III SS	Mean Square	F Value	Pr > F
grp	1	1772.410000	1772.410000	8.90	0.0080
Error	18	3583.140000	199.063333		

Source	DF	Type III SS	Mean Square	F Value	Pr > F	Adj Pr > F G - G	H - F
probe	4	3371.300000	842.825000	14.48	<.0001	<.0001	<.0001
probe*grp	4	79.940000	19.985000	0.34	0.8479	0.8068	0.8479
Error(probe)	72	4191.960000	58.221667				

with a restriction (no interaction between S and B within each level of A), $s_{(i)k} \sim IN(0, \rho\sigma^2)$ and jointly independent of the errors, $e_{ijk} \sim IN(0, (1 - \rho)\sigma^2)$. Note that for the less than full rank model the mean $\mu_{ij} = \mu + \alpha_i + \beta_j + (\alpha\beta)_{ij}$ is over parameterized. Program 3_5_2.sas is used to analyze the data set. The results are presented in the Output 3.5.5. Only PROC GLM using REPEATED command is displayed. PROC GLM using RANDOM and PROC MIXED commands are also listed in the program. The SAS code for this example is similar to that provided in Program 3_5_1.sas, except we have excluded from the code the analysis using the restricted full rank linear model.

3.5.3 Test for Equal Covariance Matrices and for Circularity

In mixed linear models, F tests that involve the repeated measures dimension are only valid if the within subjects covariance matrix has a certain structure. While the compound symmetry structure is the only necessary condition for exact F test, the more general assumption of circularity is both a necessary and sufficient condition. For a single group model, the condition for tests that involve the repeated measures condition is that there exists an orthonormal hypothesis test matrix C such that

$$C'\Sigma C = \sigma^2 I \tag{3.25}$$

where $C'C = I$ so that C' is an orthonormal matrix of order $(p-1) \times p$, Σ is a $p \times p$ within measures covariance matrix, σ^2 is a scalar and I is the identity matrix of order $(p - 1) \times (p - 1)$. Thus, under the transformation, the covariance structure is homogeneous. With more than one group, then the multisample circularity assumption must be satisfied to use the mixed model for tests that involve the repeated measures dimension. The multisample assumption is:

$$C'\Sigma_1 C = C'\Sigma_2 C = \cdots = C'\Sigma_g C = C'\Sigma C = \sigma^2 I \tag{3.26}$$

where Σ_i is the covariance matrix for the ith group, $i = 1, \ldots, I$. Thus, for multisamples we have two assumptions: (1) homogeneity of the reduced covariance matrices and (2) circularity of the common matrix.

To test the homogeneity assumption under multivariate normality, the test due to Box (1949) called Box's M test is often used. A major shortcoming of the test is that it is severely affected by departures from multivariate normality. To test homogeneity we form the statistic

$$M = (N - I) \log |C'SC| - \sum_{i=1}^{I} \nu_i \log |C'S_iC| \qquad (3.27)$$

where I is the number of groups, $N = \sum n_i$, S is an unbiased estimate of the pooled within covariance matrix: $S = \sum_{i=1}^{I} \nu_i S_i$, S_i is an unbiased estimate of Σ_i for the ith group and $\nu_i = n_i - 1$. The quantity $X_B^2 = (1 - C_*)M$ converges to a chi-square distribution with degrees of freedom $\nu = q(q+1)(I-1)/2$ where $q = (p-1)$ when the hypothesis is true as $N \to \infty$, and

$$C_* = \frac{2q^2 + 3q - 1}{6(q+1)(I-1)} \left(\sum_{i=1}^{I} \frac{1}{\nu_i} - \frac{1}{N-I} \right).$$

Equality is rejected at the level α if $X_B^2 > \chi_\nu^2(1 - \alpha)$, the upper $(1 - \alpha)$ critical value of the chi-square distribution. This approximation works well under normality when each $n_i > 20$ and $q < 6$ and $I < 6$.

If the sample sizes are small or q or I are larger than 6, the F distribution may be used. Then we calculate

$$F = \left(\frac{1 - C_* - \nu/\nu_0}{\nu} \right) M$$

where

$$C_0 = \frac{(q-1)(q+2)}{6(I-1)} \left(\sum_{i=1}^{I} \frac{1}{\nu_i^2} - \frac{1}{(N-I)^2} \right)$$

$$\nu_0 = \frac{\nu + 2}{C_0 - C_*^2},$$

and $F \sim F(\nu, \nu_0)$, under the null hypothesis. The hypothesis of homogeneity is rejected as the level α if $F > F_{(\nu, \nu_0)}^{(1-\alpha)}$.

The first sections of Program 3_5_3.sas contain the PROC IML code to perform the above hypothesis test on the Ammon data analyzed in Section 3.5.2. The results are displayed in the Output 3.5.6. Since the P-value for the test of equality of reduced covariance matrices larger than $\alpha = 0.10$, we do not reject equality.

The null hypothesis of circularity is:

$$H : C'\Sigma C = \sigma^2 I \qquad (3.28)$$

Mauchly (1940) developed a chi-square statistic to test this hypothesis. The statistic is

$$X_C^2 = - \left[(N - I) - \frac{2q^2 + q + 2}{6q} \right] \left[\log |C'SC| - q \log \left(\frac{\text{tr}(C'SC)}{q} \right) \right]. \qquad (3.29)$$

Output 3.5.6: Tests of Covariance Structures.

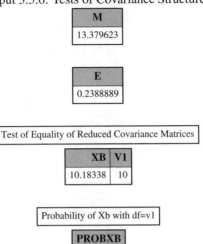

M
13.379623

E
0.2388889

Test of Equality of Reduced Covariance Matrices

XB	V1
10.18338	10

Probability of Xb with df=v1

PROBXB
0.4245548

Output 3.5.7: Mauchley's Test for Ammon Data.

Sphericity Tests				
Variables	DF	Mauchly's Criterion	Chi-Square	Pr > ChiSq
Transformed Variates	9	0.31758	18.830336	0.0267
Orthogonal Components	9	0.5692268	9.250404	0.4145

The test statistic converges to a chi-square distribution with degrees of freedom $\nu = [q(q+1)/2 - 1]$ when the null hypothesis is true. Returning to Program 3_5_2.sas, Mauchley test is executed in SAS using PROC GLM with the REPEATED statement using the option PRINTE (Output 3.5.7). $X_C^2 = 9.250404$ with $p = 0.4145$ for the Ammon data associated with the variables "Orthogonal Components".

Another test for circularity which has been found to be more powerful than the Mauchley test, at least for small sample sizes, is the locally best invariant (LBI) test that was proposed by John (1971) and Sugiura (1972). The test depends on the trace criterion $V = \mathrm{tr}(C'\Sigma C)^2 / [\mathrm{tr}(C'\Sigma C)]^2$. The test statistic is

$$V^* = \frac{q^2(N-I)}{2}\left[V - q^{-1}\right]. \tag{3.30}$$

When the null hypothesis is true, V^* converges to a chi-square distribution with degrees of freedom $\nu = q(q+1)/2 - 1$ (see, Cornell, Young, Seaman, & Kirk, 1992).

Program 3_5_3.sas contains the PROC IML code for both tests. The results are displayed in the Output 3.5.8; observe that the P-value for Mauchley's test agrees

Output 3.5.8: Mauchley and LBI Tests of Circularity.

Tests of Circularity (Mauchly)

XS	DFXS
9.250404	9

Probability of Xs with df=dfxs

PROBXS
0.4144883

Test of Circularity (L.B.I.)

VS
8.9501179

DFXS
9

Probability of Vs with df=dfxs

PROBVS
0.4418915

with the results in the Output 3.5.7 and that the P-value for the LBI test is larger, 0.4419. Thus, the circularity test is not rejected for the Ammon data.

In this example, we first tested for equality and then for circularity. To control the overall familywise level for the tests near the nominal level α, each test is tested using the nominal level $\alpha/2$. Alternatively, Mendoza (1980) developed a test that controls both tests exactly at the nominal level α; details are provided in Timm (2002, p. 142). The difficulty with the joint approach is that upon rejection you do not know whether it is due to lack of homogeneity or circularity. Thus, a two step process is recommended. Timm (2002, p. 307) reviews tests of location with unequal covariance matrices.

3.6 ANALYSIS OF COVARIANCE

In Chapter 2, we discussed the general intraclass model, an unrestricted model that allows the slopes of the regression lines within separate populations to differ. We now place restrictions on the general model to develop the analysis of covariance (ANCOVA) model. In the one-factor ANCOVA model we require the slopes of the regression lines within different populations to be equal; thus, a restriction on the intraclass model is: $\gamma_1 = \gamma_2 = \cdots = \gamma_I$ for I populations. We discussed this test in Chapter 2. With fixed effects, it does not make sense to test hypotheses regarding intercepts with unequal slopes. The ANCOVA model may be represented as a restricted intraclass model or as an unrestricted full rank model. As a restricted model we have:

$$y_{ij} = \alpha_i + \gamma_i z_{ij} + e_{ij} \quad i = 1, \ldots, I; j = 1, 2, \ldots, n_i \qquad (3.31)$$

with a restriction $(R : \gamma_1 = \gamma_2 = \cdots = \gamma_I = \gamma)$ and where γ is the common slope and α_i represent the intercepts for the linear regression of y on z. Using vector notation, the linear model matrices are:

$$X = \begin{pmatrix} 1_1 & 0_1 & \cdots & 0_1 & Z_1 & 0_1 & \cdots & 0_1 \\ 0_2 & 1_2 & \cdots & 0_2 & 0_2 & Z_2 & \cdots & 0_2 \\ \vdots & \vdots & \ddots & \vdots & \vdots & \vdots & \ddots & \vdots \\ 0_I & 0_I & \cdots & 1_I & 0_I & 0_I & \cdots & Z_I \end{pmatrix}$$

$$\beta' = \begin{pmatrix} \alpha_1 & \alpha_2 & \cdots & \alpha_I & \gamma & \gamma & \cdots & \gamma \end{pmatrix}$$

where 1_i is a $n_i \times n_i$ matrix of 1's, 0_i is a $n_i \times n_i$ matrix of 0's, and we have incorporated the $(I - 1)$ restrictions into the model.

The error variance estimate for this model is based on the variation about regression functions with equal slopes fitted to each sample. The degrees of freedom for error is:

$$df_{error} = n - \text{(No. Columns in Design Matrix)} + \text{(No. of Restrictions)}$$
$$= n - 2I + (I - 1)$$
$$= n - I - 1$$

where $n = \sum_{i=1}^{I} n_i$.

We may also test the hypothesis that the common slope is zero

$$H_{reg} : \gamma = 0. \qquad (3.32)$$

This test may be thought of as a test of the effectiveness of the covariate, much like the test of a blocking factor in a randomized blocking design. In an ANCOVA model, we expect to reject the regression hypothesis.

While in most applications, a linear regression equation in the variables is assumed for the analysis, the only requirement for the application of the model is that the regression function be linear in the parameters. For example, one may assume the relation between y and z is of the form:

$$y = \log(z)$$
$$y = z^2$$
$$y = \sqrt{z}$$

or other functions that are linear in the parameters. For example, we assume a model of the form

$$y_{ij} = \alpha_i + \gamma z_{ij} + \xi z_{ij}^2 + e_{ij}$$

where γ and ξ represent (common) linear and quadratic terms of the regression equation across the I populations. Then, the restriction for the intraclass model becomes

$$R : \begin{pmatrix} \gamma_1 = \gamma_2 = \cdots = \gamma_I = \gamma \\ \xi_1 = \xi_2 = \cdots = \xi_I = \xi \end{pmatrix}$$

and the error degrees of freedom for the restricted model is

$$
\begin{aligned}
df_{error} &= n - (\text{No. Columns in Design Matrix}) + (\text{No. of Restrictions}) \\
&= n - 3I + 2(I - 1) \\
&= n - I - 2.
\end{aligned}
$$

Thus, an additional degree of freedom is lost by the inclusion of the quadratic term. And, one may test the effectiveness of the linear term, the quadratic term or both. In general complex models and additional covariates should only be included in the model if there is a significant reduction in the residual variance since for each additional covariate that is added to the model, one degree of freedom is lost.

3.6.1 ANCOVA with One Covariate

Returning to the data examined in Section 2.10, we now analyze the data using an ANCOVA model by placing restrictions on the intraclass model. To test for differences in the intercepts, the restriction matrix

$$R = \begin{pmatrix} 0 & 0 & 0 & 1 & 0 & -1 \\ 0 & 0 & 0 & 0 & 1 & -1 \end{pmatrix}$$

must be incorporated into the model. Returning to Program 2_10.sas, the Parallelism test now becomes the restriction. And, the TEST option is used to test for intercepts and no regression; Program 3_6_1.sas contains the code. The hypothesis test matrices for the test of intercepts (A) and regression (REG) are:

$$
\begin{aligned}
C_A &= \begin{pmatrix} 1 & 0 & -1 & 0 & 0 & 0 \\ 0 & 1 & -1 & 0 & 0 & 0 \end{pmatrix} \\
C_{REG} &= \begin{pmatrix} 0 & 0 & 0 & 0 & 0 & 1 \end{pmatrix}.
\end{aligned}
$$

For the parameter vector:

$$\beta' = \begin{pmatrix} \alpha_1 & \alpha_2 & \alpha_2 & \gamma & \gamma & \gamma \end{pmatrix}.$$

The results are displayed in the Output 3.6.1-3.6.2. Thus, for this example the test of regression is significant, indicating that the covariate should be retained. If the hypothesis is not significant, then one would perform an ANOVA for the data.

Output 3.6.1: ANCOVA with One Covariate Using PROC REG.

Test A Results for Dependent Variable y				
Source	DF	Mean Square	F Value	Pr > F
Numerator	2	10.21357	20.60	<.0001
Denominator	14	0.49570		

Test REG Results for Dependent Variable y				
Source	DF	Mean Square	F Value	Pr > F
Numerator	1	21.51852	43.41	<.0001
Denominator	14	0.49570		

Output 3.6.2: ANCOVA with One Covariate Using PROC GLM.

Source	DF	Type III SS	Mean Square	F Value	Pr > F
a	2	20.42714260	10.21357130	20.60	<.0001
z	1	21.51852314	21.51852314	43.41	<.0001

With the rejection of the intercepts, differences in adjusted means for the three groups, one would again investigate contrasts in the adjusted means. The contrast estimate between treatment 1 and treatment 2 is

$$\hat{\psi} = \hat{\alpha}_1 - \hat{\alpha}_2 = 6.658 - 9.201 = -2.543.$$

To compute the standard error estimate for this contrast using the restricted full rank model, the estimated variance-covariance matrix is used:

$$\hat{\sigma}_{\hat{\psi}} = \sqrt{\hat{\sigma}_{\hat{\alpha}_1}^2 + \hat{\sigma}_{\hat{\alpha}_2}^2 - 2\text{cov}(\hat{\alpha}_1, \hat{\alpha}_2)}$$
$$= \sqrt{0.240 + .178 - 2(.130)}$$
$$= 0.398.$$

Using the critical $\alpha = 0.05$ critical F-value for the S-method,

$$c_0 = \sqrt{\nu_h F_{(2,14)}^{(1-.95)}} = \sqrt{2(3.74)} = 2.73,$$

the 95% confidence set is:

$$\hat{\alpha}_1 - \hat{\alpha}_2 \pm c_0 \hat{\sigma}_{\hat{\psi}} = -2.544 \pm 2.73(.398) = \{-3.63, -1.46\}.$$

This is readily obtained from the output of the less than full rank model with the LSMEANS statement as shown in the (Output 3.6.3). Note, the adjusted means for the ANCOVA model are obtained using the option OUT on the LSMEANS statement,

Output 3.6.3: ANCOVA with One Covariate — Adjusted Mean Differences and 95% CI.

		Difference Between Means	Simultaneous 95% Confidence Limits for LSMean(i)-LSMean(j)	
i	j			
1	2	-2.543457	-3.628614	-1.458299
1	3	-1.528222	-2.833010	-0.223433
2	3	1.015235	-0.252122	2.282592

(Table title: **Least Squares Means for Effect a**)

while the weighted unadjusted means are produced by using MEANS statement. The 95% confidence sets for other contrasts can also be computed in a similar way,

$$-2.833 \leq \hat{\alpha}_1 - \hat{\alpha}_3 \leq -0.223$$
$$-0.252 \leq \hat{\alpha}_2 - \hat{\alpha}_3 \leq 2.283.$$

3.6.2   ANCOVA with Two Covariates

In the data file 3_6_2.dat, we add a second covariate to the data set. Thus, we have a two covariate example. For the intraclass model, the parameter vector is:

$$\beta' = \begin{pmatrix} \alpha_1 & \alpha_2 & \alpha_3 & \gamma_{11} & \gamma_{12} & \gamma_{13} & \gamma_{21} & \gamma_{22} & \gamma_{23} \end{pmatrix}.$$

Using the intraclass model, we first test for parallelism using the unrestricted full rank model. The hypothesis test matrix is:

$$C_{Parallelism} = \begin{pmatrix} 0 & 0 & 0 & 1 & 0 & -1 & 0 & 0 & 0 \\ 0 & 0 & 0 & 0 & 1 & -1 & 0 & 1 & 0 \\ 0 & 0 & 0 & 0 & 0 & 0 & 1 & 0 & -1 \\ 0 & 0 & 0 & 0 & 0 & 0 & 0 & 1 & -1 \end{pmatrix}.$$

The SAS code is found in Program 3_6_2.sas. The test is not significant (Output 3.6.4).

Next, we use the parallelism test matrix as a restriction matrix to form the AN-COVA model. And, using this model we test for differences in intercepts and for no regression for both covariates. The hypothesis test matrices follow:

$$C_A = \begin{pmatrix} 1 & -1 & 0 & 0 & 0 & 0 & 0 & 0 & 0 \\ 0 & 1 & -1 & 0 & 0 & 0 & 0 & 0 & 0 \end{pmatrix}$$

$$C_{reg} = \begin{pmatrix} 0 & 0 & 0 & 0 & 0 & 1 & 0 & 0 & 0 \\ 0 & 0 & 0 & 0 & 0 & 0 & 0 & 0 & 1 \end{pmatrix}.$$

Output 3.6.4: Test of Parallelism.

Test parallel Results for Dependent Variable y				
Source	DF	Mean Square	F Value	Pr > F
Numerator	4	0.63534	1.35	0.3247
Denominator	9	0.47121		

Output 3.6.5: ANCOVA Tests Two Covariates.

Test A Results for Dependent Variable y				
Source	DF	Mean Square	F Value	Pr > F
Numerator	2	9.71232	18.62	0.0002
Denominator	13	0.52172		

Test REG Results for Dependent Variable y				
Source	DF	Mean Square	F Value	Pr > F
Numerator	2	10.83802	20.77	<.0001
Denominator	13	0.52172		

The results of these tests are displayed in the Output 3.6.5. Observe that the error variance for two covariates, $MS_{error} = .52$ (Output 3.6.5), is larger than the value of $MS_{error} = 0.50$ (Output 3.6.1) for one covariate. This is an indication that the additional variate is not effective in reducing the error variance. This is verified by investigating the separate tests for each covariate. The second covariate test is nonsignificant: P-value $= 0.5920$ (Output 3.6.6).

Finally, the code for the analysis of this example using the less than full rank model is provided in Program 3_6_2.sas. First observe that PROC GLM does not allow one to test the hypothesis of parallelism jointly. It only allows one to perform the test for each covariate separately by using the interaction terms: a* z1 and a* z2 (Output 3.6.7). While the procedure provides both a Type I SS and Type III SS, one must use the Type III SS for the tests (Output 3.6.8) which agree with the restricted model tests: Reg1 and Reg2 (Output 3.6.6). For additional details, see Milliken and Johnson (2001).

3.6.3   ANCOVA Nested Designs

To further illustrate the principles involved in constructing full rank models for designs having covariates, we now discuss methods used in the analysis of a nested design with covariates. An example is displayed in the Table 3.13 where factor B is

Output 3.6.6: Separate Tests for Each Covariate.

Test Reg1 Results for Dependent Variable y				
Source	DF	Mean Square	F Value	Pr > F
Numerator	1	10.99423	21.07	0.0005
Denominator	13	0.52172		

Test Reg2 Results for Dependent Variable y				
Source	DF	Mean Square	F Value	Pr > F
Numerator	1	0.15751	0.30	0.5920
Denominator	13	0.52172		

Output 3.6.7: Test of Parallelism Using PROC GLM.

Source	DF	Type III SS	Mean Square	F Value	Pr > F
a	2	2.36746801	1.18373401	2.51	0.1359
z1	1	9.23132928	9.23132928	19.59	0.0017
z2	1	0.64638942	0.64638942	1.37	0.2716
z1*a	2	0.63396351	0.31698176	0.67	0.5342
z2*a	2	2.46564467	1.23282233	2.62	0.1272

Output 3.6.8: ANCOVA with Two Covariates Using PROC GLM.

Source	DF	Type III SS	Mean Square	F Value	Pr > F
a	2	19.42463585	9.71231792	18.62	0.0002
z1	1	10.99423200	10.99423200	21.07	0.0005
z2	1	0.15751059	0.15751059	0.30	0.5920

Table 3.13: ANCOVA Nested Design.

A_1				A_2					
$B_{(1)1}$		$B_{(1)2}$		$B_{(2)1}$		$B_{(2)2}$		$B_{(2)3}$	
α_{11}	β_{11}	α_{12}	β_{12}	α_{21}	β_{21}	α_{22}	β_{22}	α_{23}	β_{23}

nested within factor A. The intraclass model for the design in Table 3.13 is

$$y_{ijk} = \alpha_{ij} + \beta_{ij} z_{ijk} + e_{ijk} \quad i = 1, \ldots, I; j = 1, \ldots, J_i; k = 1, \ldots, n_{ij} \quad (3.33)$$
$$e_{ijk} \sim IN(0, \sigma^2).$$

Since there are 10 parameters in this design, the design matrix has 10 columns.
We may test the equality of all five slopes:

$$H_0 : \text{all } \beta_{ij} \text{ are equal} \quad (3.34)$$

by using the procedures discussed with the other designs. If all slopes are equal, the intraclass model becomes a fixed effect nested ANCOVA design having the form:

$$y_{ijk} = \alpha_{ij} + \beta_p z_{ijk} + e_{ijk} \quad i = 1, \ldots, I; j = 1, \ldots, J_i; k = 1, \ldots, n_{ij} \quad (3.35)$$
$$e_{ijk} \sim IN(0, \sigma^2)$$

where the hypothesis of the equality of the 5 slopes becomes the matrix of restrictions.

Suppose however that all 5 slopes are not equal, but that regression lines for factor B within factor A are equal:

$$\beta_{11} = \beta_{12} = \beta_1$$
$$\beta_{21} = \beta_{22} = \beta_{23} = \beta_2.$$

Then the model becomes:

$$y_{ijk} = \alpha_{ij} + \beta_i z_{ijk} + e_{ijk} \quad i = 1, \ldots, I; j = 1, \ldots, J_i; k = 1, \ldots, n_{ij} \quad (3.36)$$
$$e_{ijk} \sim IN(0, \sigma^2).$$

and we would perhaps test hypotheses about factor B within factor A. Differences in A would not be of interest since this is an intraclass model with different slopes for factor A and interpretation is difficult as discussed in Chapter 2.

To illustrate, consider the design in Table 3.13 with the intraclass parameter vector:

$$\beta' = \begin{pmatrix} \alpha_{11} & \alpha_{12} & \alpha_{21} & \alpha_{22} & \alpha_{23} & \beta_{11} & \beta_{12} & \beta_{21} & \beta_{22} & \beta_{23} \end{pmatrix}.$$

The hypothesis test matrix of homogeneous slopes, all β_{ij} equal, is:

$$C_{Parallelism} = \begin{pmatrix} 0 & 0 & 0 & 0 & 0 & 1 & 0 & 0 & 0 & -1 \\ 0 & 0 & 0 & 0 & 0 & 0 & 1 & 0 & 0 & -1 \\ 0 & 0 & 0 & 0 & 0 & 0 & 0 & 1 & 0 & -1 \\ 0 & 0 & 0 & 0 & 0 & 0 & 0 & 0 & 1 & -1 \end{pmatrix}.$$

The analysis of covariance (assuming equality of the five slopes) may be performed using the matrix $C_{Parallelism}$ as a restriction and the matrices:

$$C = \begin{pmatrix} \frac{1}{2} & \frac{1}{2} & -\frac{1}{3} & -\frac{1}{3} & -\frac{1}{3} & 0 & 0 & 0 & 0 & 0 \end{pmatrix},$$

to test for factor A differences, and

$$C_{B(A)} = \begin{pmatrix} 1 & -1 & 0 & 0 & 0 & 0 & 0 & 0 & 0 & 0 \\ 0 & 0 & 1 & 0 & -1 & 0 & 0 & 0 & 0 & 0 \\ 0 & 0 & 0 & 1 & -1 & 0 & 0 & 0 & 0 & 0 \end{pmatrix}$$

to test for differences among the levels of B within A. If, however, we only assume that the slopes of the regression lines for levels of B within A are equivalent, then the restriction matrix would take the form:

$$R = \begin{pmatrix} 0 & 0 & 0 & 0 & 0 & 1 & -1 & 0 & 0 & 0 \\ 0 & 0 & 0 & 0 & 0 & 0 & 0 & 1 & 0 & -1 \\ 0 & 0 & 0 & 0 & 0 & 0 & 0 & 0 & 1 & -1 \end{pmatrix},$$

and one would again use the matrix $C_{B(A)}$ for testing for differences among the levels of B within A.

CHAPTER 4

Weighted General Linear Models

4.1 INTRODUCTION

In Chapter 2 and 3, we discussed the general linear model (GLM) assuming a simple structure for Ω, namely $\Omega = \sigma^2 I_n$, where the observations are independent with common variance σ^2. There we used ordinary least square estimation (OLSE). Two additional structures that Ω may have are

$$\Omega = \sigma^2 V \qquad (V \text{ known and nonsingular}) \qquad (4.1)$$

$$\Omega = \Sigma \qquad (\Sigma \text{ unknown}). \qquad (4.2)$$

When V is unknown, then (4.1) is equivalent to (4.2). Alternatively when Σ is known, (4.2) reduces to (4.1) by letting $\sigma^2 = 1$ and $V \equiv \Sigma$.

When V is known, case (4.1) is often called generalized least squares (GLS), weighted least squares estimation (WLSE) or the Gauss-Markov (GM) model. In WLSE, V is a diagonal weight matrix or a correlation matrix, $V = R = (\rho_{ij})$. With a correlation matrix V, a common structure for Ω is

$$\Omega = \sigma^2 V = \sigma^2 [(1 - \rho)I + \rho J]$$

$$= \sigma^2 \begin{pmatrix} 1 & \rho & \cdots & \rho \\ \rho & 1 & \cdots & \rho \\ \vdots & \vdots & \ddots & \vdots \\ \rho & \rho & \cdots & 1 \end{pmatrix} \qquad (4.3)$$

where J is a matrix of 1's, I is an identity matrix, and ρ is the population correlation coefficient. Ω in (4.3) is known as compound symmetry, common in repeated measurement design which we discussed in Chapter 3. Since V is known and nonsingular then it may be factored such that $V = FF'$, so that Ω may be reduced to spherical structure since

$$(F'F)^{-1} F' \Omega F (F'F)^{-1} = A' \Omega A = \sigma^2 I. \qquad (4.4)$$

When (4.2) can be reduced to (4.4), the matrix Ω is said to be a type H matrix or Hankel matrix; (4.4) is the circularity assumption in repeated measurement design discussed in Section 3.5.3.

Case (4.2), when Σ is unknown with no special structure and thus must be estimated by $\widehat{\Sigma}$, is often called feasible generalized least squares (FGLS) estimation or the general Gauss-Markov (GGM) model. If σ^2 and V are both unknown, then this is a special case of (4.2) since Σ has structure $\Sigma = \sigma^2 V$.

In this chapter, we show how to estimate the parameter vector β given that more general structure of the covariance matrix shown in (4.1) and (4.2). With both known and unknown covariance structures, we then show how to test linear hypotheses. OLSE and FGLS estimates are compared. In addition, basic asymptotic theory and Fisher's information matrix are developed to establish large sample tests. Examples in this chapter include weighted regression analysis, and categorical data analysis using a linear model.

4.2 ESTIMATION AND HYPOTHESIS TESTING

For the moment, assume that $\Omega = \Sigma$ and that Σ is known. Let

$$y_0 = \Sigma^{-1/2} y \qquad X_0 = \Sigma^{-1/2} X \tag{4.5}$$

where $\Sigma^{1/2} = P\Lambda^{1/2}$, P is an orthogonal matrix $(P'P = PP' = I)$ such that $P'\Sigma P = I$ (by the spectral decomposition theorem, Theorem 7.3). Then, substituting y_0 for y in the GLM (1.1), the structure for y_0 becomes $\sigma^2 I_n$. Thus, the OLSE = best linear unbiased estimator (BLUE) for β is

$$\hat{\beta} = (X_0' X_0)^{-1} X_0' y_0 = (X'\Sigma^{-1} X)^{-1} X'\Sigma^{-1} y. \tag{4.6}$$

The estimate (4.6) is called the GLS estimator of β or the Gauss-Markov estimator of β. The covariance matrix of $\hat{\beta}$ is

$$\text{cov}(\hat{\beta}) = (X'\Sigma^{-1} X)^{-1}. \tag{4.7}$$

If $\Omega = \sigma^2 V$ and V is known and nonsingular, the WLSE of β is

$$\hat{\beta} = (X_0' X_0)^{-1} X_0' y_0 = (X'V^{-1} X)^{-1} X'V^{-1} y \tag{4.8}$$

and does not involve the unknown common variance σ^2. Alternatively, the estimators given in (4.6) and (4.8) may be obtained by minimizing the more general least squares criterion

$$(y' - X\beta)'\Sigma^{-1}(y - X\beta) = \text{tr}(\Sigma^{-1} ee') \tag{4.9}$$

with regard to β, which requires Σ to be nonsingular. For a more general treatment, see Rao (1973, p. 297-302). When $\Omega = \sigma^2 V$, V known and nonsingular, the Gauss-Markov theory applies so that a GLS estimator (GLSE) of β is obtained without making distributional assumptions regarding y. However, to test the hypothesis H : $C\beta = \xi$ we make the distributional assumption $y \sim N_n(X\beta, \Omega = \sigma^2 V)$ so that $y_0 = V^{-1/2} y$ is normal with mean $V^{-1/2} X\beta$ and covariance matrix $\Omega = \sigma^2 I_n$. Hence, the general test theory of Chapter 2 and 3 may be used for GLS.

When $\Omega = \Sigma$ and Σ is known and Gramian (nonnegative definite, represented as $\Sigma \geq 0$), one can apply the general Gauss-Markov theory to estimate β without making the distributional assumptions. Usually we assume that $y \sim N_n(X\beta, \Omega = \Sigma)$. The maximum likelihood (ML) estimate of β is

$$\hat{\beta}_{ML} = (X'\Sigma^{-1}X)^{-1}X'\Sigma^{-1}y. \tag{4.10}$$

Comparing (4.9) with the density of the multivariate normal distribution (1.26), observe that (4.9) is included in the exponent of the distribution indicating that OLSE = BLUE = ML estimator (MLE) of β under multivariate normality and known Σ. With known Σ, the statistic

$$X^2 = (C\hat{\beta}_{ML} - \xi)'(C(X'\Sigma^{-1}X)^{-1}C')^{-1}(C\hat{\beta}_{ML} - \xi) \tag{4.11}$$

has a noncentral χ^2 distribution. The likelihood ratio test of $H : C\beta = \xi$ is to reject H if the $P(X^2 > c|H) = \alpha$ where X^2 has a central χ^2 distribution when H is true and degrees of freedom $\nu = \text{rank}(C) = g$.

When Σ is unknown, estimation of β and hypothesis testing is more complicated since it requires a consistent estimator of Σ. To see how we might proceed, we know that the ML estimate of β in (4.10) is normally distributed since y is normally distributed. Σ is unknown, but suppose that we can find a (weakly) consistent estimator of $\hat{\Sigma}$ of Σ that converges in probability of $\hat{\Sigma}$, $\text{plim}\hat{\Sigma} = \Sigma$. Let the FGLS estimator of β be $\hat{\beta}_{FGLS}$ or

$$\tilde{\beta}_{FGLS} = \tilde{\beta}_{CAN} = (X'\hat{\Sigma}^{-1}X)^{-1}X'\hat{\Sigma}^{-1}y \tag{4.12}$$

so that $\tilde{\beta}_{CAN}$ is a consistent asymptotically normal (CAN) estimator of β since the $\text{plim}\tilde{\beta}_{CAN} = \beta$. If $\hat{\Sigma}$ is the ML estimate of Σ, then by maximum likelihood estimation theory, the estimator in (4.12) is the best asymptotically normal (BAN) estimator of β.

Definition 4.1. A CAN estimator $\tilde{\beta}_k$ of order k based on a sample size n is said to be a BAN estimator for β if $n^{1/2}(\tilde{\beta}_k - \beta)$ converges in distribution to a multivariate normal distribution,

$$n^{1/2}(\tilde{\beta}_k - \beta) \xrightarrow{d} N_k(0, F^{-1}(\beta))$$

where $F^{-1}(\beta)$ is the inverse of the $k \times k$ Fisher's information matrix (the inverse of expectations of second order partial derivatives of the log of the likelihood of β evaluated at its true value times $1/n$).

In general, maximum likelihood estimates are BAN provided the density of the estimator satisfies certain general regularity conditions, Stuart and Ord (1991, p. 649-682). The information matrix for $\tilde{\beta}$ in (4.12) is given by

$$F_{n,\beta} = \frac{X'\Sigma^{-1}X}{n} \tag{4.13}$$

where Σ is unknown. The asymptotic covariance matrix for $\tilde{\beta}$ is then

$$\frac{1}{n} F_{n,\beta}^{-1} = (X'\Sigma^{-1}X)^{-1} \tag{4.14}$$

since X is of full rank and Σ is nonsingular.

To test the hypothesis $H : C\beta = \xi$ when Σ is unknown, we may use the test statistic due to Wald (1943). Wald showed that

$$W = (C\tilde{\beta}_{BAN} - \xi)'(C(X'\widehat{\Sigma}^{-1}X)^{-1}C')^{-1}(C\tilde{\beta}_{BAN} - \xi) \tag{4.15}$$

converges asymptotically to a central χ^2 distribution with degrees of freedom $g = \text{rank}(C)$, when H is true. The null hypothesis H is rejected at the nominal level α if $W > c$ where c is the critical value of the central χ^2 distribution such that

$$P(W > c|H) = \alpha. \tag{4.16}$$

When n is small, W/g may be approximated by an F distribution with degrees of freedom $g = \text{rank}(C)$ and $\nu_e = n - \text{rank}(X)$, Theil (1971, p. 402-403). Relating W/g to Hotelling T^2 type statistic (see, for example, Timm, 2002, p. 99), the scaled statistic (for known covariance matrix) has an exact F distribution. Replacing the covariance matrix with $\widehat{\Sigma}$, the statistic

$$T^2 = \left(\frac{\nu_e}{\nu_e + g - 1}\right)\left(\frac{W}{g}\right) \xrightarrow{d} F(g, \nu_e). \tag{4.17}$$

To estimate Σ in practice, one usually has multivariate data with multiple vector observations for y. In univariate analysis, the covariance matrix Σ is diagonal, $\Sigma = \text{diag}(\sigma_i^2)$. Then the observations y_i have unequal, heterogeneous variances, σ_i^2. If the observations have a common mean, μ, then the WLSE or ML estimate of μ is

$$\hat{\mu} = \frac{\sum_{i=1}^{n} w_i y_i}{\sum_{i-1}^{n} w_i} \tag{4.18}$$

where $w_i = 1/\sigma_i^2$. If $y_i \sim N(\mu_i, \sigma_i^2)$, to test the hypothesis $H : \mu_1 = \mu_2 = \cdots = \mu_n = \mu$ the statistic X^2 in (4.11) becomes

$$Q = \sum_{i=1}^{n} w_i(\hat{\mu}_i - \hat{\mu})^2 \tag{4.19}$$

where $\hat{\mu}_i$ is the unbiased estimator of μ_i. From (4.11), Q has a central χ^2 distribution when H is true provided σ_i^2 are known. For unknown variance σ_i^2, the statistic Q becomes W in (4.15) which converges asymptotically to a χ^2 distribution. Finding the distribution of W for small sample sizes is known as the generalized Behren-Fisher problem. For a review of the two sample cases, one may consult S. H. Kim and Cohen (1998). The statistic Q is also used in meta-analysis, Hedges and Vevea (1998).

4.3 OLSE VERSUS FGLS

To estimate β and to test hypotheses of the form $H : C\beta = \xi$ using the GLM (1.1) having unknown, nonsingular covariance structure for $\Omega = \Sigma$, we obtained the GM estimator of β and substituted a consistent estimator for Σ into the GM estimator for the Σ to obtain the FGLS (4.12). Using asymptotic theory, we are able to estimate β optimally and to test hypotheses. The FGLS is used when Σ is unknown and we have no knowledge about the structure of Σ. When we do know knowledge of the structure of Σ, $\Sigma = \sigma^2 I_n$, or $\Sigma = \sigma^2 V$ and V is known, using (4.5) we can use the OLSE, and it is the BLUE.

Now suppose we let $\Sigma = \sigma^2 V$ where V is not known. The WLSE of β given in (4.8) required V to be known. For the WLSE of β to be the BLUE, the matrix V must be eliminated from the WLSE or

$$(X'V^{-1}X)^{-1}X'V^{-1} = (X'X)^{-1}X'. \tag{4.20}$$

McElroy (1967) showed that a necessary and sufficient condition for the OLSE of β to be the BLUE is that V must have the structure of compound symmetry given in (4.3) with $0 \leq \rho \leq 1$ which ensures the existence of V^{-1} for all values of n. For a given n, the correlation ρ can be replaced by $-1/(n-1) < \rho < 1$; for more detail, see Arnold (1981, p. 232-238). He called this situation the exchangeable linear model. The optimal structure of compound symmetry for Σ is a special case of the more general class of covariance matrices of the form

$$\Sigma = ZVZ' + \sigma^2 I \tag{4.21}$$

which arises from the mixed model given in (2.82)

$$y = X\beta + Zb + e \tag{4.22}$$

where b is a random vector, $b \sim N(0, V = I_n \otimes D)$, $e \sim N(0, \sigma^2 I)$, and $\text{cov}(b, e) = 0$.

Rao (1967) compared the OLSE with the FGLS estimate under the more general assumption that Σ has the general form given in (4.21). Rao showed that the

$$\text{cov}[(X'X)^{-1}X'y] = \text{cov}(\hat{\beta}_{OLSE}) = (X'\Sigma^{-1}X)^{-1} \tag{4.23}$$

$$\text{cov}[(X'\hat{\Sigma}^{-1}X)^{-1}X'\hat{\Sigma}^{-1}y] = \text{cov}(\tilde{\beta}_{FGLS}) = \frac{n-1}{n-k-1}(X'\Sigma^{-1}X)^{-1} \tag{4.24}$$

so that the effect of using a ML estimate, $\hat{\Sigma}$ of Σ when Σ is a member of the class of matrices defined by (4.21), is to reduce the estimator's efficiency; thus OLSE is better than the FGLS. Rao further showed that if Σ is not a member of the class, then the

$$\text{cov}(\hat{\beta}_{OLSE}) = (X'X)^{-1}X'\Sigma X(X'X)^{-1} \tag{4.25}$$

and that the $\text{cov}(\tilde{\beta}_{FGLS})$ is smaller than this only if

$$|(X'X)^{-1}(X'\Sigma X)(X'X)^{-1}| > |(X'\Sigma^{-1}X)^{-1}| \tag{4.26}$$

thus compensating for the multiplication factor in (4.24). Note that the $\text{cov}(\tilde{\beta}_{FGLS})$ is given by (4.24) whether or not Σ is a member of the class in (4.21).

More generally Rao (1967) also found the class of covariance matrices Σ for the FGLS estimate to equal the OLSE. That is, to determine Σ such that

$$(X'\Sigma^{-1}X)^{-1}X'\Sigma^{-1} = (X'X)^{-1}X'. \tag{4.27}$$

Theorem 4.1. *Given the linear model $y = X\beta + e$ where $\mathcal{E}(e) = 0$ and $\text{cov}(e) = \Sigma$, the necessary and sufficient condition that the OLSE (Gauss-Markov estimator) of this model to be equivalent to that obtained for the model with $\text{cov}(e) = \sigma^2 I_n$ is that Σ has the structure*

$$\Sigma = X\Delta X' + Z\Gamma Z' + \sigma^2 I_n \tag{4.28}$$

where $X_{n\times k}$, $\text{rank}(X) = k$, $ZX = 0$ and Δ, Γ, and σ^2 are unknown.

This theorem implies that y has the mixed model form

$$y = X\beta + X\delta + Z\gamma + e \tag{4.29}$$

where δ, γ, and e are uncorrelated random vectors with covariance matrices Δ, Γ, and $\sigma^2 I_n$ and means 0.

4.4   GENERAL LINEAR MIXED MODEL CONTINUED

The general structure of the general linear mixed model (GLMM) was introduced in Section 2.6, equation (2.82). There we restricted the model to be linear in the random components as defined in (2.92) and the covariance matrix of the random vectors b and e to be diagonal. More generally, we again assume the model to be linear

$$\Omega_0 : y_{n\times 1} = X_{n\times k}\beta_{k\times 1} + Z_{n\times h}b_{h\times 1} + e_{n\times 1}, \tag{4.30}$$

but allow the covariance matrices for b and e to have the general structure $R_{n\times n}$ and $G_{h\times h}$, respectively. We again assume that the $\text{cov}(e, b) = 0$ where

$$e \sim N_n(0, R) \tag{4.31}$$

$$b \sim N_n(0, G) \tag{4.32}$$

so that $y \sim N_n(X\beta, \Omega)$ and the

$$\text{cov}(y) = \Omega = ZGZ' + R. \tag{4.33}$$

If G and R are known, then the ML estimate of the fixed vector β is

$$\hat{\beta}_{ML} = (X'\Omega^{-1}X)^{-1}X'\Omega^{-1}y \tag{4.34}$$

which is equal to the GLSE of β. Any estimable function $\psi = c'\beta$ is estimated by the BLUE $\hat{\psi} = c'\hat{\beta}_{ML}$. Under normality, the distribution of the BLUE is normal,

$$\hat{\psi} \sim N(\psi, c'(X'\Omega^{-1}X)^{-1}c). \qquad (4.35)$$

However, in practice neither G or R are in general known and must be estimated. Replacing the covariance matrices G and R with consistent estimators $\widehat{G}$ and $\widehat{R}$, the FGLS estimator of the fixed parameter vector β is

$$\tilde{\beta}_{FGLS} = (X'\widehat{\Omega}^{-1}X)^{-1}X'\widehat{\Omega}^{-1}y \qquad (4.36)$$

where $\widehat{\Omega}^{-1} = (Z\widehat{G}Z' + \widehat{R})^{-1}$. The parametric function $\tilde{\psi} = c'\tilde{\beta}$ is the FGLS estimate of ψ. While $\tilde{\psi}$ is an unbiased estimator of ψ for most balanced designs, this is not the case for unbalanced experimental designs. The covariance of $\tilde{\psi}_{FGLS}$ is

$$\mathrm{cov}(\tilde{\psi}) = c'(X'\Omega^{-1}X)c. \qquad (4.37)$$

While the OLSE of ψ, $\hat{\psi}_{OLSE} = c'\hat{\beta}_{OLSE} = c'(X'X)^{-1}X'y$, is always an unbiased estimate and has covariance

$$\mathrm{cov}(\hat{\psi}_{OLSE}) = c'(X'X)^{-1}(X'\Omega X)(X'X)^{-1}c, \qquad (4.38)$$

the covariance of each estimate is unknown since the Ω is unknown. However, a reasonable estimate of each is realized by replacing Ω with $\widehat{\Omega}$. Because the bias of the FGLS estimate is usually small and because for most balanced designs the estimates are equal, the FGLS estimate of $\psi = c'\beta$ is calculated by the procedure PROC MIXED in SAS.

To estimate the fixed parameter vector β and the random vector b in the GLMM (2.82), Henderson's mixed model normal equations

$$\begin{pmatrix} X'\widehat{R}^{-1}X & X'\widehat{R}^{-1}Z \\ Z'\widehat{R}^{-1}Z & Z'\widehat{R}^{-1}Z + \widehat{G}^{-1} \end{pmatrix} \begin{pmatrix} \hat{\beta} \\ \hat{b} \end{pmatrix} = \begin{pmatrix} X'\widehat{R}^{-1}y \\ Z'\widehat{R}^{-1}y \end{pmatrix} \qquad (4.39)$$

are solved. The solution to the normal equations is

$$\hat{\beta} = (X'\widehat{\Omega}^{-1}X)^{-1}X'\widehat{\Omega}^{-1}y \qquad (4.40)$$

$$\hat{b} = \widehat{G}Z'\widehat{\Omega}^{-1}(y - X\hat{\beta}). \qquad (4.41)$$

The estimate of the parameter vector β is seen to be equal to the FGLSE. The estimate of the random vector b is called the empirical best linear unbiased predictor (EBLUP) of the random vector b. If G and Ω are known, then the EBLUP is called the best linear unbiased predictor (BLUP).

Hypotheses for the fixed and random vectors in the GLMM have the general structure

$$H : C \begin{pmatrix} \beta \\ b \end{pmatrix} = 0. \qquad (4.42)$$

Letting the matrix

$$\widehat{A} = \begin{pmatrix} X'\widehat{R}^{-1}X & X'\widehat{R}^{-1}Z \\ Z'\widehat{R}^{-1}Z & Z'\widehat{R}^{-1}Z + \widehat{G}^{-1} \end{pmatrix}^{-1} = \begin{pmatrix} \widehat{C}_{11} & \widehat{C}'_{21} \\ \widehat{C}_{21} & \widehat{C}_{22} \end{pmatrix}, \qquad (4.43)$$

where $\widehat{C}_{11} = (X'\widehat{\Omega}^{-1}X)^{-1}$, $\widehat{C}_{21} = -\widehat{G}Z'\widehat{\Omega}^{-1}Z\widehat{C}_{11}$, and $\widehat{C}_{22} = (Z'\widehat{R}^{-1}Z + \widehat{G}^{-1})^{-1} - \widehat{C}_{21}X'\widehat{\Omega}^{-1}Z\widehat{G}$, the F statistic for testing H is

$$F = \frac{\begin{pmatrix} \hat{\beta} \\ \hat{b} \end{pmatrix}' C(C'\widehat{A}C)^{-1}C \begin{pmatrix} \hat{\beta} \\ \hat{b} \end{pmatrix}}{\text{rank}(C)}. \qquad (4.44)$$

While the degrees of freedom for the hypothesis is $\nu_h = \text{rank}(C)$, the error degrees of freedom for error $\hat{\nu}_e$ must be estimated. Unlike fixed effect ANOVA models, the F statistic does not exactly follow an F distribution. In most applications, one is most interested in fixed and random effects. For these tests, the SAS procedure PROC MIXED offers three exact methods for estimating the error degrees of freedom and two approximate methods. We will say more about how this is done shortly.

To estimate the GLMM parameters and to test hypotheses regarding the parameters requires one to obtain estimates for th unknown covariance matrices G and R. For the variance components model, these matrices are both diagonal. When this is the case, one may obtain estimates of the variance components using the ANOVA method, as discussed in Chapter 2. While this procedure results in unbiased estimates of the variance components, they may be negative. And, if some of the random effects in the model depend on the fixed effects, as in unbalanced designs, the procedure may not be used. Two methods that overcome these problems are the ML and restricted maximum likelihood (REML) methods. To estimate the covariance matrices requires the minimization of the log-likelihood functions under normality. These functions are

$$\ell_{ML}(G, R) = -\tfrac{1}{2}\log|\Omega| - r'\Omega^{-1}r - \tfrac{1}{2}n\log(2\pi) \qquad (4.45)$$

$$\ell_{REML}(G, R) = -\tfrac{1}{2}\log|\Omega| - \tfrac{1}{2}\log|X'\Omega^{-1}X| - r'\Omega^{-1}r - \tfrac{1}{2}(n-k)\log(2\pi) \qquad (4.46)$$

where the residual vector $r = y - X(X'\Omega^{-1}X)^{-1}X'\Omega^{-1}y$ and $k = \text{rank}(X)$. Minimizing -2 times the log of the likelihoods yields the ML and REML estimates of the covariance matrices G and R. The ML and REML estimates of the variance components are always nonnegative. In general, the REML estimates are less biased and for balanced designs agree with the ANOVA method estimates. Thus, the REML method is usually used to estimate the unknown covariance matrices.

The GLMM is a very general model that may be used to analyze multivariate data, split-plot, split-split plot, strip plot, and repeated measures designs with either complete or incomplete data. We illustrate how the model is used to analyze the repeated measures design discussed in Chapter 2 using the restricted linear model.

4.4.1 Example: Repeated Measures Design

The linear model for the repeated measures design is written as

$$y_{ijk} = \mu + \alpha_i + \beta_j + \gamma_{ij} + e_{ijk} \tag{4.47}$$

for $i = 1, \ldots, I$, $j = 1, \ldots, J$, and $k = 1, \ldots, n_i$. Let

$$e'_{ik} = \begin{pmatrix} e_{i1k} & e_{i2k} & \ldots & e_{iJk} \end{pmatrix}$$

and assume that $e_{ik} \sim N(0, \Sigma)$. Then, the GLMM for the design is $y = X\beta + Z_1 b_1 + e$ where $Z_1 = 0$ so that the covariance matrix $G = 0$ and $\Omega = R$. Hence, the covariance structure for design has block diagonal structure

$$R = \text{cov}(e) = \text{diag}(\Sigma) \tag{4.48}$$

for the observation vector $y' = (y_{111}, y_{121}, \ldots, y_{1J1}, \ldots, y_{I1n}, y_{I2n}, \ldots, y_{IJn})$. The homogeneous covariance matrices may have any number of structures. For example, it may be

1. Unrestricted

$$\Sigma = (\sigma_{ij}) \tag{4.49}$$

2. Compound symmetry

$$\Sigma = \sigma^2 \begin{pmatrix} 1 & \rho & \cdots & \rho \\ \rho & 1 & \cdots & \rho \\ \vdots & \vdots & \ddots & \vdots \\ \rho & \rho & \cdots & 1 \end{pmatrix} \tag{4.50}$$

3. Circularity (Huyhn-Feldt)

$$\Sigma = \sigma^2 \begin{pmatrix} \lambda + 2\delta_1 & \delta_1 + \delta_2 & \cdots & \delta_1 + \delta_J \\ \delta_2 + \delta_1 & \lambda + 2\delta_2 & \cdots & \delta_2 + \delta_J \\ \vdots & \vdots & \ddots & \vdots \\ \delta_J + \delta_1 & \delta_J + \delta_2 & \cdots & \lambda + 2\delta_J \end{pmatrix} \tag{4.51}$$

for some λ and δ_j.

4. AR(1) structure

$$\Sigma = \sigma^2 \begin{pmatrix} 1 & \rho & \cdots & \rho^{J-1} \\ \rho & 1 & \cdots & \rho^{J-2} \\ \vdots & \vdots & \ddots & \vdots \\ \rho^{J-1} & \rho^{J-2} & \cdots & 1 \end{pmatrix} \tag{4.52}$$

among others including heterogeneous compound symmetry structure, SAS (1999).

Repeated measures designs have structures identical to some split-plot, split-split plot, strip-plot, and multilevel (hierarchical) designs. For these designs in general, there are two structures in the model; the treatment structure and the design structure which are assumed not to interact. The treatment structure consists of those treatments in the study selected for analysis and are directly related to the dependent variable. The design structure are the factors used to form groups or blocks of the experimental units. While the treatment structure parameters may be fixed or random, the design structure effects are almost always random. Thus, the models are mixed models and analyzed using the GLMM, Littell and Portier (1992) and Raudenbush and Bryk (2002). While we address these designs more in Chapter 11, Milliken (2003) provides an excellent overview.

4.4.2 Estimating Degrees of Freedom for F Statistics in GLMMs

To estimate the error degrees of freedom for the F statistic for the fixed parameter vector β, one may use the residual option. The degrees of freedom for error is $\hat{\nu}_e = n - \text{rank}(X, Z)$. This option assumes that the random components in the model are fixed. Since the method ignores the covariance structure for the mixed model, it usually requires large sample sizes and assumes the structure of $R = \sigma^2 I$. It is the default method when one or more random statements are used to specify the covariance structure for the model, even if one uses a repeated statement. With no random statement, the default method used by SAS is the `betwithin` option. It should only be used if the repeated measures design has compound symmetry structure. When this is not the case, the betwithin option should not be used. Instead, an approximate method is preferred.

For complicated mixed models with complex covariance structures and/or unbalanced data the distribution of the F statistic is unknown. Assuming that it can be approximated by an F distribution, SAS currently provides two approximation methods: Fai-Cornelius (FC) and Kenward-Roger (KR) methods. Using Theorem 7.3, Fai and Cornelius (1996) represented the matrix $\widehat{C}_{11}$ as $P'\widehat{C}_{11}P = \text{diag}(\lambda_m)$ where λ_m are the eigenvalues of $\widehat{C}_{11}$. Then, the approximate F statistic may be represented as a sum of ν_h squared t variables

$$F = \sum_{m=1}^{\nu_h} \frac{(p'_m C \hat{\beta})^2}{\lambda_m} = \sum_{m=1}^{\nu_h} t_{\hat{\nu}_m}^2 \qquad (4.53)$$

where p'_m is the mth eigenvector of p and $\hat{\nu}_m$ is Satterthwaite (1941) approximate degrees of freedom for the t variable. Since $\mathcal{E}(F) = \nu_e/(\nu_e - 2) = \sum_{m=1}^{\nu_h} \hat{\nu}_m/(\hat{\nu}_m - 2) \equiv E$ (for $\hat{\nu}_m > 2$) because the F statistic is the sum of independent Student's t distributions, the approximate error degrees of freedom for the test is

$$\hat{\nu}_e = \frac{2E}{E - \nu_h}. \qquad (4.54)$$

This approximation method is called the sattherthwaite option in SAS.

The Kenward and Roger (1997) approximation method first implements an adjustment to covariance matrix: $\text{cov}(\hat{\beta}) = (X'\widehat{\Omega}^{-1}X)$ to account for small sample bias that incorporates the variability in the estimation of R and G. Let this covariance matrix be defined as $\widehat{\Omega}^*$. They then calculate a scale factor s to approximate the degrees of freedom for the statistic

$$F_{KR} = \frac{s(C\hat{\beta})'(C'\widehat{\Omega}^*C)^{-1}(C\hat{\beta})}{\text{rank}(C)}. \tag{4.55}$$

The approximate moments of the statistic F_{KR} are calculated and equated to the moments of an F distribution to obtain the scale factor and the approximate degrees of freedom for error. In SAS this method of approximation is called kenwardroger option.

SAS documentation warns that the approximate methods have not been fully studied for complex covariance structures and/or unbalanced designs. However, McBride (2000) carried out an extensive Monte Carlo study of the two approximation methods and showed that the KR method works as well or better than the FC method with both simple and complex covariance structures that are not severely unbalanced. In general, the observed P-value of the test using the KR approximation method may be considered an approximate lower bound for the test.

4.5 MAXIMUM LIKELIHOOD ESTIMATION AND FISHER'S INFORMATION MATRIX

The principle of maximum likelihood states that the best value to assign to unknown parameters is those that maximize the likelihood of the observed sample. That is, the values of the parameters that are most likely, given the data.

To obtain ML estimates, let us first consider the random vector y with mean zero and density $f(y_i|\theta)$ and the joint density $f(y|\theta)$ where

$$f(y|\theta) = f(y_1, y_2, \ldots, y_n|\theta) = \prod_{i=1}^{n} f(y_i|\theta). \tag{4.56}$$

The likelihood function of θ is defined as the joint density $f(y|\theta)$ where the observations are given and θ is a vector of parameters. Thus the likelihood is a function of θ

$$L(\theta|y) \propto f(y|\theta). \tag{4.57}$$

The principle of ML estimation consists of finding $\hat{\theta}$ as an estimate of θ, where

$$L(\hat{\theta}|y) = \max_{\theta} L(\theta|y). \tag{4.58}$$

If $\hat{\theta}$ is the ML estimator and $\hat{\delta}$ is any other estimator, the following inequality holds

$$L(\hat{\theta}|y) > L(\hat{\delta}|y). \tag{4.59}$$

If $\theta_{m \times 1}$ is continuous and $L(\theta|y)$ is differentiable for all values of θ, the maximum of $L(\theta|y)$ can be obtained by equating the partial derivatives to zero so that

$$\frac{\partial L(\theta|y)}{\partial \theta_j} = 0 \qquad j = 1, 2, \ldots, m. \tag{4.60}$$

Since the $\log L(\theta|y)$ is a monotone increasing function of $L(\theta|y)$, both attain their maximum value at the same value of θ. Therefore, we may write (4.60) as

$$\frac{\partial \log L(\theta|y)}{\partial \theta_j} = 0$$

or

$$\sum_{i=1}^{n} \frac{\partial \log f(y_i|\theta)}{\partial \theta_j} = 0 \qquad j = 1, 2, \ldots, m. \tag{4.61}$$

The system of equation in (4.61) is called the ML equations or the normal equations, and a solution to the system of equations yields the ML estimate, $\hat{\theta}$.

If $\hat{\theta}$ is the ML estimator of θ, then $\hat{\theta}$ is invariant under single-valued transformations of the parameters so that if $T(\theta) = \delta$, then the ML estimate of δ is $\hat{\delta} = T(\hat{\theta})$. More importantly, ML estimates are asymptotically efficient (BAN), under general regularity conditions on the likelihood, Stuart and Ord (1991).

Expanding (4.61) in a Taylor series about $\theta = \theta_0$, we have that the

$$\frac{\partial \log L(\theta_0|y)}{\partial \theta_j} + (\hat{\theta} - \theta_0) \left[\frac{\partial^2 \log L(\theta_0|y)}{\partial \theta_j \partial \theta_k} \right] + \cdots = 0 \qquad j, k = 1, 2, \ldots, m. \tag{4.62}$$

Neglecting the higher order terms, observe that

$$(\hat{\theta} - \theta_0) = - \left[\sum_{i=1}^{n} \frac{\partial^2 \log f(y_i|\theta)}{\partial \theta_j \partial \theta_k} \right]^{-1} \left[\sum_{i=1}^{n} \frac{\partial \log f(y_i|\theta)}{\partial \theta_j} \right]. \tag{4.63}$$

Multiplying (4.63) by the $n^{1/2}$, we have that

$$n^{1/2}(\hat{\theta} - \theta_0) = - \left[\frac{1}{n} \sum_{i=1}^{n} \frac{\partial^2 \log f(y_i|\theta)}{\partial \theta_j \partial \theta_k} \right]^{-1} \left[\frac{1}{n^{1/2}} \sum_{i=1}^{n} \frac{\partial \log f(y_i|\theta)}{\partial \theta_j} \right]. \tag{4.64}$$

The second term to the right of the equal sign in (4.64) is the sum of n independent and identically distributed random variables. The random vector $[\partial \log f(y_i|\theta)]/[\partial \theta_j]$ has zero mean and covariance matrix

$$\text{cov}\left(\frac{\partial \log f(y_i|\theta)}{\partial \theta_j} \right) = \mathcal{E} \left[\left(\frac{\partial \log f(y_i|\theta)}{\partial \theta_j} \right) \left(\frac{\partial \log f(y_i|\theta)}{\partial \theta_k} \right) \right] \qquad \text{for all } j \text{ and } k \tag{4.65}$$

since the $\mathcal{E}\{\partial \log f(y_i|\theta)/[\partial\theta_j]\} = 0$.

Taking the expected value of the matrix of second derivatives results in the following

$$-\mathcal{E}\left[\frac{\partial^2 \log f(y_i|\theta)}{\partial\theta_j\partial\theta_k}\right] = \text{var}\left[\frac{\partial \log f(y_i|\theta)}{\partial\theta_j}\right] \qquad (4.66)$$

which is defined as the information matrix F_θ (Casella & Berger, 1990, p. 308-312) and

$$F_\theta = \text{cov}\left[\frac{\partial \log f(y_i|\theta)}{\partial\theta_j}\right] = -\mathcal{E}\left[\frac{\partial^2 \log f(y_i|\theta)}{\partial\theta_j\partial\theta_k}\right]. \qquad (4.67)$$

We now consider the two terms to the right of the equality in (4.64). By the central limit theorem, the random vector $n^{-1/2}\sum_{i=1}^{n} [\partial \log f(y_i|\theta)]/[\partial\theta_j]$ as $n \to \infty$ has a multivariate normal distribution with mean 0 and covariance matrix F_θ. Furthermore, by the weak law of large numbers, the matrix

$$n^{-1}\left[\sum_{i=1}^{n} \frac{\partial^2 \log f(y_i|\theta)}{\partial\theta_j\partial\theta_k}\right] \qquad (4.68)$$

is asymptotically equal (converges in probability) to $-F_\theta$ so that

$$-\left[\frac{1}{n}\sum_{i=1}^{n} \frac{\partial^2 \log f(y_i|\theta)}{\partial\theta_j\partial\theta_k}\right]^{-1} \approx F_\theta^{-1}. \qquad (4.69)$$

Hence for large n, the quantity $n^{1/2}(\hat{\theta} - \theta_0)$ has a multivariate normal distribution with mean 0 and covariance matrix F_θ^{-1} so that the

$$\text{cov}(\hat{\theta} - \theta_0) = \text{cov}(\hat{\theta}) \approx \frac{F_\theta^{-1}}{n} \qquad (4.70)$$

which for large n means that the covariance matrix of the ML estimator attains the Cramer-Rao bound and is hence BAN which establishes Definition 4.1. This result implies that an estimate of the asymptotic covariance matrix of the estimators can be derived by evaluating the negative inverse of Fisher's information matrix at the point estimate corresponding to the maximum of the likelihood function and then multiplied by the factor $1/n$.

4.6 WLSE FOR DATA HETEROSCEDASTICITY

WLSE is most often used in the analysis of linear models to stabilize the variance of the dependent variable when the variance-covariance matrix $\Omega \neq \sigma^2 I$. For the classical linear model we assume that the variance of each variable is constant. When this is not the case we have heteroscedasticity; the variance of the dependent variable y_i is

$$\text{var}(y_i) = \sigma_i^2. \qquad (4.71)$$

To eliminate heteroscedasticity one usually has two options: transform the data to re-move the effect of the lack of homogeneity variance, or use WLSE where the weights are, for example, $w_i = \frac{1}{\sigma_i^2}$. The weights in the diagonal weight matrix are selected proportional to the reciprocal of the variances, thereby discounting unreliability that is due to large variability.

Assuming a simple linear model $(y_i = \alpha + \beta x_i + e_i)$, we estimate the parameters with estimators that minimize the weighted sum of squares

$$\sum_i \frac{1}{\sigma_i^2}(y_i - \alpha - \beta x_i)^2 = \sum_i \frac{1}{w_i}(y_i - \alpha - \beta x_i)^2 \qquad (4.72)$$

to obtain a WLS estimate of the model parameters. For example, suppose that the variance of y_i is proportional to x_i^2 so that σ_i is proportional to x_i:

$$\text{var}(y_i) = \sigma_i^2 = c^2 x_i^2$$
$$\sigma_i = c x_i.$$

When one fits a simple linear regression model using OLSE and plots the resid-uals $(\hat{e} = y_i - \hat{y}_i)$ versus the fitted values $(\hat{y}_i)$ a fan pattern will result, suggesting that σ_i may be proportional to x_i. If this is the case, one may transform the original data by dividing the dependent variable by the independent variable, and perform a simple linear regression using OLSE with the transformed variable:

$$y_i^* = y_i / x_i = \alpha(1/x_i) + \beta + (e_i / x_i).$$

The variance of the transformed variable is a constant: $\text{var}(y_i^2) = c^2$. Alternatively, we may minimize (4.72) directly using WLSE. Using matrix notation, the weight matrix is:

$$W = \text{diag}(w_i) = \text{diag}\left(\frac{1}{\sigma_i^2}\right) = V^{-1}. \qquad (4.73)$$

For our simple linear regression example, the weight matrix for $c = 1$ is

$$W = \text{diag}\left(\frac{1}{s_i}\right) = V^{-1}. \qquad (4.74)$$

In applications of multiple linear regression, one often finds that the dependent variable is related in a known way to one of several independent variables, for exam-ple x_1, and then $\sigma_i = c x_{1i}$ as described above. Other patterns of heteroscedasticity may include the relations,

$$\text{var}(y_i) = \sigma_i^2 = c^2 \sqrt{x_{1i}}$$
$$\text{var}(y_i) = \sigma_i^2 = c^2 x_{1i}.$$

To illustrate the OLSE and WLSE using PROC REG, we analyze a problem given in (Neter, Wasserman, & Kutner, 1990, p. 430, problem 11.22). In this application,

Graph 4.6.1: Scatterplot of residual by age.

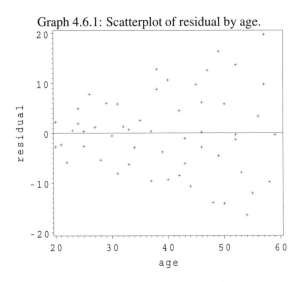

a health educator is interested in studying the relation between diastolic blood pressure (PBLPRESS) and age among healthy adult women aged 20 to 60 years (AGE), using a random sample of 54 women. The SAS code for the analysis is provided in Program 4_6.sas.

The data are read from the file 4_6.dat and then the mean and standard deviation of blpress are computed for each of four age groups using the PROC MEANS. The Output Delivery System (ODS) is used to store the summary data in the data set OUT1. It will be used for obtaining WLSE. To obtain the OLSE solution for the simple linear model ($y = \alpha + \beta x$), the MODEL statement is used for the linear model: blpress=age. The options CLB tells SAS to generate confidence intervals for the model parameters using $\alpha = 0.025$ to obtain confidence sets with an overall family-wise error rate near the nominal level of $\alpha = 0.05$ and the option P outputs predicted values for the dependent variable. The PLOT statement plots OLSE residuals versus age (see Graph 4.6.1). The plot is seen to be FAN SHAPED. Thus, we do not have homogeneity of the variances. The OLS equation for the example is

$$\widehat{blpress} = 56.1569 + 0.580 * age.$$

Upon examining the four subgroups of age: < 30, < 40, < 50 and 55+, one observes that the standard deviations are 4.9, 7.11, 9.82, and 11.54, respectively. To obtain WLSE solution for these data, the weight vector is created in the data step called wlse. Note that OUT1 of the ODS procedure is merged with the data by weighting the data in each group by $\frac{1}{s_i}$. The WEIGHT statement in PROC REG causes the SAS system to perform a WLS analysis. The weights were created in the

Graph 4.6.2: Scatterplot of weighted residual by age.

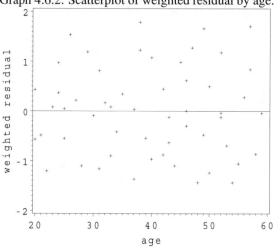

data step labeled wlse. The WLS solution for the data is:

$$\widehat{blpress} = 56.208 + 0.582 * age.$$

While this is close to the OLS solution, the standard errors of the regression coefficients are smaller for the WLS solution. More importantly, the plot of the WLSE residuals versus the fitted values no longer shows a fan shape; the pattern is now random (Graph 4.6.2). And, the data are normally distributed. Finally, the program ends with SAS code used to create the high resolution .eps files. These files may be imported into text files to provide high resolution graphics.

4.7   WLSE FOR CORRELATED ERRORS

WLSE was used to modify a regression equation when the variance of the dependent variable increased with the independent variable resulting in heteroscedasticity. The GLSE procedure may also be used to modify the regression equation in the study of longitudinal change where the errors are correlated over time. This occurs frequently in the behavioral sciences − in particular in studies of learning, maturation and development.

To illustrate, here is the multiple regression model for time-series data

$$y_t = x_t'\beta + e_t \qquad\qquad (4.75)$$

where

$$e_t = \rho e_{t-1} + \nu_t \qquad\qquad t = \ldots, -2, -1, 0, 1, 2, \ldots,$$

$\nu_t \sim N(0, \sigma^2)$ and $|\rho| < 1$, the so-called Markov chain model. Markov models have no "memory" of past states of the process. The probability of the system being in state y_t depends only on the state of the system at time $t - 1$ and no prior times. The error terms ν_t are assumed to satisfy the usual assumptions, $\mathcal{E}(\nu) = 0$ and $\text{cov}(\nu) = \sigma^2 I$. These assumptions regarding ν ensure that the $\mathcal{E}(e) = 0$. To see this, we write e_t as

$$
\begin{aligned}
e_t &= \rho e_{t-1} + \nu_t \\
&= \rho(\rho e_{t-2} + \nu_{t-1}) + \nu_t \\
&= \rho[\rho(\rho e_{t-3} + \nu_{t-2}) + \nu_{t-1}] + \nu_t \\
&= \nu_t + \rho \nu_{t-1} + \rho^2 \nu_{t-2} + \rho^3 \nu_{t-3} + \cdots \\
&= \sum_{s=0}^{\infty} \rho^s \nu_{t-s}
\end{aligned}
$$

so that $\mathcal{E}(e_t) = 0$, for all t. And provided $|\rho| < 1$, the

$$
\text{var}(e_t) = \sigma^2 \sum_{s=0}^{\infty} \rho^{2s} = \frac{\sigma^2}{(1 - \rho^2)}
$$

since we can sum the infinite series. To obtain the covariance between e_t and e_{t+1}, observe that the

$$
\begin{aligned}
\text{cov}(e_t, e_{t+1}) &= \mathcal{E}(e_t e_{t+1}) = \mathcal{E}[e_t(\rho e_t + \nu_{t+1})] = \mathcal{E}(\rho e_t^2) + \mathcal{E}(e_t \nu_{t+1}) \\
&= \rho \text{var}(e_t) + 0,
\end{aligned}
$$

since the $\mathcal{E}(e_t \nu_{t+1}) = 0$ for all t. Furthermore,

$$
\begin{aligned}
e_t &= \rho e_{t+1} + \nu_{t+2} \\
&= \rho(\rho e_t + \nu_{t+1}) + \nu_{t+2}
\end{aligned}
$$

so that the

$$
\begin{aligned}
\text{cov}(e_t, e_{t+2}) &= \mathcal{E}(e_t e_{t+2}) \\
&= \mathcal{E}[e_t(\rho^2 e_t + \rho \nu_{t+1} + \nu_{t+2})] \\
&= \rho^2[\mathcal{E}(e_t^2) + \rho \mathcal{E}(e_t \nu_{t+1}) + \mathcal{E}(e_t \nu_{t+2}) \\
&= \rho^2 \text{var}(e_t) + 0 + 0.
\end{aligned}
$$

More generally, the

$$
\text{cov}(e_t, e_{t+s}) = \rho^s \text{var}(e_t) = \rho^2 \left(\frac{\sigma^2}{1 - \rho^2} \right) \qquad t = 0, 1, 2, \ldots \qquad (4.76)
$$

so that the autocorrelation parameter ρ is the correlation between two successive error terms:

$$
\rho(e_t, e_{t+1}) = \frac{\rho \left[\sigma^2 / (1 - \rho^2) \right]}{\sqrt{\sigma^2 / (1 - \rho^2)} \sqrt{\sigma^2 / (1 - \rho^2)}} = \rho.
$$

The correlation between e_t and e_{t+s} is then ρ^s.

Returning to the Markov chain model, we see that the covariance matrix of the errors is

$$\Sigma = \left(\frac{\sigma^2}{1-\rho^2}\right)\begin{pmatrix} 1 & \rho & \rho^2 & \cdots & \rho^\ell \\ \rho & 1 & \rho & \cdots & \rho^{\ell-1} \\ \rho^2 & \rho & 1 & \cdots & \rho^{\ell-2} \\ \vdots & \vdots & \vdots & \ddots & \vdots \\ \rho^\ell & \rho^{\ell-1} & \rho^{\ell-2} & \cdots & 1 \end{pmatrix}$$

and the inverse of the matrix is

$$\Sigma^{-1} = \left(\frac{1}{\sigma^2}\right)\begin{pmatrix} 1+\rho^2 & -\rho & 0 & \cdots & 0 \\ -\rho & 1+\rho^2 & -\rho & \cdots & 0 \\ 0 & -\rho & 1+\rho^2 & \cdots & 0 \\ \vdots & \vdots & \vdots & \ddots & \vdots \\ 0 & 0 & 0 & \cdots & 1+\rho^2 \end{pmatrix}.$$

The GLSE of β is obtain using the general expression

$$\hat{\beta} = (X'\Sigma^{-1}X)^{-1}X'\Sigma^{-1}y \tag{4.77}$$

for the Markov chain model.

Alternatively, we can transform the observation vector y so the errors are uncorrelated, which in the analysis of serially correlated time series data is called the analysis of the "first differences". It involves calculating the generalized differences

$$y_t = y_t - \rho y_{t-1}$$
$$x_t = \rho x_t$$

and

$$y_1 = y_1\sqrt{1-\rho^2}$$
$$x_1 = x_1\sqrt{1-\rho^2}$$

and performing OLSE. The matrix of differences Dy and Dx are defined using the matrix

$$D = \begin{pmatrix} \sqrt{1-\rho^2} & -\rho & 0 & \cdots & 0 \\ -\rho & \sqrt{1-\rho^2} & 0 & \cdots & 0 \\ 0 & 0 & \sqrt{1-\rho^2} & \cdots & 0 \\ \vdots & \vdots & \vdots & \ddots & \vdots \\ 0 & 0 & 0 & \cdots & \sqrt{1-\rho^2} \end{pmatrix},$$

motivated by the observation that $-\rho e_{t-1} + e_t = \nu_t$. In forming the matrix product $D'D$, observe that $D'D = \frac{\Sigma^{-1}}{\sigma^2}$ and the the constant σ^2 cancels using the GLSE formula. Thus, using matrix of differences transformations yields the GLSE exactly. This is only the case if ρ is known. When ρ is unknown, iterative methods must be employed (Neter et al., 1990, p. 469).

4.8 FGLS FOR CATEGORICAL DATA

We have shown how WLSE is used to improve estimates when the variance of the dependent variable y depends on the independent variable x and when the error of a linear model has known serial correlation. To illustrate the application of FGLS and the Wald statistic, we discuss the analysis of categorical data. We begin with a brief overview of the theory, and then discuss tests of marginal homogeneity, homogeneity of proportions, and independence.

4.8.1 Overview of the Categorical Data Model

Assume that the categorical data to be analyzed are samples of size n_{i+}, $i = 1, \ldots, s$ from s independent populations of r mutually exclusive and exhaustive categories. Letting N_{ij} represent the number of observations falling into the cell j for population i, the multinomial distribution of N_{ij} is

$$P(N_{i1} = n_{i1}, N_{i2} = n_{i2}, \ldots, N_{ir} = n_{ir}) = n_{i+}! \prod_{j=1}^{r} \pi_{ij}^{n_{ij}} \qquad (4.78)$$

where π_{ij} is the probability of a single observation fall in cell j for population i, the $\sum_{j=1}^{r} \pi_{ij} = 1$, and $\sum_{j=1}^{r} n_{ij} = n_{i+}$. Given i, the probability of an observation falling into cell j follows a binomial distribution so that the $\mathcal{E}(N_{ij}) = n_{i+}\pi_{ij}$ and the $\mathrm{var}(N_{ij}) = n_{i+}\pi_{ij}(1 - \pi_{ij})$. Furthermore, we can think of the random variables N_{ij} as an experiment of n_{i+} independent Bernoulli trials so that $N_{ij} = \sum_{u=1}^{n_{i+}} Y_u$ and $N_{ij'} = \sum_{v=1}^{n_{i+}} X_v$ where $Y_u = 1$ if trial t results in class j, and $X_v = 1$ if trial t results in class j'. Hence, the

$$
\begin{aligned}
\mathrm{cov}(N_{ij}, N_{ij'}) &= \sum_{u=1}^{n_{i+}} \sum_{v=1}^{n_{i+}} \mathrm{cov}(Y_u, X_v) \\
&= \sum_{u=1}^{n_{i+}} \mathrm{cov}(Y_u, X_v) + 2\sum_{u<v} \mathrm{cov}(Y_u, X_v) \\
&= \sum_{u=1}^{n_{i+}} (-\pi_{iu}\pi_{iv}) \\
&= -n_{i+}\pi_{iu}\pi_{iv}
\end{aligned}
$$

so that the N_{ij} are not independent. Since n_{ij} represent the observed number of observations falling into each cell, $p_{ij} = \frac{n_{ij}}{n_{i+}}$ is an estimate of the population proportion π_{ij}. The $\mathcal{E}(p_{ij}) = \pi_{ij}$ and the $\mathrm{var}(p_{ij}) = \pi_{ij}(1 - \pi_{ij})$. Grizzle, Starmer, and Koch (1969) proposed analyzing the population proportions using a linear model.

For a $s \times r$ contingency table, we let

$$
\begin{aligned}
\pi_i' &= \begin{pmatrix} \pi_{i1} & \pi_{i2} & \ldots & \pi_{ir} \end{pmatrix} \\
\pi' &= \begin{pmatrix} \pi_1' & \pi_2' & \ldots & \pi_s' \end{pmatrix} \\
p_i' &= \begin{pmatrix} p_{i1} & p_{i2} & \ldots & p_{ir} \end{pmatrix} \\
p' &= \begin{pmatrix} p_1' & p_2' & \ldots & p_s' \end{pmatrix}.
\end{aligned}
$$

Since the N_{ij} are not independent, the variance-covariance matrix for the proportions in the ith population is

$$
\text{var}(p_i) = \Sigma_{r \times r} = \frac{1}{n_{i+}}
\begin{pmatrix}
\pi_{i1}(1 - \pi_{i1}) & -\pi_{i1}\pi_{i2} & \cdots & -\pi_{i1}\pi_{ir} \\
-\pi_{i2}\pi_{i1} & \pi_{i2}(1 - \pi_{i2}) & \cdots & -\pi_{i2}\pi_{ir} \\
\vdots & \vdots & \ddots & \vdots \\
-\pi_{ir}\pi_{i1}) & -\pi_{ir}\pi_{i2} & \cdots & \pi_{ir}(1 - \pi_{ir})
\end{pmatrix}
\tag{4.79}
$$

and the $\text{var}(p) = \Omega = I_s \otimes \Sigma$.

Furthermore, we let $f_m(\pi)$ represent any function(s) of the elements of the vector π that have partial derivatives up to second order with respect to the elements π_{ij} for $m = 1, 2, \ldots, q < (r-1)s$. And let $f_m(p) = f_m(\pi)$ evaluated at $\pi = p$. And using these results we form the following functions:

$$
\begin{aligned}
[F(\pi)]' &= \begin{bmatrix} f_1(\pi) & f_2(\pi) & \ldots & f_q(\pi) \end{bmatrix} \\
[F(p)]' &= \begin{bmatrix} f_1(p) & f_2(p) & \ldots & f_q(p) \end{bmatrix} \equiv \widehat{F}' \\
H_{q \times rs} &= \left[\frac{\partial f_m(\pi)}{\partial \pi_{ij}} \Big|_{\pi_{ij}} \right] = p_{ij}.
\end{aligned}
$$

By the multivariate δ-method, $\widehat{F}$ converges in distribution to a multivariate normal distribution with covariance matrix $\Sigma_F = H\Omega H'$ where $\Omega = I_s \otimes \Sigma_{r \times r}$. The approximate large sample estimate of $\Sigma_F = H\Omega H'$ is

$$
S_{q \times q} = H\widehat{\Omega} H', \quad \text{where} \quad \widehat{\Omega} = I_s \otimes \widehat{\Sigma}_{r \times r}
\tag{4.80}
$$

where $\widehat{\Sigma}_{r \times r} = \text{var}(\hat{p}_i)$ with sample proportions substituted for the population values π_{ij} in (4.79), see Agresti (1990, p. 424) or Bishop, Fienberg, and Holland (1975, p. 493). When $f_m(p)$ are linear functions of p, then S is the exact covariance matrix. Finally, we assume that the functions of the elements of π are chosen such that they are jointly independent of each other and of the population constraint $\sum_{j=1}^{r} \pi_{ij} = 1$, for $i = 1, 2, \ldots, s$, then the matrices H and $H\widehat{\Omega}H'$ are of full rank q.

With these preliminaries, Grizzle et al. (1969) proposed a linear model for the functions of π

$$
F_{q \times 1}(\pi) = X_{q \times k}\beta_{k \times 1}
\tag{4.81}
$$

where β is the parameter vector and X is a known design matrix of rank $k \leq q$. For the parameter vector β in the linear model, the FGLS estimator is

$$\tilde{\beta} = (X'S^{-1}X)^{-1}X'S^{-1}\widehat{F}. \tag{4.82}$$

This estimate minimizes the quadratic form

$$Q = (\widehat{F} - X\beta)'S^{-1}(\widehat{F} - X\beta). \tag{4.83}$$

The FGLS estimate has an asymptotic normal distribution with estimated covariance matrix

$$\text{cov}(\tilde{\beta}) = (X'S^{-1}X)^{-1}. \tag{4.84}$$

To evaluate the fit of the model, one examines the residual sum of squares

$$Q = (\widehat{F} - X\tilde{\beta})'S^{-1}(\widehat{F} - X\tilde{\beta}) = \widehat{F}'S^{-1}\widehat{F} - \tilde{\beta}'(X'S^{-1}X)\tilde{\beta} \tag{4.85}$$

sometimes called the lack of fit or goodness of fit test (Timm, 1975, p. 291). Under the hypothesis $H : F(\pi) - X\beta = 0$, Q has an asymptotic chi-square distribution with degrees of freedom $q - k$, the difference between the number of independent functions and the number of estimated parameters in the vector β. Given the linear model, the test of the hypothesis $H : (C\beta = \beta)$ may be tested by Wald statistic given in (4.15),

$$W = (C\tilde{\beta} - \xi)'(X'S^{-1}X)^{-1}(C\tilde{\beta} - \xi) \tag{4.86}$$

which has an asymptotic χ^2 distribution with degrees of freedom $\nu = \text{rank}(C)$, when H is true.

In applying the general least squares theory presented above, two classes of functions cover most of the problems of interest in categorical data analysis. The first class of functions are linear functions of the cell probabilities. In this case,

$$F(\pi_{q \times 1}) = A_{q \times rs}\pi_{rs \times 1} = X_{q \times k}\beta_{k \times 1} \tag{4.87}$$

where the rank of $A = q \leq (r-1)s$. The element of the matrix $A = (a_{ijk})$ where $i = 1, 2, \ldots, q; j = 1, 2, \ldots, s$ and $k = 1, 2, \ldots, r$. For the linear model in (4.87), $H = \frac{\partial F}{\partial \pi} = A$, and $S = A\widehat{\Omega}A'$ where $\widehat{\Omega}_{rs \times rs}$ is block diagonal.

The second class of functions include logarithmic functions of the cell probabilities. For this case,

$$F(\pi_{t \times 1}) = K_{t \times q}\log(A_{q \times rs}\pi_{rs \times 1}) = X_{q \times k}\beta_{k \times 1}. \tag{4.88}$$

For this model, $H = \frac{\partial F}{\partial \pi} = KD^{-1}A$, $S = KD^{-1}A\widehat{\Omega}A'D^{-1}K'$. The matrix $D = \text{diag}(a_i'p)$ where a_i' is the ith row of A. The model in (4.81) proposed by Grizzle et al. (1969) for the analysis of categorical data is a special case of a more general class of models called generalized linear models (Agresti, 1990; Dobson, 1990; McCullagh & Nelder, 1989).

The FGLS procedure and the Wald statistic are also used in the analysis of longitudinal data when observations are correlated, missing and normally distributed, Kleinbaum (1973) and is discussed in Chapter 12. However, when normality cannot be assumed as with binary data, we must develop additional theory. Liang and Zeger (1986) extended the estimating equations for the generalized linear model to the generalized estimating equations (GEE) to analyze binary longitudinal data. An elementary introduction is provided by Dunlop (1994). Growth curve models with continuous and categorical variables are discussed in Chapter 12.

4.8.2 Marginal Homogeneity

A common problem in the analysis of categorical data in the behavioral, social and medical sciences is to test the equality of two dependent proportions for longitudinal data consisting of two time points, before and after. Using a contingency table, this is the test of marginal homogeneity (McNemar's test for change for 2×2 tables or Stuart's test for $I \times I$ tables). For a 3×3 table, the null hypothesis of marginal homogeneity is

$$H : \pi_{i+} = \pi_{+j} \qquad i = 1, 2, 3 \tag{4.89}$$

where $\pi_{i+} = \sum_j \pi_{ij}$ and $\pi_{+j} = \sum_i \pi_{ij}$. The sample estimates of the population proportions are $p_{i+} = \frac{n_{i+}}{n_{++}}$ and $p_{+j} = \frac{n_{+j}}{n_{++}}$. To represent the marginal homogeneity hypothesis as a linear model, observe that

$$\pi_{1+} = \pi_{+1} \implies \pi_{12} + \pi_{13} - \pi_{21} - \pi_{31} = 0$$
$$\pi_{2+} = \pi_{+2} \implies \pi_{21} + \pi_{23} - \pi_{12} - \pi_{32} = 0$$
$$\pi_{3+} = \pi_{+3} \implies \pi_{31} + \pi_{32} - \pi_{13} - \pi_{23} = 0.$$

Thus, the hypothesis is a linear function of the probabilities. Letting

$$\pi'_{1 \times rs} = \begin{pmatrix} \pi_{11} & \pi_{12} & \pi_{13} & \pi_{21} & \pi_{22} & \pi_{23} & \pi_{31} & \pi_{32} & \pi_{33} \end{pmatrix},$$

the null hypothesis has the linear model form:

$$F(\pi_{q \times 1}) = A_{q \times rs} \pi_{rs \times 1} = \begin{pmatrix} 0 & 1 & 1 & -1 & 0 & 0 & -1 & 0 & 0 \\ 0 & -1 & 0 & 1 & 0 & 1 & 0 & -1 & 0 \\ 0 & 0 & -1 & 0 & 0 & -1 & 1 & 1 & 0 \end{pmatrix} \pi. \tag{4.90}$$

The matrix A is singular since (-1) times the first row plus (-1) times the second row yields the third row. The matrix A must be of full rank for S to be of full rank. Since the chi-square statistic is invariant to the row basis of the matrix A, we may delete any row of A to obtain a full rank matrix. A suitable matrix is

$$A_* = \begin{pmatrix} 0 & 1 & 1 & -1 & 0 & 0 & -1 & 0 & 0 \\ 0 & -1 & 0 & 1 & 0 & 1 & 0 & -1 & 0 \end{pmatrix}. \tag{4.91}$$

Table 4.1: Racial Composition of Schools.

| | | 1955 | | | |
		(a) Black (90% − 100%)	(b) Black/White (11% − 89%)	(c) White (90% − 100%)	Total
	(a)	50	9	10	69
1965	(b)	2	6	29	37
	(c)	0	0	18	18
	Total	52	15	57	124

Using the matrix A_*, the equivalent hypothesis becomes:

$$H : F_* = A_* \pi = X\beta = 0. \tag{4.92}$$

Since $A_* p = 0$ if p is an estimate of π, and $X\beta = 0$, the test statistic for testing H is from (4.85):

$$Q = \widehat{F}' S^{-1} \widehat{F} = \widehat{F}'_* (A_* \widehat{\Omega} A'_*)^{-1} \widehat{F}_* = (A_* p)' (A_* \widehat{\Omega} A'_*)^{-1} (A_* p) \tag{4.93}$$

which has a chi-square distribution with degrees of freedom $\nu = \text{rank}(A_*) = q - 1 = 2$. The estimate of the covariance matrix for p is

$$\text{var}(p) = \frac{1}{n_{++}} [D(p) - pp'] = \widehat{\Omega} \tag{4.94}$$

where $D(p)$ is a diagonal matrix of proportions since we have only one population.

As an example of the application of the test of marginal homogeneity, we use data from Marascuilo and McSweeney (1977, p. 174). The data are from a study of school desegregation in which the racial composition of a city's schools in 1955 is compared to the racial composition in 1965. The racial composition of a sample of 124 school is given in Table 4.1.

To analyze the categorical data in Table 4.1 for the equality of marginal proportions, we use the CATMOD procedure and also illustrate how the IML procedure may be used for the analysis. Other SAS procedures for the analysis of categorical data include the procedures LOGISTIC and GENMOD. PROC LOGISTIC is a procedure used to analyze logistic regression models, while PROC GENMOD is used in the analysis of generalized linear models (Stokes, Davis, & Koch, 1995). The SAS code for the analysis is provided in Program 4_8_2.sas.

The analysis of the school desegregation data using PROC IML is straightforward. We have only to evaluate the matrix expression for the $\text{var}(p)$ given in (4.94) and A_* as defined in (4.91). The IML code first reads the data into a matrix named freq, prints it, calculates the covariance matrix, defined by S, prints it, and calculates the test statistic Q and the associated P-value for the test (Output 4.8.1). To test the hypothesis, the chi-square statistic Q is compared to the chi-square critical value of 5.99 for the chi-square distribution with 2 degrees of freedom for a test of size $\alpha = 0.05$. The test statistic $Q = 62.73$ is clearly significant. Thus, the racial composition of the schools in 1965 differs from that in 1955.

Program 4_8_2.sas also contains the code to perform the analysis using the PROC CATMOD. For this procedure the frequency data are input using the INPUT statement. The variables input are T55 (category of desegregation in 1955), and T65 (category of desegregation in 1965), and COUNT (number of schools). The @@ symbol allows for the input of more than one cell frequency in a line.

The four statements used in PROC CATMOD are WEIGHT, RESPONSE, MODEL, and REPEATED. The WEIGHT statement includes the name of the variable that contains the cell frequencies. The RESPONSE statement defines the response function F, and the term MARGINAL specifies that the response functions are marginal proportions. Only the marginal estimates of the population parameters π_{1+}, π_{2+}, π_{+1}, π_{+2} are calculated and associated with the response function $\hat{F}$ in PROC CATMOD since we are using marginal proportions. The RESPONSE statement directs the CATMOD procedure to build a full rank factorial design for the response function with an intercept and slope parameter to predict the year variable:

$$\hat{f} = X\beta,$$

$$\begin{pmatrix} p_{1+} \\ p_{2+} \\ p_{+1} \\ p_{+2} \end{pmatrix} = \begin{pmatrix} 1 & 0 & 1 & 0 \\ 0 & 1 & 0 & 1 \\ 1 & 0 & -1 & 0 \\ 0 & 1 & 0 & -1 \end{pmatrix} \begin{pmatrix} \beta_1 \\ \beta_2 \\ \beta_3 \\ \beta_4 \end{pmatrix}. \tag{4.95}$$

The design matrix X is generated from the full rank design matrix

$$X_0 = \begin{pmatrix} 1 & 1 \\ 1 & -1 \end{pmatrix}$$

to obtain the design matrix for X for the repeated measures. The linear model for the function $\hat{f}$ is specified using the MODEL and REPEATED statements. Because we have two repeated factor, the REPEATED statement identifies the factor and the number of independent levels. The MODEL statement says that the response variables are crossed.

PROC CATMOD provides an estimate of the marginal response function. The ANOVA table for the factor YEAR tests that the two contrasts $\psi_1 = \pi_{+1} - \pi_{1+}$ and $\psi_2 = \pi_{+2} - \pi_{2+}$ are simultaneously zero. The test statistic for YEAR in the ANOVA table is 62.73 as calculated using PROC IML (Output 4.8.2).

4.8.3 Homogeneity of Proportions

Another common problem in the analysis of contingency tables is to determine whether proportions are homogeneous across independent populations, a one-way ANOVA. This example, taken from Grizzle et al. (1969), deals with a study of

Output 4.8.1: PROC IML of Racial Composition of Schools.

FREQ		
50	9	10
2	6	29
0	0	18

P
0.4032258
0.0725806
0.0806452
0.016129
0.0483871
0.233871
0
0
0.1451613

S	
0.0012142	-0.000912
-0.000912	0.0023476

INVS	
1162.4717	451.37977
451.37977	601.2337

F
0.1370968
0.1774194

Chi-square Test of Marginal Homogeneity

Q
62.733059

p-value

PV
2.387E-14

Output 4.8.2: PROC CATMOD of Racial Composition of Schools.

Analysis of Variance			
Source	DF	Chi-Square	Pr > ChiSq
Intercept	2	505.23	<.0001
year	2	62.73	<.0001
Residual	0	.	.

Table 4.2: Severity of Dumping Syndrome.

	Clinical Evaluation of Dumping	Surgical Procedure			
		A	B	C	D
Hospital 1	None	23	23	20	24
	Slight	7	13	10	10
	Moderate	2	5	5	6
	Total N	32	38	38	40
	Ave Score	1.3	1.5	1.6	1.6
Hospital 2	None	18	18	13	9
	Slight	6	6	13	15
	Moderate	1	2	2	2
	Total N	25	26	28	26
	Ave Score	1.3	1.4	1.6	1.7
Hospital 3	None	8	12	11	7
	Slight	6	4	6	7
	Moderate	3	4	2	4
	Total N	17	20	19	18
	Ave Score	1.7	1.6	1.5	1.8
Hospital 4	None	12	18	14	13
	Slight	9	3	8	6
	Moderate	1	2	3	4
	Total N	22	20	25	23
	Ave Score	1.5	1.4	1.6	1.6

the effects of four different procedures on the dumping syndrome, an undesirable aftermath of duodenal ulcers. The data are provided in Table 4.2.

There are several hypotheses of interest in this study. The primary analysis is concerned with fitting an additive linear model to the observed cell probabilities for the four hospitals and the different surgical procedures: A, B, C, and D. The four surgical procedures consist of operations which remove differing amounts of the stomach: A =none, $B = \frac{1}{4}$, $C = \frac{1}{2}$, and $D = \frac{3}{4}$. If the additive model fits, one can then test hypotheses concerning the model parameters. Of primary interest is to determine if the severity of the dumping syndrome is linearly related to the amount of the stomach removed.

For each independent group, the proportions:

$$p'_i = \begin{pmatrix} p_{i1} & p_{i2} & p_{i3} \end{pmatrix}$$

so that

$$p'_1 = \begin{pmatrix} 23/32 & 7/32 & 2/32 \end{pmatrix}$$
$$p'_2 = \begin{pmatrix} 23/28 & 10/28 & 5/28 \end{pmatrix}$$
$$\vdots$$
$$p'_{16} = \begin{pmatrix} 13/23 & 16/23 & 4/23 \end{pmatrix}$$

where the estimated covariance matrix is

$$\text{var}(p_i) = \widehat{\Sigma}_{3\times3} = \frac{1}{n_{i+}} \begin{pmatrix} p_{i1}(1-p_{i1}) & -p_{i1}p_{i2} & -p_{i1}p_{i3} \\ -p_{i2}p_{i1} & p_{i2}(1-p_{i2}) & -p_{i2}p_{i3} \\ -p_{i3}p_{i1}) & -p_{i3}p_{i2} & p_{i3}(1-p_{i3}) \end{pmatrix} \quad (4.96)$$

so that

$$\widehat{\Omega} = I_s \otimes \widehat{\Sigma}_{r\times r} = \text{diag}[\text{var}(p_i)] = \bigoplus_{i=1}^{16} \text{var}(p_i).$$

For each of the categories of the response (none, slight, and moderate), the scores of 1, 2, and 3 are assigned. Hence, the mean score of the 16 populations is calculated as $1p_{i1} + 2p_{i2} + 3p_{i3}$. It is this mean score within each treatment combination that is used in the analysis. To calculate these mean, the matrix A is postmultiplied by p where the matrix

$$A = \text{diag}\begin{pmatrix} 1 & 2 & 3 \end{pmatrix}.$$

The function of the population proportions of interest for the study is

$$F_{16\times1}(\pi) = Ap = X\beta \quad (4.97)$$

is an additive model. The design matrix and parameter vector for the example are:

$$
X = \begin{pmatrix}
1 & 1 & 1 & 0 & 1 & 0 & 0 \\
1 & 1 & 1 & 0 & 0 & 1 & 0 \\
1 & 1 & 1 & 0 & 0 & 0 & 1 \\
1 & 1 & 1 & 0 & -1 & -1 & -1 \\
1 & 0 & 0 & 0 & 1 & 0 & 0 \\
1 & 0 & 0 & 0 & 0 & 1 & 0 \\
1 & 0 & 0 & 0 & 0 & 0 & 1 \\
1 & 0 & 0 & 0 & -1 & -1 & -1 \\
1 & 0 & 0 & 1 & 1 & 0 & 0 \\
1 & 0 & 0 & 1 & 0 & 1 & 0 \\
1 & 0 & 0 & 1 & 0 & 0 & 1 \\
1 & 0 & 0 & 1 & -1 & -1 & -1 \\
1 & -1 & -1 & -1 & 1 & 0 & 0 \\
1 & -1 & -1 & -1 & 0 & 1 & 0 \\
1 & -1 & -1 & -1 & 0 & 0 & 1 \\
1 & -1 & -1 & -1 & -1 & -1 & -1
\end{pmatrix}, \quad \beta = \begin{pmatrix}
\mu \\
\alpha_1 \\
\alpha_2 \\
\alpha_3 \\
\lambda_1 \\
\lambda_2 \\
\lambda_3
\end{pmatrix}
$$

where μ is the overall mean, α_k is the effect of the kth hospital, $k = 1, 2, 3$, and λ_m is the effect of the mth treatment condition. To make the design matrix full rank, side conditions are added to the model: $\sum_{k=1}^{4} \alpha_k = 1$, $\sum_{m=1}^{4} \lambda_m = 1$. Hence, the last effect α_4 and λ_4 are estimated from the prior estimates: $\hat{\alpha}_4 = -\sum_{k=1}^{3} \hat{\alpha}_k$, $\hat{\lambda}_4 = -\sum_{m=1}^{3} \hat{\lambda}_m$. For the model, $H = \frac{\partial F}{\partial \pi} = A$, so that

$$
S = A\hat{\Omega}A'
$$
$$
\tilde{\beta} = (X'S^{-1}X)^{-1}X'S^{-1}F
$$
$$
Q = (F - X\tilde{\beta})'S^{-1}(F - X\tilde{\beta}).
$$

Using Program 4_8_3.sas and PROC IML, the estimate of β for the model is

$$
\tilde{\beta} = \begin{pmatrix}
1.54 \\
-0.04 \\
-0.04 \\
0.11 \\
-0.11 \\
-0.07 \\
0.05
\end{pmatrix}
$$

and $\hat{\alpha}_4 = -\sum_{k=1}^{3} \hat{\alpha}_k = -0.03$, $\hat{\lambda}_4 = -\sum_{m=1}^{3} \hat{\lambda}_m = 0.13$. Because the value of $Q = 6.32$, the P-value for the lack of model fit test is 0.71 (Output 4.8.3). Hence, the additive model fits the data.

Given an additive model, we may test hypotheses concerning the model parameters using the Wald statistic. To test for hospital and treatment effects, the hypothesis

Output 4.8.3: PROC IML of Severity of Dumping Syndrome.

BETA
1.5448619
-0.040816
-0.035645
0.1061082
-0.110471
-0.072985
0.0496295

Chi-square Test of Fit

Q
6.3262338

p-value for fit

PVQ
0.7068753

Output 4.8.4: PROC CATMOD of Severity of Dumping Syndrome.

Analysis of Variance			
Source	DF	Chi-Square	Pr > ChiSq
Intercept	1	1999.88	<.0001
hosp	3	2.33	0.5065
treat	3	8.90	0.0307
Residual	9	6.33	0.7069

test matrices:

$$C_H = \begin{pmatrix} 0 & 1 & 0 & 0 & 0 & 0 & 0 \\ 0 & 0 & 1 & 0 & 0 & 0 & 0 \\ 0 & 0 & 0 & 1 & 0 & 0 & 0 \end{pmatrix}$$

$$C_T = \begin{pmatrix} 0 & 0 & 0 & 0 & 1 & 0 & 0 \\ 0 & 0 & 0 & 0 & 0 & 1 & 0 \\ 0 & 0 & 0 & 0 & 0 & 0 & 1 \end{pmatrix}$$

are used, resulting in the ANOVA table shown in Output 4.8.4.

The degrees of freedom in the ANOVA table for the main effect tests are the number of independent rows in the hypothesis test matrix. Using $\alpha = 0.05$ for each test we find that the treatment effect is marginally significant. As usual, Scheffé type confidence sets may be constructed for contrasts of the form: $\psi = c'\beta$. The

Output 4.8.5: PROC CATMOD Contrast of Severity of Dumping Syndrome.

Analysis of Contrasts			
Contrast	DF	Chi-Square	Pr > ChiSq
Linear Trend - Treatment	1	8.74	0.0031

confidence set has the form:

$$c'\tilde{\beta} \pm [\chi^2_{(\nu,1-\alpha)} c'(X'S^{-1}X)^{-1}c]^{1/2} \tag{4.98}$$

where $\nu = \text{rank}(C)$ for the test and $\chi^2_{(\nu,1-\alpha)}$ is the upper $(1-\alpha)$ critical value for the chi-square distribution.

To test that the severity of syndrome is linearly related to the amount of stomach removed, we test that the orthogonal linear polynomial

$$H_L : -3\lambda_1 - \lambda_2 + \lambda_3 + 3\lambda_4 = 0.$$

However, for the reparameterized model, this becomes: $-3\lambda_1 - \lambda_2 + \lambda_3 + 3(-\lambda_1 - \lambda_2 - \lambda_3) = -6\lambda_1 - 4\lambda_2 - 2\lambda_3 = 0$ or $-3\lambda_1 + 2\lambda_2 + \lambda_3 = 0$. Hence, the contrast matrix is

$$L = \begin{pmatrix} 0 & 0 & 0 & 0 & -3 & 2 & 1 \end{pmatrix}$$

for the model. The Wald statistic $W = (L\tilde{\beta} - \xi)'(L(X'S^{-1}X)^{-1}(L\tilde{\beta} - \xi) = 8.74$ (Output 4.8.5). The severity of syndrome was significantly linearly related to the amount of stomach removed.

4.8.4   Independence

The examples of the linear model approach to the analysis of categorical data have assumed only linear functions of π. We now consider the hypothesis of independence which utilizes the logarithmic function. The standard expression for the test of independence for a two-way table is

$$H : \pi_{ij} = \pi_{i+}\pi_{+j} \tag{4.99}$$

for all i and j. To write this test in the form $F(\pi) = 0$, we let $\mu_{ij} = \log(\pi_{ij})$ and representing the cell mean as a linear model: $\mu_{ij} = \mu + \alpha_i + \beta_j + \gamma_{ij}$ with side conditions and using dot notations, the parameters for the reparameterized linear mode are

$$\alpha_i = \mu_{i\cdot} - \mu_{\cdot\cdot} \qquad \beta_j = \mu_{\cdot j} - \mu_{\cdot\cdot} \qquad \gamma_{ij} = \mu_{ij} - \mu_{i\cdot} - \mu_{\cdot j} + \mu_{\cdot\cdot}$$

For a 2×2 table, the linear model parameters become:

$$\alpha_1 = \frac{(\mu_{11} + \mu_{12} - \mu_{21} - \mu_{22})}{4} = \sum_j \log\left(\frac{\pi_{1j}}{\pi_{2j}}\right)$$

$$\beta_1 = \frac{(\mu_{11} - \mu_{12} + \mu_{21} - \mu_{22})}{4} = \sum_i \log\left(\frac{\pi_{i1}}{\pi_{i2}}\right)$$

$$\gamma_{11} = \frac{(\mu_{11} - \mu_{12} - \mu_{21} + \mu_{22})}{4} = \frac{\log\left(\frac{\pi_{11}\pi_{22}}{\pi_{12}\pi_{21}}\right)}{4}.$$

Thus, the interaction term is proportional to the logarithm of the cross-product ration in the 2×2 table, which is also seen to be an odds ratio: $\frac{\pi_{11}\pi_{22}}{\pi_{12}\pi_{21}}$. Hence, for a 2×2 table, if $\lambda_{11} > 0$ the expected frequency in cell $(1,1)$ is greater than $n\pi_{1+}\pi_{+1}$ corresponding to independence. More generally, the test of independence has the general form:

$$H : \log\left(\frac{\pi_{ij}\pi_{i'j}}{\pi_{ij'}\pi_{i'j'}}\right) = 0 \quad \text{for} \quad i \neq i', j \neq j'. \tag{4.100}$$

For a 2×2 table, the test is written as:

$$H : F(\pi) = K \log A(\pi) = 0$$

$$= \log\left(\frac{\pi_{11}\pi_{22}}{\pi_{12}\pi_{21}}\right) = 0$$

$$= \log(\pi_{11}) - \log(\pi_{12}) - \log(\pi_{21}) + \log(\pi_{22}) = 0.$$

To obtain the test statistic, using (4.87), observe that $A = I_4$, $K = (1, -1, -1, 1)$,

$$\widehat{\Omega} = \left(\frac{1}{n_{++}}\right) \begin{pmatrix} p_{11}(1 - p_{11}) & -p_{11}p_{12} & -p_{11}p_{21} & -p_{11}p_{22} \\ -p_{12}p_{11} & p_{12}(1 - p_{12}) & -p_{12}p_{21} & -p_{12}p_{22} \\ -p_{21}p_{11} & -p_{21}p_{12} & p_{21}(1 - p_{21}) & -p_{21}p_{22} \\ -p_{22}p_{11} & -p_{22}p_{12} & p_{22}p_{21} & p_{22}(1 - p_{22}) \end{pmatrix}$$

$$D = \begin{pmatrix} p_{11} & 0 & 0 & 0 \\ 0 & p_{12} & 0 & 0 \\ 0 & 0 & p_{13} & 0 \\ 0 & 0 & 0 & p_{14} \end{pmatrix}$$

so that

$$KD^{-1}A = \left(\frac{1}{p_{11}} \quad \frac{-1}{p_{12}} \quad \frac{-1}{p_{21}} \quad \frac{1}{p_{22}}\right)$$

$$KD^{-1}A\widehat{\Omega} = \frac{1}{n_{++}}\left(1 \quad -1 \quad -1 \quad 1\right)$$

$$S = KD^{-1}A\widehat{\Omega}A'D^{-1}K' = \frac{1}{n_{++}}\left(\frac{1}{p_{11}} + \frac{1}{p_{12}} + \frac{1}{p_{21}} + \frac{1}{p_{22}}\right).$$

Table 4.3: Teacher Performance and Holding an Office.

		Teacher Performance		
		Successful	Unsuccessful	Total
Held Office	Yes	50	20	70
	No	10	20	30
	Total	60	40	100

And since $X\beta = 0$ for the test of independence, the test statistic is

$$Q = [K \log(Ap)]' S^{-1} [K \log(Ap)]$$

$$= n_{++} \left[\frac{(\log p_{11} - \log p_{12} - \log p_{21} + \log p_{22})^2}{\frac{1}{p_{11}} + \frac{1}{p_{12}} + \frac{1}{p_{21}} + \frac{1}{p_{22}}} \right]$$

which has a chi-square distribution with one degree of freedom. For a general $I \times J$ table, Bhapkar (1966) showed that the Wald statistic is equal to Neyman (1949) modified chi-square statistic:

$$X_N^2 = \sum_i \sum_j \frac{(n_{ij} - n_{++}\hat{\pi}_{ij})^2}{n_{ij}}.$$

In general, this is not equal to Karl Pearson's chi-square statistic:

$$X_{KP}^2 = \sum_i \sum_j \frac{(n_{ij} - n_{++}\hat{\pi}_{ij})^2}{n_{++}\hat{\pi}_{ij}}$$

which is also generally not the same as the Neyman-Pearson likelihood ratio statistic:

$$X_{NP}^2 = -2 \sum_i \sum_j n_{ij} \log \left(\frac{n_{++}\hat{\pi}_{ij}}{n_{++}} \right).$$

These statistics are calculated in the SAS procedure PROC FREQ for the test of independence.

Our example of the test of independence is taken from Marascuilo and Mc-Sweeney (1977, p. 205). The study is designed to determine if there is any association between how a teacher performs and whether the teacher held an elected office while a student. The data are shown in Table 4.3.

Program 4_8_4.sas includes the SAS code to perform the necessary calculations for the test of independence using PROC CATMOD (Output 4.8.6). Using the formula for the 2×2 table derived above,

$$Q = \frac{100(\log 0.50 - \log 0.20 - \log 0.10 + \log 0.20)^2}{\frac{1}{0.50} + \frac{1}{0.20} + \frac{1}{0.10} + \frac{1}{0.20}} = \frac{100(1.609)^2}{22} = 11.77.$$

The same result is obtained using PROC CATMOD. Hence, there appears to be association between teacher performance and holding office.

Output 4.8.6: PROC CATMOD of Teacher Performance and Holding an Office.

Analysis of Variance			
Source	DF	Chi-Square	Pr > ChiSq
office	1	9.81	0.0017
perform	1	3.42	0.0643
Residual	1	11.77	0.0006

Output 4.8.7: PROC MIXED of Real Data Example.

Kenward–Roger Method

Type 3 Tests of Fixed Effects						
Effect	Num DF	Den DF	Chi-Square	F Value	Pr > ChiSq	Pr > F
grp	1	16.9	18.74	18.74	<.0001	0.0005
prob	4	72.2	57.76	14.44	<.0001	<.0001
grp*prob	4	72.2	1.37	0.34	0.8495	0.8485

Satterthwaite Method

Type 3 Tests of Fixed Effects						
Effect	Num DF	Den DF	Chi-Square	F Value	Pr > ChiSq	Pr > F
grp	1	16.9	18.81	18.81	<.0001	0.0005
prob	4	72.2	57.76	14.44	<.0001	<.0001
grp*prob	4	72.2	1.37	0.34	0.8495	0.8485

4.8.5 Univariate Mixed Linear Model, Less Than Full Rank Representation

Program 3_5_2.sas was used to analyze the univariate mixed linear model for a simple split plot design where the covariance matrix had compound symmetry structure. In Program 4_9.sas, we analyze the same data using PROC MIXED, but allow the variance-covariance matrix to have structure AR(1). And, we use both the Kenward-Roger (KR) and Satterthwaite methods to estimate error degrees of freedom for the fixed parameter F-tests. The structure of the variance-covariance matrix is specified on the RANDOM statement using the TYPE option. The results are displayed in Output 4.8.7. The error degrees of freedom for both procedure are the same for the split-plot design. The CHISQ option on the model statement generates the Wald statistic.

CHAPTER 5

Multivariate General Linear Models

5.1 INTRODUCTION

The multivariate general linear model (MGLM) without restrictions has wide applicability in regression analysis, analysis of variance (ANOVA), analysis of covariance (ANCOVA), analysis of repeated measurements, and more broadly defined, experimental designs involving several dependent variables. Timm (1975, 1993a, 2002) provides numerous examples of applications in education and psychology.

In this chapter, we develop the MGLM, estimate the model parameters and test general linear hypotheses. The multivariate normal distribution is extended to the matrix normal distribution and the general multivariate linear mixed model (GM-LMM) is introduced. A special case of the GMLMM, known as Scheffé's mixed model, is used to analyze repeated measurement data. We then illustrate the general theory using examples in multivariate regression, multivariate analysis of variance (MANOVA), and multivariate analysis of covariance (MANCOVA). We show the relationship between univariate and multivariate analysis of repeated measurement data and discuss how one may analyze extended linear hypotheses.

5.2 DEVELOPING THE MODEL

The univariate GLM was obtained from (1.1) by assuming the covariance structure $\Omega = \sigma^2 I_n$ where y was a vector of n independent observations for one dependent variable. To extend the model to p variables, we write a linear model for each variable

$$y_i = X\beta_i + e_i \qquad\qquad i = 1, 2, \ldots, p \qquad\qquad (5.1)$$
$$\text{var}(y_i) = \sigma_{ii} I_n$$
$$\text{cov}(y_i, y_j) = \sigma_{ij} I_n \qquad\qquad \text{for } i \neq j.$$

In (5.1), we assume $\mathcal{E}(e_i) = 0$. The p n-variate observation vectors $y_{i(n \times 1)}$ are related to the p k-variate parameter vectors β_i by a common design matrix $X_{n \times k}$ where the $\text{rank}(X) = k$. The p separate univariate models are related by the $[p(p-1)]/2$ covariances.

The classical matrix representation for (5.1) is to let

$$Y_{n \times p} = \begin{pmatrix} y_1 & y_2 & \cdots & y_p \end{pmatrix} \qquad B_{k \times p} = \begin{pmatrix} \beta_1 & \beta_2 & \cdots & \beta_p \end{pmatrix} \qquad (5.2)$$

where $U_{n \times p} = (e_1, e_2, \ldots, e_p)$ is the matrix of errors. Then (5.1) is written as

$$\Omega_0 : Y_{n \times p} = X_{n \times k} B_{k \times p} + U_{n \times p} \qquad (5.3)$$

$$\Omega = \text{cov}(Y) = I_n \otimes \Sigma. \qquad (5.4)$$

Here $\Sigma = (\sigma_{ij})$ is the common $(p \times p)$ covariance matrix and $\otimes$ is a Kronecker product defined as $A \otimes B = (a_{ij} B)$, so that each row of the data matrix Y has a common covariance structure Σ.

Stacking the p n-vectors y_i into a single $(np \times 1)$ vector y, the p k-vectors β_i into a $(kp \times 1)$ vector β, and the p n-vectors e_i into a $(np \times 1)$ vector e, we may write (5.1) in vector form as follows

$$\begin{pmatrix} y_1 \\ y_2 \\ \vdots \\ y_p \end{pmatrix} = D \begin{pmatrix} \beta_1 \\ \beta_2 \\ \vdots \\ \beta_p \end{pmatrix} + \begin{pmatrix} e_1 \\ e_2 \\ \vdots \\ e_p \end{pmatrix}$$

$$y = D\beta + e \qquad (5.5)$$

where $D = (I_p \otimes X)$ and $\Omega = \Sigma \otimes I_n$. Thus, (5.5) has the GLM structure (1.1). Expression (5.5) may also be obtained from (5.3) using the $\text{vec}(\cdot)$ operator. The $\text{vec}(\cdot)$ operator creates a vector from a matrix by stacking the columns of a matrix sequentially (Magnus & Neudecker, 1988); hence, $\text{vec}(Y) = y$ and $\text{vec}(B) = \beta$. Using this notation, (5.5) becomes

$$\mathcal{E}[\text{vec}(Y)] = D\text{vec}(B) = (I_p \otimes X)\text{vec}(B) \qquad (5.6)$$

$$\text{cov}[\text{vec}(Y)] = \Omega = \Sigma \otimes I_n. \qquad (5.7)$$

Alternatively, the ith row $y_{i(p \times 1)}$ of Y has the structure

$$y_i = (I_p \otimes x_i')\text{vec}(B) + e_i \qquad (5.8)$$

where x_i' and e_i' are the ith rows of X and E, respectively. The GLM includes the multivariate regression model, MANOVA, MANCOVA, and numerous other multivariate linear models.

Instead of using the classical matrix form (5.3), we may use the responsewise orientation. Then (5.3) is written as

$$Y_{p \times n} = \begin{pmatrix} y_1 & y_2 & \cdots & y_n \end{pmatrix} = B_{p \times k} X_{k \times n} + U_{p \times n} \qquad (5.9)$$

where now each column of Y and U in (5.9) has common covariance structure Σ for $j = 1, 2, \ldots, n$. Or one may use the $\text{vec}(\cdot)$ operator on the rows of the classical matrices to obtain the responsewise linear model. Letting $\tilde{y} = \text{vec}(Y')$, $\tilde{\beta} = \text{vec}(B')$ and $\tilde{e} = \text{vec}(U')$ for the matrices in (5.3), the responsewise model is represented as $\tilde{y} = \tilde{D}\tilde{\beta} + \tilde{e}$ where $\tilde{D} = X \otimes I$ and the $\text{cov}(\tilde{y}) = I \otimes \Sigma$.

5.3   ESTIMATION THEORY AND HYPOTHESIS TESTING

To estimate B or $\beta = \text{vec}(B)$, we have from (5.5) and the generalized least square result (4.6) that the best linear unbiased estimator (BLUE) of β is

$$\hat{\beta} = (D'\Omega^{-1}D)^{-1}D'\Omega^{-1}y \qquad (5.10)$$

where $\Omega = \Sigma \otimes I$ is known and $D = (I_p \otimes X)$. However, even when Ω is unknown, (5.10) may be used because the estimator of β does not depend on Σ; Σ drops out of (5.10) since the design matrix X is the same for each variable. To see this, observe that on substituting Ω and D into (5.10) that

$$
\begin{aligned}
\text{vec}(\widehat{B}) = \hat{\beta} &= [(I_p \otimes X)'(\Sigma \otimes I_n)^{-1}(I_p \otimes X)]^{-1}(I_p \otimes X)'(\Sigma \otimes I_n)^{-1}y \\
&= (\Sigma^{-1} \otimes X'X)^{-1}(\Sigma^{-1} \otimes X')y \\
&= (I_p \otimes (X'X)^{-1}X')y \\
&= (I_p \otimes (X'X)^{-1}X')\text{vec}(Y) \\
&= \text{vec}[(X'X)^{-1}X'Y] \qquad (5.11)
\end{aligned}
$$

where the last step in (5.11) uses Theorem 5.1 with $A = (X'X)^{-1}X'$ and $C' = I_p$. From (5.11), we have for the classical model (5.3) that

$$\widehat{B} = (X'X)^{-1}X'Y. \qquad (5.12)$$

Partitioning B and Y and equating columns, the BLUE of β_i is the ordinary least squares estimator (OLSE), $\beta_i = (X'X)^{-1}X'y_i$. To obtain the covariance matrix for $\hat{\beta}$, formula (4.7) is used

$$
\begin{aligned}
\text{cov}(\hat{\beta}) &= (D'\Omega^{-1}D)^{-1} \\
&= [(I_p \otimes X)'(\Sigma^{-1} \otimes I_n)(I_p \otimes X)]^{-1} \\
&= \Sigma \otimes (X'X)^{-1}. \qquad (5.13)
\end{aligned}
$$

We have obtained the BLUE of B using the vector version of the GLM with covariance structure $\Omega = \Sigma \otimes I_n$. To directly obtain the OLSE = BLUE of B one may minimize the matrix error sum of squares using the classical model (5.3),

$$\text{tr}(U'U) = \text{tr}[(Y - XB)'(Y - XB)] \qquad (5.14)$$

to obtain the matrix normal equations

$$(X'X)\widehat{B} = X'Y \qquad (5.15)$$

and directly obtain result (5.12).

Theorem 5.1. *Let A, B, and C be three matrices such that the matrix product ABC is defined, then*

$$\text{vec}(ABC) = (C' \otimes A)\text{vec}(B). \qquad (5.16)$$

To test hypotheses regarding the parameter matrix B or the vector $\beta = \text{vec}(B)$, we again make distributional assumptions. Using (5.3), we assume that the n rows $(e_1', e_2', \ldots, e_n')$ of U are independently normally distributed

$$e_i' \sim IN_p(0, \Sigma) \tag{5.17}$$

$$e = \text{vec}(U) \sim N_{np}(0, \Omega = \Sigma \otimes I_n). \tag{5.18}$$

Alternatively using the observation vector y, we assume that

$$y = \text{vec}(Y) \sim N_{np}(D\beta, \Omega = \Sigma \otimes I_n) \tag{5.19}$$

where the design matrix $D = I_p \otimes X$. Letting $N = np$, by (1.26) the density for y is

$$(2\pi)^{N/2} |\Omega|^{-1/2} \exp\left[-\frac{1}{2}(y - D\beta)' \Omega^{-1}(y - D\beta) \right]. \tag{5.20}$$

While (5.20) may be used to find the maximum likelihood (ML) estimate of β and Ω and hence Σ, it is convenient to use the matrix normal distribution for the data matrix $Y_{n \times p}$. By simplification of (5.20), observe first that the

$$|\Omega|^{-1/2} = |\Sigma \otimes I_n|^{-1/2} = |\Sigma|^{-n/2} |I_n|^{-p/2} = |\Sigma|^{-n/2} \tag{5.21}$$

using the identity that the $|A_m \otimes B_n| = |A|^n |B|^m$. Next we use the results from Theorem 5.1 and the identity $\text{tr}(A'B) = [\text{vec}(A)]'[\text{vec}(B)]$ so that

$$
\begin{aligned}
(y - D\beta)' \Omega^{-1}(y - D\beta) &= [\text{vec}(Y - XB)]'(\Sigma \otimes I_n)^{-1}[\text{vec}(Y - XB)] \\
&= [\text{vec}(Y - XB)]'[\text{vec}(Y - XB)]\Sigma^{-1} \\
&= \text{tr}[(Y - XB)'(Y - XB)\Sigma^{-1}] \\
&= \text{tr}[\Sigma^{-1}(Y - XB)'(Y - XB)]. \tag{5.22}
\end{aligned}
$$

The matrix normal distribution for $Y_{n \times p}$ is

$$(2\pi)^{-np/2} |\Sigma|^{-n/2} \text{etr}\left[-\frac{1}{2}\Sigma^{-1}(Y - XB)'(Y - XB) \right]. \tag{5.23}$$

Using (5.23), we immediately observe from the exponent that the ML estimate of B is equal to the BLUE. Furthermore, the ML estimate of Σ is

$$
\begin{aligned}
\widehat{\Sigma} &= \frac{Y'(I - X(X'X)^{-1}X')Y}{n} \\
&= \frac{(Y - X\widehat{B})'(Y - X\widehat{B})}{n}. \tag{5.24}
\end{aligned}
$$

Result (5.23) is a special case of definition (5.1) with $W \equiv I$.

Definition 5.1. The data matrix $Y_{n \times p}$ has a matrix normal distribution with parameters μ, Σ, and W written as $Y \sim N_{np}(\mu, \Sigma \otimes W)$ if the density for Y is

$$(2\pi)^{-np/2} |\Sigma|^{-n/2} |W|^{-p/2} \text{etr} \left[-\frac{1}{2} \Sigma^{-1} (Y - \mu)' W^{-1} (Y - \mu) \right] \qquad (5.25)$$

or $\text{vec}(Y) \sim N_{np}[\text{vec}(\mu), \Omega = \Sigma \otimes W]$.

From the matrix normal distribution, the ML estimate of μ and Σ are

$$\hat{\mu} = (X'W^{-1}X)^{-1} X'W^{-1}Y \qquad (5.26)$$

$$\widehat{\Sigma} = \frac{Y'W^{-1}Y - Y'X(X'W^{-1}X)^{-1}X'Y}{n}$$

$$= \frac{(Y - X\hat{\mu})'(Y - X\hat{\mu})}{n}. \qquad (5.27)$$

To test hypotheses using (5.3), we are usually interested in bilinear parametric functions of the form $\psi = c'Ba$ or linear sets of functions $\Psi = CBA$ where $C_{g \times k}$ and $A_{p \times u}$ are known matrices, $c_{k \times 1}$ is a row in C and $a_{i(p \times 1)}$ is a column of A. Letting

$$\beta' = \begin{pmatrix} \beta'_1 & \beta'_2 & \cdots & \beta'_p \end{pmatrix} = \text{vec}(B) \qquad (5.28)$$

where β_i is the ith column of B, $a' = (a_1, a_2, \ldots, a_p)$, and $c'_i = a_i C'$ with $i = 1, 2, \ldots, p$, observe that

$$\psi = c'Ba = \sum_{i=1}^{p} c'_i \beta_i = c'_* \beta \qquad (5.29)$$

where $c'_* = (c'_1, c'_2, \ldots, c'_p)$. Hence we may consider the more general problem of estimating $c'_* \beta$ where c'_* is arbitrary and apply the result to the special case $\psi = c'Ba$. In a similar manner, instead of estimating $\Psi = CBA$, we may investigate the more general case of estimating $\Psi = C_* \beta$, where $C_* = (C'_1, C'_2, \ldots, C'_p)$ is a known matrix. To apply the general result to the special case $\Psi = CBA$, we let $C'_{gu \times k} = a_i \otimes C$ for $i = 1, 2, \ldots, p$, a_i equal to the ith row of A, and $C_* = A'_{u \times p} \otimes C_{g \times k}$. Using the classical form of the model, hypotheses of the form

$$H : CBA = \Gamma \Leftrightarrow (A' \otimes C)\text{vec}(B) = \text{vec}(\Gamma) \Leftrightarrow (A' \otimes C)\beta = \gamma \qquad (5.30)$$

are of general interest, where a hypothesis is testable if and only if B is estimable. Alternatively, general hypothesis H becomes $H : C_* \beta = \gamma$. Under multivariate normality of $\hat{\beta}$, the ML estimate of $\gamma = \text{vec}(\Gamma)$ is $\hat{\gamma} = C_* \hat{\beta}$ where $\hat{\beta} = \text{vec}(\widehat{B})$ and $\widehat{B}$ is defined in (5.12), the ML estimate of B. Then, the distribution of $\hat{\gamma}$ is multivariate normal and represented as

$$\hat{\gamma} \sim N_{gu}[\gamma, (C_*(D'\Omega^{-1}D)^{-1}C'_*)]. \qquad (5.31)$$

Simplifying the structure of the covariance matrix, observe that

$$
\begin{aligned}
C_*(D'\Omega^{-1}D)^{-1}C_*' &= (A' \otimes C)[(I \otimes X')(\Sigma \otimes I)^{-1}(I \otimes X)]^{-1}(A \otimes C') \\
&= (A' \otimes C)[\Sigma \otimes (X'X)^{-1}](A \otimes C') \\
&= A'\Sigma A \otimes C(X'X)^{-1}C'.
\end{aligned}
\tag{5.32}
$$

When $\Phi = A'\Sigma A$ is known, following (4.11), the likelihood ratio test for

$$
H : CBA = \Gamma \Leftrightarrow H : \gamma = \gamma_0
\tag{5.33}
$$

where C, A, and Γ are known and where $\widehat{B}$ is the ML estimate of B is to reject H if $X^2 > c$ where c is chosen such that the $P(X^2 > c | H \text{ true}) = \alpha$, and X^2 is defined as

$$
\begin{aligned}
X^2 &= (\hat{\gamma} - \gamma_0)'[(A'\Sigma A) \otimes (C(X'X)^{-1}C')]^{-1}(\hat{\gamma} - \gamma_0) \\
&= \mathrm{tr}[(C\widehat{B}A - \Gamma)'(C(X'X)^{-1}C')^{-1}(C\widehat{B}A - \Gamma)(A'\Sigma A)^{-1}]
\end{aligned}
\tag{5.34}
$$

by using Theorem 5.1 and the $\mathrm{tr}(A'B) = [\mathrm{vec}(A)]'[\mathrm{vec}(B)]$. X^2 has a central χ^2 distribution with $\nu = gu$ degrees of freedom when H is true.

To develop the likelihood ratio test of $H : CBA = \Gamma$ with unknown Σ, we let $\mu = XB$ in the matrix normal distribution and obtain the maximum likelihood estimates of B and Σ under Ω_0 and ω. With $\mu = XB$, the ML estimate of B and Σ is given in (5.26) (Seber, 1984, p. 406). Following (2.12), the corresponding ML estimates under Ω_0 are

$$
\widehat{B} = \widehat{B} - (X'X)^{-1}C'[C(X'X)^{-1}C']^{-1}(C\widehat{B}A - \Gamma)
\tag{5.35}
$$

$$
\widehat{\Sigma} = \frac{(Y - X\widehat{B}_\Omega)'(Y - X\widehat{B}_\Omega)}{n}.
\tag{5.36}
$$

Following the univariate procedure,

$$
\begin{aligned}
\lambda &= \frac{L(\hat{\omega})}{L(\widehat{\Omega}_0)} = \frac{|\widehat{\Sigma}_\omega|^{-n/2}}{|\widehat{\Sigma}|^{-n/2}} = \left[\frac{|\widehat{\Sigma}|}{|\widehat{\Sigma}_\omega|} \right]^{n/2} \\
&= \left[\frac{|(Y - X\widehat{B})'(Y - X\widehat{B})|}{|(Y - X\widehat{B}_\omega)'(Y - X\widehat{B}_\omega)|} \right]^{-1/2}
\end{aligned}
\tag{5.37}
$$

so that

$$
\Lambda = \lambda^{2/n} = \frac{|E|}{|E + H|}
\tag{5.38}
$$

where

$$
E = A'Y'(I - X(X'X)^{-1}X')YA
\tag{5.39}
$$

$$
\begin{aligned}
H &= (C\widehat{B}A - \Gamma)'(C(X'X)^{-1}C')^{-1}(C\widehat{B}A - \Gamma) \\
&= (\widehat{\Gamma} - \Gamma_0)'(C(X'X)^{-1}C')^{-1}(\widehat{\Gamma} - \Gamma_0).
\end{aligned}
\tag{5.40}
$$

The matrices E and H have independent Wishart distributions

$$E \sim W_u(\nu_e = n - k, A'\Sigma A, \Delta = 0) \qquad (5.41)$$

$$H \sim W_u(\nu_h = g, A'\Sigma A, (A'\Sigma A)^{-1}\Delta) \qquad (5.42)$$

where the noncentrality matrix

$$\Delta = (\Gamma - \Gamma_0)'[C(X'X)^{-1}C']^{-1}(\Gamma - \Gamma_0). \qquad (5.43)$$

The likelihood ratio test is to reject $H : CBA = \Gamma$ if

$$\Lambda = \frac{|E|}{|E + H|} = \prod_{i=1}^{s}(1 + \lambda_i) \leq U_{u,g,n-k}^{1-\alpha} \qquad (5.44)$$

where $U_{u,g,n-k}^{1-\alpha}$ is the upper $(1 - \alpha)100\%$ of the Λ-statistic, λ_i are the roots of $|H - \lambda E| = 0$, $s = \min(g, u)$, $g = \operatorname{rank}(C)$, $u = \operatorname{rank}(A)$, and $k = \operatorname{rank}(X)$.

Theorem 5.2. *The null distribution of the likelihood ratio criterion or more generally the U-statistic*

$$U = \frac{|E|}{|E + H|} = \frac{1}{|I + HE^{-1}|} \qquad (5.45)$$

when $\nu_e \geq u$ and $\nu_h \geq u$, $H \sim W_u(\nu_h, \Sigma)$, $E \sim W_u(\nu_e, \Sigma)$ and H and E are independent in the distribution of the product of u independent beta random variables V_i, $\prod_{i=1}^{u} V_i$ where V_i is beta$[(\nu_e - i + 1)/2, \nu_h/2]$.

Theorem 5.3. *When $\nu_e \geq u > \nu_h$, the null distribution of U_{ν_h,u,ν_e} is the same as U_{u,ν_h,ν_e}.*

Theorem 5.4. *The asymptotic distribution of U is given by*

$$X^2 = -\left[\nu_e - \frac{(u - \nu_h + 1)}{2}\right]\log U \sim \chi^2(u\nu_h), \qquad (5.46)$$

a χ^2 distribution with $\nu = u\nu_h$ degrees of freedom, as the sample size n tends to infinity.

Box's asymptotic expansion may be used, to obtain p-value for the U-statistic (T. W. Anderson, 1984, p. 317), using Oh notation where $O(n^{-\gamma})$ represent terms in the expansion that are bounded when divided by $n^{-\gamma}$.

Theorem 5.5. *The*

$$P(\rho n \log U_{u,g,n-k} > c) = P(X_\nu^2 > c) + \gamma[P(X_{\nu+4}^2 > c) - P(X_\nu^2 > c)] + O[(\rho n)^{-3}] \qquad (5.47)$$

where $\nu = ug$, $\gamma = [\nu(u^2 + g^2 - 5)]/[48(\rho n)^2]$, $\rho = 1 - n^{-1}[k - g + (u + g + 1)/2]$ and X_ν^2 represents χ^2 random variable with ν degrees of freedom.

Anderson states that the expression in Theorem 5.5 is accurate to three decimal places if $u^2 + g^2 \leq \rho n/3$. An alternative to the χ^2 approximation for obtaining p-value is to use Rao's F approximation (Seber, 1984, p. 41).

Theorem 5.6. *The*

$$P\left[(U^{-a} - 1)\left(\frac{\nu_2}{\nu_1}\right)\right] = P(F_{\nu_1,\nu_2} > c) \tag{5.48}$$

where

$$a = \left[\frac{u^2 + \nu_h^2 - 5}{(\nu_h u)^2} - 4\right]^{1/2} \tag{5.49}$$

$$\nu_1 = \nu_h u \tag{5.50}$$

$$\nu_2 = a^{-1}\left[\nu_e - \frac{u - \nu_h + 1}{2}\right] - \frac{\nu_h u - 2}{2} \tag{5.51}$$

and $u = \text{rank}(A)$, $\nu_h = \text{rank}(C)$ and $\nu_e = n - \text{rank}(X)$. If u or ν_h is 1 or 2, the approximation is exact.

The Wishart distribution is a multivariate distribution that generalized the χ^2 distribution, a distribution that generalizes the F distribution is the multivariate beta distribution.

Theorem 5.7. *Let H and E be independently distributed Wishart distributions: $H \sim W_u(\nu_h, \Sigma)$ and $E \sim W_u(\nu_e, \Sigma)$ with $\nu_h, \nu_e \geq u$, then*

$$F = (H + E)^{-1/2} H (H + E)^{-1/2} \tag{5.52}$$

where $(H + E)^{1/2}$ is any nonsingular factorization of $(H + E)$, then the pdf of F is a multivariate beta type I distribution

$$c|F|^{(\nu_h - u - 1)/2}|I_u - F|^{(\nu_e - u - 1)/2} \tag{5.53}$$

or

$$F \sim M\beta_{I(u, \nu_h/2, \nu_e/2)}. \tag{5.54}$$

The constant c is chosen such that the total probability is equal to one in the sample space.

Inspection of the likelihood ratio criterion suggests that the null hypothesis should be rejected if the roots of $|H - \lambda E| = 0$ are large. Several proposed criteria are dependent on the roots of the following determinant equations.

$$|H - \lambda E| = 0 \tag{5.55}$$
$$|E - \nu(H + E)| = 0 \tag{5.56}$$
$$|H - \theta(H + E)| = 0. \tag{5.57}$$

Using the roots in (5.55)-(5.57), Wilks' likelihood ratio criterion (Wilks, 1932) is

$$\Lambda = \frac{|E|}{|E + H|} = \prod_{i=1}^{s} \nu_i = \prod_{i=1}^{s}(1 + \lambda_i)^{-1} = \prod_{i=1}^{s}(1 - \theta_i) = |I_u - F| \quad (5.58)$$

where $s = \min(\nu_h, u)$, $\nu_h = \text{rank}(C)$, and $u = \text{rank}(A)$. H is rejected if $\Lambda > U_{u,\nu_h,\nu_e}^{1-\alpha}$. Bartlett-Nanda-Pillai suggested the trace criterion

$$V = \text{tr}(H(H + E)^{-1}) = \sum_{i=1}^{s} \theta_i = \sum_{i=1}^{s} \frac{\lambda_i}{1 + \lambda_i} = \sum_{i=1}^{s}(1 - \nu_i) \quad (5.59)$$

where H is rejected if $V > V_{s,M,N}^{1-\alpha}$ where $M = (|\nu_h - u| - 1)/2$, $N = (\nu_e - u - 1)/2$. Employing his Union-Intersection principle, Roy (1953) recommended the largest root statistic

$$\theta_1 = \frac{\lambda_1}{1 + \lambda_1} = 1 - \nu_1 \quad (5.60)$$

where H is rejected if $\theta_1 > \theta_{s,M,N}^{1-\alpha}$. Another trace criterion developed by Bartlett-Lawley-Hotelling (Bartlett, 1939; Hotelling, 1931, 1975; Lawley, 1938) is the trace criterion

$$T_0^2 = \nu_e \text{tr}(HE^{-1}) = \nu_e \sum_{i=1}^{s} \lambda_i = \nu_e \sum_{i=1}^{s} \left(\frac{1 - \nu_i}{\nu_i} \right) = \nu_e \sum_{i=1}^{s} \left(\frac{\theta_i}{1 - \theta_i} \right) = \nu_e U_0. \quad (5.61)$$

H is rejected if T_0^2 is larger than some constant k^* to attain a predetermined Type I error level. The statistic is also called Hotelling's generalized T_0^2. An alternative form for T_0^2 is

$$T_0^2 = (\hat{\gamma} - \gamma_0)'[(A'\Sigma A) \otimes C(X'X)^{-1}C']^{-1}(\hat{\gamma} - \gamma_0)$$
$$= \text{tr}[(C\widehat{B}A - \Gamma)'(C(X'X)^{-1}C')^{-1}(C\widehat{B}A - \Gamma)(A'\widehat{\Sigma}A)^{-1}] \quad (5.62)$$

where $\widehat{\Sigma} = E/n$ is the ML estimator of Σ which is identical to the LR statistic given in (5.34) with $\Sigma = \widehat{\Sigma}$. To derive (5.62), observe that the

$$\text{vec}(ABC) = (C' \otimes A)\text{vec}(B) \quad (5.63)$$
$$\text{tr}(AZ'BZC) = \text{tr}(Z'BZCA)$$
$$= [\text{vec}(Z)]'(CA \otimes B')[\text{vec}(Z)]$$
$$= [\text{vec}(Z)]'(A'C' \otimes B)[\text{vec}(Z)]. \quad (5.64)$$

Letting $\text{vec}(Z) = \hat{\gamma} - \gamma_0$, $B = (C(X'X)^{-1}C')^{-1}$, $A = I$ and $C = (A'\widehat{\Sigma}A)^{-1}$ in (5.64), the result follows.

Having rejected an overall test of $H : CBA = \Gamma$, one may obtain $1 - \alpha$ simultaneous confidence intervals for bilinear parametric functions of the form $\psi = c'Ba$ by evaluating the expression

$$\hat{\psi} - c_0 \hat{\sigma}_{\hat{\psi}} \leq \psi \leq \hat{\psi} + c_0 \hat{\sigma}_{\hat{\psi}} \tag{5.65}$$

where $\hat{\psi} = c'Ba$, $\hat{\sigma}_{\hat{\psi}} = (a'Sa)(c'(X'X)^{-1}c)$, $\mathcal{E}(S) = \Sigma$ and c_0 is defined as

$$c_0^2 = \nu_e \left(\frac{1 - U^{1-\alpha}}{U^{1-\alpha}} \right) \qquad \text{Wilks} \tag{5.66}$$

$$c_0^2 = \nu_e \left(\frac{\theta^{1-\alpha}}{1 - \theta^{1-\alpha}} \right) \qquad \text{Roy} \tag{5.67}$$

$$c_0^2 = U_0^{1-\alpha} = \frac{T_{0;1-\alpha}^2}{\nu_e} \qquad \text{Bartlett-Lawley-Hotelling} \tag{5.68}$$

$$c_0^2 = \nu_e \left(\frac{V^{1-\alpha}}{1 - V^{1-\alpha}} \right) \qquad \text{Bartlett-Nanda-Pillai.} \tag{5.69}$$

The value $U^{1-\alpha}, \theta^{1-\alpha}, U_0^{1-\alpha}, V^{1-\alpha}$ correspond to tabled upper $(1-\alpha)100\%$ critical values for the test statistics for testing $H : CBA = \Gamma$. Tables are provided in Timm (1975, 1993b). Alternatively, one may also use F approximations for these statistics (see for example, Timm, 2002, p. 103; Rencher, 1998, p. 126-131). When the null hypothesis is true, all four test statistics maintain the size of the test at the nominal level α.

The power of the multivariate test for a specific noncentrality matrix in (5.43) is readily calculated using F distribution approximations to the noncentral Wishart distribution (Muller, LaVange, Ramey, & Ramey, 1992).

5.4 MULTIVARIATE REGRESSION

Multivariate linear regression analysis is not widely used in practice. Instead most researchers use multiple linear regression procedures one dependent variable at a time. However, simulation studies by Breiman and Friedman (1997) suggest that this practice is ill advised. If the primary goal of a study is prediction accuracy, they recommend the use of multivariate linear regression analysis. Multivariate procedures take into account the correlations among the dependent variables ignored by univariate analysis. This is also the case for multivariate procedures used to select the best subset of variables when building multivariate models.

Multivariate linear regression procedures are used in practice to explain variation in a vector of dependent (criterion) variables by employing a set of independent (predictor) variables using observational (nonexperimental) data. In these settings, the goal of the researcher is usually to select a subset or all of the predictors that accounts for the variation in the dependent variables. The theory associated with a discipline is usually used to define the initial set of predictor variables. In such studies, the goal is to discover the relationship between the dependent variables and the "best" subset of predictor variables. Multivariate linear regression procedures are also used with experimental data. In these situations, the regression coefficients are employed to evaluate the marginal or partial effect of a predictor on the dependent variable given the other predictor variables in the model. In both of these situations one is usually concerned with estimation of the model parameters, model specification and variable selection or in general model calibration and sample model fit.

Regression models are also developed to predict some outcome variable such as job performance. In this situation, predictor variables are selected to maximize the predictive power of the linear model. Here one is concerned with measures of multivariate association, population model fit, predictive accuracy, and variable selection that reduces the mean square error of prediction.

Regression methods may be applied using random dependent variables and several fixed predictors making few assumptions regarding the random errors: (i) the classical multivariate linear regression (CMR) model. Or, the method is applied for fixed predictors, assuming that the random errors have normal structure: (ii) the classical, normal multivariate linear regression (CNMR) model. In neither of these models is the set of predictors random, they remain fixed under repeated stratified sampling of the dependent variables. If the independent variables and the set of predictors are obtained from a multivariate population as a random sample of independent and identically distributed jointly multivariate normal observations, we have: (iii) the jointly normal multivariate linear (JNMR) model. Another model closely related to the JNMR model is the (iv) random, classical multivariate normal (RCMR) model. For the RCMR model, the conditional distribution of the dependent variables is assumed to be multivariate normal and the marginal distribution of the variables is unknown.

Given the design of the study and the associated model assumptions, we review model selection and model fit criteria, measures of multivariate association, and prediction accuracy in multivariate regression analysis to facilitate the transition from univariate analysis to multivariate analysis.

5.5　CLASSICAL AND NORMAL MULTIVARIATE LINEAR REGRESSION MODELS

Estimation

The CMR model is used to explore the relationship between q dependent variables $y_1, y_2, \ldots, y_q$ and p fixed independent variables $x_1, x_2, \ldots, x_p$. When $q = 1$, the CMR model reduces to the classical multiple linear regression (CR) model. The linear model for the ith variable is $y_i = \beta_{0i} + \beta_{1i}x_1 + \cdots + \beta_{pi}x_p + e_i$ for $i = 1, 2, \ldots, q$. Letting $Y = (y_1, y_2, \ldots, y_q)$ represent a $n \times q$ matrix of observations, $B = (\beta_1, \beta_2, \ldots, \beta_q)$ represent the $k \times q$ matrix of unknown parameters where each column β_i is associated with the coefficients for the ith dependent variable, X the matrix of fixed independent variables, and $U = (e_1, e_2, \ldots, e_q)$ the $n \times q$ matrix of random errors, the matrix expression for the CMR model is

$$\Omega_0 : Y_{n \times q} = X_{n \times k} B_{k \times q} + U_{n \times q} \tag{5.70}$$

where k is the rank of the matrix X and $k = p + 1$ for p variables. While the design matrix may be of less than full rank in designed experiments, we assume that one may always reparametrize the model to full rank. As we shall see later, the rank index k will be allowed to vary as the number of variables in a model change. The error matrix U is assumed to have mean 0, $\mathcal{E}(U) = 0$, and the covariance of the random matrix U is defined as $\text{cov}(U) = \text{cov}[\text{vec}(U)] = \Omega_{nq \times nq} = \Sigma_{q \times q} \otimes I_n = [\Omega_{ij}] = [\text{cov}(e_i, e_j)]$.

From (5.11), the OLSE of the matrix B is

$$\widehat{B} = (X'X)^{-1} X'Y \tag{5.71}$$

where the ith column is $\beta_i = (X'X)^{-1} X'y_i$ for $i = 1, 2, \ldots, q$. Assuming the matrix Y follows a matrix normal distribution, the OLSE is the ML estimate B. Using the properties of expectations of matrix quadratic forms, Seber (1984, p. 7), the mean squared error of the estimator $\widehat{B}$ is the risk matrix

$$\mathcal{E}[(B - \widehat{B})'(B - \widehat{B})] = \Sigma \cdot \text{tr}[(X'X)^{-1}]. \tag{5.72}$$

For $q = 1$, the estimate reduces to the univariate result with Σ replaced by σ^2.

For the multivariate linear model, the matrix $\widehat{Y} = X\widehat{B}$ is a matrix of fitted values and the matrix $\widehat{U} = Y - \widehat{Y}$ is a matrix of estimated errors or residuals. Having fit the CMR model, one may desire a measure of sample fit. However, because B is a matrix and not a scalar, a natural measure does not exist. There are many ways of reducing a matrix to a scalar. Before discussing model fit for the CMR model, we consider model selection criteria. To develop model selection criteria, an unbiased estimate of the unknown covariance matrix Ω which contains the unknown symmetric error covariance matrix Σ is needed. Given the OLSE of B, an unbiased estimate of the common covariance matrix Σ is

$$\begin{aligned} S &= \frac{(Y - X\widehat{B})'(Y - X\widehat{B})}{n - k} \\ &= \frac{Y'Y - \widehat{B}'X'Y}{n - k} \\ &= \frac{Y'(I - X(X'X)^{-1}X')Y}{n - k} \\ &= \frac{SSE}{n - k} = \frac{E}{n - k} \\ &= MSE. \end{aligned} \tag{5.73}$$

In general, the matrix S is nonsingular with probability one for independent observations, which are not necessarily normal or identically distributed. For the CNMR model, the ML estimate of Σ is

$$\widehat{\Sigma} = \frac{SSE}{n} = \frac{E}{n}. \tag{5.74}$$

As in univariate statistics, given $\widehat{\Sigma}$ one obtains the minimum variance unbiased estimate for the covariance matrix using the relation ${(n\widehat{\Sigma})}/{(n-k)} = S$. This replaces the biased ML estimate of the error covariance matrix with the unbiased estimator $S = (s_{ij})$ where $s_{ij} = {(n\hat{\sigma}_{ij})}/{(n-k)}$. The matrix of errors for each row of the random matrix U and each row of the matrix Y shares the common error covariance matrix Σ. Finally, under normality, the joint distribution of $\widehat{B}$ and S is

$$\widehat{B} \sim N_{k,q}(B, \Sigma \otimes (X'X)^{-1}) \tag{5.75}$$

$$(n-k)S \sim W_q(n-k, \Sigma), \tag{5.76}$$

where $\nu = n - k \geq q$, Ghosh and Sinha (2002).

Model Selection

Having fit a multivariate regression model with fixed X, the matrix X may represent an over fitted model. We let X^t represent the true model and X^u an under fitted model with the nesting condition that $X^u \subseteq X^t \subseteq X$ and corresponding parameter matrices B^U, B^t, and B, respectively. Assuming $Y = XB + U$ for n new observations, the average mean squared error of prediction for fixed X is the matrix

$$\Delta_f^2 = \frac{\mathcal{E}[(Y - \widehat{Y})'(Y - \widehat{Y})]}{n}$$
$$= \Sigma \cdot \left(1 + \frac{k}{n}\right). \tag{5.77}$$

The quantity Δ_f^2 is called the multivariate final prediction error (MFPE) , a parameter to be estimated. As one varies the number of variables $p_j = k_j - 1$ in the multivariate model where $k_j = 1, 2, \ldots, k$ and $j = 1, 2, \ldots, (2^p - 1)$, changes in the MFPE matrix measure the balance between the dispersion of changes U and $X\widehat{B}$. However, unlike the univariate case, the MFPE is a matrix and not a scalar parameter. The unbiased and ML estimates of MFPE are, respectively,

$$\widehat{\Delta}_u^2 = S_j \left(1 + \frac{k_j}{n}\right) \tag{5.78}$$

$$\widehat{\Delta}_{MLE}^2 = \widehat{\Sigma}_j \left(\frac{n + k_j}{n - k_j}\right). \tag{5.79}$$

Following Schott (1997, p. 403) or Bilodeau and Brenner (1999, p. 76), and given multivariate normality, the variance of the unbiased estimator $\widehat{\Delta}_u^2$ is

$$\mathrm{var}(\widehat{\Delta}_u^2) = \left(\frac{(I_{q^2} + k)(\Sigma \otimes \Sigma)}{n - k_j}\right) \left(1 + \frac{2k_j}{n} + \frac{k_j^2}{n^2}\right). \tag{5.80}$$

To use the MFPE as a criterion for model selection, the symmetric covariance matrix
or its associated dispersion matrix must be transformed to a scalar value. Given any
$q \times q$ symmetric matrix A, one can show that there are many functions of A that
map a matrix A to a scalar. However, several functions are represented in terms of
the eigenvalues λ_i for $i = 1, 2, \ldots, q$ of the characteristic equation $|A - \lambda I| = 0$
for a symmetric matrix A, Harville (1997, Chapter 13). Common functions useful in
statistics are the determinant of A $|A| = \prod_{i=1}^{q} \lambda_i$, the trace of the matrix A, $\text{tr}(A) =$
$\sum_{i=1}^{q} \lambda_i$, where the roots are ordered: $\lambda_1 \geq \lambda_2 \geq \cdots \geq \lambda_q$, the largest root of A, λ_1,
and the (Frobenius or Euclidean) norm squared of A, $\|A\|^2 = \text{tr}(A'A) = \sum_{i=1}^{q} \zeta_i^2$,
where ζ_i represent the roots of $A'A$, a special case of the more general L_p-norm. In
statistics, the determinant of the covariance matrix is called the generalized variance
(Wilks, 1932). Thus, using the determinant function, the MFPE determinant criterion
is

$$MFPE_j = |S_j| \left(\frac{n+k_j}{n} \right)^q = |\widehat{\Sigma}_j| \left(\frac{(n+k)j}{n-k_j} \right)^q = |E_j| \left(\frac{n+k_j}{n(n-k_j)} \right)^q.$$
(5.81)

The best candidate models are selected that minimize the MFPE criterion. Although
the $|S_j|$ is not an unbiased estimator of $|\Sigma_j|$, the unbiased estimator is $|\tilde{S}_j| =$
$|E_j|/\prod_{i=1}^{q}(n - k_j - q + i)$, Muirhead (1982, p. 100), we continue to replace
Σ_j by S_j. One may of course use the other function of the matrix S_j, however,
since the matrix E_j has a known Wishart distribution for the CNMR model, the de-
terminant and $|E_j|/|\Sigma|$ is related to the product of independent χ^2 random variables
$\left(\prod_{i=1}^{q} \chi_{(n-k_j-q+i)}^2 \right)$. Under multivariate normality, the distribution of the $\text{tr}(E_j)$ is
a weighted sum of χ^2 random variables $\left(\sum_{i=1}^{q} \lambda_i \chi_{(n-k_j)}^2 \right)$ where the weights λ_i are
the eigenvalues of the covariance matrix Σ, (Chattogpadhyay & Pillai, 1971). The
mean of the trace of E_j is $\mathcal{E}[\text{tr}(E_j)] = (n - k_j)\text{tr}(\Sigma)$. Because distributions of the
other matrix functions are less understood, the determinant and trace functions are
most often used in practice. Sparks, Coutsourides, Troskie, and Sugiura (1983) generalized Mallows' (1973, 1995) C_p criterion to the multivariate model. The criterion
is

$$C_p = (n - k)\widehat{\Sigma}^{-1}\widehat{\Sigma}_j + (2k_j - n)I_q.$$
(5.82)

Again, we may use the determinant, trace, norm squared, or largest root functions,
among others to reduce the criterion to a scalar. Because $2k_j - n$ may be negative,
Sparks et al. (1983) recommended a modification to the $|C_p|$. Instead of using this
determinant, they suggest using the $|E^{-1}E_j|$. When the bias is zero, the quantity
$|E^{-1}E_j| = \left(\frac{n-k_j}{n-k} \right)^q$ is always positive. Using the most popular scalar functions,

candidate models may be selected based upon the criteria

$$\text{determinant} : \min |E^{-1}E_j| \leq \left(\frac{n - k_j}{n - k}\right)^q, \tag{5.83}$$

$$\text{trace} : \min \text{tr}(C_p) \leq k_j p, \tag{5.84}$$

$$\text{root} : \text{largest root of } C_p, \text{ and} \tag{5.85}$$

$$\text{norm squared} : \min \|C_p' C_p\|^2 = \text{tr}(C_p' C_p). \tag{5.86}$$

These criteria may be applied to CMR or CNMR models to facilitate locating the best candidate model. Monte Carlo studies performed by McQuarrie and Tsai (1998) suggest that the determinant criterion out performs the others.

We now consider selection criteria that require the model errors $\text{vec}(E')$ to be multivariate normal, the CMNR model. To generalize the AIC criterion developed by Akaike (1969, 1973, 1974, 1978) and Davisson (1965) to the CNMR model, one evaluates AIC=-2log(likelihood)+2(number of parameters) where the likelihood is that of the multivariate normal density. Then the AIC criterion becomes

$$AIC = n \log(|\Sigma|) + 2d \tag{5.87}$$

where $d = (k_j q) + q(q + 1)/2$ is the number of total parameters to be estimated: $k_j q$ in B and $q(q + 1)/2$ unknown parameters in the error covariance matrix Σ. Scaling the AIC criterion by $1/n$ to express it as a rate and substituting a MLE or an unbiased estimate for the population covariance matrix, the AIC are represented as

$$AIC_{MLE} = \log(|\widehat{\Sigma}_j|) + \frac{2d}{n} = \log(|\widehat{\Sigma}_j|) + \frac{2k_j q + q(q + 1)}{n}, \tag{5.88}$$

$$AIC_u = \log(|S_j|) + \frac{2d}{n} = \log(|S_j|) + \frac{[2k_j q + q(q + 1)]}{n}. \tag{5.89}$$

Models with small values for the criterion are potential candidate models. Akaike's AIC criterion tends to overfit when there are too many parameters relative to the sample size n, $(n/d < 40)$. For this situation, Fujikoshi and Satoh (1997, eq. 7) modified Sugiura (1978) correction for the AIC (CAIC) criterion used in multiple linear regression models. Following McQuarrie and Tsai (1998, p. 148), the multivariate CAIC criteria are

$$CAIC_j = \log(|\Sigma_j|) + \frac{2d}{n - k_j - q - 1}, \tag{5.90}$$

$$CAIC_u = \log(|S_j|) + \frac{2d}{n - k_j - q - 1}. \tag{5.91}$$

The criterion $CAIC_u$ is doubly corrected in that it includes a modified penalty function and an unbiased estimate for the covariance matrix. The penalty factor is $n/(n - k_j - q - 1)$. For large sample sizes the AIC and $CAIC_j$ are equivalent.

Replacing the penalty factor $2d$ of the AIC criterion with the penalty factor $d \log(n)$ and $2d \log[\log(n)]$, respectively, extension of the information criteria BIC

proposed by Schwarz (1978) and the HQ information criterion developed by Hannan and Quinn (1979) is defined for the CNMR model as

$$BIC_j = n \log(|\Sigma_j|) + d \log(n), \qquad (5.92)$$

$$HQ_j = n \log(|\Sigma_j|) + 2d \log[\log(n)], \qquad (5.93)$$

where we may substitute for the unknown covariance matrix Σ_j with the ML estimate or the minimum variance unbiased estimates. When unbiased estimates are substituted for the error covariance matrix, the criteria are represented as BIC_u and HQ_u, respectively.

Following the development of the $CAIC$ criterion, McQuarrie and Tsai (1998, p. 156) showed that one may also correct the HQ (CHQ) criterion. They suggested the criteria

$$CHQ_j = \log(|\Sigma_j|) + \frac{2k_j q \log[\log(n)]}{n - k_j - q - 1}, \qquad (5.94)$$

$$CHQ_u = \log(|S_j|) + \frac{2k_j q \log[\log(n)]}{n - k_j - q - 1}. \qquad (5.95)$$

A large Monte Carlo study conducted by McQuarrie and Tsai (1998, p. 402) indicated that among the parametric procedures investigated that the information criteria $CAIC_u$ and CHQ_u performed best overall. Generally speaking, criteria with stronger penalty functions tend to underfit least; however, they may do so at the expense of some under fitting when the parameters are weak. The information criteria may be used to rank order multivariate models to facilitate locating the best candidate models, Burnham and Anderson (1998, p. 63).

An alternative procedure for obtaining a best subset of the variables in X that takes into account the correlations among the dependent variables was proposed by Reinsel and Velu (1998, p. 57-64) using the multivariate reduced rank regression (RRR) model. A Monte Carlo study conducted by Al-Subaihi (2000) concluded that the RRR method performed as well as information criteria in most situations. A major advantage of their procedure is that it directly takes into account inter-correlations among the dependent variables which is only indirectly accounted for by information criteria. However, two major problems with the method are that a test of significance is required to evaluate the rank of the unknown "reduced rank" parameter matrix B and in order to apply the RRR variable selection method the rank of the parameter matrix must be greater than zero. The effect of setting the rank to one when it is really zero, in order to apply the method, is unknown.

While many stepwise methods have also been suggested for the analysis of multivariate models, they tend to exclude many models for further study and do not control the experimentwise error rate, Mantel (1970). Hence, we have excluded the procedures in this text. Other all-possible multivariate regression and bootstrap techniques are discussed by McQuarrie and Tsai (1998).

Model Selection, Likelihood Ratio Tests

For CNMR models, one may also test the hypothesis that the last $h = p - m$ rows of the parameter matrix are zero. The test statistics depend on the roots of the characteristic equation $|H - \lambda E| = 0$ where H is the hypothesis test matrix and E is the error matrix. To develop the simple matrix expressions, the CNMR model is partitioned to form two sub-models, the reduced model and the component to be tested. The CNMR model is represented as

$$\Omega_0 : Y_{n \times q} = X_{n \times k} B_{k \times q} + U_{n \times q}$$
$$= X_r B_r + X_h B_h + U_{n \times q} \tag{5.96}$$

where $X_r B_r$ is the reduced component and $X_h B_h$ is the hypothesis test component. The matrices X_r and B_h have h columns and h rows, respectively. To test the null hypothesis $H : B_h = 0$ versus the alternative $H_A : B_h \neq 0$, we let $\widehat{B} = (X'x)^{-1}X'Y$ represent the ML estimate of B for the full model, and $\widehat{B}_r = (X_r'X_r)^{-1}X_r'Y$ the ML estimate for the reduced model. Then, the hypothesis and error matrices for the test are, respectively,

$$E = Y'Y - \widehat{B}'X'Y \qquad\qquad H = \widehat{B}'X'Y - \widehat{B}_r'X_r'Y. \tag{5.97}$$

One may use Wilks' likelihood ratio test, represented as Λ, the union-intersection test developed by Roy, represented as $\theta^{(s)}$, or tests using the trace criterion developed by Bartlett-Lawley-Hotelling, represented as $V^{(s)}$, or the Bartlett-Nanda-Pillai criterion, represented as $U_0^{(s)}$. Letting $\nu_h = h$, $\nu_e = n - p - 1$, $s = \min(\nu_h, q)$, $M = (|\nu_h - q| - 1)/2$, $N = (\nu_e - q - 1)/2$, the test statistic becomes

$$\Lambda = \frac{|E|}{|H + E|} = \frac{|Y'Y - \widehat{B}'X'Y|}{|Y'Y - \widehat{B}_r'X_r'Y|}$$

$$= \left(\frac{|Y'Y - \widehat{B}'X'Y|}{|Y'Y - n\bar{y}\bar{y}'|} \right) \left(\frac{|Y'Y - n\bar{y}\bar{y}'|}{|Y'Y - \widehat{B}_r'X_r'Y|} \right) = \Lambda_F \left(\frac{1}{\Lambda_r} \right)$$

$$= \prod_{i=1}^{s}(1 + \lambda_i)^{-1} \sim U_{q,h,\nu_e} \tag{5.98}$$

$$V^s = \text{tr}[H(E + H)^{-1}]$$

$$= \sum_{i=1}^{s} \frac{\lambda_i}{1 + \lambda_i} \sim V_{s,M,N} \tag{5.99}$$

$$T_0^2 = \nu_e \text{tr}(HE^{-1})$$

$$= \sum_{i=1}^{s} \lambda_i \sim \nu_e U_{0(s,M,N)} \tag{5.100}$$

$$\theta^{(s)} = \frac{\lambda_1}{1 + \lambda_1} \sim \theta_{s,M,N}. \tag{5.101}$$

Again one may also use F approximations to test the overall hypothesis. When the null hypothesis is true, all four test statistics maintain the size of the test at the nominal level α.

Since $h = p - m$, setting $m = 0$ tests the hypothesis that all the regression coefficients associated with all the predictors in the model are zero. Setting $h = 1$, the contribution of the ith variable may be evaluated. For this situation $\nu_h = 1$, $s = 1$, and there is only one nonzero eigenvalue, λ_1. In this case, all four test statistics are functions of each other and give identical results. The marginal contribution of the ith variable can be evaluated by using the partial F statistic

$$F = \left(\frac{\nu_e - q - 1}{q} \right) \lambda_1 \sim F(q, \nu_e - q - 1). \qquad (5.102)$$

Rejecting the overall test that all regression coefficients associated with the p independent variables in the multivariate linear model are simultaneously zero at some nominal level α, Rencher and Scott (1990) showed that if the p partial F tests are performed at the same nominal α-level as the overall test that the familywise error rate remains near nominal level. Given that the overall test is rejected, these tests are called protected partial F tests.

If the P-value of a protected F test is larger than α one may consider the variable a candidate for removal from the model, however, as in univariate analysis the procedure must be used with extreme caution since each test is only a marginal test considering the "additional" effect of a variable given that all other variables are included in the model. As a model selection procedure, it tends to exclude too many variables from the set of potential variables.

Model Fit

Given that one has located a potential candidate model using the sample data, one may want an overall measure of model fit using the estimates obtained from the calibration sample for either the CMR or the CMNR model. The proportion of the variation about the sample mean vector can be accounted for by the multivariate model. As a generalization of Fisher's correlation-ratio, Wilks (1932) proposed the η^2-like goodness of fit statistic for a multivariate model as

$$\begin{aligned} \tilde{\eta}_W^2 &= \frac{|\widehat{Y}'\widehat{Y} - n\bar{y}\bar{y}'|}{|Y'Y - n\bar{y}\bar{y}'|} = \frac{|H_j|}{|T|} \\ &= 1 - \frac{|E_j|}{|T|} = 1 - \frac{|E_j|}{|E_j + H_j|} = 1 - \Lambda_W. \end{aligned} \qquad (5.103)$$

This is a multivariate extension of the analysis of the matrix variation about the mean vector that the total sum of squares and cross products matrix about the mean (T) minus the error sum of squares and cross products matrix (E_j) is equal to the between sum of squares matrix, $H_j = \|\widehat{Y}'\widehat{Y} - n\bar{y}\bar{y}'\|^2$. Thus, $T = H_j + E_j$ for fixed X where k_j is the rank of the matrix X. Relating Wilks' criterion to the eigenvalues of the determinant equation $|H_j - \lambda E_j| = 0$, Wilks' multivariate η^2-like correlation ratio becomes for $\tilde{r}_i^2 = \lambda_i/(1 + \lambda_i)$ and $s = \min(p, q)$ is

$$
\begin{aligned}
\tilde{\eta}_W^2 &= 1 - \prod_{i=1}^{s} \left(\frac{1}{1 + \lambda_i} \right) = 1 - \prod_{i=1}^{s} (1 - \tilde{r}_i^2) = 1 - \Lambda_W \\
&= 1 - \frac{|Y'Y - \widehat{B}'X'Y|}{|Y'Y - n\bar{y}\bar{y}'|}.
\end{aligned}
\tag{5.104}
$$

The parameters $\tilde{r}_i^2$ are not estimates of the square of the population canonical correlation coefficients since the matrix X is fixed. Often the "sample-like" covariance matrix for the vector, $z'_{(q+p)} = (y', x')$ where y is random and x is fixed, is partitioned as

$$
\tilde{\Sigma}_{(q+p)\times(q+p)} = \begin{pmatrix} \tilde{\Sigma}_{yy} & \tilde{\Sigma}_{yx} \\ \tilde{\Sigma}_{xy} & \tilde{\Sigma}_{xx} \end{pmatrix},
\tag{5.105}
$$

then $\tilde{\eta}_W^2$ may be expressed as a "sample-like" covariance formula

$$
\begin{aligned}
\tilde{\eta}_W^2 &= 1 - \frac{|Y'Y - \widehat{B}'X'Y|}{|Y'Y - n\bar{y}\bar{y}'|} = 1 - \frac{|\tilde{\Sigma}|}{|\tilde{\Sigma}_{yy}||\tilde{\Sigma}_{xx}|} \\
&= 1 - \frac{|\tilde{\Sigma}_{yy} - \tilde{\Sigma}_{yx}\tilde{\Sigma}_{xx}^{-1}\tilde{\Sigma}_{xy}|}{|\tilde{\Sigma}_{yy}|} = 1 - \Lambda_W
\end{aligned}
\tag{5.106}
$$

since the $|\tilde{\Sigma}| = |\tilde{\Sigma}_{xx}||\tilde{\Sigma}_{yy} - \tilde{\Sigma}_{yx}\tilde{\Sigma}_{xx}^{-1}\tilde{\Sigma}_{xy}|$. While $\tilde{\eta}_W^2$ may be expressed as a "sample-like" formula, the matrices are not estimates of population covariance matrices because the matrix X is fixed and not random. In particular, because the matrix X is fixed and not random for either the CMR model and the CNMR, observe that the total matrix $T = (Y'Y - n\bar{y}\bar{y}')$ divided by $(n-1)$ is not an unbiased estimate of Σ_{yy}. Following Goldberger (1991, p.179-180), the $\mathcal{E}(T)/(n-1) = \Sigma + B'_\ell(X'_\ell P_1 X_\ell)B_\ell/(n-1)$ for $X = (1, X_\ell)$ where 1 is a vector of n 1's and $P_1 = I - 1(1'1)^{-1}1'$ is a projection matrix. The total matrix divided by $(n-1)$ is only an estimate of Σ if $B_\ell = 0$ in the population.

Substituting mean square matrices for the sum of square matrices in (5.104), Jobson (1991, p. 218) proposed the less biased η^2-like correlation ration index of fit

$$
\tilde{\eta}_J^2 = 1 - \frac{n\Lambda_W/(n-k)}{1 + \Lambda_W/(n-k)}.
\tag{5.107}
$$

Using standard test statistics in MANOVA, due to Bartlett-Lawley-Hotelling, Roy, and Bartlett-Nanda-Pillai, Muller and Peterson (1984) defined model fit criteria for the CMR and CNMR models using the roots λ_i of the matrix $|H_j - \lambda E_j| = 0$ where

$s = \min(p, q)$. Their criteria are defined by

$$\tilde{\eta}^2_{B-L-H} = \frac{\sum_{i=1}^{s} \frac{\lambda_i}{s}}{1 + \sum_{i=1}^{s} \frac{\lambda_i}{s}} = \frac{\sum_{i=1}^{s} \left(\frac{\tilde{r}_i^2}{1-\tilde{r}_i^2}\right)\left(\frac{1}{s}\right)}{1 + \sum_{i=1}^{s} \left(\frac{\tilde{r}_i^2}{1-\tilde{r}_i^2}\right)\left(\frac{1}{s}\right)} \tag{5.108}$$

$$\tilde{\eta}^2_R = \left(\frac{\lambda_1}{1+\lambda_1}\right) = \tilde{r}_1^2 \tag{5.109}$$

$$\tilde{\eta}^2_{B-N-P} = \sum_{i=1}^{s} \frac{\frac{\lambda_i}{1+\lambda_i}}{s} = \sum_{i=1}^{s} \frac{\tilde{r}_i^2}{s}. \tag{5.110}$$

Other adjustments are suggested by Huberty (1994) and Tatsuoka and Lohnes (1988).

Cramer and Nicewander (1979) proposed using one minus the geometric mean of Λ_W; Coxhead (1974) proposed one minus the harmonic mean of the elements in Λ_W, and we may, for example, also use the matrix norm. These measures of sample multivariate fit are

$$\tilde{\eta}^2_G = 1 - \left(\prod_{i=1}^{s}(1 - \tilde{r}_i^2)\right)^{1/s} = 1 - \Lambda^{1/s} \tag{5.111}$$

$$\tilde{\eta}^2_H = 1 - \left(\frac{s}{\sum_{i=1}^{s} \frac{1}{1-\tilde{r}_i^2}}\right) = 1 - \frac{1}{\text{tr}[H_j^{-1}(H_j + E_j)]/s} \tag{5.112}$$

$$\tilde{\eta}^2_N = 1 - \frac{\|E_j\|^2}{\|T\|^2} = 1 - \frac{\text{tr}(E_j^2)}{\text{tr}(T^2)}. \tag{5.113}$$

Thus, for the CMR model we have several alternative criteria of sample model fit from the data since there are many ways of transforming the matrices E_j and T to a scalar.

Wilks' R^2-like multivariate coefficient of determination for CMR models may be defined as

$$\tilde{R}^2_W = \frac{|\tilde{\Sigma}_{yx}\tilde{\Sigma}_{xx}^{-1}\tilde{\Sigma}_{xy}|}{|\tilde{\Sigma}_{yy}|} \neq 1 - \frac{|\tilde{\Sigma}_{yy} - \tilde{\Sigma}_{yx}\tilde{\Sigma}_{xx}^{-1}\tilde{\Sigma}_{xy}|}{|\tilde{\Sigma}_{yy}|} = 1 - \Lambda_W = \tilde{\eta}^2_W \tag{5.114}$$

where for a single dependent variable $\tilde{R}^2_W = R^2 = \tilde{\eta}^2_W$. For the multivariate model, when $q \neq 1$, $\tilde{R}^2_W \neq \tilde{\eta}^2_W$ since we are using determinants. The coefficient $\tilde{R}^2_W$ is not a measure of multivariate model fit.

For the CMR and CNMR models one may calculate any number of measures of sample model fit and hence also employ them to locate a candidate model. However, to convert any of the sample fit measures into measures of association, the sampling scheme must be performed in such a manner that the observations (y_i', x_i') are independently and identically distributed to form the random pair (Y, X). The matrix X must be random or stochastic. Or, one may use the RCMR model where the conditional distribution of $\tilde{y} = \text{vec}(Y|x)$ is multivariate normal.

Additional Comments

Power for the multivariate tests for evaluating nested models has been extensively studied for fixed X where it has been shown that power depends on the configuration of the population eigenvalues. No test is uniformly most powerful; however, the asymptotic power ranking of the tests when at least two eigenvalues in the population are nonzero is as follows: $V^{(s)} \geq \Lambda \geq U_0^{(s)} \geq \theta_{(s)}$, Breusch (1979). If only one population eigenvalue is nonzero, the inequalities of the test statistics are reversed. Using the noncentral F distribution, approximate power calculations for the tests and procedures for calculating approximate sample sizes have been developed by Muller et al. (1992). While Robinson (1974) developed the asymptotic theory for estimates of the coefficients in the CNMR model, Sampson (1974) has developed the general theory for calculating power with a random matrix X. Glueck and Muller (2003) showed how to evaluate power when a design matrix is random, fixed or both; however, they did not evaluate the power of the multivariate criteria for a random design matrix. Any definition of effect size in multivariate regression depends upon the multivariate test statistic employed in the analysis, Muller and Peterson (1984). Unique to multivariate models are methods of variable selection that are based upon simultaneous test procedures. Several methods are reviewed by McKay (1977). Schmidhammer (1982) discussed a finite intersection test (FIT) procedure. Most of the test procedures applied to the multivariate models require nested models; however, the procedure developed by Timm and Al-Subaihi (2001) may be used for nonnested multivariate models with fixed X.

5.6 JOINTLY MULTIVARIATE NORMAL REGRESSION MODEL

Estimation and Model Fit

Given that the pair of matrices (Y, X) are random and assuming the sample of observation vectors (y_i', x_i') for $i = 1, 2, \ldots, n$ where $y_i' = (y_1, y_2, \ldots, y_q)$ and $x_i' = (x_1, x_2, \ldots, x_p)$ are jointly multivariate normal with mean vector $\mu' = (\mu_y', \mu_x')$ and nonsingular covariance matrix

$$\Sigma_{(q+p) \times (q+p)} = \begin{pmatrix} \Sigma_{yy} & \Sigma_{yx} \\ \Sigma_{xy} & \Sigma_{xx} \end{pmatrix}, \tag{5.115}$$

we desire normalized linear functions $U = \alpha' y$ and $V = \beta' x$ of unit variance such that the correlation between U and V is maximum for $s = \min(q, p)$. Letting V_i and W_i represent the normalized eigenvector of the determinant equations,

$$|\Sigma_{yx} \Sigma_{xx}^{-1} \Sigma_{xy} - \rho^2 \Sigma_{yy}| = 0 \qquad |\Sigma_{xy} \Sigma_{yy}^{-1} \Sigma_{yx} - \rho^2 \Sigma_{xx}| = 0 \tag{5.116}$$

respectively, with eigenvalues $\rho_1^2 \geq \rho_2^2 \geq \cdots \geq \rho_s^2 \geq 0$. The canonical variates U_i and V_i are the linear combinations with maximum correlation ρ_i called the canonical correlation between the canonical variates. Thus, the square of the canonical correlations is the eigenvalues of the matrices

$$C_{xy} = \Sigma_{yy}^{-1/2}\Sigma_{yx}\Sigma_{xx}^{-1}\Sigma_{xy}\Sigma_{yy}^{-1/2} \quad C_{yx} = \Sigma_{xx}^{-1/2}\Sigma_{xy}\Sigma_{yy}^{-1}\Sigma_{yx}\Sigma_{xx}^{-1/2} \quad (5.117)$$

provided the matrices Σ_{yy} and Σ_{xx} are nonsingular. Solving the determinant equation $|\Sigma_{yy} - \gamma I| = 0$, there exists an orthogonal matrix O, $(O'O = I)$, such that $O'\Sigma_{yy}O = \Gamma$. Then, the matrix $\Sigma_{yy} = (O\Gamma^{1/2}O')(O\Gamma^{1/2}O') = \Sigma_{yy}^{1/2}\Sigma_{yy}^{1/2}$. So, the square root matrix $\Sigma_{yy}^{-1/2} = (O\Gamma^{1/2}O')^{-1} = O\Gamma^{-1/2}O'$. The matrix $\Sigma_{yy}^{1/2}$ is called the square root matrix of the matrix Σ_{yy}. The matrix $\Sigma_{xx}^{-1/2}$ may be similarly constructed. The determinant

$$|C_{yx}| = |C_{xy}| = \prod_{i=1}^{s} \rho_i^2 \qquad (5.118)$$

is called the generalized alienation coefficient, Hooper (1959). Or, since canonical correlations are invariant to changes in scales, the population covariance matrices may be replaced by population correlation matrices P_{ij} where the indexes i and j assume the values x and/or y. If all vectors in the space spanned by X are orthogonal to the vectors spanned by Y, then $\Sigma_{yy} = 0$ and the determinant $|C_{yx}|$ is zero. If $y = Ax$, then $C_{xy} = I$ and the $|C_{xy}| = 1$. Or if any ρ_i^2 is zero, the $|C_{yx}|$ is zero and if all canonical correlations are one, then the quantity is one. Thus, the generalized alienation coefficient is an overall measure of nonassociation. This led Hotelling (1936) to propose the quantity $\prod_{i=1}^{s}(1 - \rho_i^2)$, which he called the vector alienation coefficient.

Since one minus a quantity not explained must be explained and since the eigenvalues of $(I_q - A)$ are equal to $(1 - \delta_i)$ where δ_i is an eigenvalue of the matrix A, Rozeboom (1965) defined a measure of multivariate association which he called the multivariate correlation ratio defined as

$$\eta_{yx}^2 = 1 - \prod_{i=1}^{s}(1 - \rho_i)^2 \qquad (5.119)$$

$$= 1 - |I_q - \Sigma_{yy}^{-1/2}\Sigma_{yx}\Sigma_{xx}^{-1}\Sigma_{xy}\Sigma_{yy}^{-1/2}| \qquad (5.120)$$

$$= 1 - \frac{|\Sigma_{yy} - \Sigma_{yx}\Sigma_{xx}^{-1}\Sigma_{xy}|}{|\Sigma_{yy}|}. \qquad (5.121)$$

It is one minus Hotelling's vector alienation coefficient. If the matrix X is fixed, the multivariate correlation ratio reduces to multivariate η^2-like correlation ratio; however, because the matrix X is random and not fixed, $\hat{\eta}_{yx}^2$, is a measure of model fit in the population and not the sample. As in the case of one dependent variable, if the relationship between each dependent variable and the set of independent variables is linear in the dependent variable, it becomes the vector correlation coefficient and we set $\eta_{yx}^2 = R_{yx}^2$. If $q = 1$, the vector correlation coefficient is the square of the multiple correlation coefficient in multiple linear regression. Following M. W. Browne (1975b), these are measures of fit in the population or measures of internal precision within the calibration sample and are not measures of predictive validity in the population. A generalization of predictive validity, a multivariate extension of ρ_{rc}^2 (the shrunken estimator of the coefficient of determination) has yet to be constructed.

Since the eigenvalues of the determinant equation $|\lambda_i(\Sigma_{yy} - \Sigma_{yx}\Sigma_{xx}^{-1}\Sigma_{xy}) - \Sigma_{yx}\Sigma_{xx}^{-1}\Sigma_{xy}| = 0$ are the same as the eigenvalues of the determinant equation $|\rho_i^2\Sigma_{yy} - \Sigma_{yx}\Sigma_{xx}^{-1}\Sigma_{xy}| = 0$. Or, $\rho_i^2 = \lambda_i/(1 + \lambda_i)$, we observe that the test of independence, $H : \Sigma_{yx} = 0$ or $H : \rho_1 = \rho_2 = \cdots = \rho_s = 0$ is equivalent to the hypothesis that the regression coefficients associated with the independent variables are simultaneously zero. Thus, $R_{yx}^2 = 1 - \prod_{i=1}^s (1 + \lambda_i)^{-1} = 1 - \prod_{i=1}^s (1 - r_i^2) = 1 - \Lambda_W$. The two sets of variables are linearly dependent if $R_{yx}^2 = 1$ and independent if $R_{yx}^2 = 0$. However, observe that the vector correlation coefficient may attain the value one if the square of the maximum canonical correlation is unity, irrespective of the other values, $\rho_2^2, \ldots, \rho_s^2$. This led Yanai (1974) to introduce a quantity he called the generalized coefficient of determination.

Instead of using the determinant of a matrix, Yanai (1974), following Hooper (1959) and Srikantan (1970), suggested using the trace criterion. His coefficient is defined as

$$D_{yx}^2 = \frac{\text{tr}(C_{yx})}{s} = \frac{\text{tr}(C_{xy})}{s} = D_{xy}^2 = \sum_{i=1}^s \frac{\rho_i^2}{s} \quad (5.122)$$

the mean of the squared canonical correlations, or the Bartlett-Nanda-Pillai test statistic for the test of independence divided by s. The coefficient was so named since it is a generalization of the square of the multiple correlation coefficient in multiple linear regression. Clearly, the quantity lies between zero and one. The primary advantage of the generalized coefficient of determination over the vector correlation coefficient is that if the maximum canonical correlation ρ_i is one, and all remaining canonical correlations are less than one, the generalized coefficient of determination will always be less than one while the vector correlation coefficient will always be equal to one. For $q = 1$, both measures of association are identical.

An alternative measure of multivariate association or population fit based upon calibration sample was proposed by Coxhead (1974) and Shaffer and Gillo (1974). Their measure is one minus the harmonic mean of $1 - \rho_i^2$. Or by rescaling the ratio in (5.116) using the matrix of partial covariances and the trace operator to relate the eigenvalues of $|\lambda_i(\Sigma_{yy} - \Sigma_{yx}\Sigma_{xx}^{-1}\Sigma_{xy}) - \Sigma_{yx}\Sigma_{xx}^{-1}\Sigma_{xy}| = 0$, $|\rho_i^2\Sigma_{yy} - \Sigma_{yx}\Sigma_{xx}^{-1}\Sigma_{xy}| = 0$, and $|\theta_i\Sigma_{yy} - (\Sigma_{yy} - \Sigma_{yx}\Sigma_{xx}^{-1}\Sigma_{xy})| = 0$, their measure is defined as

$$M_{xy}^2 = \frac{\text{tr}[(\Sigma_{yy} - \Sigma_{yx}\Sigma_{xx}^{-1}\Sigma_{xy})^{-1}\Sigma_{yx}\Sigma_{xx}^{-1}\Sigma_{xy}]}{\text{tr}[(\Sigma_{yy} - \Sigma_{yx}\Sigma_{xx}^{-1}\Sigma_{xy})^{-1}\Sigma_{yy}]}$$

$$= 1 - \frac{s}{\sum_{i=1}^s 1/(1 - \rho_i^2)} = 1 - \frac{s}{\sum_{i=1}^s \theta_i}. \quad (5.123)$$

Cramer and Nicewander (1979) showed that if ρ_i^2 are the ordered eigenvalues of C_{xy} that the several measures of association are related as

$$\rho_s^2 \le D_{xy}^2 \le M_{xy}^2 \le \rho_1^2 \le R_{xy}^2. \quad (5.124)$$

Using the norm squared function, Robert and Escoufier (1976) proposed the measure of association

$$RV^2 = \frac{\text{tr}(\Sigma_{xy}\Sigma_{yx})}{\sqrt{\|\Sigma_{xx}\|^2\|\Sigma_{yy}\|^2}} = \frac{\text{tr}(\Sigma_{xy}\Sigma_{yx})}{\sqrt{\text{tr}(\Sigma_{xx}^2)\text{tr}(\Sigma_{yy}^2)}}. \tag{5.125}$$

Their measure is the ratio of the matrix norm squared of the intra-covariances divided by the product of the matrix norm squared of the inter-covariances. Ogasawara (1999) suggested the overall symmetric composite redundancy index

$$RD^2 = \frac{1}{2}\left[\frac{\text{tr}(P_{yx}P_{xx}^{-1}P_{xy})}{q} + \frac{\text{tr}(P_{xy}P_{yy}^{-1}P_{yx})}{p}\right] \tag{5.126}$$

where the matrices P denote population correlation matrices. Each term in Ogasawara's measure is the Stewart and Love (1968) redundancy index for the set $y|x$ and $x|y$, respectively. The redundancy measures represent an average of the squared multiple correlation coefficients. Alternatively, one may construct a weighted average by using the variance of the individual variables as weights.

To obtain estimates of the measures of multivariate association, one replaces the population values with their corresponding sample estimates using either sample covariance matrices (S_{ij}), or correlation matrices (R_{ij}) where the subscripts represent the appropriate set of variables x and/or y, and for population canonical correlation their sample estimates r_i. Using the δ-method, Ogasawara (1999) has developed formula to obtain the standard errors and hence confidence intervals for several of the multivariate measures of association.

Prediction, Joint Normality

Assuming the sample observation vectors (y_i', x_i') are jointly multivariate normal, then the conditional normal distribution of $y|x$ has mean

$$\begin{aligned}\mu_{y \cdot x} &= \mu_y + \Sigma_{yx}\Sigma_{xx}^{-1}(x - \mu_x) \\ &= (\mu_y - \Sigma_{yx}\Sigma_{xx}^{-1}\mu_x) + \Sigma_{yx}\Sigma_{xx}^{-1}x \\ &= \beta_0 + B_\ell' x \end{aligned} \tag{5.127}$$

where

$$B_\ell = \Sigma_{xx}^{-1}\Sigma_{xy} \tag{5.128}$$
$$\beta_0 = \mu_y - \Sigma_{yx}\Sigma_{xx}^{-1}\mu_x \tag{5.129}$$

and the conditional covariance matrix is

$$\Sigma_{yy \cdot x} = \Sigma_{yy} - \Sigma_{yx}\Sigma_{xx}^{-1}\Sigma_{xy}. \tag{5.130}$$

An unbiased estimate of the conditional covariance matrix is given by $S_{yy \cdot x} = S_{yy} - S_{yx} S_{xx}^{-1} X_{xy}$. Given that the matrix X is random, the multivariate random mean squared error of prediction (MRPE) is

$$D_r^2 = \mathcal{E}\left[\left(y - \hat{\beta}_0 - \hat{B}_\ell' x\right)\left(y - \hat{\beta}_0 - \hat{B}_\ell' x\right)'\right]$$
$$= \Sigma_{yy \cdot x} + (\beta_0 - \mu_y + B_\ell' \mu_x)(\beta_0 - \mu_y + B_\ell' \mu_x)' + (B_\ell' - \hat{B}_\ell')\Sigma_{xx}(B_\ell' - \hat{B}_\ell')' \tag{5.131}$$

where the maximum likelihood estimates of the parameter β_0 and B_ℓ are

$$\hat{B}_\ell = \hat{\Sigma}_{xx}^{-1} \hat{\Sigma}_{xy} = S_{xx}^{-1} S_{xy} \tag{5.132}$$
$$\hat{\beta}_0 = \bar{y} - \hat{B}_\ell' \bar{x} = \bar{y} - S_{yx} S_{xx}^{-1} \bar{x}. \tag{5.133}$$

Kerridge (1967) showed that the expected value of D_r^2 is

$$\Delta_r = \mathcal{E}(D_r^2) = \Sigma_{yy \cdot x}\left(1 + \frac{1}{n}\right)\left(\frac{n-2}{n-p-2}\right) \tag{5.134}$$

which depends on both the sample size n and the number of predictors p. An unbiased estimate of Δ_r is

$$\hat{\Delta}_r = \frac{S_{yy \cdot x}(n+1)(n-2)}{(n-k)(n-p-2)} \tag{5.135}$$

where $S_{yy \cdot x} = E/(n-k)$. As in the multiple linear regression model, $\hat{\Delta}_r$ is not an unbiased estimate of the MRPE matrix. However, over all calibration samples, we expect that the $\mathcal{E}(\hat{\Delta}_r - D_r^2) = 0$.

To obtain an unbiased estimator, a new sample of n^* observations must be available. Then, letting the cross-validation estimator of D_r^2 be defined as

$$\hat{D}_r^{2*} = \sum_{i=1}^n \frac{(y_i^* - \hat{\beta}_0 - \hat{B}_\ell' x_i^*)(y_i^* - \hat{\beta}_0 - \hat{B}_\ell' x_i^*)'}{n^*}, \tag{5.136}$$

and following the argument given by M. W. Browne (1975a) using the Wishart distribution, one can show that $\hat{D}_r^{2*}$ is a conditionally unbiased estimator of D_r^2. As in univariate analysis, data splitting tends to reduce precision. Hence, the use of a single large sample usually results in better estimates of $\Delta_r = \mathcal{E}(D_r^2)$.

Any discussion of predictive validity or predictive precision for the multivariate linear regression model, outside the calibration sample, is a complex problem since it involves the evaluation of the moments of the matrix variate multivariate t-distribution, Dickey (1967) and Wegge (1971). The predictive validity matrix would have as diagonal elements quantities of the form $\rho_{rc_i}^2$ for $i = 1, 2, \ldots, q$ and associated covariances as off diagonal elements.

Model Selection

To develop a model selection procedure with random X and assuming that the q-variate and p-variate pairs (y_i', x_i') are jointly multivariate normal, we are interested in evaluating whether or not using m predictors instead of p will result in no increase in the expected value of the mean square error of prediction. Or, will the inclusion of $h = p - m$ predictor increase the accuracy of prediction. From Wegge (1971, eq. 4), the density of the random matrix of parameters involves the determinant of D_r^2 based upon a model with p independent variables. Partitioning the model, following (5.96), into two subsets: $\widehat{B}_\ell' = (\widehat{B}_m', \widehat{B}_h')$, the conditional density of $\widehat{B}^* = (\widehat{B}_h | \widehat{B}_m)$ also follows a matrix variate t-distribution where the MRPE matrix depends on the $h \times h$ submatrix of $\Sigma_{yy \cdot x}$ and the estimate of the matrix $\widehat{B}^*$. While one might construct a likelihood ratio test to evaluate the MRPE matrices for the two models with m and p variables, respectively, it involves the evaluation of the matrix variate t-distribution, Gupta and Nagar (2000, p. 133). Following M. W. Browne (1975a), one might also consider that statistic $r^2 = (r_p^2 - r_m^2)$ where r_p^2 is the squares of the largest sample canonical correlation based upon a subset p variables in X and r_m^2 is the square of the largest canonical correlation for a subset of m variables. Although, the asymptotic distribution of a squared sample canonical correlation is normal, the exact distribution is complicated, Muirhead (1982, Theorem 10.4.2).

To construct a test, we partition the vector x into the two vectors: $x_p' = (x_1', x_2')$ where the vector x_1' contains m variables and the vector x_2' contains $h = p - m$ variables. In addition, suppose the mean and covariance matrices are partitioned as

$$\mu_x = \begin{pmatrix} \mu_1 \\ \mu_2 \end{pmatrix} \tag{5.137}$$

$$\Sigma_{xx} = \begin{pmatrix} \Sigma_{11} & \Sigma_{12} \\ \Sigma_{21} & \Sigma_{22} \end{pmatrix} \tag{5.138}$$

$$\Sigma_{xy} = \begin{pmatrix} \Sigma_{1y} \\ \Sigma_{2y} \end{pmatrix}. \tag{5.139}$$

Then, $\mu_{y \cdot x_1} = \mu_y + \Sigma_{y1} \Sigma_{11} (x_1 - \mu_1)$ and we may predict the vector y by using

$$\hat{y} = \bar{y} + S_{y1} S_{11}^{-1} (x_1 - \bar{x}_1) \tag{5.140}$$

where $\bar{x}' = (\bar{x}_1', \bar{x}_2')$. Following Fujikoshi (1982), the hypothesis that x_2 provides no additional information about y in the presence of x_1 implies that y and x_2 given x_1 are independent or that

$$\Sigma_{2y} = 0 \Leftrightarrow \text{tr}(\Sigma_{yy}^{-1/2} \Sigma_{y1} \Sigma_{11}^{-1} \Sigma_{1y} \Sigma_{yy}^{-1/2}) = \text{tr}(M_{xy}^2) \tag{5.141}$$

where

$$\begin{pmatrix} \Sigma_{22 \cdot 1} & \Sigma_{2y \cdot 1} \\ \Sigma_{y2 \cdot 1} & \Sigma_{yy \cdot 1} \end{pmatrix} = \begin{pmatrix} \Sigma_{22} & \Sigma_{2y} \\ \Sigma_{y2} & \Sigma_{yy} \end{pmatrix} - \begin{pmatrix} \Sigma_{21} \\ \Sigma_{y1} \end{pmatrix} \Sigma_{xx}^{-1} \begin{pmatrix} \Sigma_{12} & \Sigma_{1y} \end{pmatrix}. \tag{5.142}$$

Then by defining

$$\Lambda_m = \begin{vmatrix} S_{22\cdot1} & S_{2y\cdot1} \\ S_{y2\cdot1} & S_{yy\cdot1} \end{vmatrix} |S_{22\cdot1}||S_{yy\cdot1}|$$

$$= \left(\frac{|S|}{|S_{11}|}\right)\left(\frac{|S_{11}|}{|S_{xx}|}\right)\left(\frac{1}{|S_{yy\cdot1}|}\right)$$

$$= \frac{|S_{yy\cdot x}|}{|S_{yy\cdot1}|}, \tag{5.143}$$

Fujikoshi (1985) showed that one should select the subset of variables $m < p$ that maximizes the criterion

$$AIC(p) - AIC(m) = n\log\Lambda_m + 2q(p - m). \tag{5.144}$$

The proper subset of variables ensures that the coefficient of determination $D^2_{xy}(m) < D^2_{xy}(p)$. And, hence that the subset of variables so selected will result in no increase in the trace of the expected mean squared error of prediction. Alternatively, one may also use determinants, for example, the log of $|\widehat{\Delta}_p|/|\widehat{\Delta}_m|$. However, to select the appropriate subset, one must approximate the critical value of the sum of independent log-Beta distributions, McQuarrie and Tsai (1998, p. 237).

Prediction, Shrinkage

When estimating the regression coefficients for the multivariate regression model, the estimate for the matrix B did not make direct use of the correlations among the dependent variables to improve predictive accuracy. Breiman and Friedman (1997) showed that prediction accuracy can be improved by considering the correlations among the dependent variables. Following Helland and Almoy (1994), we approach the problem using the "centered" random multivariate linear model

$$\Omega_0 : Y_{n\times q} = X_{n\times k}B_{k\times q} + U_{n\times q} = \begin{pmatrix} 1 & X_d \end{pmatrix} B_{k\times q} + U_{n\times q} \tag{5.145}$$

where

$$U_{n\times q} \sim N_{nq}(0, \Sigma \otimes I_n)$$
$$X_d \sim N_{np}(0, \Sigma_{xx} \otimes I_n).$$

For a given value of the predictor variables x', one wants an estimate of y'. Using the OLSE of B, we assume that the prediction model has the linear form

$$y' = x'\widehat{B}_{k\times q} = \begin{pmatrix} x'\hat{\beta}_1 & \dots & x'\hat{\beta}_q \end{pmatrix} \tag{5.146}$$

so that prediction of the ith dependent variable is accomplished by considering only the ith multiple linear regression equation. Assuming the predicted future observation follows the same prediction model, $y' = x'B + e$, where $x \sim N_p(0, \Sigma_{xx})$ and $e \sim N_q(0, \Sigma)$, following Stein (1960), Bilodeau and Brenner (1999, p. 156) defined the SPER (sum of squares of prediction error when independent variables are random) by subtracting q from the expected value of Mahalanobis' distance, a matrix norm scalar risk function defined as

$$
\begin{aligned}
\mathcal{E}[(\hat{y} - y)'\Sigma^{-1}(\hat{y} - y)] &= \mathcal{E}\{[x'(\widehat{B} - B) - e']\Sigma^{-1}[(\widehat{B} - B)'x - e]\} \\
&= \mathcal{E}\{\text{tr}[(\widehat{B} - B)\Sigma^{-1}(\widehat{B} - B)'\Sigma_{xx}]\} + q \\
&= \mathcal{E}\{\text{tr}[\Sigma_{xx}^{1/2}(\widehat{B} - B)\Sigma^{-1/2}\Sigma^{-1/2}(\widehat{B} - B)'\Sigma_{xx}^{1/2}]\} + q \\
&= \mathcal{E}\{\text{tr}[(\Sigma_{xx}^{1/2}U)(U'\Sigma_{xx}^{1/2})]\} \\
&= \mathcal{E}\{\|V\|^2\} + q
\end{aligned}
\tag{5.147}
$$

where $V = \Sigma_{xx}^{1/2}U$ and the $\text{tr}(VV') = \|V\|^2$ is the square of the Euclidean matrix norm. Then, the SPER risk criterion for the OLS estimator is defined as

$$
\begin{aligned}
R_{SPER}(\widehat{B}) &= \mathcal{E}\{\text{tr}[(\widehat{B} - B)\Sigma^{-1}(\widehat{B} - B)'\Sigma_{xx}]\} \\
&= \text{tr}\{\mathcal{E}[\Sigma_{xx}^{1/2}(\widehat{B} - B)\Sigma^{-1}(\widehat{B} - B)'\Sigma_{xx}^{1/2}]\} \\
&= \text{tr}\{\mathcal{E}[\Sigma_{xx}^{1/2}(\widehat{B} - B)\Sigma^{-1/2}\Sigma_{-1/2}(\widehat{B} - B)'\Sigma_{xx}^{1/2}]\} \\
&= \text{tr}\{\mathcal{E}[\Sigma_{xx}^{1/2}UU'\Sigma_{xx}^{1/2}]\}.
\end{aligned}
\tag{5.148}
$$

Letting $U = (\widehat{B} - B)\Sigma^{-1/2}$, $\mathcal{E}(U'U|X) = \text{tr}(I_q)(X'X)^{-1} = q(X'X)^{-1}$ and using properties of the inverted Wishart distribution (Muirhead, 1982, p. 97): for $x \sim N_p(0, \Sigma_{xx})$ then $X'X \sim W_p(n-1, \Sigma_{xx})$ and $\mathcal{E}(X'X)^{-1} = [(n-1)-p-1]^{-1}\Sigma_{xx}^{-1}$, the SPER risk criterion for the OLSE is

$$
R_{SPER}(\widehat{B}) = \frac{qp}{n - p - 2} = qr
\tag{5.149}
$$

where $r = {p}/{(n - p - 2)}$. For SPER risk to be defined, $n > p + 2$.

An alternative risk function used in prediction is SPE risk proposed by Dempster, Schatzoff, and Wermuth (1977). It is obtained by replacing Σ_{xx} in $R_{SPER}(\widehat{B})$ with $X'X$. The SPE risk function is defined as

$$
R_{SPE}(\widehat{B}) = \mathcal{E}\{\text{tr}[(\widehat{B} - B)\Sigma^{-1}(\widehat{B} - B)'X'X]\}.
\tag{5.150}
$$

SPE risk for the OLSE is $R_{SPE}(\widehat{B}) = qp$, Bilodeau and Kariya (1989, eq. 3.26).

Assuming each row $x \sim N_p(0, \Sigma)$ and that the triple (B, Σ, Σ_{xx}) is known, Breiman and Friedman (1997) using the linear model of the form $y = \mu_y + A(x - \mu_x) + e$ showed that the optimal matrix A that minimizes the matrix norm risk criterion

$$
\min_A \mathcal{E}[(A'\hat{y} - y)'\Sigma^{-1}(A'\hat{y} - y)]
\tag{5.151}
$$

is the matrix

$$
A_{OPT} = [(B'\Sigma_{xx}B) + r\Sigma]^{-1} \qquad \text{where } r = \frac{p}{n - p - 2}.
\tag{5.152}
$$

Setting $\widetilde{B}_{OPT} = \widehat{B} A_{OPT}$ reduces the SPER risk criterion more than the OLSE. Using the matrix A_{OPT}, the OLSE has been shrunken and the corresponding predicted values in the vector $\tilde{y} = A'_{OPT} y$ have been flattened. The matrix A_{OPT} is called the multivariate flattening matrix following Stone (1974). To estimate the matrix $\widetilde{B}_{OPT}$ and hence the matrix A_{OPT}, Breiman and Friedman (1997) proposed a complex computer-based, ordinary cross-validation (CV) iterative procedure and a generalized cross-validation (GCV) procedure that is an approximation to CV that is less computer intensive.

For small values of r, $A_{OPT} \approx I_q - r(B'\Sigma_{xx}B)^{-1}\Sigma$. Using this result, Merwe and Zidek (1980) proposed an A defined as

$$\widehat{A}_{FICYREG} = I_q - r^*(n-k)(\widehat{B}'X'X\widehat{B})^{-1}S \tag{5.153}$$

and called their estimate the filtered canonical y-variate regression (FICYREG) estimate, $\widetilde{B}_{FICYREG} = \widehat{B}\widehat{A}_{FICYREG}$. They showed that SPE risk for the FICYREG estimate is less than the SPE risk for the OLSE provided $(n-k) > (q+1)$, $(q+1) < p$ and $r^* = (p-q-1)/(n-k+q+1)$. SPE risk for the FICYREG estimator is

$$R_{SPE}(\widehat{B}_{FICYREG}) = qp - r^*(p-q-1)\text{tr}[(n-r)S(\widehat{B}'X'X\widehat{B})]. \tag{5.154}$$

If $r^* = 0$, then the FICYREG estimate is equal to the OLSE.

Bilodeau and Kariya (1989, Theorem 3.4) developed an Efron-Morris empirical Bayes estimator for the matrix B that has smaller SPE risk than the FICYREG estimator. Their proposed estimate for the multivariate flattening matrix A is

$$\widehat{A}_{EM} = I_q - r^*(n-k)(\widehat{B}X'X\widehat{B})^{-1}S - \frac{b^*(n-k)}{\text{tr}(\widehat{B}'X'X\widehat{B})} \tag{5.155}$$

where $b^* = (q-1)/(n-k+q+1)$ provided $r^* > 0$ and $b^* > 0$. If $r^* < 0$ one may set $r^* = 0$ provided $b^* > 0$, then $\widetilde{B}_{EM} = \widehat{B}\widehat{A}_{EM}$. For $b^* = 0$, their estimate reduces to the FICYREG estimate. SPE risk for the Efron-Morris estimate is

$$R_{SPE}(\widehat{B}_{EM}) = R_{SPE}(\widehat{B}_{FICYREG}) - b^*(q-1)\text{tr}[(n-k)S][\text{tr}(\widehat{B}'X'X\widehat{B})]^{-1} \tag{5.156}$$

which is seen to be less than or equal to the SPE risk of the FICYREG estimator.

Bilodeau (2002) obtained an expression for the exact SPE risk for Breiman and Friedman (1997) GCV estimator for the parameter matrix B.

5.7 MULTIVARIATE MIXED MODELS AND THE ANALYSIS OF REPEATED MEASUREMENTS

When employing a mixed model, the number of independent parameters is reduced in Σ by assuming a linear structure for $\Phi = A'\Sigma A$. The classical MANOVA model makes no assumptions about Φ, other than positive definiteness. The general multivariate mixed model corresponding to the multivariate general linear model is

$$\Omega_0 : Y_{n \times t} = X^*_{n \times k} B^*_{k \times t} + Z^*_{n \times r} \Lambda_{r \times t} + E_{0(n \times t)}$$
$$= XB + U \tag{5.157}$$

where Λ and E_0 are random matrices,

$$\text{vec}(Y) \sim N[(I_t \otimes X)\text{vec}(B), Z^* D^* Z^{*'} + \Psi], \tag{5.158}$$

D^* is positive definite block diagonal matrix, Z^* is a matrix of fixed covariates, and $\Psi = \Sigma_e \otimes I_n$. Alternatively, from the GLM $Y = XB + U$, the ith row of Y by (5.8) is

$$y_i = (I_t \otimes x'_i)\text{vec}(B) + u_i. \tag{5.159}$$

Then, a simple multivariate linear mixed model is constructed by modeling u_i as

$$u_i = Z_i \theta_i + e_i \qquad\qquad i = 1, 2, \ldots, n \tag{5.160}$$

where Z_i is a $(t \times h)$ within-subject design matrix from random effects, the rank$(Z) = h$,

$$\theta_i \sim N_h(0, D_h) \tag{5.161}$$
$$e_i \sim N(0, \sigma^2 I_t) \tag{5.162}$$

and θ_i and e_i are independent. Thus,

$$\Sigma_i = Z_i D_h Z'_i + \sigma^2 I_t \tag{5.163}$$

which is of the same form as (2.92) for the univariate mixed model. Comparing (5.159) and (5.163) with (5.8), we see that the multivariate mixed model is identical to the standard multivariate linear model if we rearrange the data matrix, assume homogeneity of $\Sigma = \Sigma_i$ for $i = 1, 2, \ldots, n$ and impose Hankel structure on $\Phi = A'\Sigma A$.

Using (5.159), the ML estimator of Γ is $\widehat{\Gamma} = C\widehat{B}A$ or

$$\text{vec}(\widehat{\Gamma}) = (A' \otimes C) \left[\sum_{i=1}^{n} (\widehat{\Sigma}_i^{-1} \otimes x_i x'_i) \right]^{-1} \left[\sum_{i=1}^{n} (\widehat{\Sigma}_i^{-1} \otimes x_i) y_i \right] \tag{5.164}$$

where $\widehat{\Sigma}_i = Z_i \widehat{D}_h Z'_i + \hat{\sigma}^2 I_t$, $\widehat{D}_h$ is the ML estimate of D_h and $\hat{\sigma}^2$ is the ML estimate of σ^2. To test hypotheses about Γ, one usually employs large sample theory, except if Σ_i has special structure. If the Σ_i are not allowed to vary among subjects, (5.163) becomes

$$\Sigma_i = \Sigma = ZD_h Z' + \sigma^2 I_t. \tag{5.165}$$

The simplest mixed model, attributed to Scheffé (1959), is obtained by associating Z with a vector of t 1's, $Z = 1_t$ so that $h = 1$, then

$$\Sigma = \sigma_s^2 11' + \sigma^2 I_t. \tag{5.166}$$

Postmultiplication of (5.157) by A yields

$$YA = XBA + UA \tag{5.167}$$

$$Z_{n \times u} = X_{n \times k} \Theta_{k \times u} + \xi_{n \times u}. \tag{5.168}$$

The distribution of $\text{vec}(\xi)$ is normal, $\text{vec}(\xi) \sim N(0, \sigma^2 I_t)$ if $\Phi = A'\Sigma A = \sigma^2 I_u$ has Hankel structure. This is also known as circularity or sphericity which Huynh and Feldt (1970) showed was a necessary and sufficient condition for a univariate mixed model analysis to be valid. Hence, one should be able to recover univariate mixed model F-statistics from the multivariate linear model. To see this, consider the example in Scheffé (1959, Chapter 8). The model with side condition is

$$y_{ij} = \mu + \alpha_j + e_{ij} \qquad i = 1, 2, \ldots, n \tag{5.169}$$

$$\sum_j \alpha_j = 0 \qquad j = 1, 2, \ldots, t$$

$$\mathcal{E}(e_{ij}) = 0$$

$$\text{cov}(e_{ij}, e_{i'j'}) = \delta_{ii'} \sigma_{jj'}$$

where $\delta_{ii'} = 1$ if $i = i'$ and 0 otherwise. Thus, for $n = 3$ and $t = 3$ the linear model is

$$\begin{pmatrix} y_{11} \\ y_{12} \\ y_{13} \\ y_{21} \\ y_{22} \\ y_{23} \\ y_{31} \\ y_{32} \\ y_{33} \end{pmatrix} = \begin{pmatrix} 1 & 1 & 0 \\ 1 & 0 & 1 \\ 1 & -1 & -1 \\ 1 & 1 & 0 \\ 1 & 0 & 1 \\ 1 & -1 & -1 \\ 1 & 1 & 0 \\ 1 & 0 & 1 \\ 1 & -1 & -1 \end{pmatrix} \begin{pmatrix} \mu \\ \alpha_1 \\ \alpha_2 \end{pmatrix} + \begin{pmatrix} e_{11} \\ e_{12} \\ e_{13} \\ e_{21} \\ e_{22} \\ e_{23} \\ e_{31} \\ e_{32} \\ e_{33} \end{pmatrix}$$

$$y = X\beta + e \tag{5.170}$$

where $\mathcal{E}(y) = 0$ and $\text{cov}(y) = I + n \otimes \Sigma$. The random effects are contained within the covariance matrix Σ. The problem with specification (5.170) is that the pattern of the elements in Σ is unspecified. The univariate model assumes a certain sphericity pattern for Σ. We shall show how a univariate repeated measures design may be represented as a multivariate linear model and, with additional restrictions on Σ, how to recover univariate tests.

We can rearrange the univariate vector into a data matrix with t occasions and n

subjects as

$$\begin{pmatrix} y_{11} & y_{21} & y_{31} \\ y_{12} & y_{22} & y_{32} \\ y_{13} & y_{23} & y_{33} \end{pmatrix} = \begin{pmatrix} 1 & 1 & 0 \\ 1 & 0 & 1 \\ 1 & -1 & -1 \end{pmatrix} \begin{pmatrix} \mu \\ \alpha_1 \\ \alpha_2 \end{pmatrix} \begin{pmatrix} 1 & 1 & 1 \end{pmatrix} + \begin{pmatrix} e_{11} & e_{21} & e_{31} \\ e_{12} & e_{22} & e_{32} \\ e_{13} & e_{23} & e_{33} \end{pmatrix}$$
$$Y' = X_W B' X_B' + U'. \tag{5.171}$$

Letting

$$A' = \begin{pmatrix} \frac{1}{3} & \frac{1}{3} & \frac{1}{3} \\ \frac{2}{3} & -\frac{1}{3} & -\frac{1}{3} \\ -\frac{1}{3} & \frac{2}{3} & -\frac{1}{3} \end{pmatrix}, \tag{5.172}$$

notice that $A' X_w = I$. Hence, multiplying (5.171) on the left by A' reduces (5.171) to the transpose of a standard multivariate linear model

$$A'Y' = B'X_B' + A'U' \tag{5.173}$$
$$Z' = B'X_B' + \xi' \tag{5.174}$$

or taking transposes, the standard multivariate linear model results

$$YA = X_B B + UA \tag{5.175}$$
$$Z = X_B B + \xi \tag{5.176}$$

where $A = (X_W')^{-1}$ and $Z = Y(X_W')^{-1}$. Finally, observe that $X = X_B \otimes X_W$ or $X' = X_B' \otimes X_W'$. The condition that the univariate design matrix X can be represented as the product $X_B \otimes X_W$ is called the condition of separability. Using (5.175), the BLUE of B is

$$\widehat{B} = (X_B'X_B)^{-1}X_B'Z$$
$$= (X_B'X_B)^{-1}X_B'Y(X_B')^{-1}. \tag{5.177}$$

Recalling Theorem 5.1 and applying the $\text{vec}(\cdot)$ operator to both sides of (5.177),

$$\hat{\beta}' = \text{vec}(\widehat{B}')[(X_B'X_B)^{-1}X_B \otimes X_W^{-1}]\text{vec}(Y'). \tag{5.178}$$

Next observe that the general form of the univariate repeated measures model (5.170) is

$$\mathcal{E}(y) = X\beta \qquad\qquad \text{cov}(y) = I_n \otimes \Sigma = \Omega. \tag{5.179}$$

Further, assume that $X = X_B \otimes X_W$ so that the design matrix is separable. Then, by (4.6)

$$\hat{\beta} = (X'\Omega^{-1}X)^{-1}X'\Omega^{-1}y$$
$$= [(X_B \otimes X_W)'(I_n \otimes \Sigma)^{-1}(X_B \otimes X_W)]^{-1}(X_B \otimes X_W)'(I_n \otimes \Sigma)^{-1}y$$
$$= [(X_B'X_B)^{-1}X_B' \otimes (X_W'\Sigma^{-1}X_W)^{-1}X_W'\Sigma^{-1}]y. \tag{5.180}$$

Comparing (5.180) with (5.178), we have equivalence of the univariate and multivariate estimates if and only if

$$X_W^{-1} = (X_W' \Sigma^{-1} X_W)^{-1} X_W' \Sigma^{-1} \tag{5.181}$$

so that X_W^{-1} must exist or the within design matrix must be of full rank, $\text{rank}(X_W) = t$, or saturated. Hence, the univariate and multivariate estimates are identical if and only if the univariate design matrix is separable ($X = X_B \otimes X_W$) and the within design matrix is saturated ($\text{rank}(X_W) = t$ or X_W^{-1} exists). In conclusion, the two models are really one model, arranged in two ways, Davidson (1988).

To demonstrate the equivalence of the two models for hypothesis testing is more complicated. Using the reduced form of the model (5.167), $\text{vec}(\xi) \sim N(0, \Phi \otimes I_n)$ where $\Phi = A' \Sigma A = \sigma^2 I_u$. The ML estimator of σ^2 is

$$\hat{\sigma}^2 = \frac{\hat{\xi}' \hat{\xi}}{n} = \frac{\text{tr}(\hat{\xi}\hat{\xi}')}{n} \tag{5.182}$$

so that the ML estimate of Φ, assuming sphericity, is

$$\widetilde{\Phi} = \frac{I_u \text{tr}(E)}{un} \tag{5.183}$$

where $u = \text{rank}(A)$. The usual unbiased estimator of Φ is

$$\widehat{\Phi} = \frac{I_u \text{tr}(E)}{u\nu_e} \tag{5.184}$$

where $\nu_e = n - \text{rank}(X) = n - k$ and E is the SSP matrix, $E = Y'(I - X(X'X)^{-1}X')Y$. Substituting $[(nu\nu_h)/(n-k)]\Phi$ into (5.62) for $\Phi = A' \Sigma A$ results in the likelihood ratio statistic for testing $H : C\Theta = 0$:

$$F = \frac{(\nu_e/\nu_h)\{[\text{vec}(\Theta)]'[I_u \otimes C'(C(X'X)^{-1}C')^{-1}C][\text{vec}(\Theta)]\}}{\text{tr}(E)} \tag{5.185}$$

which under sphericity has a F distribution with degrees of freedom $u\nu_h$ and $u\nu_e$, and noncentrality parameter

$$\delta = \frac{\theta'\{I_u \otimes C'[C(X'X)^{-1}C']^{-1}C\}\theta}{\sigma^2} \tag{5.186}$$

where $\theta' = \text{vec}(\Theta)$ and $\Theta = BA$.

An alternative approach to establish the equivalence of the models for hypothesis testing is to observe that the multivariate hypothesis is equivalent to the univariate hypothesis,

$$H : C_B BA_W = 0 \Leftrightarrow C\beta = 0 \tag{5.187}$$

if and only if $C = C_B' \otimes A_W$ where $\text{vec}(B') = \beta$ so that the univariate hypothesis is separable. If both the univariate design matrix X and hypothesis test matrix

Table 5.1: 2×2 Repeated Measures Design.

		Conditions	
		1	2
	I	A	B
Groups		μ_{11}	μ_{12}
	II	B	A
		μ_{21}	μ_{22}

C are separable, and $\Phi = A'_W \Sigma A_W = \sigma^2 I$ has spherical structure, the univariate hypothesis and error sum of squares defined in (2.15)-(2.16) have the following representation

$$SS_H = \mathrm{tr}(H) \qquad\qquad SS_E = \mathrm{tr}(E) \qquad (5.188)$$

where

$$H = (C_B \widehat{B} A_W)'[C_B(X'_B X_B)^{-1} C'_B]^{-1}(C_B \widehat{B} A_W) \qquad (5.189)$$

$$E = A'_W Y'[I - X_B(X'_B X_B)^{-1} X'_B] Y A_W. \qquad (5.190)$$

The F-statistic for testing $H : C\beta = 0$ is

$$F = \frac{SS_H/\nu_h^*}{SS_E/\nu_e^*} = \frac{MS_h}{MS_e} \qquad (5.191)$$

where $\nu_e^* = \mathrm{rank}(E)\mathrm{rank}(A_W) = \nu_e u$ and $\nu_h^* = \mathrm{rank}(H)\mathrm{rank}(A_W) = \nu_h u$ as discussed by Boik (1988), Davidson (1988), and Timm (1980b).

5.8   EXTENDED LINEAR HYPOTHESES

In our discussion of the MGLM model, we demonstrated how one tests standard hypotheses of the form $H : CBA = \Gamma$ and obtains simultaneous confidence intervals for bilinear parametric functions of the form $\psi = c'Ba$; however, not all potential hypotheses may be represented as $H : CBA = \Gamma$. To illustrate, consider a simple repeated measures design for two groups (I and II), two conditions (1 and 2), and two stimuli (A and B) as shown in Table 5.1.

For the design in Table 5.1, the parameter matrix B is

$$B = \begin{pmatrix} \mu_{11} & \mu_{12} \\ \mu_{21} & \mu_{22} \end{pmatrix}. \qquad (5.192)$$

The overall test for differences in conditions is to select the matrices $C = C_0$ and $A = A_0$ such that

$$C_0 = \begin{pmatrix} 1 & 0 \\ 0 & 1 \end{pmatrix} \qquad\qquad A_0 = \begin{pmatrix} 1 \\ -1 \end{pmatrix} \qquad (5.193)$$

so that the test becomes

$$H : \begin{pmatrix} \mu_{11} \\ \mu_{21} \end{pmatrix} = \begin{pmatrix} \mu_{12} \\ \mu_{22} \end{pmatrix}. \tag{5.194}$$

Having rejected the overall test of H using one of the multivariate test criteria, one investigates parametric functions $\psi = c'Ba$ for significance. For this example, one may be interested in the contrasts

$$\psi_1 : \mu_{11} - \mu_{22} = 0 \qquad\qquad \psi_2 : \mu_{21} - \mu_{12} = 0 \tag{5.195}$$

that the differences in conditions is due either to stimulus A or stimulus B. However, these contrasts are not of the standard classical bilinear form, $\psi = c'Ba$, since the pattern of significance is not the same for each variable.

Using the more general form for ψ_1 and ψ_2, $H_\psi : C_*\beta = \gamma$, the contrasts in (5.195) become

$$\begin{pmatrix} 1 & 0 & 0 & -1 \\ 0 & 1 & -1 & 0 \end{pmatrix} \begin{pmatrix} \mu_{11} \\ \mu_{21} \\ \mu_{12} \\ \mu_{22} \end{pmatrix} = \begin{pmatrix} 0 \\ 0 \end{pmatrix}. \tag{5.196}$$

Because $C_* = A_0' \otimes C_0$ for the matrices C_0 and A_0 in (5.193), the overall test and the contrasts may be combined and written as

$$C_*\beta = \begin{pmatrix} 1 & 0 & 0 & -1 \\ 0 & 1 & -1 & 0 \\ 1 & 0 & -1 & 0 \\ 0 & 1 & 0 & -1 \end{pmatrix} \begin{pmatrix} \mu_{11} \\ \mu_{21} \\ \mu_{12} \\ \mu_{22} \end{pmatrix} = \begin{pmatrix} 0 \\ 0 \\ 0 \\ 0 \end{pmatrix} \tag{5.197}$$

where the first two rows of C_* in (5.197) are the contrasts associated with ψ_1 and ψ_2 and the last two rows represent the overall test H. The LR test for testing (5.197) is to use the χ^2 statistic given in (5.34)

$$X_\nu^2 = (\hat{\gamma} - \gamma_0)'[C_*(D'\Omega^{-1}D)^{-1}C_*']^{-1}(\hat{\gamma} - \gamma_0) \tag{5.198}$$

where $\nu = \operatorname{rank}(C_*)$, $\Omega = \Sigma \otimes I_n$ and $D = I_p \otimes X$. Because Σ is unknown, we must replace Σ by a consistent estimator so that (5.198) is the Wald statistic given in (4.15). When $C_* = A' \otimes C$, (5.198) is T_0^2 by (5.62).

Alternatively, observe that (5.197) may be written as the joint test of

$$\operatorname{tr}\left[\begin{pmatrix} 1 & 0 \\ 0 & -1 \end{pmatrix} \begin{pmatrix} \mu_{11} & \mu_{12} \\ \mu_{21} & \mu_{22} \end{pmatrix} \right] = 0 \tag{5.199}$$

$$\operatorname{tr}\left[\begin{pmatrix} 0 & 1 \\ -1 & 0 \end{pmatrix} \begin{pmatrix} \mu_{11} & \mu_{12} \\ \mu_{21} & \mu_{22} \end{pmatrix} \right] = 0 \tag{5.200}$$

$$\operatorname{tr}\left[\begin{pmatrix} 1 & 0 \\ -1 & 0 \end{pmatrix} \begin{pmatrix} \mu_{11} & \mu_{12} \\ \mu_{21} & \mu_{22} \end{pmatrix} \right] = 0 \tag{5.201}$$

$$\operatorname{tr}\left[\begin{pmatrix} 0 & 1 \\ 0 & -1 \end{pmatrix} \begin{pmatrix} \mu_{11} & \mu_{12} \\ \mu_{21} & \mu_{22} \end{pmatrix} \right] = 0 \tag{5.202}$$

where (5.201)-(5.202) are constructed from the fact that $\psi = c'Ba = \text{tr}(ac'B) = \text{tr}(MB)$. More generally, this suggests a union-intersection representation for the multivariate hypothesis ω_H

$$\omega_H = \cap_{M \in M_0} H_M : \text{tr}(MB) = \text{tr}(M\Gamma) \tag{5.203}$$

where M_0 is some set of matrices of order $p \times k$. Letting M_0 be defined as the family of matrices

$$M_0 = \left(M | M = \sum_{j=1}^{q} \theta_j C_j' \right) \qquad \theta_j\text{'s are real} \tag{5.204}$$

the set of q linearly independent matrices form a matrix decomposition of the multivariate hypothesis called the extended linear hypothesis of order q by Mudholkar, Davidson, and Subbaiah (1974) who used Roy's UI principle and symmetric gauge functions to develop test statistics for testing (5.203), a generalized T_0^2 statistic.

As we saw in our example, the family ω_H includes the general linear hypotheses implied by the maximal hypothesis H so that a test procedure constructed for ω_H includes H and each minimal hypothesis ω_M: $\text{tr}(MB) = \text{tr}(M\Gamma)$. By the UI principle, suppose we can construct a test statistic $t_\psi(M)$ for each minimal hypothesis $\omega_M \subseteq \omega_H$ where the maximum hypothesis $H : C_0 B A_0 = \Gamma$ is a subfamily of ω_H so that a test procedure for ω_H will include H. Constructing a test statistic $\omega_M \subseteq \omega_H$ independent of $M \in M_0$, the extended linear hypothesis ω_H is rejected if

$$t(M) = \sup_{M \in M_0} t_\psi(M) \geq c_\psi(1 - \alpha) \tag{5.205}$$

is significant for some minimal hypothesis ω_M where the critical value $c_\psi(1 - \alpha)$ is chosen such that the $P(t(M) \leq c_\psi(1 - \alpha)|\omega_H) = 1 - \alpha$.

To construct a test of ω_H, Mudholkar et al. (1974) related $t_\psi(M)$ to symmetric gauge function (SGF) to generate a class of invariant tests containing both Roy's largest root test and the Bartlett-Lawley-Hotelling trace test statistics. Informally, a SGF is a mapping from a vector space defined by M of order $p \times k$, $p \leq k$ and $p = \text{rank}(M)$ to the real numbers. The mapping is invariant under nonsingular transformations of M and generates a matrix norm $\|M\|_\psi = \psi\left(\lambda_1^{1/2}, \lambda_2^{1/2}, \ldots, \lambda_k^{1/2}\right)$ where $\lambda_1 > \lambda_2 > \cdots > \lambda_k$ are the eigenvalues of MM'. Formally, a SGF is a mapping from a normed vector space (a vector space on which a norm is defined) to the real numbers (a Banach space or complete normed vector space) that is invariant to permutations and arbitrary sign changes of the coordinates. For a normed vector space of order r, the SGF is

$$t_{\psi_r}(M) = \left(\sum_i \sum_j |m_{ij}|^r \right)^{1/r}. \tag{5.206}$$

For $r = 2$, the SGF defines the familiar Euclidean (Frobenius) matrix norm

$$t_{\psi_2}(M) = \left(\sum_i \sum_j m_{ij}^2\right)^{1/2} \qquad \|M\|_{\psi_2} = [\text{tr}(MM')]^{1/2} = \left(\sum_{i=1}^{p} \lambda_i\right)^{1/2}.$$

(5.207)

Associated with a SGF is a normed conjugate SGF. For $t_{\psi_r}(M)$, the associated conjugate SGF is

$$t_{\psi_s}(M) = \left(\sum_i \sum_j |m_{ij}|^s\right)^{1/2} \qquad \frac{1}{r} + \frac{1}{s} = 1.$$

(5.208)

Hence for $r = s = 2$, the Euclidean norm is self-conjugate. For $r = 1$, $t_{\psi_1}(M) = \sum_i \sum_j |m_{ij}|$, the matrix norm for ψ_1 is $\|M\|_{\psi_1} = \sum_{i=1}^{p} \lambda_i^{1/2}$. The corresponding conjugate SGF is $t_{\psi_\infty}(M) = \max(m_{ij})$ with matrix norm $\|M\|_{\psi_\infty} = \lambda^{1/2}$, also called the spectral norm. To construct a test statistic for ω_H, a result due to Mudholkar et al. (1974) that related two conjugate norms ψ and ϕ is required.

Theorem 5.8. *For real matrices A and B of order $p \times k$, $p \leq k$, the*

$$\|B\|_\psi = \sup_{A \neq 0} \left(\frac{\text{tr}(AB')}{\|A\|_\phi}\right).$$

(5.209)

With (5.205) and (5.209), we can develop a general class of tests based on a SGF ψ to test ω_H. We consider first the subfamily of maximal hypotheses $H : C_0 B A_0 = \Gamma$; from (5.39) let

$$E_0 = A_0' Y'(I - X(X'X)^{-1}X')Y A_0 \qquad (5.210)$$
$$H_0 = (\widehat{B}_0 - \Gamma)' W_0^{-1}(\widehat{B}_0 - \Gamma) \qquad (5.211)$$

where

$$W_0 = C_0(X'X)^{-1}C_0' \qquad (5.212)$$
$$\widehat{B}_0 = C_0 \widehat{B} A_0 \qquad (5.213)$$
$$\widehat{B} = (X'X)^{-1}X'Y \qquad (5.214)$$

for testing H. Then, if we associate matrices A and B in (5.209) with the matrices in (5.210)-(5.214) for

$$A = E_0^{1/2} M W_0^{1/2} \qquad B = E_0^{-1/2} \widehat{B}_0' W_0^{-1/2} \qquad (5.215)$$

the $\text{tr}(A'B) = \text{tr}(M\widehat{B}_0)$, M is of order $(u \times g)$ and the $\|B\|_\psi$ depends only on the eigenvalues of $H_0 E_0^{-1}$. Comparing (5.209) with (5.205), a general test statistic for the minimal hypothesis $\omega_H : \text{tr}(MB_0) = \text{tr}(M\Gamma)$ is

$$|t_\psi(M)| = \frac{\text{tr}[M(\widehat{B}_0 - \Gamma)]}{\|W_0^{1/2}M'E_0^{1/2}\|_\phi} \tag{5.216}$$

with the critical value $c_\psi(1 - \alpha)$ for $\psi\left(\lambda_1^{1/2}, \lambda_2^{1/2}, \ldots, \lambda_s^{1/2}\right)$, $s = \min(g, u)$, $g = \text{rank}(C_0)$, $u = \text{rank}(A_0)$, λ_i are the roots of $H_0 E_0^{-1}$, and where ϕ and ψ are conjugate SGF's. Since the denominator in (5.126) is a matrix norm,

$$\|W_0^{1/2}M'E_0^{1/2}\|_\phi = \phi\left(\lambda_1^{1/2} \quad \ldots \quad \lambda_s^{1/2}\right) \tag{5.217}$$

where λ_i represent the characteristic roots of $MW_0M'E_0$. When $\psi = \psi_2$ is the self conjugate SGF, the test statistic for $\omega_M : \text{tr}(MB_0) = \text{tr}(M\Gamma)$ becomes

$$t_\psi(M) = \frac{|\text{tr}[M(\widehat{B}_0 - \Gamma)]|}{[\text{tr}(MW_0M'E_0)]^{1/2}}, \tag{5.218}$$

a generalized T_0^2 statistic. The hypothesis ω_H is rejected if the supremum of $t_\psi(M)$, $t(M)$, exceeds the critical constant $c_\psi(1 - \alpha)$. To compute the supremum, let

$$M = \theta_1 M_1 + \theta_2 M_2 + \cdots + \theta_q M_q \tag{5.219}$$

for a matrix hypothesis ω_H. Furthermore, let

$$\tau_i = \text{tr}[M_i(\widehat{B}_0 - \Gamma)] \qquad T = (t_{ij}) \qquad t_{ij} = \text{tr}(M_i W_0 M_j' E_0), \tag{5.220}$$

then by (5.34), $t(M) = [t_\psi(M)]^2 = {(\theta'\tau)^2}/{(\theta'T\theta)}$ is the supremum over all vectors θ. To evaluate this supremum, recall from Timm (1975, p. 107) that the sup ${(\theta'A\theta)}/{(\theta'B\theta)}$ is the largest root of the determinant equation $|A - \lambda B| = 0$ for symmetric matrices A and B, λ_1 equals the largest root of AB^{-1}. Letting $A = \tau\tau'$ and $B = T$, λ_1 equals $\tau'T^{-1}\tau$, a quadratic form. Hence,

$$t(M) = \tau'T^{-1}\tau \tag{5.221}$$

and ω_H is rejected if $t(M)$ exceeds $c_\psi^2(1-\alpha)$. More importantly, $1-\alpha$ simultaneous confidence intervals may be constructed for contrasts of the parametric function $\psi = \text{tr}(MB_0) = \sum_i \sum_j \beta_{0ij}$ estimated by $\hat{\psi} = \text{tr}(M\widehat{B}_0)$. The simultaneous confidence intervals again have the general form

$$\hat{\psi} - c_0\hat{\sigma}_{\hat{\psi}} \leq \psi \leq \hat{\psi} + c_0\hat{\sigma}_{\hat{\psi}} \tag{5.222}$$

where

$$\hat{\sigma}_{\hat{\psi}}^2 = \text{tr}(MW_0M'A_0'SA_0) \tag{5.223}$$

$\mathcal{E}(S) = \Sigma$ and the critical constant is $c_0^2 = \nu_e U_{0(s,M,N)}^{1-\alpha}$ for the overall hypothesis H. Letting the SGF be $t_{\psi_\infty}(M)$, one obtains Roy's extended largest root statistic. For this criterion, the $1 - \alpha$ simultaneous confidence intervals in (5.222) are defined with

$$c_0^2 = \frac{\theta_{s,M,N}^{1-\alpha}}{1 - \theta_{s,M,N}^{1-\alpha}} \tag{5.224}$$

$$\hat{\sigma}_{\hat{\psi}} = \sum_{i=1}^{s} \lambda_i^{1/2} \tag{5.225}$$

where $\lambda_i^{1/2}$ are same as in (5.217). A computer program for tests of extended linear hypotheses and the establishment of confidence intervals is included in Krishnaiah, Mudholkar, and Subbaiah (1980).

In (5.198), we showed how the Wald statistic X_ν^2 may be used to test the general hypotheses $H : C_* \beta = \gamma_0$ by substituting for the unknown covariance matrix Σ with any consistent estimate. Hecker (1987) derived a test for H in (5.197) for situations in which a Wishart distributed estimator of Σ is available and independent of X_ν^2. To test H, let $\widehat{\Sigma} \sim W_p(\nu_e, \Sigma)$ independent of X_ν^2 ($\nu = \text{rank}(C_*)$) and

$$\lambda = \lambda_1 \lambda_2 \tag{5.226}$$

where

$$\lambda_1 = \frac{\text{tr}(\widehat{\Sigma})}{p\nu_e} \qquad \lambda_2 = \frac{|\widehat{\Sigma}|^{\nu_e/2}}{\left\{ [\text{tr}(\widehat{\Sigma})]/_{[p]} \right\}^{p\nu_e/2}}. \tag{5.227}$$

The hypothesis $H : C_* \beta = \gamma_0$ is rejected if the

$$P\left(\frac{X_\nu^2}{\lambda} > k \right) = \alpha \tag{5.228}$$

for some constant k. To find the constant k, the expression

$$C(\nu, \nu_e, p) = \frac{X_\nu^2}{\lambda_1 \lambda_2} \tag{5.229}$$

must be evaluated where X_ν^2 and λ_1 have χ^2 distribution with ν and $p\nu_e$ degrees of freedom, respectively, and

$$k^* = \lambda_2^{2/\nu_e} \tag{5.230}$$

is distributed as a product of $p-1$ independent beta random variables V_i with $(\nu_e - i)/2$ and $i(1/2 + p^{-1})$ degrees of freedom for $i = 1, 2, \ldots, p - 1$ (Srivastava & Khatri, 1979, p. 209). The distribution of the test statistic X_ν^2/λ proposed by Hecker (1987) is unknown; however, its value is between the distribution of two known distributions

$$F = \frac{X_\nu^2}{\lambda_1} \sim F(\nu, p\nu_e) \qquad \qquad X_\nu^2 \sim \chi_\nu^2 \tag{5.231}$$

when $\widehat{\Sigma}$ is the pooled within covariance matrix for the standard MANOVA model that is independent of the mean.

Hence, we may test extended linear hypotheses using the generalized T_0^2 statistic defined in (5.218), Roy's extended largest root procedure using (5.224), Wald's statistic defined in (5.198) with Σ replaced by a consistent estimator $\widehat{\Sigma}$, or Hecker (1987) statistic. While all the proposed methods are asymptotically equivalent, the performance of each is unknown for small sample sizes.

5.9 Multivariate Regression: Calibration and Prediction

For our first example of the multivariate linear model, we illustrate the variable selection process, model fit criteria, and the development of a multivariate prediction model. Two programs contain the code for this example: Multsel.sas and program 5_9.sas. The first program is used to select the best subset of variables for the multivariate regression model, the calibration model, and the second program is used to calculate the fit criteria and to develop a multivariate prediction equation. The classical data found in the file tobacco.dat on the chemical components of $n = 25$ tobacco leaf samples analyzed by Sparks et al. (1983) are utilized. There are three dependent variables and six independent variables in the data set. The vector of dependent variables $y' = (y_1, y_2, y_3)$ represent y_1: rate of cigarette burn in inches per 1000 seconds, y_2: percentage of sugar in the leaf, and y_3: percentage of nicotine in the leaf. The vector of independent variables $x' = (x_1, x_2, \ldots, x_6)$ represents the percentage of x_1: nitrogen, x_2: chlorine, x_3: potassium, x_4: phosphorus, x_5: calcium, and x_6: magnesium, used in crop cultivation.

5.9.1 Fixed X

Assuming a MR model for these data, suppose the primary goal of the study is to find the best candidate model or to discover the relationship among the dependent and fixed independent variables from the sample data. While the errors are assumed to be identically distributed, the observations are not since the data are collected at fixed values of the independent variables. For the six independent variables, one may fit all-possible regression models or a total of $2^6 - 1 = 64$ candidate models. For these data, $n = 25$, $k = 7$, and $p = 6$. The CMR model for the example is $Y_{25 \times 3} = X_{25 \times 7} B_{7 \times 3} + U$. Centering the matrix X, the MLE of the unknown parameter matrix B is

$$\widehat{B} = \begin{pmatrix} \hat{\alpha}_0' \\ B_1 \end{pmatrix} = \begin{pmatrix} \bar{y}_1' \\ B_1 \end{pmatrix} = \begin{pmatrix} 1.688 & 16.616 & 2.161 \\ 0.062 & -4.319 & 0.552 \\ -0.160 & 1.326 & -0.279 \\ 0.292 & 1.590 & 0.218 \\ -0.658 & 13.958 & -0.723 \\ 0.173 & 0.555 & 0.323 \\ -0.428 & -3.502 & 2.005 \end{pmatrix}. \tag{5.232}$$

The intercept for the noncentered data is obtained using the expression: $\hat{\beta}_0 = \bar{y} - \hat{B}_1\bar{x}$. Using (5.73)-(5.74), ML and minimum variance biased estimate of the error covariance matrix Σ are

$$\hat{\Sigma} = \begin{pmatrix} .008 & -.008 & .005 \\ -.008 & 1.090 & .003 \\ .005 & .003 & .051 \end{pmatrix} \quad S = \begin{pmatrix} .011 & -.011 & .077 \\ -.011 & 1.520 & .005 \\ .077 & .005 & .072 \end{pmatrix}. \quad (5.233)$$

The correlation matrices for the dependent and independent variables are calculated to evaluate the inter- and intra-correlations among the variables,

$$R_{yy} = \begin{pmatrix} 1.00 & -.32 & .22 \\ -.32 & 1.00 & -.72 \\ .22 & -.72 & 1.00 \end{pmatrix} \quad (5.234)$$

$$R_{yx} = \begin{pmatrix} .23 & -.62 & .49 & -.32 & -.12 & -.31 \\ -.71 & .43 & .19 & .24 & -.48 & -.53 \\ .77 & -.27 & -.30 & -.05 & .69 & .73 \end{pmatrix} \quad (5.235)$$

$$R_{xx} = \begin{pmatrix} 1.00 & -.09 & -.10 & -.11 & .63 & .60 \\ -.09 & 1.00 & -.09 & .07 & .14 & .12 \\ -.01 & -.09 & 1.00 & -.21 & -.59 & -.61 \\ -.11 & .07 & -.21 & 1.00 & .01 & .15 \\ .63 & .14 & -.59 & .01 & 1.00 & .76 \\ .60 & .12 & -.61 & .15 & .76 & 1.00 \end{pmatrix}. \quad (5.236)$$

Averaging the entries in the matrices, the sample average intra-correlations among the x's and y's are .05 and $-.27$, and the average of the inter-correlations is .06. Using the absolute values of the correlations, the values are .41, .28, and .41, respectively.

To evaluate the candidate models for these data, the following multivariate model selection criteria: det(MFPE), $|E^{-1}E_j|$, the corrected information criteria CHQ_u and $CAIC_u$, and the MLE information criteria BIC_{MLE}, AIC_{MLE}, and HQ_{MLE} are used. The candidate models selected using the criteria are provided in Table 5.2. Given the small sample size, the ratio: $\{n/[2qk+q(q+1)] = 25/52 < 40\}$ is significantly smaller than 40; hence, the ML information criteria may have located an over fitted model. While the doubly corrected information criteria generally compensate for the small ratio, they appear to be overly restrictive for these data. The MFPE and C_p criteria both locate four variable models.

Most statistical routines in the analysis of a CMNR model provide an overall test of no significant linear relationship: the null hypothesis that $B_1 = 0$. In practice, this test has limited value since the test is most often rejected. For this example, the P-values for all of the multivariate criteria are $< .001$. Given that the overall hypothesis is rejected, one may use (with caution) the protected F statistics defined in (5.102), with $\alpha = .05$, to include/remove variables from the model. The significant P-values suggest a model that includes only two variables: $(2, 6)$.

Assuming fixed independent variables, the fit criteria for the full model are shown in Table 5.3. As the entries in Table 5.3 indicate, there is a great variability among the

Table 5.2: Model Selection Criteria — Candidate Models.

Model selection criteria	Independent variables
$\|MFPE\|$	$(1, 2, 4, 6)$
$\|E^{-1}E_j\|$	$(1, 2, 4, 6)$
$AIC_{MLE}, HQ_{MLE}, BIC_{MLE}$	$(1, 2, 6)$
$CHQ_u, CAIC_u$	(6)
Protected F tests $(\alpha = .05)$	$(2, 6)$

Table 5.3: Sample Fit Criteria — Full MR Model.

Criterion	Value	Criterion	Value
$\tilde{\eta}_W^2$	.9674	$\tilde{\eta}_{B-N-P}^2$	.5726
$\tilde{\eta}_J^2$	.9548	$\tilde{\eta}_G^2$	.6806
$\tilde{\eta}_{B-L-H}^2$	.7560	$\tilde{\eta}_H^2$	.7561
$\tilde{\eta}_R^2$	.8702	$\tilde{\eta}_N^2$	.9321

fit criteria, ranging from a low of .57 to a high of .97. In multiple linear regression, fit criteria are employed to evaluate the proportion of the variation in the dependent variables associated with the independent variables. Such an interpretation may also be applied to the CNMR model; however, the criteria are not single valued. Thus, the practice of converting these indices to effect sizes must be questioned.

Using the models in Table 5.2 as a basis for the relationship between the dependent and independent variables, the test statistics in (5.98)-(5.101) may be used to test the hypothesis that the coefficients associated with the variables not in the model are zero. Generally, one should find the tests to be nonsignficant. Nonsignificance tends to support the exclusion of the subset of variables from the model and indirectly substantiates the validity of the candidate model. The P-values for the joint tests and the selected subsets of variables are provided in Table 5.4.

The P-values for the test statistics in the Table 5.4 seem to support the candidate

Table 5.4: Tests of Significance for (Variables Excluded) by Criteria.

Test	$MLE - IC$	C_p	Corrected $- IC$	Protected F
Statistic	$(3 - 5)$	$(3, 5)$	$(1 - 5)$	$(1, 3 - 5)$
Λ	.5111	.5792	.0033	.0353
$V^{(s)}$	.4576	.5638	.0407	.0682
$U^{(s)}$	.5427	.5792	.0007	.0277
$\theta^{(s)}$	.2003	.2521	.0001	.0010

models selected using the modified C_p criterion, and the MLE information criteria. The models selected using the doubly corrected criteria and the protected F test are clearly rejected. Because it is better to over fit than under fit, the selection procedures seem to support a model with only three independent variables: $(1, 2, 6)$. In general, if one is unsure when fitting two models that differ by one or two variables, it is often better to select the model that is wrong with one more variable than one that is wrong with one less variable. One cannot estimate the coefficient associated with a model when a variable has been excluded incorrectly. It is sometimes better to be approximately right than exactly wrong.

5.9.2 Random X

Assuming that the data are collected so that the vector pairs are jointly multivariate normal, the observation vectors are identically distributed and we may perform a canonical analysis of the data. Because the data vectors are jointly normal, the fit criteria are estimates of internal precision in the population. And, may be viewed as estimates of the population parameter. If the primary goal of the analysis is to discover the best candidate model with random independent variables, one may still use the information criteria to evaluate candidate models. However, with random predictors, the goal of the data analysis is often to find the best subset of predictors. When this is the case, to locate the best candidate models, one should evaluate the reduction in the mean square error of prediction.

To initiate the analysis, the correlation matrices are substituted into (5.117) since the eigenvalues of the matrix C_{xy} are identical to R_{xy}. The values of the matrices for the tobacco leaf data follow

$$R_{yy}^{-1/2} = \begin{pmatrix} 1.0426902 & .1617986 & -.028089 \\ .1617986 & 1.3374387 & .5288663 \\ -.028089 & .5288663 & 1.3005564 \end{pmatrix} \qquad (5.237)$$

$$R_{xy} = \begin{pmatrix} .671353 & .161800 & -.028090 \\ -.105689 & .451065 & .528866 \\ -.080075 & .528860 & .590478 \end{pmatrix}. \qquad (5.238)$$

Solving the deterimental equation, the sample canonical correlations are: $r_1^2 = .0870178$, $r_2^2 = .708677$, and $r_3^2 = .138823$. Alternatively, solving the deterimental equation $|H - \lambda E| = 0$ for testing the null hypothesis $H_0 : B_1 = 0$, the eigenvalues are $\lambda_1 = 6.7029$, $\lambda_2 = 2.4326$, and $\lambda_3 = .1612$. The relationship between the square of the canonical correlations and the eigenvalues is as claimed: $r_i^2 = \lambda_i/(1 + \lambda_i)$ where $s = \min(q, p) = 3$.

Evaluating (5.124) for the sample data, we have the following result for the measures of internal precision or population fit since the matrix X is random

$$\rho_s^2 \leq D_{xy}^2 \leq M_{xy}^2 \leq \rho_1^2 \leq \rho_{xy}^2$$

$$.1388 < .5726 < .7571 < .8702 < .9674 \qquad (5.239)$$

illustrating that the inequality is satisfied. The population measures of precision RV^2 and RD^2 for the tobacco leaf data are: $RV^2 = .5906$ and $RD^2 = (.7437 + .5015)/_2 = .6226$ which appear on the lower end of multivariate association. There does not appear to be any best measure of population model fit for a jointly normal multivariate regression model.

5.9.3   Random X, Prediction

Given joint multivariate normality, primary interest of an analysis is often not about model precision, but model prediction. Because it is difficult to estimate the multivariate random mean square error of prediction, D_r^2, the proxy estimate $\widehat{\Delta}_r^2$ is used. Assuming the independent variables are specified beforehand and are not chosen from a larger set of predictors and that any estimate of model parameters may be applied to other members of population, we illustrate the AIC criterion suggested by Fujikoshi (1985) to obtain a best subset of random predictors for the tobacco leaf data. The subsets of variables that maximize the criterion in (5.143) for the 62 possible models for $m = 1, 2, 3, 4, 5 < 6$ are reviewed. The maximum over all subsets occurs for the subset of variables $(1, 2, 6)$, a maximum value of 17.76, almost twice the value of the other subset $(1, 2, 3)$, and three times the value of the criterion for subsets $(2, 3, 6)$ and $(2, 4, 6)$, with values 9.50, 5.28, and 4.76, respectively. The next best candidate model includes the four variables $(1, 2, 4, 6)$, a value of 12.94, and the variables $(1, 2, 3, 6)$, a value of 10.88. The best two variable model includes the variables $(2, 6)$, and the single variable model includes only variable (6).

5.9.4   Overview — Candidate Model

For these data, the centered three variable candidate model is

$$Y_{25\times 3} = X_{25\times 4}B_{4\times 3} + U_{25\times 3}.$$

The OLS estimate for the parameter matrix B is

$$\widehat{B} = \begin{pmatrix} \widehat{\alpha}_0' \\ \widehat{B}_1 \end{pmatrix} = \begin{pmatrix} \bar{y}' \\ \widehat{B}_1 \end{pmatrix} = \begin{pmatrix} 1.688 & 16.616 & 2.161 \\ .317 & -3.760 & .830 \\ -.160 & 1.414 & -.252 \\ -.754 & -4.387 & 2.093 \end{pmatrix}. \tag{5.240}$$

The MLE and minimum variance unbiased estimate of the error covariance matrix Σ are

$$\widehat{\Sigma} = \begin{pmatrix} .010 & -.010 & .007 \\ -.010 & 1.310 & -.003 \\ .007 & -.003 & .056 \end{pmatrix} \quad S = \begin{pmatrix} .012 & -.019 & .009 \\ -.019 & 1.560 & .004 \\ .009 & .004 & .067 \end{pmatrix}. \tag{5.241}$$

Table 5.5: Sample Fit Criteria — Reduced Model.

Criterion	Value	Criterion	Value
$\tilde{\eta}_W^2$	.9478	$\tilde{\eta}_{B-N-P}^2$	.4995
$\tilde{\eta}_J^2$	.9377	$\tilde{\eta}_G^2$	.6262
$\tilde{\eta}_{B-L-H}^2$	.7175	$\tilde{\eta}_H^2$	.7175
$\tilde{\eta}_R^2$	.8530	$\tilde{\eta}_N^2$	.9020

And, the correlation matrices for the dependent and independent variables follow

$$R_{yy} = \begin{pmatrix} 1.00 & -.32 & .22 \\ -.32 & 1.00 & -.72 \\ .22 & -.72 & 1.00 \end{pmatrix} \tag{5.242}$$

$$R_{yx} = \begin{pmatrix} .23 & -.62 & -.31 \\ -.71 & .43 & -.53 \\ .77 & -.27 & .73 \end{pmatrix} \tag{5.243}$$

$$R_{xx} = \begin{pmatrix} 1.00 & -.09 & .60 \\ -.09 & 1.00 & .12 \\ .60 & .12 & 1.00 \end{pmatrix}. \tag{5.244}$$

Averaging the entries in the sample correlation matrices, the sample average intra-correlations among the revised x's is .21, and average inter-correlations is $-.03$. The corresponding values for the full model were: .05 and .06, respectively.

The sample fit criteria for the reduced model with fixed X are shown in Table 5.5.

Comparing the fit criteria for the full and reduced models, the differences do not appear to be large. The corresponding values of association or population model fit for random X follow

$$\rho_s^2 \leq D_{xy}^2 \leq M_{xy}^2 \leq \rho_1^2 \leq \rho_{xy}^2$$
$$.0009 < .4995 < .7175 < .8530 < .9478. \tag{5.245}$$

The population measures of association for RV^2 and RD^2 for the tobacco leaf data are: $RV = .67$ and $RD^2 = .66$. These two measures of precision are larger for the reduced model than the full model.

5.9.5 Prediction and Shrinkage

The sum of square of prediction error when the independent variables are random for the reduced model, under normality, is $R_{SPER}(\widehat{B}) = .45$ while the corresponding value for the full model is 1.056. Thus, the reduced model has better predictive precision than the full model. We cannot calculate the $FICYREG$ estimate for the matrix B for the reduced model since $p < q + 1$ and $r^* < 0$. Setting $r^* = 0$, the

$FICY REG$ estimate is identical to the OLS estimate. However, $b^* = .08$ so that the Efron-Morris estimate may be calculated by setting $r^* = 0$. The estimate of the multivariate flattening matrix is

$$\widehat{A}_{EM} = \begin{pmatrix} .99972 & .00027 & -.00021 \\ .00027 & .96465 & .00010 \\ -.03341 & .00010 & .99847 \end{pmatrix}. \tag{5.246}$$

Using this estimate, the matrix of estimated regression coefficients becomes

$$\widehat{B}_{EM} = \begin{pmatrix} \bar{y}' \\ \widehat{B}_c \end{pmatrix} = \begin{pmatrix} 1.6916 & 16.0290 & 2.1591 \\ .3161 & -3.6268 & .8286 \\ -.1398 & 1.3636 & -.2513 \\ -.7553 & -4.2312 & 2.0893 \end{pmatrix}. \tag{5.247}$$

And, the SPE risk for the estimate is $8.9257 < 9$, the SPE risk for the OLS estimate.

Continuing, for the full model both estimates $\widehat{B}_{FICY REG}$ and $\widehat{B}_{EM}$ may be using (5.153) and (5.155) since the values of r^* and b^* are both equal to .09091 for the full model. The estimate of the multivariate flattening matrices is

$$\widehat{A}_{FICY REG} = \begin{pmatrix} .93006 & -.94557 & -.18441 \\ -.01003 & .66725 & -.05832 \\ -.03341 & -1.09107 & .78781 \end{pmatrix} \tag{5.248}$$

$$\widehat{A}_{EM} = \begin{pmatrix} .92983 & -.94534 & -.18456 \\ -.01002 & .63620 & -.05841 \\ -.03356 & -1.09116 & .78633 \end{pmatrix}. \tag{5.249}$$

Using the matrix $\widehat{A}_{EM}$, the OLS estimate $\widehat{B}$ is flattened by the matrix $\widehat{A}_{EM}$ to create the matrix $\widehat{B}_{EM} = \widehat{B}\widehat{Z}_{EM}$, given below:

$$\widehat{B}_{EM} = \begin{pmatrix} \bar{y}' \\ \widehat{B}_c \end{pmatrix} = \begin{pmatrix} 1.331 & 6.617 & .417 \\ .082 & -3.409 & .675 \\ -.153 & 1.299 & -.267 \\ .248 & .498 & .024 \\ -.727 & 10.288 & -1.262 \\ .145 & -.164 & .190 \\ -.403 & -4.011 & 1.860 \end{pmatrix}. \tag{5.250}$$

The matrix $\widehat{A}_{EM}$ "flattened" the matrix $\widehat{B}$ and the fitted values $\widehat{Y} = X\widehat{B}_{EM}$ have increased precision in that the SPE risk for the flattened estimate is smaller than the OLS estimate. While the value of $R_{SPER}(\widehat{B}) = 18$ for the OLS estimate, the corresponding value for the $FICY REG$ and Efron-Morris estimates are 16.7702 and 16.7047, respectively. Thus, one can improve upon the predictive precision of the model by using a multivariate flattening matrix which takes into account the correlations among the dependent variables. This fact is typically ignored by practitioners who continue to use univariate multiple linear regression methods to estimate regression coefficients that do not take into account the correlations among the dependent variables.

5.10   MULTIVARIATE REGRESSION: INFLUENTIAL OBSERVATIONS

In our previous example, the data followed a multivariate distribution. Having examined the data for normality, one often wants to also determine whether or not certain observations are influential on the parameter estimates and whether there are outliers in the data. Multivariate extensions of some univariate measures have been proposed by Barrett and Ling (1992); these include, for example, Cook's distance, DFFITS, DFBETA, CVR indices for a single observation a or a subset of observations. If we let $\widehat{B}$ be the OLSE of B, and $\widehat{B}_{(I)}$ the estimate with I observations in the data matrix Y excluded (the complement of $B_{(I)}$ is represented as B_I) then the multivariate extension of Cook's distance measure which may be used to evaluate the effect of I observations on the estimation of B and is given by

$$D_I^2 = \frac{[\text{vec}(\widehat{B} - \widehat{B}_{(I)})]'[S^{-1} \otimes (X'X)][\text{vec}(\widehat{B} - \widehat{B}_{(I)})]}{k} \qquad (5.251)$$

where $S = \frac{E}{(n-k)}$ is the unbiased estimate of Σ and $\widehat{B}_{(I)} = (X'_{(I)}X_{(I)})^{-1}X'_{(I)}Y_{(I)}$. Letting $V = kS$ and $M = (X'X)$, Cook's distance has the quadratic form structure:

$$(\hat{\gamma} - \gamma_0)'(V \otimes M)^{-1}(\hat{\gamma} - \gamma_0)$$

which is similar to (5.34). While D_I^2 does not have a chi-square distribution, like the univariate case, one may relate its value to the corresponding chi-square distribution with $\nu = kp$ degrees of freedom. If the percentile value is larger than 50 critical value, the distance between $\widehat{B}$ and $\widehat{B}_{(I)}$ is large so that the I observations affect the estimate of the model parameters. Since the 50th percentile for large data sets is near ν, if $D_I^2 > \nu$ the I observations are considered influential and some remedial action may have to be taken.

Barrett and Ling (1992) show that Cook's distance may be written as a function of the leverage or "hat" matrix $H_I = X_I(X'X)^{-1}X'_I$ and the residual matrix $Q_I = U_{(I)}(U'U)^{-1}U'_{(I)}$ where the matrix $U_{(I)} = Y_{(I)} - X_{(I)}\widehat{B}_{(I)}$ and $Y_{(I)}$, $\widehat{B}_{(I)}$ and $X_{(I)}$ are the complements of Y_I, $\widehat{B}_I$ and X_I, respectively, with the I indexed observations included. After some manipulation, Cook's distance measure for the multivariate regression model may be expressed:

$$D_I^2 = [(n - k)k]\text{tr}[(I - H_I)^{-2}H_IQ_I]. \qquad (5.252)$$

We see that D_I^2 has the general form $f(\cdot)\text{tr}(L_IR_I)$ where $L_I = (I - H_I)^2$ and $R_I = (I - H_I)^{-1/2}Q_I(I - H_I)^{-1/2}$ where L_I and R_I represent the leverage and residual matrices, respectively. From vector geometry, the law of cosines states that the $\|x\| \|y\| \cos\theta = x'y = \text{tr}(yx')$ where $0° \leq \theta \leq 180°$ is the angle between x and y. Hence, the

$$\begin{aligned} \text{tr}(L_IR_I) &= \|vec(L_I)\|\|vec(R_I)\| \cos\theta_I \\ &= [\|L_I\|(\cos\theta_I)^{1/2}][\|R_I\|(\cos\theta)^{1/2}] \\ &= L_I^*R_I^* \end{aligned}$$

where L_I^* and R_I^* represent the relative contribution of leverage and residual to Cook's total influence measure. Barrett and Ling (1992) recommend plotting the logarithms of L_I^* and R_I^*, for various values of I to assess the joint influence of multiple observation on the estimation of B for the multivariate linear regression model.

Hossain and Naik (1989) and Srivastava and Rosen (1999) generalize Studentized residuals to the multivariate linear model by forming statistics that are the square of $r_i = \frac{\hat{e}_i}{\sqrt{s^2(1-h_{ii})}}$ and $r_{(i)} = \frac{\hat{e}_{(i)}}{\sqrt{s_{(i)}^2(1-h_{ii})}} = t_i \sim t(n-k-1)$ for the univariate linear model. The internally and externally Studentized residuals are defined as:

$$r_i^2 = \frac{\hat{e}_i' S^{-1} \hat{e}_i}{(1-h_{ii})} \text{ and } T_i^2 = \frac{\hat{e}_i' S_{(i)}^{-1} \hat{e}_i}{(1-h_{ii})} \qquad (5.253)$$

for $i = 1, \ldots, n$ where $\hat{e}_i$ is the ith row of $\widehat{E} = Y - X\widehat{B}$. Because T_i^2 has Hotelling's T^2 distribution and $r_i^2 \sim \text{Beta}[p/2, (n-q-p-2)/2)]$, assuming no outliers, an observation vector y_i' may be considered an outlier if the statistic

$$\frac{(n-q-p)}{p} \frac{T_i^2}{(n-q-1)} > F_{(p,n-q-1)}^{(1-\alpha^*)} \qquad (5.254)$$

where α^* is selected to control the familywise error rate for n tests near the nominal level α. Hossain and Naik (1989) also define multivariate extensions of the statistics DFFITS, DFBETA, and the CVR for the multivariate regression model. However, SAS procedures do not currently calculate any of these statistics. Hence, users must provide their own code using PROC IML. Finally, we may use the matrix of residuals to construct plots of residuals versus predicted values or variables not in the model to evaluate potential multivariate models.

While one may use PROC GLM to analyze classical multivariate regression models in SAS, from Section 5.9 we saw that PROC IML code is often used to analyze multivariate regression models; even though PROC REG may be used to fit multivariate models, it has severe limitations. The primary thrust of the software is toward testing multivariate hypotheses and not in fitting multivariate equations to data or in exploratory residuals for multivariate models. PROC REG uses an F approximation to obtain P-values of the multivariate test criteria which by Theorem 5.5 is often exact for Wilks' Λ criterion. The relationship between the criteria and the F-distribution is provided in the SAS/STAT *User's Guide* (1990, pp. 17-19).

To evaluate the effect of a single observation on the estimate of B for the classical multivariate linear regression model, Barrett and Ling (1992) used William D. Rohwer's data given in Timm (2002, p. 213). We use the same data in this section to fit a classical regression model to the data, calculate Cook's distance and evaluate the data for outliers using Chi-square plots. The data are in the data set Rohwer.dat. Rohwer was interested in predicting the performance on three standardized tests: Peabody Picture Vocabulary (y1), the Raven Progressive Matrices test (y2) Student Achievement (y3), and in using five paired-associate, learning-proficiency tasks: Named (N:x1), Still (S:x2), Named Still (NS:x3), Named Action (NA:x4), and

Sentence Still (SS:x5) for 32 school children in an upper-class, white residential school. The SAS code for the example is included in Program 5_10.sas.

5.10.1 Results and Interpretation

Fitting a full model to the data, the chi-square plot indicates that the data are multivariate normal; and Mardia's test of skewness and Kurtosis are not significant. The residual plots do not indicate the presence of outliers. Calculating Cook's distance for the data and comparing D^2 with the value $\nu = kp = (6)(3) = 18$, we see that none of the observed values are even close to the value of 18; however, observation 5 has the largest influence value (0.8467); this is consistent with the results reported by Barrett and Ling (1992) who found case 5 to have the largest influence and the largest leverage component. Barrett and Ling (1992) suggest investigating subsets of observations that have large joint-leverage or joint-residual components, but small influence. They found that subset $\{5,14\}$ and $\{5,25\}$ have the largest joint-leverages, although not sufficiently large to suggest data modification.

Having determined that the data are well behaved, we next move to the model refinement phase by trying to reduce the set of independent variables to explain the relationship among the set of dependent variables: model calibration. When using SAS for this phase we have to depend on univariate selection methods, e.g. C_p-plots and stepwise methods. While SAS provides a measure of the strength of the model fit using R^2 a variable at a time, among others, no multivariate measure is provided. The set of independent variables appear to predict y3 best.

We combine univariate methods and multivariate tests of hypotheses regarding the elements of the parameter matrix $B = \begin{pmatrix} \beta'_0 \\ \Gamma \end{pmatrix}$ using MTEST statements. The MTEST statements are used to test that model parameters associated with the independent variables are zero by separating the independent variable names with commas. For example, to test $H : \Gamma = 0$, the MTEST statement is mtest x1, x2, x3, x4, x5 /print; and to include the intercept, the key word intercept must be included in the list. A statement with only one variable name is test $H : \gamma_i = 0$. Thus mtest x5/print; test $H : \gamma_5 = 0$. The PRINT option prints the hypothesis test matrix H and the error matrix E for the test.mtest x1, x2, x3, x4, x5 /print; and to include the intercept, the key word intercept must be included in the list. A statement with only one variable name is test $H : \gamma_i = 0$. Thus mtest x5/print; test $H : \gamma_5 = 0$. The PRINT option prints the hypothesis test matrix H and the error matrix E for the test.

Fitting the full model and reviewing the univariate tests for each dependent variable suggests that only the variable SS can be excluded from the model. However, the joint MTEST suggest excluding both SS (x5) and N (x1).

Fitting the reduced model, the measure of model fit defined η^2-like correlation ration index given in (5.107), over 62 of the variations in the dependent variables are accounted for by the variables, compared to 70 for the full model. Thus, we use the reduced model for the data. Thus, we are led to fitting a reduced model that only includes the variables S (x2), NS (x3), NA (x4). The same conclusion is reached using the multivariate selection criterion HQIC defined in (5.93) using program MulSubSelrev.sas, for the Rohwer data.

Fitting the reduced model, we may again test that the coefficients associated with the independent variables are all zero. Following the rejection of the overall test, we may want to find simultaneous confidence sets for the regression coefficients. Because PROC REG in SAS does not generate the confidence sets for the parameter matrix, PROC IML is used. To illustrate the method, an approximate simultaneous confidence interval for β_{43} is provided using $\alpha = 0.05$ and Roy's criterion. While the confidence set does not include zero, recall that it is a lower bound for the true interval. Using planned comparisons for nine parameters, the critical value for the multivariate t (Studentized Maximum Modulus) distribution is 2.918, very near the Roy criterion.

In this example we have shown one how to find a multivariate regression equation using a calibration sample. To validate the equation, one must use a validation sample or use the methods illustrated in our first example to develop a prediction equation.

In our discussion of the GLM theory and in our examples of the multivariate linear regression, we have assumed that the design matrix has full rank or that it is reparameterized to full rank. When using PROC REG, the design matrix must be of full rank. If the matrix is not of full rank, one may use PROC GLM; however, the normal equations no longer have a unique solution. Using a generalized inverse of $X'X$ a solution may take the form:

$$\widehat{B} = (X'X)^- X'Y. \tag{5.255}$$

Even though the parameter matrix is not unique, certain linear functions of the estimate are unique. These are called estimable functions. Using the following theorem, one may easily determine whether or not a parametric function is estimable.

Theorem 5.9. *A parametric function $\psi = c'Ba$ is uniquely estimable if and only if $c'H = c'$ where $H = (X'X)^-(X'X)$. For a proof see Timm (1975, p. 174). A linear hypothesis ψ is testable if and only if the parameter is uniquely estimable.*

5.11 NONORTHOGONAL MANOVA DESIGNS

For this example, we consider a completely randomized two factor factorial design with $n_{ij} > 0$ observations per cell, factors A and B, and all interactions in the model. Hence, there are no empty cells. The overparameterized parametric model for the two-factor design is:

$$y_{ijk} = \mu + \alpha_i + \beta_j + \gamma_{ij} + e_{ijk} \text{ where } e_{ijk} \sim IN(0, \Sigma) \tag{5.256}$$

where y_{ijk} is a p-variate random vector, $i = 1, 2, \ldots, a$; $j = 1, 2, \ldots, b$; $k = 1, 2, \ldots, n_{ij} > 0$. The subscripts i and j are associated with the levels of factors A and B. The model parameters α_i and β_j are main effects and γ_{ij} is the vector of interactions. Letting $n_{ij} = 2$, $a = 3$, $b = 2$, we present the design matrix X and the fixed parameter matrix B for an orthogonal design:

$$
B = \begin{pmatrix}
\mu_{11} & \mu_{12} & \cdots & \mu_{1p} \\
\alpha_{11} & \alpha_{12} & \cdots & \alpha_{1p} \\
\alpha_{21} & \alpha_{22} & \cdots & \alpha_{2p} \\
\alpha_{31} & \alpha_{32} & \cdots & \alpha_{3p} \\
\beta_{11} & \beta_{12} & \cdots & \beta_{1p} \\
\beta_{21} & \beta_{22} & \cdots & \beta_{2p} \\
\gamma_{111} & \gamma_{112} & \cdots & \gamma_{11p} \\
\gamma_{121} & \gamma_{122} & \cdots & \gamma_{12p} \\
\vdots & \vdots & \ddots & \vdots \\
\gamma_{321} & \gamma_{322} & \cdots & \gamma_{32p}
\end{pmatrix}
$$

$$
X = \begin{pmatrix}
1 & 1 & 0 & 0 & 1 & 0 & 1 & 0 & 0 & 0 & 0 & 0 \\
1 & 1 & 0 & 0 & 1 & 0 & 1 & 0 & 0 & 0 & 0 & 0 \\
1 & 1 & 0 & 0 & 0 & 1 & 0 & 1 & 0 & 0 & 0 & 0 \\
1 & 1 & 0 & 0 & 0 & 1 & 0 & 1 & 0 & 0 & 0 & 0 \\
1 & 0 & 1 & 0 & 1 & 0 & 0 & 0 & 1 & 0 & 0 & 0 \\
1 & 0 & 1 & 0 & 1 & 0 & 0 & 0 & 1 & 0 & 0 & 0 \\
1 & 0 & 1 & 0 & 0 & 1 & 0 & 0 & 0 & 1 & 0 & 0 \\
1 & 0 & 1 & 0 & 0 & 1 & 0 & 0 & 0 & 1 & 0 & 0 \\
1 & 0 & 0 & 1 & 1 & 0 & 0 & 0 & 0 & 0 & 1 & 0 \\
1 & 0 & 0 & 1 & 1 & 0 & 0 & 0 & 0 & 0 & 1 & 0 \\
1 & 0 & 0 & 1 & 0 & 1 & 0 & 0 & 0 & 0 & 0 & 1 \\
1 & 0 & 0 & 1 & 0 & 1 & 0 & 0 & 0 & 0 & 0 & 1
\end{pmatrix}.
$$

Thus, we see that each column of B is associated with a row of the design matrix. Unfortunately, this design matrix X is not of full column rank, so a unique inverse does not exist; the $\operatorname{rank}(X) = ab$. While one may reparameterize the matrix to full rank by adding side conditions this is not the approach taken in SAS using PROC GLM; instead PROC GLM obtains a generalized inverse and provides a general form of estimable functions using Theorem 5.9. For the two-factor factorial design, the general form of parametric functions that are estimable and hence testable are given in (Timm, 1975, p. 404):

$$
\psi = c'Ba = a'[\sum_{i=1}^{a} \sum_{j=1}^{b}(\mu + \alpha_i + \beta_j + \gamma_{ij})] \tag{5.257}
$$

$$
\hat{\psi} = c'\widehat{B}a = a'(\sum_{i=1}^{a} \sum_{j=1}^{b} t_{ij} y_{ij\cdot}) \tag{5.258}
$$

where $y_{ij\cdot}$ is a vector of cell means and the $t' = (t_{ij})$ is a vector of arbitrary constants.

As in a univariate analysis, ψ's involving the individual effects of α and β are not directly estimable. One can see that no constants t_{ij} exist that would isolate an individual main effect. However, if we select the constants t_{ij} such that the $\sum_j t_{ij} = \sum_j t_{i'j} = 1$, then

$$\psi = \alpha_{is} - \alpha_{i's} + \sum_{j=1}^{b} t_{ij}(\beta_{js} + \gamma_{ij}) - \sum_{j=1}^{b} t_{i'j}(\beta_{js} + \gamma_{ij}) \qquad (5.259)$$

for $s = 1, 2, \ldots, p$ is estimated by

$$\hat{\psi} = \sum_{j=1}^{b} t_{ij} y_{ij\cdot}^{(s)} - \sum_{j=1}^{b} t_{i'j} y_{i'j\cdot}^{(s)}. \qquad (5.260)$$

where $y_{ij\cdot}^{(s)}$ is a cell mean for variable s. A similar expression exists for the β's. Thus, the hypothesis for the main effects becomes:

$$H_{A^*} : \alpha_i + \sum_{j=1}^{b} t_{ij}(\beta_j + \gamma_{ij}) \text{ are equal} \qquad (5.261)$$

$$H_{B^*} = \beta_j + \sum_{i=1}^{a} t_{ij}(\alpha_i + \gamma_{ij}) \text{ are equal} \qquad (5.262)$$

for arbitrary t_{ij} that sum to unity. For example, letting $t_{ij} = \frac{1}{b}$ for H_{A^*} and $t_{ij} = \frac{1}{a}$ for H_{B^*}, the hypothesis becomes:

$$H_A : \alpha_i + \sum_{j=1}^{b} \frac{\gamma_{ij}}{b} \text{ are equal} \qquad (5.263)$$

$$H_B : \beta_j + \sum_{i=1}^{a} \frac{\gamma_{ij}}{a} \text{ are equal.} \qquad (5.264)$$

Hence, tests of main effects are confounded by interaction, whether the design is orthogonal or nonorthogonal.

To test hypotheses involving interactions, observe that all functions (tetrads) of the form

$$\psi = \gamma_{ij} - \gamma_{i'j} - \gamma_{ij'} + \gamma_{i'j'} \qquad (5.265)$$

are estimable and estimated by

$$\hat{\psi} = y_{ij\cdot} - y_{i'j\cdot} - y_{ij'\cdot} + y_{i'j'\cdot}. \qquad (5.266)$$

The interaction hypothesis tests that all tetrads are zero; when this is the case, the parameters γ_{ij} may be removed from the model (the model becomes additive) and tests of main effects are no longer confounded by the interaction.

When the factorial design is nonorthogonal, the coefficients t_{ij} are design dependent. For example, letting the coefficients t_{ij} depend on the cell frequencies n_{ij}, we have the weighted tests:

$$H_{A^*} : \alpha_i + \sum_{j=1}^{b} n_{ij} \frac{(\beta_j + \gamma_{ij})}{n_{i+}} \text{ are equal} \tag{5.267}$$

$$H_{B^*} : \beta_j + \sum_{i=1}^{a} n_{ij} \frac{(\alpha_i + \gamma_{ij})}{n_{+j}} \text{ are equal} \tag{5.268}$$

where n_{ij} is a cell frequency, n_{i+} and n_{+j} are sums over column and row frequencies, respectively. Choosing weights proportional to the number of levels in a design yields the unweighted tests:

$$H_A : \alpha_i + \sum_{j=1}^{b} \frac{\gamma_{ij}}{b} \text{ are equal} \tag{5.269}$$

$$H_B : \beta_j + \sum_{i=1}^{a} \frac{\gamma_{ij}}{a} \text{ are equal.} \tag{5.270}$$

Other more complicated functions are also estimable and hence testable. One function tested in PROC GLM for a factorial design is the function:

$$\psi_i = a' \left[\sum_{j=1}^{b} n_{ij}(\mu + \alpha_i + \beta_j + \gamma_{ij}) - \sum_{k=1}^{a} \frac{n_{ij} n_{kj}}{n_{+j}} (\mu + \alpha_i + \beta_j + \gamma_{ij}) \right] . \tag{5.271}$$

The inclusion or exclusion of the interaction parameter also determines which functions of the model parameters are estimable and their estimates. For additional details see Timm (1975, p. 511-526) for multivariate models and Timm and Carlson (1975) or Milliken and Johnson (1992, p. 138) for the univariate case.

The orthogonal decomposition of the hypothesis test space is not unique for a nonorthogonal design (Timm & Carlson, 1975). Because of this, PROC GLM evaluates the reduction in the error sum of squares for the parameters in the model for some hierarchical sequence of the parameters. For example, one may fit α in the model in the orders:

$$I : \alpha \text{ after } \mu$$

$$II : \alpha \text{ after } \mu \text{ and } \beta$$

$$III : \alpha \text{ after } \mu, \beta, \gamma$$

These are termed the Type I, Type II, and Type III sum of squares, respectively, in SAS. The Type I sum of squares evaluates the effect of the parameter added to the model in the order defined by the MODEL statement. The Type II sum of squares evaluates the effect after adjusting for all other terms in the model that do not include it. The Type III sum of squares evaluates the effect of the last term given that all other terms are in the model. For nonorthogonal designs, the Type I, Type II, and Type III sum of squares do not add to the total sum of squares about the mean. This was illustrated in Section 3.3 for univariate designs where we also saw that the test of interaction is invariant to the weighting scheme.

For a two-factor orthogonal design, the Type I, Type II, and Type III are equal. SAS also computes a Type IV sum of squares for use with designs having empty cells; for a discussion of Type IV sum of squares, consult Milliken and Johnson (1992, p. 186). In general, the various sum of squares for a nonorthogonal design are testing very different hypotheses. For a nonorthogonal two-factor design, the Type I sum of squares is a weighted test of A, H_{A*} that depends on the cell frequencies. The Type III sum of squares is an unweighted test of A, H_A that depends on the number of levels of the factor. The Type II sums of squares is testing (5.266) that all ψ_i are equal. For most nonorthogonal design, the most appropriate test is that which uses the Type III sum of squares (Timm & Carlson, 1975; Searle, 1994) since the unequal cell frequencies do not depend upon the treatment. If one starts with a design that has equal cell frequencies, but where the unequal cell frequencies are due to treatment (subjects die), then the weighted tests are appropriate. Additional advice on the utilization of the various sum of squares is provided by Milliken and Johnson (1992, p. 150).

To illustrate the general theory of this chapter for an experimental design and to extend the discussion of nonorthogonality to a more complex design, we will analyze a completely randomized fixed effect nonorthogonal three-factor design with $n_{ijk} > 0$ observations per cell. The overparameterized model for the design is:

$$y_{ijk} = \mu + \alpha_i + \beta_j + \lambda_k + (\alpha\beta)_{ij} + (\alpha\lambda)_{ik} + (\beta\lambda)_{jk} + (\alpha\beta\lambda)_{ijk} + e_{ijkm}$$
$$\text{where } e_{ijkm} \sim IN(0, \Sigma)$$

$$(5.272)$$

and $i = 1, 2, \ldots, a$; $j = 1, 2, \ldots, b$; $k = 1, 2, \ldots, c$; $m = 1, 2, \ldots, n_{ijk} > 0$. The subscripts i, j, and k are associated with the levels of factors A, B, and C, respectively. The data for the example are hypothetical where factor A has three levels, factor B has three levels, and factor C has two levels.

Before analyzing the data for a nonorthogonal three factor design, one must determine the appropriate sum of squares to test for main effects and interactions. For a two-factor design, the interaction test was invariant to the design weights. For a three factor design, this is also the case for the highest order interaction. Now, both main effect tests and tests of two-way interactions depend on the design weights. If the unequal cell frequencies are the result of treatment effects so that upon replication of the study one would expect the same proportion of cell frequencies, then the most appropriate design weights are cell frequencies, yielding weighted tests or Type I sums of squares. If this is not the case, the cell frequencies do not depend upon the treatment then unweighted tests are most appropriate or Type III sums of squares. For most designs, Type II sums of squares are not appropriate for testing; however, they are appropriate in exploratory studies to evaluate the contribution of model parameters, adjusting for the parameters already in the model. The SAS code for the example is included in Program 5_11.sas.

First the DATA step reads in the data. Next PROC GLM, using the less-than-full-rank model, performs an unweighted analysis of the data. Then an unweighted analysis using the full-rank model is performed. Next, weighted analyses are performed, first using the less-than-full-rank model and then using the full-rank model. Finally, simultaneous confidence sets are constructed using PROC IML and the TRANSGEN procedure.

5.11.1   Unweighted Analysis

To analyze the data for this example of the three factor design, we first assume that the unequal sample sizes are not due to treatment and that the levels for the design exhaust the categories of interest so that the Type III sum of squares is most appropriate, or unweighted tests. Unweighted tests are obtained by specifying the option SS3 in the MODEL statement. ESTIMATE statements generate contrasts $\hat{\psi}$ while CONTRAST statements perform the tests that $\psi = 0$. To illustrate that differences in main effects are dependent on other model parameters we include an ESTIMATE statement that compares the average of $A1$ with the average of $A3$; the average is obtained with the option DIVISOR=3. The option E in the ESTIMATE statement causes the coefficients for the estimable function to be output. We also include the contrast and estimate of the average of $C1$ vs. $C2$ which is the same as the overall test. The contrast $A1$ vs. $A3$ performs the multivariate test of the null hypothesis that the contrast is significantly different from zero. If one includes all linearly independent rows in the CONTRAST statement: contrast "Test of A' a 1 0 -1, a 0 1 -1; the overall test of the contrast would be identical to the overall test of A. The MANOVA statement can be used to perform all unweighted tests of the model parameters. The statement associated with the MANOVA analysis in Program 5_11.sas generates all Type III F-tests and the Type III multivariate tests for all factors in the MODEL statement. Because the design is nonorthogonal, the sum of squares do not add to the total sum of squares about the mean.

The unweighted test of the interaction AB is testing that all tetrads of the form

$$\psi = (\alpha\beta)_{ij} - (\alpha\beta)_{i'j} - (\alpha\beta)_{ij'} + (\alpha\beta)_{i'j'} = 0 \qquad (5.273)$$

where the estimates are

$$\hat{\psi} = y_{ij\cdot} - y_{i'j\cdot} - y_{ij\cdot} + y_{i'j'\cdot}, \qquad (5.274)$$

tetrads in cell means. Finally, the three-factor test of interaction is testing that all differences in two-way tetrads are simultaneously zero:

$$\begin{aligned} &[(\alpha\beta)_{ijk} - (\alpha\beta)_{i'jk} - (\alpha\beta)_{ijk'} + (\alpha\beta)_{i'j'k}] \\ &- [(\alpha\beta)_{ijk'} - (\alpha\beta)_{i'jk'} - (\alpha\beta)_{ij'k'} + (\alpha\beta)_{i'j'k'}] = 0 \end{aligned} \qquad (5.275)$$

Output 5.11.1: MANOVA Test of Three-Way Interaction of ABC.

MANOVA Test Criteria and F Approximations for the Hypothesis of No Overall a*b*c Effect H = Type III SSCP Matrix for a*b*c E = Error SSCP Matrix S=2 M=0.5 N=9.5					
Statistic	Value	F Value	Num DF	Den DF	Pr > F
Wilks' Lambda	0.24745566	5.30	8	42	0.0001
Pillai's Trace	0.99325167	5.43	8	44	<.0001
Hotelling-Lawley Trace	2.06839890	5.29	8	27.776	0.0004
Roy's Greatest Root	1.34538972	7.40	4	22	0.0006
NOTE: F Statistic for Roy's Greatest Root is an upper bound.					
NOTE: F Statistic for Wilks' Lambda is exact.					

for all indices. When analyzing a three-factor design, one analyzes the factor in a hierarchical manner, highest order to lowest order. The process is similar to the univariate strategy suggested by Milliken and Johnson (1992, p. 198). Reviewing the unweighted output, the three-way interactions are significant for each variable and, jointly, the multivariate test is significant. The output for the multivariate test is displayed in Output 5.11.1. Because there is a significant three-way interaction, one may choose to analyze two-way effects (say AB) at selected levels of the other factor, usually the factor of least interest (here C). While one may use the overpa-rameterized model for this analysis it would involve finding the contrast using the general form of estimable functions as discussed by Milliken and Johnson (1992, p. 142). It is generally easier to use the cell means (μ_{ijk}) full-rank model. Then, one merely has to specify cell mean tetrads of the form, for example,

$$\psi = \mu_{ij1} - \mu_{i'j1} - \mu_{ij'1} + \mu_{i'j'1}. \tag{5.276}$$

In Program 5_11.sas we use PROC GLM with the NOINT option in the MODEL statement to specify a full-rank cell means model. The CONTRAST statement is used to test the AB interaction at the first level of C. The output is displayed in Output 5.11.2.

Using Wilks' Λ criterion, the P-Value for the multivariate test is 0.0146; thus the interaction AB is significant at $C1$. One may also test AB at $C2$. Because the test of interaction is significant, one would not investigate main effects since they are confounded with interaction. If all interaction are nonsignificant, contrasts in main effects can be studied. Depending on the number of interest, one would use either an overall multivariate criterion or a planned comparison method. PROC MULTTEST may be used for a planned comparison analysis since it can be used to obtain adjusted P-values.

5.11.2   Weighted Analysis

To obtain weighted tests using PROC GLM one must run PROC GLM several times, reordering the sequence of the effects to obtain the correct Type I sums of

Output 5.11.2: Nonorthogonal Three Factor MANOVA: Full Rank Model with Un-weighted Contrast.

MANOVA Test Criteria and F Approximations for the Hypothesis of No Overall a*b at c1 Effect H = Contrast SSCP Matrix for a*b at c1 E = Error SSCP Matrix S=2 M=0.5 N=9.5					
Statistic	Value	F Value	Num DF	Den DF	Pr > F
Wilks' Lambda	0.42748067	2.78	8	42	0.0146
Pillai's Trace	0.60141358	2.37	8	44	0.0327
Hotelling-Lawley Trace	1.27169514	3.25	8	27.776	0.0096
Roy's Greatest Root	1.21611491	6.69	4	22	0.0011
NOTE: F Statistic for Roy's Greatest Root is an upper bound.					
NOTE: F Statistic for Wilks' Lambda is exact.					

Table 5.6: ANOVA Summary Table for Variable $Y1$.

Hypothesis	Reduction Notation	SS	df	MS	F	P-Value
H_{A*}	$R(A\|\mu)$	4.89	2	2.44	2.76	0.0853
H_{B*}	$R(B\|\mu)$	9.10	2	4.55	5.13	0.0148
H_{C*}	$R(C\|\mu)$	1.94	1	1.94	2.19	0.1531
H_{AB*}	$R(AB\|\mu, A, B)$	33.24	4	8.31	9.37	0.0001
H_{AC*}	$R(AC\|\mu, A, C)$	8.25	2	4.13	4.56	0.0206
H_{BC*}	$R(BC\|\mu, B, C)$	5.86	2	2.93	3.91	0.0555
H_{ABC*}	$R(ABC\|\mu, A, B, AB, AC, BC)$	25.87	4	6.47	7.30	0.0007
$Error$		19.50	22	0.89		

squares for each main effect and interaction. While we have not stopped printing the univariate tests with the NOUNI option on the MODEL statement, one must be careful when using the output since all tests are printed; one may control the output of the multivariate tests with the MANOVA statement. The weighted tests for variable $Y1$ are provided in Table 5.6. A similar table may be constructed for the dependent variable $Y2$.

For the multivariate tests the MANOVA statement is used. For the first sequence of factors we request MANOVA weighted tests for A, B, and ABC. Only the weighted MANOVA test of ABC is equal to the unweighted MANOVA test. The weighted tests of AB or other two-way interactions are not equal to the unweighted tests. This is because, for example, the parameter

$$\psi = (\alpha\beta)_{ij} - (\alpha\beta)_{i'j} - (\alpha\beta)_{ij'} + (\alpha\beta)_{i'j'} \tag{5.277}$$

is estimated by

$$\hat{\psi} = \bar{y}_{ij\cdot} - \bar{y}_{i'j\cdot} - \bar{y}_{ij'\cdot} + \bar{y}_{i'j'\cdot}, \tag{5.278}$$

which now involve weighted cell means $\bar{y}_{ij\cdot}$. Similarly, the weighted test of A is not the same for the two models. The MANOVA statements in the other runs of PROC GLM are used to generate multivariate tests of B, C, CB, and AC.

There are no `LSMEANS` statements included with the Type I analysis since we are interested in weighted averages of cell means to construct marginal means which is accomplished using the `MEANS` statement. The `ESTIMATE` statement by default uses the `LSMEANS` to obtain estimates of marginal means which are unweighted estimates of cell means. To obtain weighted estimates, it is more convenient to use the cell means full rank model. The code for the contrast of the average of $A1$ vs $A3$ and $C1$ vs $C2$ is included in Program 5_11.sas. The coefficients are constructed from the cell frequencies. For example, the coefficients for $C1$ vs $C2$ are

$$\left(\frac{3}{21} \quad -\frac{2}{19} \quad \frac{2}{21} \quad -\frac{3}{19} \quad \frac{2}{21} \quad -\frac{2}{19} \quad \frac{2}{21} \quad -\frac{1}{19} \quad \frac{1}{21} \quad -\frac{2}{19} \quad \frac{3}{21} \quad -\frac{2}{19} \quad \frac{2}{21} \quad -\frac{1}{19} \cdots \right.$$
$$\left. \frac{3}{21} \quad -\frac{3}{19} \quad \frac{3}{21} \quad \frac{3}{19} \right).$$

The test of AB at $C1$ for the weighted analysis is identical to the unweighted analysis, since at fixed levels of C, the test of AB is invariant to the design weights. This is also the case for two-way interactions at each level of the other factors. The output with the results for the multivariate tests of these contrasts is displayed in Output 5.11.3. We have illustrated the analysis of the three-factor design using both the overparameterized model and a cell mean full-rank model using Program 5_11.sas. When all of the parameters are not included in the model, PROC GLM may not be used to generate a cell means model since currently restrictions cannot be imposed on the model parameters. With this being the case we may use either the overparameterized GLM procedure or the REG procedure, as illustrated in Chapter 3. The multivariate restricted model analysis is discussed in Chapter 7.

Following the rejection of an overall multivariate test, currently no syntax is available in SAS to generate simultaneous confidence intervals. In Program 5_11.sas we use the TRANSREG procedure to generate the design matrix for the model and the Roy criterion to develop a confidence set for the comparison $A1$ vs $A3$ for the weighted contrast using variable $Y2$. Observe that the estimate $\hat{\psi} = -11.2002$ and standard error $\hat{\sigma}_{\hat{\psi}} = 3.5517$ are identical to that obtained using PROC GLM for the full rank model. However, the approximate critical constant for the Roy criterion is obtained using PROC IML code.

5.12 MANCOVA Designs

5.12.1 Overall Tests

We have illustrated how one may apply the general theory of this chapter to analyze multivariate regression models and a nonorthogonal MANOVA design. Combining regression with experimental design into a single model is called multivariate analysis of covariance (MANCOVA). When analyzing MANCOVA designs, X contains values for the fixed, categorical design variables and Z the fixed regression variables (also called the covariate or concomitant variables). The multivariate linear mode for the MANCOVA design is

$$Y_{n \times p} = X_{n \times k} B_{k \times p} + Z_{n \times q} \Gamma_{q \times p} + U_{n \times p} \text{ where } u_i' \sim IN(0, \Sigma). \qquad (5.279)$$

Output 5.11.3: Nonorthogonal Three Factor MANOVA: Full Rank Model with Weighted Contrast.

MANOVA Test Criteria and F Approximations for the Hypothesis of No Overall a*b at c1 Effect H = Contrast SSCP Matrix for a*b at c1 E = Error SSCP Matrix S=2 M=0.5 N=9.5					
Statistic	Value	F Value	Num DF	Den DF	Pr > F
Wilks' Lambda	0.42748067	2.78	8	42	0.0146
Pillai's Trace	0.60141358	2.37	8	44	0.0327
Hotelling-Lawley Trace	1.27169514	3.25	8	27.776	0.0096
Roy's Greatest Root	1.21611491	6.69	4	22	0.0011
NOTE: F Statistic for Roy's Greatest Root is an upper bound.					
NOTE: F Statistic for Wilks' Lambda is exact.					

MANOVA Test Criteria and Exact F Statistics for the Hypothesis of No Overall a1 vs a3 Effect H = Contrast SSCP Matrix for a1 vs a3 E = Error SSCP Matrix S=1 M=0 N=9.5					
Statistic	Value	F Value	Num DF	Den DF	Pr > F
Wilks' Lambda	0.68637959	4.80	2	21	0.0192
Pillai's Trace	0.31362041	4.80	2	21	0.0192
Hotelling-Lawley Trace	0.45691977	4.80	2	21	0.0192
Roy's Greatest Root	0.45691977	4.80	2	21	0.0192

MANOVA Test Criteria and Exact F Statistics for the Hypothesis of No Overall c1 vs c2 Effect H = Contrast SSCP Matrix for c1 vs c2 E = Error SSCP Matrix S=1 M=0 N=9.5					
Statistic	Value	F Value	Num DF	Den DF	Pr > F
Wilks' Lambda	0.53998843	8.94	2	21	0.0015
Pillai's Trace	0.46001157	8.94	2	21	0.0015
Hotelling-Lawley Trace	0.85189153	8.94	2	21	0.0015
Roy's Greatest Root	0.85189153	8.94	2	21	0.0015

Each row vector in the error matrix U is multivariate normal with constant covariance matrix Σ. The matrix Y is the data matrix and B and Γ are fixed parameter matrices. Letting $X_0 = \begin{pmatrix} X \\ Z \end{pmatrix}$ and $B_0 = \begin{pmatrix} B \\ \Gamma \end{pmatrix}$, we see that the MANCOVA model has the general form (5.3) or $Y = X_0 B_0 + U$, a multivariate linear model.

The normal equations for the MANCOVA design is

$$\begin{pmatrix} X'X & X'Z \\ Z'X & Z'Z \end{pmatrix} \begin{pmatrix} B \\ \Gamma \end{pmatrix} = \begin{pmatrix} X'Y \\ Z'Y \end{pmatrix}. \tag{5.280}$$

Solving these equations, we have using (3.12) that

$$\widehat{B} = (X'X)^{-1}X'Y - (X'X)^{-1}X'Z\widehat{\Gamma} \tag{5.281}$$

$$\widehat{\Gamma} = (Z'QZ)^{-1}Z'QY = E_{ZZ}^{-1}E_{ZY} \tag{5.282}$$

where $Q = I - X(X'X)^{-1}X$, a projection matrix.

To test the general hypothesis $H : C \begin{pmatrix} B \\ \Gamma \end{pmatrix} A = 0$, one may select $C = I$ and $A = \begin{pmatrix} 0 \\ I \end{pmatrix}$ to test the null hypothesis $H : \Gamma = 0$. Alternatively, to test hypotheses regarding the elements of B, we select $C = (C_1, 0)$. Using the general theory in this chapter, the residual sum of squares for the MANCOVA model is

$$E = (Y - X_0\widehat{B})'(Y - X_0\widehat{B}) = Y'QY - \widehat{\Gamma}'(Z'QZ)\widehat{\Gamma} = E_{YY} - E_{YZ}E_{ZZ}^{-1}E_{ZY} \tag{5.283}$$

with degrees of freedom $\nu_e = n - \text{rank}(X) - \text{rank}(Z)$. To test the hypothesis $H : \Gamma = \Gamma_0$, that the covariates have no effect on the dependent variables, the hypothesis sum of squares is

$$H_\Gamma = (\widehat{\Gamma} - \Gamma_0)'(Z'QZ)(\widehat{\Gamma} - \Gamma_0) \tag{5.284}$$

with degrees of freedom $\nu_h = \text{rank}(Z)$. Any of the multivariate criteria may be used to test H. The MANCOVA model, like the MANOVA model, assumes that the regression matrix Γ is common across groups, the assumption of parallelism. A test of multivariate parallelism is given in Timm (2002, p. 227-228) using Wilks' Λ criterion.

To test hypotheses regarding the matrix B, $H : B = B_0$, we may construct the hypothesis test matrix directly by using properties of inverses for partitioned matrices. The hypothesis test matrix is

$$
\begin{aligned}
H_B = {} &(C_1\widehat{B}A - C_1B_0A)'[C_1(X'X)^-C_1' \\
&+ C_1(X'X)^-X'Z(Z'QZ)^{-1}Z'X(X'X)^{-1}C_1]^{-1}(C_1\widehat{B}A - C_1B_0A)
\end{aligned}
\tag{5.285}
$$

with degrees of freedom $\nu_h = \text{rank}(C_1)$. Given H_B and the error matrix E tests using the multivariate criteria are easily constructed.

An alternative method for calculating the hypothesis sum of square is to perform a two-step least squares procedure, Seber (1984, p. 465). For this method, one performs the calculations:

1. Calculate $\widehat{B} = (X'X)^{-1}X'Y$ and $Y'QY (= E_{YY})$ for the MANOVA model $\mathcal{E}(Y) = XB$.

2. $\widehat{\Gamma} = (Z'QZ)^{-1}Z'QY = E_{ZZ}^{-1}E_{ZY}$.

3. $E = (Y - X_0\widehat{B})'(Y - X_0\widehat{B}) = Y'QY - \widehat{\Gamma}'(Z'QZ)\widehat{\Gamma} = E_{YY} - E_{YZ}E_{ZZ}^{-1}E_{ZY}$.

4. Replace Y with the reduced matrix $Y - Z\widehat{\Gamma}$ in $\widehat{B}$ to calculate $\widehat{B}_R = (X'X)^{-1}X'(Y - Z\widehat{\Gamma})$.

5. To obtain E_ω where $\omega : C_1B = 0$, obtain $Y'Q_\omega Y (= E_{HYY}$, say) for the model $\Omega_0 : \mathcal{E}(Y) = XB$ and $\omega : \mathcal{E}(Y) = XB$ and $C_1B = 0$, and steps (2) and (3) are repeated. From (3),

$$E_\omega = Y1Q_\omega Y - Y'Q_\omega Z(Z'Q_\omega Z)^{-1}Z'Q_\omega Y = E_{HYY} - E_{HYY}E_{HZZ}^{-1}E_{HZY}.$$

Then Wilks' criterion for the test is

$$\Lambda = \frac{|E|}{|E_\omega|} = \frac{|E_{YY} - E_{YZ}E_{ZZ}^{-1}E_{ZY}|}{|E_{HYY} - E_{HYY}E_{HZZ}^{-1}E_{HZY}|} \sim U_{p,\nu_h,\nu_e}. \tag{5.286}$$

5.12.2   Tests of Additional Information

A test closely associated with the MANCOVA design is Rao's test for additional information, Rao (1973, p. 551). In many MANOVA or MANCOVA designs, one collects data on p variables and one is interested in determining whether the additional information provided by the last $(p - s)$ variables, independent of the first s variables, is significant. To develop the test procedure of this hypothesis, we begin with the linear model: $\Omega_0 : Y = XB + U$ where the usual hypothesis is $H : CB = 0$. Partitioning the data matrix: $Y = (Y_1, Y_2)$ and the parameter matrix $B : (B_1, B_2)$, we consider the alternative model

$$\Omega_1 : Y_1 = XB_1 + U_1 \tag{5.287}$$
$$H_{01} : CB_1 = 0 \tag{5.288}$$

where following Timm (1975, p. 125), the

$$\mathcal{E}(Y_2|Y_1) = XB_2 + (Y_1 - XB_1)\Sigma_{11}^{-1}\Sigma_{12}$$
$$= X(B_2 - B_1\Sigma_{11}^{-1}\Sigma_{12}) + Y\Sigma_{11}^{-1}\Sigma_{12} = X\Theta + Y_1\Gamma$$

and

$$\Sigma_{22\cdot1} = \Sigma_{21}\Sigma_{11}^{-1}\Sigma_{12}.$$

Thus, the conditional model is

$$\Omega_2 : \mathcal{E}(Y_2|Y_1) = X\Theta + Y_1\Gamma, \tag{5.289}$$

the MANCOVA model. Under Ω_2, testing

$$H_{02} : C(B_2 - B_1\Gamma) = 0 \tag{5.290}$$

corresponds to testing $H_{02} : C\Theta = 0$. If $C = I_p$, so that $\Theta = 0$, then the conditional distribution of $Y_2|Y_1$ depends only on Γ and does not involve B_1; thus Y_2 provides no additional information on B_1. Because Ω_2 is the standard MANCOVA model with $Y \equiv Y_2$ and $Z \equiv Y_1$, we may test H_{02} using the Λ criterion:

$$\Lambda_{2\cdot1} = \frac{|E|}{|E_\omega|} = \frac{|E_{22} - E_{21}E_{11}^{-1}E_{12}|}{|E_{H22} - E_{H21}E_{H11}^{-1}E_{H12}|} \sim U_{p-s,\nu_h,\nu_e} \tag{5.291}$$

where $\nu_e = n - p - s$ and $\nu_h = \text{rank}(C)$. Because $H(CB = 0)$ is true if and only if H_{01} and H_{02} are true, we may partition Λ as $\Lambda_1\Lambda_{2\cdot1}$ where Λ_1 is the test of H_{01}; this results in a stepdown test of H (Seber, 1984, p. 472).

To illustrate the analysis of a MANCOVA design, we consider a study that examines the effect of four drugs on the apgar scores for twins at birth, adjusting for the weight in kilograms of the mother. Subjects were randomly assigned to four treatment drug conditions with an equal number of subjects per treatment. The SAS code for the example is provided in Program 5_12.sas.

5.12.3   Results and Interpretation

Having ensured that the data are normally distributed with equal conditional covariance matrices, we test that the regression lines are parallel. With one covariate, this may be tested using PROC GLM by testing for the significance of an interaction between the covariate, x, and the independent variables, drugs. From the output, we see that the P-values for the test for variable `apgar1` and `apgar2` are $p = 0.6713$ and $p = 0.9698$, respectively. In addition, the overall multivariate test of interaction using, for example, Wilks' Λ criterion is nonsignificant, $p = 0.9440$. Thus, for these data we conclude that the assumption of parallelism appears tenable. Also illustrated in the code is how one may use PROC REG to test for parallelism. While the two procedures are equivalent for one covariate and many groups, this is not the case in general for more than one covariate.

Given parallelism, we next test for differences in drug treatments. Although we have an equal number of observations per cell, a balanced design, PROC GLM generates both a Type I and Type III sum of squares. These are not equal since the covariate is changing from mother to mother. The correct sum of squares for the covariance analysis is the Type III sum of squares. The P-values for the `apgar1` and `apgar2` dependent variables are $p = 0.1010$ and $p = 0.1112$, respectively. And, the overall multivariate test using any of the multivariate criteria is not significant using $\alpha = 0.05$ for the test size. Thus, there appears to be no difference in the apgar

Output 5.12.1: Multivariate MANCOVA: Adjusted Means.

Obs	_NAME_	drugs	LSMEAN	STDERR	NUMBER	COV1	COV2	COV3	COV4
1	apgar1	1	5.31013	0.28811	1	0.08301	0.03593	-0.00950	-0.07681
2	apgar1	2	5.32566	0.24134	2	0.03593	0.05825	-0.00677	-0.05478
3	apgar1	3	5.76735	0.18551	3	-0.00950	-0.00677	0.03441	0.01448
4	apgar1	4	5.09686	0.38695	4	-0.07681	-0.05478	0.01448	0.14973
5	apgar2	1	3.95570	0.66435	1	0.44136	0.19104	-0.05050	-0.40842
6	apgar2	2	4.35980	0.55651	2	0.19104	0.30970	-0.03601	-0.29125
7	apgar2	3	6.02270	0.42777	3	-0.05050	-0.03601	0.18299	0.07700
8	apgar2	4	7.16181	0.89227	4	-0.40842	-0.29125	0.07700	0.79614

scores for the four treatment drugs. If one did not take into account the weights of the mothers, the differences in apgar scores would have been judged as significant.

When one performs an analysis of covariance, PROC GLM can be directed to compute adjusted means and perform pairwise comparisons, a variable at a time, by using the LSMEANS statement. The option OUT=ADJMEANS in the LSMEANS statement creates a SAS data set with the adjusted means for each variable. To compute the adjusted means for each variable in the jth group, PROC GLM uses the formula $\bar{y}_{adj,j} = \bar{y} - \hat{\gamma}(\bar{x}_j - \bar{x})$ where $\bar{x}_j$ is the mean of the covariate for the jth treatment group, $\bar{x}$ is the overall mean for the covariate, and $\hat{\gamma}$ is the common slope estimate. The common estimate for the slope of the regression equation is output by using the SOLUTION option in the MODEL statement in PROC GLM. The values can be found in the univariate ANCOVA test output in the column labeled estimate and the row labeled x. For group 1 and apgar1, we have that

$$\bar{y}_{adj,j} = \bar{y} - \hat{\gamma}(\bar{x}_j - \bar{x}) = 2.75 - 0.167877153(39.75 - 55) = 5.31012658.$$

The standard error of the adjusted mean is

$$\hat{\sigma}_{adj,1} = \sqrt{MS_e}\sqrt{\frac{1}{n_1} + \frac{(\bar{x}_1 - \bar{x})^2}{\sum_i (x_{ij} - \bar{x})^2}}$$
$$= 0.510867\sqrt{\frac{1}{8} + \frac{(39.75 - 55)^2}{1204.75}}$$
$$= 0.28810778$$

as found in the adjusted mean output in Output 5.12.1. The other values are calculated similarly.

In Program 5_12.sas, the test of the hypothesis of no difference between drug 1 versus drug 4 was performed using the CONTRAST and ESTIMATE statements. The difference $\hat{\psi} = 0.2133$ is the difference in adjusted means for the variable apgar1. Observe that SAS performs t-tests, but does not generate confidence sets for multivariate tests. To obtain an approximate simultaneous confidence set for the multivariate test, PROC IML code must be used using the design matrix created by the TRANSREG procedure. The approximate confidence set using Roy's criterion for the difference of drug1 versus drug4 for the variable `apgar1` is illustrated in the code.

Finally, Rao's test of additional information for `apgar2` given `apgar1` is evaluated. The p-value for the test is $p = 0.7534$. Thus, given `apgar1`, `apgar2` provides no additional information.

5.13 STEPDOWN ANALYSIS

In the last section we illustrated the test of additional information for two dependent variables and one covariate. The test more generally is used in a stepdown analysis. We illustrate it with a model that involves two covariates and eleven dependent variables. The data reproduced in Seber (1984, p. 469) from Smith, Gnanadesikan, and Hughes (1962) are utilized. The data consist of urine samples from men in four weight categories; the 11 variables and two covariates are:

$$y_1 = pH$$
$$y_2 = \text{modified createnine coefficient}$$
$$y_3 = \text{pegment createnine}$$
$$y_4 = \text{phosphate } (mg/ml)$$
$$y_5 = \text{calcium } (mg/ml)$$
$$y_6 = \text{phosphorus } (mg/ml)$$
$$y_7 = \text{createnine } (mg/ml)$$

$$y_8 = \text{chloride } (mg/ml)$$
$$y_9 = \text{bacon } (\mu g/ml)$$
$$y_{10} = \text{choline } (\mu g/ml)$$
$$y_{11} = \text{copper } (\mu g/ml)$$
$$x_1 = \text{volume } (ml)$$
$$x_2 = (\text{specific gravity}-1) \times 10^3.$$

The model considered by Smith et al. (1962) is a one-way MANCOVA with two covariates:

$$y_{ij} = \mu + \alpha_i + \gamma_1 x_{ij1} + \gamma_2 x_{ij2} + e_{ij} \text{ where } e_{ij} \sim IN(0, \Sigma) \qquad (5.292)$$

with the unweighted Type III restriction that the $\sum_i \alpha_i = 0$. The hypotheses considered by Smith et al. (1962) were:

$$H_{01} : \mu = 0$$
$$H_{02} : \alpha_1 = \alpha_2 = \alpha_3 = \alpha_4 = 0$$
$$H_{03} : \gamma_1 = 0$$
$$H_{04} : \gamma_2 = 0$$

where H_{01} is the null hypothesis that the overall mean is zero, H_{02} is the null hypothesis of differences among weight groups, and H_{03} and H_{04} are testing that the covariates are each zero. The SAS code to evaluate these tests and parallelism for `x1` and `x2` a variable at a time and jointly is provided in Program 5_13.sas.

First, the data statement in the program is used to create an $I = 4$ group multivariate intraclass covariance model which allows one to test for parallelism jointly. For the example, we may test that $H_{\text{parallel}} : \Gamma_1 = \Gamma_2 = \Gamma_3 = \Gamma_4 = \Gamma$. While one may use PROC GLM to test for parallelism by investigation of an interaction with one covariate and multiple groups, or with many covariates and only two groups a variable-at-a-time, this is not the case with more than one covariate and more than

Output 5.13.1: Multivariate MANCOVA with Stepdown Analysis: Joint Test of Parallelism Using PROC REG.

Multivariate Statistics and F Approximations					
S=6 M=2 N=10.5					
Statistic	Value	F Value	Num DF	Den DF	Pr > F
Wilks' Lambda	0.09054987	1.10	66	128.53	0.3146
Pillai's Trace	1.75627665	1.05	66	168	0.3883
Hotelling-Lawley Trace	3.56870524	1.17	66	67.731	0.2609
Roy's Greatest Root	1.91813982	4.88	11	28	0.0003
NOTE: F Statistic for Roy's Greatest Root is an upper bound.					

Table 5.7: Tests of Hypotheses Using PROC GLM for the MANCOVA Model.

	Hypothesis	Λ	F	P-value
H_{01} :	$\mu = 0$	0.06920	35.46	0.0001
H_{02} :	$\alpha_1 = \alpha_2 = \alpha_3 = \alpha_4 = 0$	0.06011	4.17	< 0.0001
H_{03} :	$\gamma_1 = 0$	0.37746	4.35	0.0007
H_{04} :	$\gamma_2 = 0$	0.25385	7.75	0.0001

two groups (Timm, 2002, p. 239). Instead, the procedure PROC REG should be used.

For our example, the test of joint parallelism is nonsignificant and is displayed in Output 5.13.1. We next test the hypotheses using PROC GLM for the MANCOVA model. The test results using Wilks' Λ criterion for the hypotheses are displayed in Table 5.7. These values do not agree with those reported by Smith et al. (1962). Their values for Λ were $0.0734, 0.4718, 0.2657$, and 0.0828, respectively.

Given significance between weight groups, Smith et al. (1962) were interested in evaluating whether variables $y2$, $y3$, $y4$, $y6$, $y9$, $y10$, $y11$, (SET 2) add additional information to the analysis of weight difference over the above that provide by variables $y1$, $y5$, $y7$, $y8$, (SET 1). To evaluate the contribution of SET 2 given SET 1, we calculate $\Lambda_{2 \cdot 1}$ using PROC GLM. The test of additional information has a P-value < 0.0001 since $\Lambda_{2 \cdot 1} = 0.016614119$. We also calculate Λ_1 to verify that $\Lambda = \Lambda_1 \Lambda_{2 \cdot 1}$ where Λ is Wilks' criterion for the MANCOVA for group differences: H_{02}. From the Output 5.13.2, "Test of Gamma1", $\Lambda_1 = 0.36180325$.

Using the MULTTEST procedure, stepdown p-value adjustments can be made. The stepdown options include STEPBON for stepdown Bonferroni, STEPBOOT for stepdown for bootstrap resampling, and others.

5.14   REPEATED MEASURES ANALYSIS

As our next example of the GLM, we illustrate how to use the model to analyze repeated measurement data. Using the data in Timm (1975, p. 244), we showed

Output 5.13.2: Multivariate MANCOVA: Test of Gamma1.

MANOVA Test Criteria and F Approximations for the Hypothesis of No Overall group Effect H = Type III SSCP Matrix for group E = Error SSCP Matrix S=3 M=0 N=17					
Statistic	Value	F Value	Num DF	Den DF	Pr > F
Wilks' Lambda	0.36180325	3.73	12	95.539	0.0001
Pillai's Trace	0.77940037	3.33	12	114	0.0004
Hotelling-Lawley Trace	1.39630489	4.10	12	58.825	0.0001
Roy's Greatest Root	1.10187599	10.47	4	38	<.0001
NOTE: F Statistic for Roy's Greatest Root is an upper bound.					

in Chapter 3 how to analyze repeated measurement data using a restricted univariate linear model which required the covariance matrix Σ within each group to have Hankel structure, meet the circularity condition (Kirk, 1995, p. 525). To analyze repeated measurement data using the multivariate linear model, we allow Σ to be arbitrary but homogeneous across groups where the number of subjects within each group is at least equal to the number of repeated measurements acquired on each subject.

For a repeated measurement experiment for I groups and a p-variate response vector of observations, we assume that the observation vector follows a multivariate normal distribution

$$y'_{ij} = \begin{pmatrix} y_{ij1} & y_{ij2} & \cdots & y_{ijp} \end{pmatrix} \sim IN_p(\mu, \Sigma)$$

where $i = 1, 2, \ldots, I$; $j = 1, 2, \ldots, n_i$, and $n = \sum_{i=1}^{I} n_i$. Employing the GLM, the data matrix is $Y_{n \times p}$; the parameter matrix $B_{I \times p} = (\mu_{ij})$ contains means where the ith row vector in B has the structure

$$\mu'_i = \begin{pmatrix} \mu_{i1} & \mu_{i2} & \cdots & \mu_{ip} \end{pmatrix}.$$

The design matrix has diagonal structure $X_{n \times I} = I_n \otimes 1_{n_i} = \oplus_{i=1}^{I} 1_{n_i}$, where 1_{n_i} is a column vector of n_i 1's.

To create the multivariate linear model using PROC GLM, we may use two model statements

```
model y1-y5=group/intercept nouni;
model y1-y5=group/noint nouni;
```

Using the first statement, SAS generates a less-than full-rank model of the form:

$$\mathcal{E}(y_{ij}) = \mu + \alpha_i$$

and prints hypotheses associated with the intercept. In the second, SAS generates a full-rank model of the form

$$\mathcal{E}(y_{ij}) = \mu_i.$$

Either model may be used to test hypotheses regarding the means in the parameter matrix B. The multivariate linear model corresponds to a random effects model where the subjects are random and groups are fixed. Or, under circularity, the model corresponds to a split-plot univariate model where the p variables represent the within (condition) factor.

When one analyzes a repeated measures design, three hypotheses are of primary interest:

H_{01} : Are the profiles for the I groups parallel?

H_{02} : Are there differences among the p conditions?

H_{03} : Are there differences among the I groups?

If the profiles are parallel, or equivalently when there is no group by condition inter-action, one may alter the hypotheses H_{01} and H_{02} to test for differences in groups given parallelism or differences in conditions given parallelism. We represent these as H_{02}^* and H_{03}^*. Furthermore, if there is an unequal number of subjects in each group that is due to the treatment, one may test for differences in group means given parallelism. We represent this test as H_{03}^{**}. In terms of the model parameters, the tests are:

$$H_{01} : \begin{pmatrix} \mu_{11} - \mu_{12} \\ \mu_{12} - \mu_{13} \\ \vdots \\ \mu_{1(p-1)} - \mu_{1p} \end{pmatrix} = \cdots = \begin{pmatrix} \mu_{I1} - \mu_{I2} \\ \mu_{I2} - \mu_{I3} \\ \vdots \\ \mu_{I(p-1)} - \mu_{Ip} \end{pmatrix}$$

$$H_{02} : \begin{pmatrix} \mu_{11} \\ \mu_{21} \\ \vdots \\ \mu_{I1} \end{pmatrix} = \cdots = \begin{pmatrix} \mu_{1p} \\ \mu_{2p} \\ \vdots \\ \mu_{Ip} \end{pmatrix}$$

$$H_{03} : \mu_{11} = \cdots = \mu_{I1}$$

$$H_{02}^* : \sum_{i=1}^{I} \frac{\mu_{i1}}{I} = \cdots = \sum_{i=1}^{I} \frac{\mu_{ip}}{I}$$

$$H_{03}^* : \sum_{j=1}^{p} \frac{\mu_{1j}}{p} = \cdots = \sum_{j=1}^{p} \frac{\mu_{I}}{p}$$

$$H_{03}^{**} : \sum_{i=1}^{I} \frac{n_i \mu_{i1}}{n} = \cdots = \sum_{i=1}^{I} \frac{n_i \mu_{ip}}{n}.$$

The SAS code to perform the tests is in Program 5_14.sas.

5.14.1 Results and Interpretation

To show how SAS is used to analyze repeated measurement data, it is convenient to express the hypotheses of interest using the general matrix product form $CBA = 0$

where B is the parameter matrix. For the example, the matrix B is

$$B_{2\times5} = \begin{pmatrix} \mu_{11} & \mu_{12} & \mu_{13} & \mu_{14} & \mu_{15} \\ \mu_{21} & \mu_{22} & \mu_{23} & \mu_{24} & \mu_{25} \end{pmatrix}. \tag{5.293}$$

To test H_{03}, differences in group mean vectors, we set $C = (1, -1)$ to obtain the difference in group means and the matrix $A = I_5$. The within-matrix A is equal to the identity matrix since we are evaluating the equivalence of the means across the p variables simultaneously. In PROC GLM, statement

```
manova h=group/printe printh;
```

is used where the options PRINTE and PRINTH print H and E for hypothesis testing.

To test H_{02}, differences among treatments, the matrices

$$C = I_2 \text{ and } A = \begin{pmatrix} 1 & 0 & 0 & 0 \\ -1 & 1 & 0 & 0 \\ 0 & -1 & 1 & 0 \\ 0 & 0 & -1 & 1 \\ 0 & 0 & 0 & -1 \end{pmatrix}$$

are used. The matrix A creates differences within variables, the within-subject dimension, and C is the identity matrix since we are evaluating vectors across groups, simultaneously. To test these using PROC GLM, one uses the CONTRAST statement, the full rank model (NOINT option in the MODEL statement) and the MANOVA statement:

```
contrast 'Mult Cond' group 1 0,
                      group 0 1;
manova m= (1 -1 0 0 0,
           0 1 -1 0 0,
           0 0 1 -1 0,
           0 0 0 1 -1)
prefix = diff/ printe printh;
```

where $m = A'$ and the group matrix is the identity matrix I_2.

To test for parallelism of profiles, H_{01}, we use the matrices

$$C = \begin{pmatrix} 1 & -1 \end{pmatrix} \text{ and } A = \begin{pmatrix} 1 & 0 & 0 & 0 \\ -1 & 1 & 0 & 0 \\ 0 & -1 & 1 & 0 \\ 0 & 0 & -1 & 1 \\ 0 & 0 & 0 & -1 \end{pmatrix}.$$

The matrix A again forms differences across variables while the matrix C creates differences among the groups. For more than two groups, say four, the matrix C

would become

$$C = \begin{pmatrix} 1 & 0 & 0 & -1 \\ 0 & 1 & 0 & -1 \\ 0 & 0 & 1 & -1 \end{pmatrix}$$

which compares each group with the last and the $\text{rank}(C) = 3 = I - 1$. Of course the matrices C and A are not unique since other differences could be specified. The tests are invariant to the specific form provided they span the same test space.

To test H_{02}^*, differences in condition given parallelism, the matrices

$$C = \begin{pmatrix} \frac{1}{2} & \frac{1}{2} \end{pmatrix} \text{ and } A = \begin{pmatrix} 1 & 0 & 0 & 0 \\ -1 & 1 & 0 & 0 \\ 0 & -1 & 1 & 0 \\ 0 & 0 & -1 & 1 \\ 0 & 0 & 0 & -1 \end{pmatrix}$$

are used. Using PROC GLM, we use the statement

```
manova h =_all_ m=(1 -1 0 0 0,
                   0 1 -1 0 0,
                   0 0 1 -1 0,
                   0 0 0 1 -1)
prefix = diff/ printe printh;
```

In this statement, $h = \begin{pmatrix} 1 & -1 \\ 1 & -1 \end{pmatrix}$ and the test of "intercept" is the test of equal condition means (summed over groups) given parallelism of profiles using the less-than-full-rank model. Because the multivariate tests are invariant to scale, it does not matter whether we use $C = (1, 1)$ or $C = \begin{pmatrix} \frac{1}{2}, \frac{1}{2} \end{pmatrix}$.

The final test of interest for these data is H_{03}^*, differences in groups given parallelism, sometimes called the test of coincidence. For this hypothesis we select the matrices

$$C = \begin{pmatrix} 1 & -1 \end{pmatrix} \text{ and } A = \begin{pmatrix} 1/5 \\ 1/5 \\ 1/5 \\ 1/5 \\ 1/5 \end{pmatrix}.$$

Using PROC GLM, it is again convenient to form a contrast using either the full-rank or less-than-full-rank model:

```
contrast 'Univ Gr' group 1 -1;
manova m=(.2 .2 .2 .2 .2) prefix=Gr/ printe printh;
```

where the matrix $m = A'$ in SAS. For more than two groups, say three, one would add rows to the contrast statement. That is:

Table 5.8: Multivariate Analysis I.

Hypothesis	df	Wilks' Λ	P-Value
$H_{03} : Groups$	1	0.556	0.1083
$H_{02} : Conditions$	2	0.195	0.0008
$H_{01} : Parallelism$	1	0.839	0.5919

Table 5.9: Multivariate Analysis II.

Hypothesis	df	Wilks' Λ	P-Value
$H_{03}^* : Groups|Parallelism$	1	0.669	0.0080
$H_{02}^* : Conditions|Parallelism$	1	0.220	0.0001
$H_{01} : Parallelism$	1	0.829	0.5919

```
contrast 'Univ Gr' group 1 0 -1, 0 1 -1;
```

From Output 5.14.1-5.14.2, following Timm (1980b), it is convenient to create two MANOVA table, Multivariate Anlaysis I and II. The Multivariate Analysis I is used if the test of parallelism is significant, i.e., we reject the null hypothesis of parallelism. Given parallelism, one would use the result in the Multivariate Analysis II table. Tables 5.8 and 5.9 summarize the output.

Finally, if one can ensure a Hankel structure or circularity for Σ, that $A'\Sigma A = \sigma^2 I$, where $A'A = I$, so that the matrix A is semi-orthogonal, then one may obtain from the tests given parallelism, the F-tests for a split plot design as illustrated in Chapter 3. To accomplish this using PROC GLM, one employs the REPEATED statement as shown in Program 5_14.sas. The ANOVA table is displayed in Output 5.14.3.

Because the post matrix A used in the Multivariate Analysis II tests was not normalized, one may not directly obtain the univariate tests by averaging the diagonal elements in the multivariate test matrices as illustrated in Timm (1980b). However, PROC GLM performs the normalization when using the repeated statement.

Both the multivariate (Multivariate Analysis I and II) and the univariate test results show that there are differences among conditions. The Multivariate Analysis I analysis failed to detect group differences where as with the univariate analysis group differences were detected. This is due to the fact that the univariate result examines only one contrast of the means while the multivariate tests an infinite number of contrasts; hence, the multivariate test is more conservative.

As discussed in Chapter 3, observe that both the Greenhouse-Geisser and the Huynh-Felt Epsilon adjusted tests are calculated when using PROC GLM with the REPEATED statement. These tests do not have a power advantage over the exact multivariate tests (Boik, 1991).

Output 5.14.1: Repeated Measurement Analysis I.

H = Type III SSCP Matrix for group					
	y1	y2	y3	y4	y5
y1	396.05	293.7	373.8	298.15	511.75
y2	293.7	217.8	277.2	221.1	379.5
y3	373.8	277.2	352.8	281.4	483
y4	298.15	221.1	281.4	224.45	385.25
y5	511.75	379.5	483	385.25	661.25

MANOVA Test Criteria and Exact F Statistics for the Hypothesis of No Overall group Effect
H = Type III SSCP Matrix for group
E = Error SSCP Matrix

S=1   M=1.5   N=6

Statistic	Value	F Value	Num DF	Den DF	Pr > F
Wilks' Lambda	0.55607569	2.24	5	14	0.1083
Pillai's Trace	0.44392431	2.24	5	14	0.1083
Hotelling-Lawley Trace	0.79831634	2.24	5	14	0.1083
Roy's Greatest Root	0.79831634	2.24	5	14	0.1083

H = Contrast SSCP Matrix for Mult Cond				
	diff1	diff2	diff3	diff4
diff1	4377.7	-4062.2	3220.3	-1412.2
diff2	-4062.2	3770	-2988.2	1303.6
diff3	3220.3	-2988.2	2368.9	-1039
diff4	-1412.2	1303.6	-1039	538.4

MANOVA Test Criteria and F Approximations for the Hypothesis of No Overall Mult Cond Effect
on the Variables Defined by the M Matrix Transformation
H = Contrast SSCP Matrix for Mult Cond
E = Error SSCP Matrix

S=2   M=0.5   N=6.5

Statistic	Value	F Value	Num DF	Den DF	Pr > F
Wilks' Lambda	0.19454894	4.75	8	30	0.0008
Pillai's Trace	0.88327644	3.16	8	32	0.0094
Hotelling-Lawley Trace	3.74006487	6.78	8	19.248	0.0003
Roy's Greatest Root	3.62985954	14.52	4	16	<.0001

NOTE: F Statistic for Roy's Greatest Root is an upper bound.

NOTE: F Statistic for Wilks' Lambda is exact.

H = Type III SSCP Matrix for group				
	diff1	diff2	diff3	diff4
diff1	26.45	-20.7	19.55	-55.2
diff2	-20.7	16.2	-15.3	43.2
diff3	19.55	-15.3	14.45	-40.8
diff4	-55.2	43.2	-40.8	115.2

MANOVA Test Criteria and Exact F Statistics for the Hypothesis of No Overall group Effect
on the Variables Defined by the M Matrix Transformation
H = Type III SSCP Matrix for group
E = Error SSCP Matrix

S=1   M=1   N=6.5

Statistic	Value	F Value	Num DF	Den DF	Pr > F
Wilks' Lambda	0.83905852	0.72	4	15	0.5919
Pillai's Trace	0.16094148	0.72	4	15	0.5919
Hotelling-Lawley Trace	0.19181198	0.72	4	15	0.5919
Roy's Greatest Root	0.19181198	0.72	4	15	0.5919

Output 5.14.2: Repeated Measurement Analysis II.

H = Contrast SSCP Matrix for Univ Gr	
	Gr1
Gr1	354.482

MANOVA Test Criteria and Exact F Statistics for the Hypothesis of No Overall Univ Gr Effect on the Variables Defined by the M Matrix Transformation
H = Contrast SSCP Matrix for Univ Gr
E = Error SSCP Matrix

S=1 M=-0.5 N=8

Statistic	Value	F Value	Num DF	Den DF	Pr > F
Wilks' Lambda	0.66905173	8.90	1	18	0.0080
Pillai's Trace	0.33094827	8.90	1	18	0.0080
Hotelling-Lawley Trace	0.49465273	8.90	1	18	0.0080
Roy's Greatest Root	0.49465273	8.90	1	18	0.0080

H = Type III SSCP Matrix for Intercept

	diff1	diff2	diff3	diff4
diff1	4351.25	-4041.5	3200.75	-1357
diff2	-4041.5	3753.8	-2972.9	1260.4
diff3	3200.75	-2972.9	2354.45	-998.2
diff4	-1357	1260.4	-998.2	423.2

MANOVA Test Criteria and Exact F Statistics for the Hypothesis of No Overall Intercept Effect on the Variables Defined by the M Matrix Transformation
H = Type III SSCP Matrix for Intercept
E = Error SSCP Matrix

S=1 M=1 N=6.5

Statistic	Value	F Value	Num DF	Den DF	Pr > F
Wilks' Lambda	0.21986464	13.31	4	15	<.0001
Pillai's Trace	0.78013536	13.31	4	15	<.0001
Hotelling-Lawley Trace	3.54825288	13.31	4	15	<.0001
Roy's Greatest Root	3.54825288	13.31	4	15	<.0001

H = Type III SSCP Matrix for group

	diff1	diff2	diff3	diff4
diff1	26.45	-20.7	19.55	-55.2
diff2	-20.7	16.2	-15.3	43.2
diff3	19.55	-15.3	14.45	-40.8
diff4	-55.2	43.2	-40.8	115.2

MANOVA Test Criteria and Exact F Statistics for the Hypothesis of No Overall group Effect on the Variables Defined by the M Matrix Transformation
H = Type III SSCP Matrix for group
E = Error SSCP Matrix

S=1 M=1 N=6.5

Statistic	Value	F Value	Num DF	Den DF	Pr > F
Wilks' Lambda	0.83905852	0.72	4	15	0.5919
Pillai's Trace	0.16094148	0.72	4	15	0.5919
Hotelling-Lawley Trace	0.19181198	0.72	4	15	0.5919
Roy's Greatest Root	0.19181198	0.72	4	15	0.5919

Output 5.14.3: F-Test Summary for Split Plot Design.

Source	DF	Type III SS	Mean Square	F Value	Pr > F
group	1	1772.410000	1772.410000	8.90	0.0080
Error	18	3583.140000	199.063333		

| | | | | | | Adj Pr > F | |
Source	DF	Type III SS	Mean Square	F Value	Pr > F	G - G	H - F
cond	4	3371.300000	842.825000	14.48	<.0001	<.0001	<.0001
cond*group	4	79.940000	19.985000	0.34	0.8479	0.8068	0.8479
Error(cond)	72	4191.960000	58.221667				

Having found a significant overall multivariate effect, one may want to investigate contrasts in the means of the general form $\psi = c'Ba$. The vectors c and a are selected depending upon whether one is interested in differences among groups, differences in conditions, or an interaction. Because there is not a significant interaction in our example, the differences in conditions or groups, averaging over the other factor may be of interest. Following (5.65), the confidence interval has the structure

$$\hat{\psi} - c_0\hat{\sigma}_{\hat{\psi}} \leq \psi \leq \hat{\psi} - c_0\hat{\sigma}_{\hat{\psi}} \qquad (5.294)$$

where the critical constant c_0 depends upon the test criteria defined in (5.66)-(5.69) and the estimated variance is $\hat{\sigma}^2_{\hat{\psi}} = (a'Sa)(c'(X'X)^{-1}c)$. For our example, the covariance matrix is obtained in SAS using the error matrix $S = \frac{E}{\nu_e}$.

Because SAS does not calculate simultaneous confidence sets for the multivariate tests we provide PROC IML code for the calculations in Program 5_14.sas for test differences in conditions given parallelism. While one may use Statistical Tables in Timm (1975) to obtain the critical constants, we have provided IML code to calculate approximate critical values for c_0 for Hotelling's T_0^2 and Roy's criterion θ. From Timm (1975, p. 605), the critical constant $c_0^2 = 14.667$ when testing for differences in conditions given parallelism for $\alpha = 0.05$, as calculated in Program 5_14.sas (because the $\min(\nu_h, u) = \min(\text{rank}(C), \text{rank}(A)) = 1$, the two test criteria have the same critical values for hypothesis). Letting $c = (.5, .5)$, $a_1' = (1, 1, 0, 0, 0)$ and $a_2' == (0, 1, 0, 0, -1. -1)$, the confidence intervals for differences in conditions one versus two:$\psi_1 = \mu_1 - \mu_5$, and two versus five:$\psi_2 = \mu_2 - \mu_5$ are significant:

$$5.04 \leq \mu_1 - \mu_5 \leq 24.46$$
$$-14.62 \leq \mu_2 - \mu_5 \leq -0.28$$

where the standard errors for the estimated contrasts are $\hat{\sigma}_{\hat{\psi}_1} = 2.5349227$ and $\hat{\sigma}_{\hat{\psi}_2} = 1.8714967$, respectively, and their estimates are: $\hat{\psi}_1 = 14.75$ and $\hat{\psi}_2 = 4.75$. Comparing each of level one, three and four with level five, we find each comparison to be nonsignificant (Timm, 1975, p. 249). Note, PROC IML code for the other multivariate hypotheses is provided in Program 5_15.sas, discussed in the next section. For additional repeated measurement examples, one may also consult Khattree and Naik (1995).

5.15   EXTENDED LINEAR HYPOTHESES

When comparing means in any MANOVA design, but especially in repeated measurement designs, one frequently encounters contrasts in the means that are of interest, but are not expressible as a bilinear form $\psi = c'Ba$. When this occurs, one may use the extended linear hypotheses defined in (5.203) to test hypotheses and evaluate contrasts in cell means. To illustrate, suppose one has a one-way MANOVA design involving three groups and three variables, or three repeated measures, so that the parameter matrix for the multivariate linear model is

$$B = \begin{pmatrix} \mu_{11} & \mu_{12} & \mu_{13} \\ \mu_{21} & \mu_{22} & \mu_{23} \\ \mu_{31} & \mu_{32} & \mu_{33} \end{pmatrix}.$$

And, suppose that one is interested in testing for differences in group means

$$H_g : \begin{pmatrix} \mu_{11} \\ \mu_{12} \\ \mu_{13} \end{pmatrix} = \begin{pmatrix} \mu_{21} \\ \mu_{22} \\ \mu_{23} \end{pmatrix} = \begin{pmatrix} \mu_{31} \\ \mu_{32} \\ \mu_{33} \end{pmatrix}.$$

To test H_g one may select $C \equiv C_0$ and $A \equiv A_0$ to test the overall multivariate hypothesis. For our example, we may select

$$A_0 = I_3 \text{ and } C_0 = \begin{pmatrix} 1 & -1 & 0 \\ 0 & 1 & -1 \end{pmatrix} \tag{5.295}$$

so that the overall test is represented in the form $H_g : C_0 B A_0 = 0$.

Upon rejection of the overall hypothesis, suppose we are interested in the contrast:

$$\psi = \frac{\mu_{11} + \mu_{22} + \mu_{33} - (\mu_{12} + \mu_{21})}{2} - \frac{(\mu_{13} + \mu_{31})/2 - (\mu_{23} + \mu_{32})}{2} \tag{5.296}$$

which compares the diagonal means with the average of the off diagonals, a common situation in some repeated measures designs. This contrast may not be represented in the familiar bilinear form. However, using the $\text{vec}(\cdot)$ operator on the parameter matrix B, we may represent the contrast in the form $\psi = C_* \text{vec}(B)$ where the matrix

$$C_* = \begin{pmatrix} 1 & -.5 & -.5 & -.5 & 1 & -.5 & -.5 & -.5 & 1 \end{pmatrix}.$$

We may test that the contrast $\psi = 0$ using the chi-square statistic defined in (5.198) with $\nu = \text{rank}(C_*) = 1$. Alternatively, letting

$$M = \begin{pmatrix} 1 & .5 \\ -.5 & .5 \\ -.5 & 1 \end{pmatrix}$$

the contrast may be expressed as an extended linear hypothesis

$$\psi = \text{tr}(M B_0) = \text{tr}(M C_0 B A_0) = \text{tr}(A_0 M C_0 B) = \text{tr}(G B)$$

where $B_0 = C_0 B A_0$ and $G = A_0 M C_0$. Thus, we have that the matrix G has the form

$$G = \begin{pmatrix} 1 & -.5 & -.5 \\ -.5 & 1 & -.5 \\ -.5 & -.5 & 1 \end{pmatrix},$$

for our example where the coefficients in each row and column of the matrix G sum to one. A contrast of this type has been termed a generalized contrast by Bradu and Gabriel (1974).

Testing the contrast associated with H_g, that the diagonal elements is equal to the off diagonal elements, was achieved using a single matrix M.

Alternatively suppose we are interested in a subset of the elements of the overall hypothesis. For example,

$$H_0 : \mu_{11} = \mu_{21} = \mu_{31}$$
$$\mu_{12} = \mu_{22} = \mu_{32}$$
$$\mu_{13} = \mu_{23} = \mu_{33}.$$

This hypothesis may not be expressed in the form $C_0 B A_0 = 0$, but may be represented as an extended linear hypothesis. To see this, observe that H_0 is equivalent to testing

$$\psi_0 = \text{tr}(M_i B_0) = \text{tr}(M_i C_0 B A_0) = \text{tr}(A_0 M_i C_0) = \text{tr}(G_i B) \text{ for } i = 1, 2, 3, 4$$

where the M matrices are defined:

$$M_1 = \begin{pmatrix} 1 & 0 \\ 0 & 0 \\ 0 & 0 \end{pmatrix} \quad M_2 = \begin{pmatrix} 0 & 0 \\ 1 & 0 \\ 0 & 0 \end{pmatrix}$$

$$M_3 = \begin{pmatrix} 0 & 0 \\ 0 & 1 \\ 0 & 0 \end{pmatrix} \quad M_4 = \begin{pmatrix} 0 & 0 \\ 0 & 0 \\ 0 & 1 \end{pmatrix}.$$

H_0 requires finding the supremum in (5.218) as outlined in (5.220).

The examples we have illustrated have assumed that the post matrix $A_0 = I$. For $A_0 \neq I$, consider the test of parallelism for our example. It has the simple structure:

$$\begin{pmatrix} 1 & -1 & 0 \\ 0 & 1 & -1 \end{pmatrix} \begin{pmatrix} \mu_{11} & \mu_{12} & \mu_{13} \\ \mu_{21} & \mu_{22} & \mu_{23} \\ \mu_{31} & \mu_{32} & \mu_{33} \end{pmatrix} \begin{pmatrix} 1 & 0 & 0 \\ -1 & 1 & 0 \\ 0 & -1 & 1 \end{pmatrix} = 0$$

$$C_0 B A_0 = 0.$$

Following the overall test, suppose we are interested in the sum of the following tetrads:

$$\psi_p = (\mu_{21} + \mu_{12} - \mu_{31} - \mu_{22}) + (\mu_{32} + \mu_{23} - \mu_{13} - \mu_{22}). \tag{5.297}$$

This contrast may not be expressed as a bilinear form $\psi_p = c'Ba$. However, selecting

$$M = \begin{pmatrix} 0 & 1 \\ 1 & 0 \end{pmatrix}$$

we see that the contrast has the form

$$\psi_p = \operatorname{tr}(MB_0) = \operatorname{tr}(MC_0BA_0) = \operatorname{tr}(A_0MC_0) = \operatorname{tr}(GB)$$

for the matrix

$$G = \begin{pmatrix} 0 & 1 & -1 \\ 1 & -2 & 1 \\ -1 & 1 & 0 \end{pmatrix}$$

is a generalized contrast.

Finally, suppose we wish to test the multivariate hypothesis that the means over the repeated measures are zero, the multivariate test of conditions. For our example the hypothesis is

$$H_c : \begin{pmatrix} \mu_{11} \\ \mu_{21} \\ \mu_{31} \end{pmatrix} = \begin{pmatrix} \mu_{12} \\ \mu_{22} \\ \mu_{32} \end{pmatrix} = \begin{pmatrix} \mu_{13} \\ \mu_{23} \\ \mu_{33} \end{pmatrix}.$$

Representing the test in the linear form $H_c : C_0BA_0 = 0$, for our example, we let $C_0 = I_3$ and

$$A = \begin{pmatrix} 1 & 0 \\ -1 & 1 \\ 0 & -1 \end{pmatrix}.$$

Following the overall test, suppose we are interested in the contrast

$$\psi_c = (\mu_{11} - \mu_{12}) + (\mu_{22} - \mu_{23}) + (\mu_{31} - \mu_{33}). \tag{5.298}$$

This contrast is again an extended linear hypothesis and may be tested using the matrices

$$M = \begin{pmatrix} 1 & 0 & 1 \\ 0 & 1 & 1 \end{pmatrix} \text{ or } G = \begin{pmatrix} 1 & 0 & 1 \\ -1 & 1 & 0 \\ 0 & -1 & -1 \end{pmatrix}$$

in the expression

$$\psi_0 = \operatorname{tr}(MB_0) = \operatorname{tr}(MC_0BA_0) = \operatorname{tr}(A_0MC_0B) = \operatorname{tr}(GB).$$

To obtain a point estimate of $\psi_0 = \operatorname{tr}(MB_0) = \operatorname{tr}(GB)$, we use $\hat{\psi}_0 = \operatorname{tr}(M\widehat{B}_0) = \operatorname{tr}(MC_0\widehat{B}A_0) = \operatorname{tr}(G\widehat{B})$. A test of $\psi = 0$ is rejected if the ratio $\frac{\hat{\psi}}{\hat{\sigma}_{\hat{\psi}}} > c_o$. To construct an approximate $1 - \alpha$ simultaneous confidence interval, we utilize the

F-approximation discussed in the SAS/STAT User's Guide (1990, p. 18) and also by Timm (2002, p. 103). For Roy's test, $\frac{\lambda_1}{(1+\lambda_1)} = \theta_1$ or $\frac{\theta_1}{(1-\theta_1)} = \lambda_1$. Letting $s = \min(\nu_h, u)$, $M = \frac{(|\mu_h - u| - 1)}{2}$, $N = \frac{(\nu_e - u - 1)}{2}$ where $\nu_h = \mathrm{rank}(C_0)$ and $u = \mathrm{rank}(A_0)$, we find the critical constant (5.67) so that the critical value is

$$c_0^2 = \nu_e \left(\frac{r}{((\nu_e - r + \nu_h))} \right) F_{(r,\nu_e - r + \nu_h)}^{1-\alpha} \qquad (5.299)$$

where $r = \max(\nu_h, u)$ and

$$\hat{\sigma}_{\hat{\psi}} = \sum_{i=1}^{s} \lambda_i^{1/2} (MW_0 M' E_0) \qquad (5.300)$$

for W_0 and E_0 defined in (5.212) and (5.210), respectively.

Alternatively, using the Bartllett-Lawley Hotelling trace criterion, also called T_0^2, the critical constant is

$$c_0^2 = \nu_e \left(\frac{s(2M + s + 1)}{2(sN + 1)} \right) F_{(s(2M+s+1),2(sn+1))}^{1-\alpha} \qquad (5.301)$$

and the standard error of the estimate is

$$\hat{\sigma}_{\hat{\psi}} = [\mathrm{tr}(MW_0 M' E_0)]^{1/2}. \qquad (5.302)$$

Thus, for contrasts of the form one may construct simultaneous $1 - \alpha$ confidence sets using (5.222) for generalized contrasts.

While Krishnaiah et al. (1980) have developed a "Roots" program to test extended linear hypotheses involving a single matrix M, we illustrate the procedure using PROC IML and data from Timm (1975, p. 454) involving three groups and three repeated measurements. The SAS code for the analysis is provided in Program 5_15.sas.

5.15.1 Results and Interpretation

In Program 5_15.sas the multivariate tests for group differences, parallelism, and differences in conditions are illustrated using both PROC GLM and with PROC IML to verify the calculations of eigenvalues and eigenvectors for each overall hypothesis. PROC IML is used to test the significance of several extended linear hypotheses, both simple (those involving a single matrix M) and complex hypotheses (those using several M_i).

In Program 5_15.sas, PROC IML is used to test the overall hypothesis of no group differences. Finding the difference in groups to be significant using $\alpha = 0.05$, suppose we tested the extended linear hypothesis

$$\psi = \mu_{11} + \mu_{22} + \mu_{33} - \frac{(\mu_{12} + \mu_{21})}{2} - \frac{(\mu_{13} + \mu_{31})}{2} - \frac{\mu_{23} + \mu_{32}}{2} = 0. \quad (5.303)$$

PROC IML is used to verify the calculations of the overall test and the estimate of ψ. The estimate is

$$\hat{\psi} = \text{tr}(M\widehat{B}_0) = \text{tr}(MC_0\widehat{B}A_0) = \text{tr}(G\widehat{B}) = -2.5$$

where the matrices M, G, and $\widehat{B}$ are defined:

$$M = \begin{pmatrix} 1 & .5 \\ -.5 & .5 \\ -.5 & 1 \end{pmatrix} \quad G = \begin{pmatrix} 1 & -.5 & -.5 \\ -.5 & 1 & -.5 \\ -.5 & -.5 & 1 \end{pmatrix} \quad \widehat{B} = \begin{pmatrix} 4 & 7 & 10 \\ 7 & 8 & 11 \\ 5 & 7 & 9 \end{pmatrix}.$$

Solving the eigen equation $|H - \lambda E_0^{-1}| = 0$ with $H = MW_0M'$, $W_0 = C_0(X'X)^{-1}$ and $E_0 = A_0'EA_0 = A_0'Y'(I - X(X'X)^{-1}X')YA_0$, we solve the eigen equation using EIGVAL function for a symmetric matrix. The last eigenvalue is near zero and $\lambda_1 = 6.881$ and $\lambda_2 = 2.119$. By (5.217), let

$$\sigma_{\text{Trace}} = [\text{tr}(MW_0M'E_0)]^{1/2} = \left(\sum_{i-1}^{s}\lambda_i\right)^{1/2}$$

$$\sigma_{\text{Root}} = \sum_{i-1}^{s}\lambda_i^{1/2}$$

for the trace and root criterion $|\hat{\psi}| = 2.5$, the extended trace and root statistics for testing $\psi = 0$ are

$$\frac{|\hat{\psi}|}{\sigma_{\text{Trace}}} = 0.8333 \text{ and } \frac{|\hat{\psi}|}{\sigma_{\text{Root}}} = 0.6123.$$

Using the F distribution estimates for the critical constants, the approximate values are 4.6141561 and 3.4264693, respectively, for the trace and root criteria using $\alpha = 0.05$. Thus, we fail to reject the null hypothesis. Also illustrated in the Program are the construction of confidence intervals for the contrast using each criterion. Both intervals contain zero:

$$\text{Root Interval} : (-16.48953, 11.489534)$$
$$\text{Trace Interval} : (-16.34247, 11.342468).$$

Next we test that the subset hypothesis:

$$H_0 : \mu_{11} = \mu_{21} = \mu_{31} \tag{5.304}$$
$$\mu_{12} = \mu_{22} = \mu_{32} \tag{5.305}$$
$$\mu_{13} = \mu_{23} = \mu_{33}. \tag{5.306}$$

This test is more complicated since it involves the maximization of G over G_i. With

Output 5.15.1: Extended Linear Hypotheses: Multivariate Test of Group — Using PROC IML.

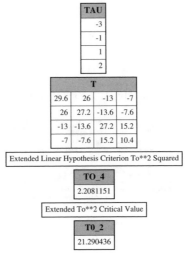

TAU
-3
-1
1
2

T			
29.6	26	-13	-7
26	27.2	-13.6	-7.6
-13	-13.6	27.2	15.2
-7	-7.6	15.2	10.4

Extended Linear Hypothesis Criterion To**2 Squared

TO_4
2.2081151

Extended To**2 Critical Value

TO_2
21.290436

$\lambda_i = \text{tr}(M_i' \widehat{B}_0)$ and $t_{ij} = \text{tr}(M_i W_0 M_j' E_0)$, we have from Output 5.15.1 that

$$\lambda' = \begin{pmatrix} -3 & -1 & 1 & 2 \end{pmatrix}$$

$$T = \begin{pmatrix} 29.6 & 26.0 & -13.0 & -7.0 \\ 26.0 & 27.2 & -13.6 & -7.6 \\ -13.0 & -13.6 & 27.2 & 15.2 \\ -7.0 & -7.6 & 15.2 & 10.4 \end{pmatrix}$$

so that $\lambda' T^{-1} \lambda = 2.208$. Comparing this with the statistic $T_0^2 = (4.614)^2 = 21.29$, the test is nonsignificant using $\alpha = 0.05$.

Finally, Program 5_15.sas contains code for testing

$$\psi_p = (\mu_{21} + \mu_{12} - \mu_{31} - \mu_{22}) + (\mu_{32} + \mu_{23} - \mu_{13} - \mu_{22}) = 0 \qquad (5.307)$$

and

$$\psi_c = (\mu_{11} - \mu_{12}) + (\mu_{22} - \mu_{23}) + (\mu_{31} - \mu_{33}) = 0 \qquad (5.308)$$

following the tests of parallelism and conditions, respectively, for the repeated measurement design. Contrasts for the two tests are displayed in Output 5.15.2-5.15.3. Because the interval for ψ_p contains zero, it is not significant. However, the test that $\psi_c = 0$ is significant since the interval does not contain zero.

Output 5.15.2: Extended Linear Hypotheses: Multivariate Test of Parallelism —
Using PROC IML.

	RL	RU
Extended Root Interval: (	-2.790005 ,	4.7900054)

	VL	VU
Extended Trace Interval: (	-2.709069 .	4.7090687)

Output 5.15.3: Extended Linear Hypotheses: Multivariate Test of Conditions —
Using PROC IML.

	RL	RU
Extended Root Interval: (	-11.03643 ,	-8.963574)

	VL	VU
Extended Trace Interval: (	-11.01909 ,	-8.980907)

CHAPTER 6

Doubly Multivariate Linear Model

6.1 INTRODUCTION

In this chapter, we extend the multivariate general linear model to the doubly multivariate linear model (DMLM) which allows one to analyze vector-valued repeated measurements or panel data. Estimation and hypothesis testing theory for the classical and the responsewise forms of the model are presented. The doubly multivariate linear mixed model is also reviewed. Finally, data are analyzed using both the DMLM and mixed model approaches.

6.2 CLASSICAL MODEL DEVELOPMENT

The multivariate linear model is extended to a doubly multivariate linear model by obtaining p-variates on n subjects at each of t occasions. If we let the number of variables $p = 1$, the repeated measurement design over t occasions results. In this chapter, we assume that all vector observations are complete so that matrix of responses $R_{i(i \times p)}$ for $i = 1, 2, \ldots, n$ have no missing values. Missing data problems are considered in Chapter 12.

The classical doubly multivariate linear model (CDMLM) for the p-variate repeated measurements over t occasions is

$$\Omega_0 : Y_{n \times pt} = X_{n \times k} B_{k \times pt} + U_{n \times pt} \tag{6.1}$$

where $Y_{n \times pt}$ is the data matrix for a sample of n subjects with $pt \leq n - k$ responses, $X_{n \times k}$ is the between subjects design matrix of full rank k, $B_{k \times pt}$ is a matrix of fixed effects, and $U_{n \times pt}$ is a matrix of random errors. For hypothesis testing, we further assume that the n rows of Y are distributed independently multivariate normal,

$$y_i \sim N(0, \Sigma_i). \tag{6.2}$$

Assuming homogeneity of Σ_i so that $\Sigma_i = \Sigma$ for all $i = 1, 2, \ldots, n$, we write (6.2) as

$$\mathcal{E}(Y) = XB \tag{6.3}$$

$$\text{cov}(Y) = I_n \otimes \Sigma_{pt \times pt}. \tag{6.4}$$

Letting the u_i denote the columns of U and $\Sigma_i = \Sigma$, $u_i \sim IN(0, \Sigma)$ or

$$\text{vec}(U) \sim N_{npt}(0, \Sigma \otimes I_n). \tag{6.5}$$

We can then write (6.1) as

$$\text{vec}(Y) = (I_{pt} \otimes X)\text{vec}(B) + \text{vec}(U) \tag{6.6}$$

which is again a general linear model with unknown Σ. The model for the ith subject is

$$y_i = (I_{pt} \otimes x_i')\text{vec}(B) + u_i \tag{6.7}$$
$$u_i \sim IN(0, \Sigma). \tag{6.8}$$

Alternatively, the ith row of Y is related to the ith response matrix $R_{i(t \times p)}$, so

$$y_{i(pt \times 1)} = \text{vec}(R_i) = (I_{pt} \otimes x_i')\text{vec}(B) + u_i. \tag{6.9}$$

To analyze contrasts among the t time periods, a post-matrix P of the form

$$P = (I_p \otimes A) \tag{6.10}$$

is constructed where the matrix A again satisfies the semiorthogonal condition that $A'A = I_u$ and the $\text{rank}(A) = u$. Using P in (6.10), the model in (6.1) that has reduced form is

$$Z = X\Theta + \xi \tag{6.11}$$

where $Z = YP$, $\Theta = BP$, and $\xi = UP$, and

$$\text{vec}(\xi) \sim N(0, \tilde{\Phi} \otimes I_n) \tag{6.12}$$

where

$$\tilde{\Phi} = (I_p \otimes A)'\Sigma(I_p \otimes A). \tag{6.13}$$

This is a general linear model (GLM) with $\Omega = \tilde{\Phi} \otimes I_n$. The matrix $\tilde{\Phi}$ may have Kronecker or direct product structure, $\tilde{\Phi} = \Sigma_e \otimes \Sigma_t$ (Galecki, 1994).

The primary goal of the model is to test hypotheses about the coefficients of B. A convenient form of H is

$$H : CB(I_p \otimes A) = \Gamma_0 \qquad \text{or} \qquad H : \Gamma = \Gamma_0 \tag{6.14}$$

where $C_{\nu_h \times k}$ has rank ν_h, and $A_{t \times u}$ has rank u. Without loss of generality, we may assume $A'A = I_u$. Assuming a CDMLM, the best linear unbiased estimator (BLUE) of Γ is

$$\hat{\Gamma} = C\hat{B}(I_p \otimes A) \qquad \text{where} \qquad \hat{B} = (X'X)^{-1}X'Y, \tag{6.15}$$

and $\widehat{\Gamma}$ is also the ML estimate of Γ. The estimator $\widehat{\Gamma}$ is the minimum variance unbiased estimator of Γ and does not depend on Σ; that is, the maximum likelihood estimator is invariant to the assumed structure of Σ. Under multivariate normality,

$$\hat{\gamma} = \text{vec}(\widehat{\Gamma}) \sim N \left[\text{vec}(\Gamma), \tilde{\Phi} \otimes C(X'X)^{-1}C' \right] \qquad (6.16)$$

where $\tilde{\Phi} = (I_p \otimes A)' \Sigma (I_p \otimes A)$. Thus, the distribution of $\widehat{\Gamma}$ depends on Σ so that inferences about Γ depend on the structure of Σ. Again if $p = 1$, the standard multivariate model results, or

$$H : CBA = \Gamma_0 \qquad \text{or} \qquad H : \Gamma = \Gamma_0 \qquad (6.17)$$

where $\widehat{\Gamma} = C\widehat{B}A$, $\widehat{B} = (X'X)^{-1}X'Y$, and $\Phi = A'\Sigma A$. Letting $\psi = c'Ba$, the $\text{var}(\hat{\psi}) = a'\Sigma a[c'(X'X)^{-1}c]$.

To test hypotheses about Γ in (6.14), the test statistics again depend on the characteristic roots λ_i of $|H - \lambda E| = 0$ where

$$E = (I_p \otimes A)'Y'[I_n - X(X'X)^{-1}X']Y(I_p \otimes A) \qquad (6.18)$$

$$H = (\widehat{\Gamma} - \Gamma_0)'[C(X'X)^{-1}C']^{-1}(\widehat{\Gamma} - \Gamma_0). \qquad (6.19)$$

The matrices E and H have independent Wishart distributions

$$E \sim W_{pu}(\nu_e, \tilde{\Phi}, 0) \qquad \text{and} \qquad H \sim W_{pu}(\nu_h, \tilde{\Phi}, \tilde{\Phi}^{-1}\Delta) \qquad (6.20)$$

where $\nu_e = n - k$ and the noncentrality matrix is

$$\Delta = (\Gamma - \Gamma_0)'[C'(X'X)^{-1}C']^{-1}(\Gamma - \Gamma_0). \qquad (6.21)$$

When $p = 1$, we have the following simplification

$$E = A'Y'[I_n - X(X'X)^{-1}X']YA \qquad (6.22)$$

$$H = (C\widehat{B}A - \Gamma_0)'[C(X'X)^{-1}C']^{-1}(C\widehat{B}A - \Gamma_0). \qquad (6.23)$$

To test H in (6.14), several test statistics are available that are functions of the roots of $|H - \lambda E| = 0$ for H and E defined in (6.18) and (6.19).

Wilks' Λ

$$\Lambda = \frac{|E|}{|E + H|} = \prod_{i=1}^{s}(1 + \lambda_i)^{-1} \sim U_{g, \nu_h, \nu_e} \qquad (6.24)$$

where $\nu_e = n - k$, $\nu_h = \text{rank}(C)$, and $g = \text{rank}(I_p \otimes A) = pu$.

Roy's Largest Root

$$\theta_1 = \frac{\lambda_1}{1 + \lambda_1} \sim \theta_{s, M, N} \qquad (6.25)$$

where $s = \min(\nu_h, u)$, $M = (|u - \nu_h| - 1)/2$, and $N = (\nu_e - u - 1)/2$.

Bartlett–Nanda–Pillai Trace

$$V = \text{tr}[H(H+E)^{-1}] = \sum_{i=1}^{s} \frac{\lambda_i}{1+\lambda_i} \sim V_{s,M,N}. \qquad (6.26)$$

Bartlett–Lawley–Hotelling Trace

$$T^2 = \frac{T_0^2}{\nu_e} = \sum_{i=1}^{s} \frac{\lambda_i}{\nu_e} \sim U_{0(s,M,N)}. \qquad (6.27)$$

An alternative expression for T_0^2 is

$$T_0^2 = [\text{vec}(\widehat{\Gamma} - \Gamma_0)]'\{\tilde{\Phi}^{-1} \otimes [C(X'X)^{-1}C']^{-1}\}[\text{vec}(\widehat{\Gamma} - \Gamma_0)] \qquad (6.28)$$

where

$$\tilde{\Phi} = \frac{E}{\nu_e} = \frac{(I_p \otimes A)'Y'[I_n \otimes X(X'X)^{-1}X']Y(I_p \otimes A)}{\nu_e}. \qquad (6.29)$$

6.3 RESPONSEWISE MODEL DEVELOPMENT

Our presentation of the DMLM so far followed the classical approach. In previous chapters, we saw that when $p = 1$, there were problems in representing the model for mixed model analysis of the data. To reduce the rearrangement of data problems, we represent (6.1) using the transpose of the data matrix Y, following Boik (1988). The reponsewise doubly multivariate linear model (RDMLM) is represented as

$$Y_{pt \times n} = B_{pt \times k} X_{k \times n} + U_{pt \times n} \qquad (6.30)$$

which is no more than the transpose of the corresponding matrices in (6.1). Using presentation in (6.30) the $pt \times 1$ vector of random errors for the jth subject is again assumed to be normally distributed,

$$u_j \sim IN(0, \Sigma). \qquad (6.31)$$

Now, however,

$$\text{vec}(U) \sim N_{npt}(0, I_n \otimes \Sigma) \qquad (6.32)$$

which has a nice block diagonal structure. Because the model has changed, the format for testing hypotheses must change; it is the transpose of (6.14) written as

$$H : (A' \otimes I_p)BC' = \Gamma_0 \qquad \text{or} \qquad H : \Gamma = \Gamma_0. \qquad (6.33)$$

To reduce (6.30) to perform a multivariate repeated measures analysis, the following reduced linear model results

$$(A' \otimes I_p)Y = (A' \otimes I_p)BX + (A' \otimes I_p)U. \qquad (6.34)$$

However, because of the alternative model representation, the form of E and H becomes

$$\tilde{E} = (A' \otimes I_p)'Y[I_n - X'(X'X)^{-1}X]Y'(A \otimes I_p) \tag{6.35}$$

$$\tilde{H} = (\widehat{\Gamma} - \Gamma_0)'[C(X'X)^{-1}C']^{-1}(\widehat{\Gamma} - \Gamma_0) \tag{6.36}$$

where

$$\widehat{\Gamma} = (A' \otimes I_p)\widehat{B}C' \qquad \text{and} \qquad \widehat{B} = (X'X)^{-1}X'Y. \tag{6.37}$$

We again have the E and H follow Wishart distributions as defined by (6.20); however,

$$\tilde{\Phi} = \tilde{\Phi}^* = (A \otimes I_p)'\Sigma(A \otimes I_p) \tag{6.38}$$

and the noncentrality parameter becomes

$$\Delta \equiv \tilde{\Delta}^* = (\Gamma - \Gamma_0)'[C(X'X)^{-1}C']^{-1}(\Gamma - \Gamma_0) \tag{6.39}$$

with Γ defined in (6.33).

6.4   THE MULTIVARIATE MIXED MODEL

Returning to the classical form of the DMLM $Y_{n \times pt}$ as defined in (6.1), the ith row of Y is

$$y_i = \text{vec}(R_i) = (I_{pt} \otimes x_i')\text{vec}(B) + u_i \tag{6.40}$$

where x_i and u_i are the ith rows of X and U. One can again construct a multivariate mixed model (MMM) by modeling

$$u_i = Z_i\theta_i + e_i \qquad\qquad i = 1, 2, \ldots, n \tag{6.41}$$

where Z_i is a $(pt \times h)$ within-subjects design matrix for the random effects θ_i, the $\text{rank}(Z_i) = h$ and the vectors θ_i and e_i are independent multivariate normal

$$\theta_i \sim N_h(0, D_h) \tag{6.42}$$

$$e_i \sim N_{pt}(0, \Sigma_e \otimes I_t) \tag{6.43}$$

as discussed by Boik (1991). Hence,

$$y_i \sim N[(I_{pt} \otimes x_i')\text{vec}(B), \Sigma_i] \tag{6.44}$$

where

$$\Sigma_i = Z_i D_h Z_i' + (\Sigma_e \otimes I_t) \qquad\qquad i = 1, 2, \ldots, n. \tag{6.45}$$

The mixed model ML estimate of Γ is $\widehat{\Gamma} = C\widehat{B}(I_p \otimes A)$ and

$$\hat{\gamma} = \text{vec}(\widehat{\Gamma}) = \left[(I_p \otimes A) \otimes C' \right]' \left[\sum_{i=1}^{n} (\widehat{\Sigma}_i^{-1} \otimes x_i x_i') \right]^{-1} \left[\sum_{i=1}^{n} (\widehat{\Sigma}_i^{-1} \otimes x_i) y_i \right]$$

(6.46)

where $\widehat{\Sigma}_i = Z_i \widehat{D}_h Z_i' + (\widehat{\Sigma}_e \otimes I_t)$, $\widehat{D}_h$ is the ML estimate of D_h and $\widehat{\Sigma}_e$ is the ML estimate of Σ_e. Using the multivariate central limit theorem, the

$$\lim_{n \to \infty} n^{1/2} \text{vec}(\widehat{\Gamma} - \Gamma_0) \sim N\left(0, \Sigma_{\widehat{\Gamma}}\right)$$

(6.47)

where

$$\Sigma_{\widehat{\Gamma}} = \lim_{n \to \infty} n \left[(I_p \otimes A) \otimes C' \right]' \left[\sum_{i=1}^{n} (\Sigma_i^{-1} \otimes x_i x_i') \right]^{-1} \left[(I_p \otimes A) \otimes C' \right]$$

(6.48)

provided $\Sigma_{\widehat{\Gamma}}$ exist. Substituting the ML estimate $\widehat{\Sigma}_i = {}^E\!/_n$ for Σ_i, inference on Γ may be obtained using (6.47) in large samples since $(\hat{\gamma} - \gamma_0)' \widehat{\Sigma}_{\widehat{\Gamma}}^{-1} (\hat{\gamma} - \gamma_0) \sim \chi_\nu^2$ and $\nu = \text{rank}(C)$; the Wald statistic defined in (4.15) with $(X'\widehat{\Sigma}^{-1}X)^{-1} \equiv \widehat{\Sigma}_{\widehat{\Gamma}}$ may be used.

The p-variate generalization of Scheffé's mixed model under sphericity is obtained by associating Z_i in (6.45) with the matrix $I_p \otimes 1_t$ so that $h = p$ and

$$\Sigma_i = \Sigma = (D_p \otimes 1_t 1_t') + (\Sigma_e \otimes I_t) \qquad i = 1, 2, \ldots, n.$$

(6.49)

This is called the Scheffé MMM and satisfies the multivariate sphericity condition that

$$\tilde{\Phi} = \Sigma_e \otimes I_u$$

(6.50)

as discussed by Reinsel (1982, 1984). It is a special case of (6.45).

Showing the equivalence of the Scheffé MMM and the standard doubly multivariate model is again complicated because of the orientation of the data matrix $Y_{n \times pt}$. To solve this problem, Boik (1988) formulated the Scheffé MMM to be rearranged to $Y_{pt \times n}$ as in (6.30). Using the RDMLM (6.30), the jth column y_j of $Y_{pt \times n}$ is the vector of responses for the jth subject. Boik (1988) shaped the vector of responses for the jth subject into a $(p \times t)$ matrix Y_j^* where the columns of Y_j^* are the t occasions and the rows are the p dependent variables. The rearranged matrix is represented as

$$Y^* = \left(Y_1^* \quad Y_2^* \ldots \quad Y_n^* \right).$$

(6.51)

The $(pt \times 1)$ response vector for the jth subject is related:

$$y_j = \text{vec}(Y_j^*)$$

(6.52)

which is a rearrangement of $R_{i(t \times p)}$ for the CDMLM. In a similar manner, the elements in $B_{pt \times k}$ and $U_{pt \times n}$ are rearranged so that

$$B^* = \begin{pmatrix} B_1^* & B_2 & \cdots & B_k^* \end{pmatrix} \qquad U^* = \begin{pmatrix} U_1^* & U_2^* & \cdots & U_n^* \end{pmatrix} \qquad (6.53)$$

where B_i^* and U_j^* are each $(p \times t)$ matrices, the ith column of B satisfies $B_i = \text{vec}(B_i^*)$ and the jth column of U satisfies $u_j = \text{vec}(U_j^*)$. Finally, $\text{vec}(Y^*) = \text{vec}(Y)$, $\text{vec}(B^*) = \text{vec}(B)$, and $\text{vec}(U^*) = \text{vec}(U)$.

This rearrangement allows one to rewrite the linear model (6.30) and the reduced linear model (6.34), respectively, as

$$Y^* = B^*(X \otimes I_t) + U^* \qquad (6.54)$$
$$Y^*(I_n \otimes A) = B^*(X \otimes A) + U^*(I_n \otimes A). \qquad (6.55)$$

Assuming $\tilde{\Phi}^*$ in (6.38) satisfies the condition of multivariate sphericity

$$\tilde{\Phi}^* = I_u \otimes \Sigma_e \qquad (6.56)$$

a MMM analysis for testing

$$H : B^*(C' \otimes A) = \Gamma_0^* \qquad \text{or} \qquad H : \Gamma^* = \Gamma_0^* \qquad (6.57)$$

is obtained by performing a multivariate analysis using the reduced model specified by (6.55).

To test (6.57), one uses the standard multivariate test statistics obtained from the reduced form of the model with the assumption that multivariate sphericity is satisfied. In testing (6.57), observe that the design matrix is separable, has the form $X \otimes A$, and that the hypothesis test matrix is also separable in that it takes the form $C' \otimes A$ as required using the standard multivariate linear model and the univariate mixed model. Establishing an expression for the error matrix, say E^*, is only complicated by the fact that the design matrix involves a Kronecker product. For the rearranged model (6.54),

$$B^{*'} = [(X \otimes I_t)(X \otimes I_t)']^{-1}(X \otimes I_p)Y^{*'} \qquad (6.58)$$
$$= [(X'X)^{-1}X \otimes I_t]Y^{*'}. \qquad (6.59)$$

For the reduced model (6.55),

$$\widehat{B}^{*'} = [(X \otimes A)(X \otimes A)']^{-1}(X \otimes A)Y^{*'}$$
$$= [(X'X) \otimes (A'A)]^{-1}(X \otimes A'A)Y^{*'}$$
$$= [(X'X)^{-1}X \otimes I_t]Y^{*'}. \qquad (6.60)$$

To find E^*, recall that in general

$$E = Z[I - M'(M'M)^{-1}M']Z' \qquad (6.61)$$

where Z and M are column forms of the data matrix and design matrix, respectively. Letting $Z = Y^*(I_n \otimes A)$ and $M = X \otimes A$,

$$
\begin{aligned}
E^* &= Y^*(I_n \otimes A)\{(I - X \otimes A)'[(X \otimes A)(X \otimes A)']^{-1}(X \otimes A)\}(I_n \otimes A)'Y^{*'} \\
&= Y^*(I_n \otimes A'A)Y^{*'} - \widehat{B}^*(X \otimes A)(X \otimes A)'\widehat{B}^* \\
&= Y^*(I_n \otimes A'A)Y^{*'} - Y^*(X'(X'X)^{-1}X \otimes A'A)Y^{*'} \\
&= Y^*\{[I_n - X'(X'X)^{-1}X] \otimes A'A\}Y^{*'}.
\end{aligned}
\tag{6.62}
$$

If we set $\widehat{\Gamma}^* = \widehat{B}^*(C' \otimes A)$ where $\widehat{B}^{*'}$ is defined by (6.59), the hypothesis matrix becomes

$$
H^* = (\widehat{\Gamma}^* - \Gamma_0)[C(X'X)^{-1}C']^{-1}(\widehat{\Gamma}^* - \Gamma_0)'.
\tag{6.63}
$$

However, because one usually tests $H : B^*(C' \otimes A) = 0$, (6.63) reduces to

$$
H^* = Y^*\{X'(X'X)^{-1}C'[C(X'X)^{-1}C']^{-1}C(X'X)^{-1}X \otimes A'A\}Y^{*'}.
\tag{6.64}
$$

Assuming $H : \Gamma^* = 0$, we can write E^* and H^* using E and H defined by the CDMLM, by using the generalized trace operator proposed by Thompson (1973). If we let W be a $(pu \times pu)$ matrix and W_{ij} be the ij^{th} $(u \times u)$ submatrix for $i, j = 1, 2, \ldots, p$, then the generalized trace operator $T_p(W)$ is a $(p \times p)$ matrix defined as

$$
T_p(W) \equiv [\text{tr}(W_{ij})].
\tag{6.65}
$$

Partitioning E, H, $\tilde{\Phi}$, and Δ into $(p \times p)$ submatrices: $E = \{E_{ij}\}$, $H = \{H_{ij}\}$, $\tilde{\Phi} = \{\tilde{\Phi}_{ij}\}$, and $\Delta = \{\Delta_{ij}\}$ for $i = 1, 2, \ldots, u$, we define

$$
E^* = T_p(E)
\tag{6.66}
$$

$$
H^* = T_p(H)
\tag{6.67}
$$

$$
\tilde{\Phi}^* = T_p(\tilde{\Phi})
\tag{6.68}
$$

$$
\Delta^* = T_P(\Delta)
\tag{6.69}
$$

and assuming multivariate sphericity

$$
E^* \sim W_p(u\nu_e, \Sigma_e, 0) \qquad H^* \sim W_p(u\nu_h, \Sigma_e, \Sigma_e^{-1}Tp(\Delta))
\tag{6.70}
$$

where $\Sigma_e = u^{-1}T_p(\tilde{\Phi})$.

To test $H : B^*(C' \otimes A) = 0$, one again uses the multivariate criteria defined by formula (6.24)-(6.27) substituting for E and H with E^* and H^*. Again the formulas simplify; for example, T_0^2 in (6.28) is T_0^2 with $\tilde{\Phi}$ replaced with $\widehat{\tilde{\Phi}} = (E^*/u\nu_e \otimes I_u)$ which is an unbiased estimator of $\tilde{\Phi}$. If $p = 1$ then

$$
T_0^2 = \frac{\text{tr}(H^*)/\nu_h^*}{\text{tr}E^*/\nu_e^*} \sim F[u\nu_h, u\nu_e, \delta = \text{tr}(\Delta)/\text{tr}(\Sigma_e)]
\tag{6.71}
$$

a noncentral F-distribution with $\nu_h^* = u\nu_h$ and $\nu_e^* = u\nu_e$ degrees of freedom (Davidson, 1988; Timm, 1980b).

6.5 DOUBLE MULTIVARIATE AND MIXED MODEL ANALYSIS OF REPEATED MEASUREMENT DATA

In Chapter 2 we illustrated the analysis of a univariate split-plot repeated measures design when the covariance matrix across groups (plots) are homogeneous and satisfy the circularity condition, have spherical structure when transformed by a normalized within-subjects matrix. When circularity does not hold, tested using Mauchley's chi-square test, the appropriate analysis is to use a multivariate model which allows for general covariance structure as illustrated in Chapter 5. However, if the number of subjects is smaller than the number of repeated measures the multivariate approach may not be used; then, one must use an approximate ε-adjusted F test.

If the repeated measurements are vector observations over time and we can assume a block spherical structure for the transformed covariance matrix (multivariate sphericity), the standard MMM should be used to analyze the data. The MMM results may be obtained from the DMLM analysis. When the number of subjects is less than the number of total repeated measures, an ε-adjusted test is used.

To illustrate the analysis of a DMLM, we use data from Timm (1980b). The data were provided by Dr. Thomas Zullo in the School of Dental Medicine, University of Pittsburgh. Nine subjects were assigned to two orthopedic treatments to study the effectiveness of three activator conditions on three variables associated with the adjustment of the mandible.

For this problem, we have $n = 18$ subjects for which we obtain $p = 3$ variables over $t = 3$ conditions (occasions or time points) for $k = 2$ groups. To analyze these data, we may organize the p-variables at t time points so that the jth subject has a $p \times t$ data matrix Y_j^*. For the MMM, we let

$$Y^* = \begin{pmatrix} Y_1^* & Y_2^* & \cdots & Y_n^* \end{pmatrix}. \tag{6.72}$$

For this data arrangement, let $y_j = \text{vec}(Y_j^*)$ where the variables are nested within time. Thus, for p variables and two time (occasions) points, the covariance matrix has a convenient block structure:

$$\Sigma = \begin{pmatrix} \Sigma_{11} & \Sigma_{12} \\ \Sigma_{21} & \Sigma_{22} \end{pmatrix}. \tag{6.73}$$

Alternatively, one may nest the time (occasions) dimension within each variable. For this case, we have for our example the following:

$$
\begin{array}{ccc|ccc|ccc}
\multicolumn{3}{c|}{v_1} & \multicolumn{3}{c|}{v_2} & \multicolumn{3}{c}{v_3} \\
t_1 & t_2 & t_3 & t_1 & t_2 & t_3 & t_1 & t_2 & t_3
\end{array} \tag{6.74}
$$

so that when the number of variables $p = 1$, the design reduces to a standard repeated measures model with occasions (repeated measures) t_1, t_2, and t_3. Most standard computer routines require data organization (6.74); however, (6.72) is more useful for recovering mixed model results. To convert (6.74) to (6.72), it is convenient to order the postmultiplication matrix by variables and thus create a block diagonal matrix as in (6.73).

Table 6.1: DMLM Organization of Mean Vector.

	So_r-Me (MM)			ANS-Me (MM)			Pal-MP Angle (degrees)		
Variables	1			2			3		
Conditions	1	2	3	1	2	3	1	2	3
Group 1	μ_{11}	μ_{12}	μ_{13}	μ_{14}	μ_{15}	μ_{16}	μ_{17}	μ_{18}	μ_{19}
Group 2	μ_{21}	μ_{22}	μ_{23}	μ_{24}	μ_{25}	μ_{26}	μ_{27}	μ_{28}	μ_{29}

Note: the mean vector for the ith treatment group is $\mu_i = (\mu_{i1}, \mu_{i2}, \ldots, \mu_{i8}, \mu_{i9})$.

Program 6_5.sas contains the SAS code to analyze the Zullo data using both a DMLM and a MMM. In addition, we explain how to obtain from a normalized DMLM the MMM results, how to test for multivariate sphericity, and how to perform ϵ-adjusted tests. The data for the example are organized using representation (6.74). The three variables are individual measurements of the vertical position of the mandible at three activator treatment conditions and two groups. The organization of the means for the analysis is presented in Table 6.1.

Results and Interpretation

The first hypothesis of interest for these data is to test whether the profiles for the two treatment groups are parallel, by testing for an interaction between conditions and groups. The hypothesis is:

$$H_{GC} : \mu_{11} - \mu_{13}, \mu_{12} - \mu_{13}, \ldots, \mu_{17} - \mu_{18}, \mu_{18} - \mu_{19})$$
$$= (\mu_{21} - \mu_{23}, \mu_{22} - \mu_{23}, \ldots, \mu_{27} - \mu_{28}, \mu_{28} - \mu_{29}) \qquad (6.75)$$

The matrices C and $P = (I_3 \otimes A)$ of the form $CBP = 0$ to test H_{GC} are

$$C = \begin{pmatrix} 1 & -1 \end{pmatrix}, P = \begin{pmatrix} A & 0 & 0 \\ 0 & A & 0 \\ 0 & 0 & A \end{pmatrix}, \text{ and } A = \begin{pmatrix} 1 & 0 \\ 0 & 1 \\ -1 & -1 \end{pmatrix}. \qquad (6.76)$$

For the DMLM analysis in SAS, the matrix P is represented as $m = P'$ so that the columns of the matrix P become the rows in m. This organization of the data and the structure for P does not result in a convenient block structure for the hypothesis test matrix H and the error matrix E. Instead, as suggested by Boik (1988), one may use the responsewise representation for the model. Alternatively, one may use the multivariate mixed model for the analysis which is convenient in SAS as illustrated in Program 6_5.sas. Using this approach, when the matrix A is normalized the MMM test may be obtained from the DMLM as illustrated by Timm (1980b).

The overall hypothesis to test for differences in Groups is

$$H_{G^*} : \mu_1 = \mu_2 \qquad (6.77)$$

and the matrices C and P to test the hypothesis are

$$C = \begin{pmatrix} 1 & -1 \end{pmatrix}, P = I_2. \tag{6.78}$$

The hypothesis to test for vector differences in conditions becomes

$$H_{C^*} : \begin{pmatrix} \mu_{11} \\ \mu_{21} \\ \mu_{14} \\ \mu_{24} \\ \mu_{17} \\ \mu_{27} \end{pmatrix} = \begin{pmatrix} \mu_{12} \\ \mu_{22} \\ \mu_{15} \\ \mu_{25} \\ \mu_{18} \\ \mu_{28} \end{pmatrix} = \begin{pmatrix} \mu_{13} \\ \mu_{23} \\ \mu_{16} \\ \mu_{26} \\ \mu_{19} \\ \mu_{29} \end{pmatrix}. \tag{6.79}$$

The test matrices for the hypothesis are

$$C = I_2, P = \begin{pmatrix} A & 0 & 0 \\ 0 & A & 0 \\ 0 & 0 & A \end{pmatrix}, \text{ and } A = \begin{pmatrix} 1 & 0 \\ -1 & 1 \\ 0 & -1 \end{pmatrix}. \tag{6.80}$$

The test for group differences, H_{G^*}, and the test for differences in conditions, H_{C^*}, do not require parallelism of profiles. However, given parallelism, tests for differences between groups and among conditions are written as:

$$H_G : \begin{pmatrix} \sum_{j=1}^{3} \frac{\mu_{1j}}{3} \\ \sum_{j=4}^{6} \frac{\mu_{1j}}{3} \\ \sum_{j=4}^{9} \frac{\mu_{1j}}{3} \end{pmatrix} = \begin{pmatrix} \sum_{j=1}^{3} \frac{\mu_{2j}}{3} \\ \sum_{j=4}^{6} \frac{\mu_{2j}}{3} \\ \sum_{j=4}^{9} \frac{\mu_{2j}}{3} \end{pmatrix} \tag{6.81}$$

and

$$H_C : \begin{pmatrix} \sum_{i=1}^{2} \frac{\mu_{i1}}{2} \\ \sum_{i=1}^{2} \frac{\mu_{i4}}{2} \\ \sum_{i=1}^{2} \frac{\mu_{i7}}{2} \end{pmatrix} = \begin{pmatrix} \sum_{i=1}^{2} \frac{\mu_{i2}}{2} \\ \sum_{i=1}^{2} \frac{\mu_{i5}}{2} \\ \sum_{i=1}^{2} \frac{\mu_{i8}}{2} \end{pmatrix} = \begin{pmatrix} \sum_{i=1}^{2} \frac{\mu_{i3}}{2} \\ \sum_{i=1}^{2} \frac{\mu_{i6}}{2} \\ \sum_{i=1}^{2} \frac{\mu_{i9}}{2} \end{pmatrix}. \tag{6.82}$$

And, the hypothesis test matrices for the two tests are:

$$C = \begin{pmatrix} 1 & -1 \end{pmatrix}, P' = \begin{pmatrix} 1 & 1 & 1 & 0 & 0 & 0 & 0 & 0 & 0 \\ 0 & 0 & 0 & 1 & 1 & 1 & 0 & 0 & 0 \\ 0 & 0 & 0 & 0 & 0 & 0 & 1 & 1 & 1 \end{pmatrix}, \tag{6.83}$$

$$C = \begin{pmatrix} \frac{1}{2} & \frac{1}{2} \end{pmatrix}, P = \begin{pmatrix} 1 & 0 & 0 & 0 & 0 & 0 \\ 0 & 0 & 0 & 1 & 0 & 0 \\ -1 & 0 & 0 & -1 & 0 & 0 \\ 0 & 1 & 0 & 0 & 0 & 0 \\ 0 & 0 & 0 & 0 & 1 & 0 \\ 0 & -1 & 0 & 0 & -1 & 0 \\ 0 & 0 & 1 & 0 & 0 & 0 \\ 0 & 0 & 0 & 0 & 0 & 1 \\ 0 & 0 & -1 & 0 & 0 & -1 \end{pmatrix}, \tag{6.84}$$

Table 6.2: DMLM Analysis for Zullo Data.

| Hypothesis | Wilks' Criterion | | F | | |
	Λ	df	value	df	P-Value
H_{GC}	0.5830	$(6, 1, 16)$	1.3114	$(6, 11)$	0.3292
H_{G^*}	0.4222	$(9, 1, 16)$	1.216	$(9, 8)$	0.3965
H_{C^*}	0.0264	$(6, 2, 16)$	9.4595	$(12, 22)$	< 0.0001
H_G	0.8839	$(3, 1, 16)$	0.6131	$(3, 14)$	0.6176
H_C	0.0338	$(6, 1, 16)$	52.4371	$(6, 11)$	< 0.0001

respectively. Normalization of the matrices P for these tests within each variable allows one to again obtain MMM results from the multivariate tests, given parallelism. This is not the case for the multivariate tests that do not assume parallelism: H_{G^*}, and H_{C^*}. Some standard statistical packages, e.g., SPSS, do not test the multivariate tests for groups and conditions, but only these tests given parallelism. The SAS code for these tests are provided in Program 6_5.sas. Table 6.2 summarizes the tests using Wilks' Λ criterion.

Because the test of parallelism is not significant, tests of conditions and group differences given parallelism are most appropriate. From these tests one may derive MMM tests of H_C and H_{CG} given multivariate sphericity. The MMM test of group differences is identical to the test of H_G using the DMLM and does not depend on the sphericity condition.

Thomas (1983) derived the likelihood ratio tests of multivariate sphericity and Boik (1988) showed that it was a necessary and sufficient condition for the MMM tests to be exact. Boik (1988) also developed ε-adjusted tests for H_C and H_{CG} when the circularity condition is not satisfied. Recall that for the simple split-plot design F-tests are exact if and only if

$$A'\Sigma A = \sigma^2 I$$

for some matrix A and $A'A = I$. For the multivariate case, using the CDMLM, this becomes

$$(I_p \otimes A)'\Sigma(I_p \otimes A) = (\Sigma_e \otimes I_u) \tag{6.85}$$

where again the matrix A is semi-orthogonal, $A'A = I$, in the submatrix P for the DMLM. Alternatively, using the RDMLM, the condition is written as

$$(A \otimes I_p)'\Sigma(A \otimes I_p) = (I_u \otimes \Sigma_e) \tag{6.86}$$

where $u \equiv q = \text{rank}(A)$.

To test for multivariate sphericity (circularity), Thomas (1983) derived the likelihood ratio statistic:

$$\lambda = |E|^{n/2}|q^{-1} \sum_{i=1}^{q} E_{ii}|^{nq/2}. \tag{6.87}$$

The asymptotic null distribution of $-2\log(\lambda)$ follows a central chi-square distribution with degrees of freedom

$$f = \frac{p(q-1)(pq+p+1)}{2}. \tag{6.88}$$

Letting $\alpha_i(i = 1, \ldots, pq)$ be the eigenvalues of E and $\beta_i(i = 1, \ldots, p)$ the eigenvalues of the $\sum_{i=1}^{q} E_{ii}$, an alternative expression for the test statistic is

$$U = -2\log(\lambda) = n\left(q\sum_{i=1}^{p}\log(\beta_i) - \sum_{i=1}^{pq}\log(\alpha_i)\right). \tag{6.89}$$

When p or q are large relative to n, the asymptotic chi-square approximation may be poor. To correct for this, Boik (1988) developed Box's correction factor for the distribution of the test statistic U. He showed that the

$$P(U \leq U_o) = P(\rho^*U \leq \rho U_o)$$
$$= (1-\omega)P(X_f^2 \leq \rho^*U_o) + \omega P(X_{f+4}^2 \leq \rho^*U_o) + O(\nu_e^{-3}) \tag{6.90}$$

where f is the degrees of freedom for the test statistic U, U_o is the critical value of the test and

$$\rho = 1 - \left(\frac{p}{12qf\nu_e}\right)[2p^2(q^4-1) + 3p(q^3-1) - (q^2-1)] \tag{6.91}$$

$$\rho^* = \frac{\rho\nu_e}{n}\omega$$
$$= (2\rho^2)^{-1}\left[\left(\frac{(pq-1)pq(pq+1)(pq+2)}{24\nu_e^2}\right)\right.$$
$$\left. - \left(\frac{(p-1)p(p+1)(p+2)}{24q^2\nu_e^2}\right)\left(\frac{f(1-\rho)^2}{2}\right)\right] \tag{6.92}$$

and $\nu_e = n - \text{rank}(X)$. Hence, the P-value for the test of multivariate sphericity using Box's correction becomes

$$P(\rho^*U \geq U_o) = (1-\omega)P(X_f^2 \leq U_o) + \omega P(X_{f+4}^2 \leq U_o) + O(\nu_e^{-3}). \tag{6.93}$$

The code for multivariate sphericity is included in Program 6_5.sas. The results are displayed in Output 6.5.1.

Since $U = -2\log(\lambda) = 74.367228$ with degrees of freedom $f = 15$ with P-value $p < 0.0001$ or using Box's correction $\rho^*U = 54.742543$ with P-value $p < 0.0001$, we reject the null hypothesis of multivariate sphericity. The MMM assumption of sphericity is not satisfied for these data; thus, one should use the DMLM.

Assuming the test of multivariate sphericity was not significant, we show how to analyze these data using the MMM. Finally, we illustrate the construction of

Output 6.5.1: Test of Multivariate Sphericity Using Chi-Square and Adjusted Chi-Square Statistics.

E					
9.6944	7.3056	-6.7972	-4.4264	-0.6736	3.7255
7.3056	8.8889	-4.4583	-3.1915	-3.2396	2.9268
-6.7972	-4.4583	18.6156	2.5772	0.8837	-10.1363
-4.4264	-3.1915	2.5772	5.3981	1.4259	-1.8546
-0.6736	-3.2396	0.8837	1.4259	18.3704	-0.7769
3.7255	2.9268	-10.1363	-1.8546	-0.7769	6.1274

E11		
9.6944	7.3056	-6.7972
7.3056	8.8889	-4.4583
-6.7972	-4.4583	18.6156

E22		
5.3981	1.4259	-1.8546
1.4259	18.3704	-0.7769
-1.8546	-0.7769	6.1274

B
18.972307
10.357742
4.2173517

A
32.29187
18.552285
10.952687
3.4969238
1.4751884
0.3258457

CHI_2	DF	PVALUE
74.367228	15	7.365E-10

RHO
0.828125

OMEGA
0.0342649

RO_CHI_2	CPVALUE
54.742543	2.7772E-6

Table 6.3: MMM Analysis for Zullo Data.

Hypothesis	Wilks' Criterion		F		
	Λ	df	value	df	P-Value
Group|Parallelism	0.8839	$(3, 1, 16)$	0.6132	$(3, 14)$	0.6176
Condition|Parallelism	0.0605	$(3, 2, 32)$	30.6443	$(6, 60)$	< 0.0001
Group $\times$ Conditions	0.8345	$(3, 2, 32)$	0.0947	$(6, 60)$	0.4687

the ϵ-adjusted MMM test statistics developed by Boik (1988) which extends the Greenhouse-Geisser type tests to the multivariate case. As in the univariate case, the adjusted tests should only be used if the multivariate test do not exist because the number of subjects is small, $n < pq$.

While one may directly derive the MMM test matrices from the normalized DMLM tests, as illustrated in Timm (1980b), it is more convenient to use PROC GLM to obtain the test statistics and hypothesis test using the multivariate mixed model which for exact tests requires multivariate sphericity. For the MMM analysis, the data must be reorganized; subjects are random and are nested within the fixed group factor, and groups are crossed with the fixed condition (repeated measures) factor. The reorganized data follow in Output 6.5.2.

To perform the multivariate mixed model analysis using PROC GLM, the RAN-DOM statement is used to calculate the expected mean squares to form the appropriate error sum of squares matrices for the multivariate tests. The Type III expected mean squares are found in Output 6.5.3 that follow.

From the expected mean square output, we see that the subject within-group effect is the error term to test for group differences given parallelism. The PROC GLM statements for the analysis from Program 6_5.sas are:

```
model y1 - y3 = group subj(group) cond cond*group;
random subj(group);
contrast 'Group' group 1 -1/ e=subj(group);
manova h = cond group*cond/ printh printe;
```

The E option in the CONTRAST statement defines the error matrix for the test of groups. The "overall" error matrix is being used to both test for conditions given parallelism and for parallel profiles. The results of the MMM analysis are summarized in Table 6.3.

To show that the MMM results may be obtained from the DMLM results, observe that for the test of conditions the hypothesis test matrix is

$$H^* = \begin{pmatrix} 152.93 & 95.79 & 14.75 \\ 95.79 & 62.73 & 8.56 \\ 14.75 & 8.56 & 1.59 \end{pmatrix}$$

and is the average of H_{11} and H_{22}, the submatrices of H for the test of conditions given parallelism. The degrees of freedom for the mixed model are $\nu_h^* =$

Output 6.5.2: Double Multivariate Model Analysis.

Obs	group	subj	cond	y1	y2	y3
1	1	1	1	117.0	59.0	10.5
2	1	1	2	117.5	59.0	16.5
3	1	1	3	118.5	60.0	16.5
4	1	2	1	109.0	60.0	30.5
5	1	2	2	110.5	61.5	30.5
6	1	2	3	111.0	61.5	30.5
7	1	3	1	117.0	60.0	23.5
8	1	3	2	120.0	61.5	23.5
9	1	3	3	120.5	62.0	23.5

Obs	group	subj	cond	y1	y2	y3
10	1	4	1	122.0	67.5	33.0
11	1	4	2	126.0	70.5	32.0
12	1	4	3	127.0	71.5	32.5
13	1	5	1	116.0	61.5	24.5
14	1	5	2	118.5	62.5	24.5
15	1	5	3	119.5	63.5	24.5
16	1	6	1	123.0	65.5	22.0
17	1	6	2	126.0	61.5	22.0
18	1	6	3	127.0	67.5	22.0
19	1	7	1	130.5	68.5	33.0
20	1	7	2	132.0	69.5	32.5
21	1	7	3	134.5	71.0	32.0
22	1	8	1	126.5	69.0	20.0
23	1	8	2	128.5	71.0	20.0
24	1	8	3	130.5	73.0	20.0
25	1	9	1	113.0	58.0	25.0
26	1	9	2	116.5	59.0	25.0
27	1	9	3	118.0	60.5	24.5
28	2	1	1	128.0	67.0	24.0
29	2	1	2	129.0	67.5	24.0
30	2	1	3	131.5	69.0	24.0
31	2	2	1	116.5	63.5	28.5
32	2	2	2	120.0	65.0	29.5

Obs	group	subj	cond	y1	y2	y3
33	2	2	3	121.5	66.0	29.5
34	2	3	1	121.5	64.5	26.5
35	2	3	2	125.5	67.5	27.0
36	2	3	3	127.0	69.0	27.0
37	2	4	1	109.5	54.0	18.0
38	2	4	2	112.0	55.5	18.5
39	2	4	3	114.0	57.0	19.0
40	2	5	1	133.0	72.0	34.5
41	2	5	2	136.0	73.5	34.5
42	2	5	3	137.5	75.5	34.5
43	2	6	1	120.0	62.5	26.0
44	2	6	2	124.5	65.0	26.0
45	2	6	3	126.0	66.0	26.0
46	2	7	1	129.5	65.0	18.5
47	2	7	2	133.5	68.0	18.5
48	2	7	3	134.5	69.0	18.5
49	2	8	1	122.0	64.5	18.5
50	2	8	2	124.0	65.5	18.5
51	2	8	3	125.5	66.0	18.5
52	2	9	1	125.0	65.5	21.5
53	2	9	2	127.0	66.5	21.5
54	2	9	3	128.0	67.0	21.6

Output 6.5.3: Multivariate Mixed Model Analysis.

Source	Type III Expected Mean Square
group	Var(Error) + 3 Var(subj(group)) + Q(group,group*cond)
subj(group)	Var(Error) + 3 Var(subj(group))
cond	Var(Error) + Q(cond,group*cond)
group*cond	Var(Error) + Q(group*cond)

Table 6.4: Univariate Analysis for Test of Conditions, Zullo Data.

Variable	F	P-Value
Group|Parallelism	162.12	< 0.0001
Condition|Parallelism	36.82	< 0.0001
Group $\times$ Conditions	1.03	0.3693

$\nu_h u = \nu_h(\mathrm{rank}(P)/p) = \nu_h \times \mathrm{rank}(A) = 1 \times 2 = 2$ and similarly, $\nu_e^* = \nu_e u = \nu_h(\mathrm{rank}(P)/p) = \nu_e \times \mathrm{rank}(A) = 16 \times 6/3 = 32$. The results for the test of Group$\times$Conditions follow similarly.

Finally, from the MMM one variable-at-a time, the univariate split-model F-ratios are obtained. The ANOVA table for the test of conditions is displayed in Table 6.4. Recall, the univariate test is valid if and only if the sphericity (circularity) condition is met by each variable.

When multivariate sphericity is not satisfied or one cannot perform a DMLM analysis, Boik (1988) developed an ε-adjustment for the tests of conditions and the test of group$\times$conditions following the original work of Box (1954). The multivariate tests may be approximated by a Wishart distribution when the null hypothesis is true:

$$E^* \simeq W_p(\nu_e^* = \varepsilon q\nu_e, \Sigma_e, 0)$$
$$H^* \simeq W_p(\nu_h^* = \varepsilon q\nu_h, \Sigma_e, 0).$$

Boik showed how one may use an estimate of ε to adjust the approximate F-tests for the multivariate criteria where

$$\hat{\varepsilon} = \frac{\mathrm{tr}\left[\left(\sum_{i=1}^q S_{ii}\right)^2\right] + \left[\mathrm{tr}\left(\sum_{i=1}^q S_{ii}\right)\right]^2}{q\sum_{i=1}^q \sum_{j=1}^q [\mathrm{tr}(S_{ij})]^2 + \mathrm{tr}\left(S_{ij}^2\right)} \tag{6.94}$$

where $q = \mathrm{rank}(A)$ and ν_h and ν_e are the hypothesis and error degrees of freedom for the DMLM tests.

Using Wilks' Λ criterion, Boik (1988) shows how one may make the $\hat{\varepsilon}$-adjustment to the associated F statistics. The results apply equally to the other criteria using Rao's F approximation. In Program 6_5.sas, we use PROC IML code with the

Output 6.5.4: PROC IML - $\hat{\varepsilon}$-adjusted Analysis, Zullo Data.

| Epsilon adjusted F-Statistics for cond and group X cond MMM tests |

EPSILON
0.7305055

F_COND	F_GXC	DF1	DF2
65.085984	0.9026246	4.3830331	30.308807

| Epsilon adjusted pvalues for MMM tests using T0**2 Criterion |

P_COND	P_GXC
8.549E-15	0.4819311

Table 6.5: $\hat{\varepsilon}$-adjusted Analysis, Zullo Data.

Hypothesis	$\nu_e T_0^2$	F	df	P-Value
Condition$\mid$Parallelism	13.7514	65.090	$(4.38, 30.31)$	< 0.0001
Group$\times$Conditions	0.1907	0.903	$(4.38, 30.31)$	0.482

Bartlett-Lawley-Hotelling trace criterion, T_0^2. Then,

$$F = \frac{2(sN+1)T_0^2}{s^2(2m+s+1)} \simeq F[s(2m+s+1), 2(sN+1)] \qquad (6.95)$$

where

$$s = \min(\nu_h^*, p)$$
$$\nu_e^* = \hat{\varepsilon} q \nu_e$$
$$\nu_h^* = \hat{\varepsilon} q \nu_h$$
$$M = \frac{(|\nu_h^* - p| - 1)}{2}$$
$$N = \frac{(\nu_e^* - p - 1)}{2}.$$

The results are given in Output 6.5.4. From the output, $\hat{\varepsilon} = 0.7305055$. The $\hat{\varepsilon}$-adjusted P-values are summarized in Table 6.5. The P-value for the unadjusted MMM was 0.4862 for the tests of parallelism using T_0^2. Using Wilks' Λ criterion, the P-values are 0.463 and 0.467, for the adjusted MMM and the unadjusted test, respectively. The nominal P-values for the two adjusted tests are approximately equal,

0.482 versus 0.463, for T_0^2. For this problem, the adjustment employing Wilks' criterion is better than that using T_0^2 since the F statistic is exact for the Λ criterion.

It appears that we have three competing strategies for the analysis of designs with vectors of repeated measures: the DMLM, the MMM, and the $\hat{\varepsilon}$-adjusted MMM. Given multivariate sphericity, the MMM is the most powerful. When multivariate sphericity does not hold, neither the adjusted MMM or the DMLM analysis is most powerful. Boik (1991) recommends using the DMLM. The choice between the two procedures depends upon the ration of the traces of the noncentrality matrices of the associated Wishart distributions which is seldom known in practice.

In this chapter we have illustrated the analysis of the DMLM with arbitrary covariance structure. When the covariance structure satisfies the multivariate sphericity condition, the Scheffé MMM analysis is optimal. However, in many repeated measurement experiments, the structure of the covariance matrix Ω may have Kronecker structure:

$$\Omega = \Sigma_t \otimes \Sigma_e \qquad (6.96)$$

where Σ_e is a $p \times p$ matrix and Σ_t is a $t \times t$ covariance matrix, and both matrices are positive definite, a structure commonly found in three-mode factor analysis (Bentler & Lee, 1978). Under Release 6.12 of the SAS System or later, the PROC MIXED can be used to test hypotheses for the DMLM with Kronecker structure. For an example, see SAS Institute Inc. (1996, Chapter 5) or Timm (2002, p. 407).

To test the hypothesis that Ω has Kronecker structure versus the alternative that Ω has a general structure,

$$H : \Omega = \Sigma_t \otimes \Sigma_e \text{ versus } A : \Omega = \Sigma, \qquad (6.97)$$

the likelihood ratio criterion is

$$\lambda = \frac{|\widehat{\Sigma}|^{n/2}}{|\widehat{\Sigma}_e|^{np/2}|\widehat{\Sigma}_t|^{nt/2}} \qquad (6.98)$$

where $\widehat{\Sigma}_e$ and $\widehat{\Sigma}_t$ are the ML estimates of Ω under H and $\widehat{\Sigma}$ is the estimate of Ω under the alternative. The asymptotic distribution of the test statistic

$$U = -2\log(\lambda) \qquad (6.99)$$

follows a chi-square distribution with degrees of freedom $f = (p-1)(t-1)[(p+1)(t+1)+1]/2$. However, to solve the likelihood equations to obtain the ML estimates of the covariance matrices requires an iterative process as outlined by Krishnaiah et al. (1980); Boik (1991); Naik and Khattree (1996). Naik and Rao provide a computer program using the IML procedure to obtain the ML estimates. Or one may use PROC MIXED. One may also test that Ω has multivariate spherical structure versus the alternative that it has Kronecker structure (Timm, 2002, p. 402).

Given that Ω has Kronecker structure, we develop a test of the extended linear hypothesis

$$H : \psi = \text{tr}(G\Gamma) = 0 \text{ for } \Gamma = CB(I_p \otimes A) \qquad (6.100)$$

when multivariate sphericity is not satisfied and Ω has Kronecker structure. Following Mudholkar et al. (1974), the test statistic is

$$X^2 = \frac{[\text{tr}(G\widehat{\Gamma})]^2}{\text{tr}[GC(X'X)^{-1}C'G(\widehat{\Sigma}_t \otimes \widehat{\Sigma}_e)]} \qquad (6.101)$$

since

$$\hat{\psi} \sim N(\psi, \hat{\sigma}_{\hat{\psi}}^2 = \text{tr}[GC(X'X)^{-1}C'G(\widehat{\Sigma}_t \otimes \widehat{\Sigma}_e)] \qquad (6.102)$$

for fixed G. The statistic X^2 converges to a chi-square distribution with one degree of freedom.

The X^2 statistic is an alternative to Boik's $\hat{\epsilon}$-adjusted multivariate test procedure for testing $H : \Gamma = 0$ when the multivariate sphericity not satisfies and Ω has Kronecker structure. Naik and Khattree (1996) have developed an alternative Satterthwaite-type approximation for the MANOVA model when multivariate sphericity is not satisfied. The DMLM may also be analyzed using PROC MIXED as illustrated by Littell, Milliken, Stroup, and Wolfinger (1996).

CHAPTER 7

Restricted MGLM and Growth Curve Model

7.1 INTRODUCTION

Repeated measurement data analysis is popular in the social and behavioral sciences and in other disciplines. A model that is useful for analyzing experiments involving repeated measurements is the restricted multivariate general linear model (RMGLM). Special cases of the RMGLM include the seemingly unrelated regression (SUR) model and the growth curve model of Potthoff and Roy (1964), also called the generalized multivariate analysis of variance (GMANOVA) model. In this chapter, we review the RMGLM and GMANOVA models. The canonical form of the GMANOVA model is also developed. While the SUR model is introduced in this chapter, it is discussed in detail in Chapter 8 along with the restricted GMANOVA model. A comprehensive discussion of linear and nonlinear growth curve models can be found in Vonesh and Chinchilli (1997).

7.2 RESTRICTED MULTIVARIATE GENERAL LINEAR MODEL

Following the classical matrix approach, the RMGLM is represented as

$$\Omega_0 : Y_{n \times p} = X_{n \times k} B_{k \times p} + U_{n \times p} \qquad (7.1)$$

$$\text{cov}(Y) = I_n \otimes \Sigma \qquad (7.2)$$

$$R_1 B R_2 = \Theta \qquad (7.3)$$

where $R_{1(r_1 \times k)}$ of rank r_1, $R_{2(p \times r_2)}$ of rank r_2 and $\Theta_{r_1 \times r_2}$ are known matrices restricting the elements of the parameter matrix B. From (7.1), we see that R_1, R_2, and Θ are similar to the hypothesis test matrices C, A, and Γ in the MANOVA model. Employing the vector version of the restricted model, we write (7.1) following (5.5)

$$\text{vec}(Y) = D\text{vec}(B) + \text{vec}(U) \qquad (7.4)$$

$$y = D\beta + e \qquad (7.5)$$

$$R\beta = \theta \qquad (7.6)$$

where $D = I_p \otimes X$, $\Omega = \Sigma \otimes I_n$, $R = R_2' \otimes R_1$ and $\theta = \text{vec}(\Theta)$, a general linear model (GLM). To estimate β in (7.5), recall that the estimator of β without restrictions from (5.10) is

$$\hat{\beta} = (D'\Omega^{-1}D)^{-1}D'\Omega^{-1}y$$
$$= (I_p \otimes (x'X)^{-1}X')y$$
$$= \text{vec}[(X'X)^{-1}X'Y]. \qquad (7.7)$$

However, with $R = R_2' \otimes R_1$, the restricted least squares (LS) estimator for β using model (7.5) is given in (3.1) and simplifying, we find that the restricted LS estimator for β is

$$\hat{\beta}_r = \hat{\beta} - (D'D)^{-1}R'[R(D'D)^{-1}R']^{-1}(R\hat{\beta} - \theta) \qquad (7.8)$$
$$\hat{B}_r = \hat{B} - (X'X)^{-1}R_1'(X'X)^{-1}R_1')^{-1}(R_1\hat{B}R_2 - \Theta)(R_2'R_2)^{-1}R_2' \qquad (7.9)$$

where $\hat{B} = (X'X)^{-1}X'Y$. See Timm (1993a, p. 158) and Timm (1980a) for details on how the LS estimate $\hat{B}_r$ is obtained by minimizing the least squares criterion $\text{tr}[(Y-XB)'(Y-XB)]$. If one assumes matrix multivariate normality, the restricted LS estimate of B is not the ML estimate. To obtain the ML estimate of B given the restriction $R_1BR_2 = \Theta$, one uses the matrix normal distribution given in Definition 5.1. As shown by Tubbs, Lewis, and Duran (1975), the restricted ML estimate of B is

$$\hat{B}_{ML} = \hat{B} - (X'W^{-1}X)^{-1}R_1'(R_1(X'W^{-1}X)^{-1}R_1')^{-1} \dots$$
$$(R_1\widehat{(B)}R_2 - \Theta)(R_2'\Sigma R_2)^{-1}R_2'\Sigma$$
$$= \hat{B} - (X'X)^{-1}R_1'(R_1(X'X)^{-1}R_1')^{-1}(R_1\hat{B}R_2 - \Theta)(R_2'\Sigma R_2)^{-1}R_2'\Sigma \qquad (7.10)$$

where $W = I$ and $\hat{B} = (X'X)^{-1}X'Y$. Hence, the restricted ML estimate of B involves the unknown matrix Σ. However, from (7.9) observe that the ML and LS estimate of parametric functions of the form

$$\Psi = C\hat{B}_rR_2 = C\hat{B}R_2 - (X'X)^{-1}R_2'(R_1(X'X)^{-1}R_1')^{-1}(R_1\hat{B}R_2 - \Theta) \quad (7.11)$$

does not contain the unknown covariance matrix Σ and are equal. Thus we should be able to develop a test of $H : CBA = \Gamma$, if the restriction takes the form $R_1BR_2 = R_1BA = \Theta$, that is independent of the unknown value for Σ in the restricted ML estimate of B. To derive a test statistic, we need an unbiased estimator of Σ under the restriction $R_1BR_2 = \Theta$. The estimated restricted error sum of products (SSP) matrix is

$$\widetilde{E} = (Y - X\hat{B}_r)'(Y - X\hat{B}_r)$$
$$= (Y - X\hat{B})'(Y - X\hat{B}) + (\hat{B} - \hat{B}_r)'(X'X)(\hat{B} - \hat{B}_r)$$
$$= E + R_2(R_2'R_2)^{-1}(R_1\hat{B}R_2 - \Theta)'(R_1(X'X)^{-1}R_1')^{-1}(R_2'R_2)^{-1}R_2' \quad (7.12)$$

where $E = Y'(I - X(X'X)^{-1}X')Y$. The matrix $S = \tilde{E}/(n - k + r_1)$ is an unbiased estimate of Σ given the restriction since $\mathcal{E}(\tilde{E}) = (n - k)\Sigma + r_1\Sigma = (n - k + r_1)\Sigma$. Under multivariate normality, $R'_2\tilde{E}R_2 \sim W_{r_2}(n - k + r_1, R'_2\Sigma R_2)$.

To test hypotheses of the form $H : CBA = \Gamma$ assuming (7.1) is complicated by the restrictions. We consider two situations. The first situation is to assume that either $R_2 = A$ and A arbitrary or that $R_2 = A = I$ where no row of C is equal to or dependent on any row of R_1 or other rows of C. Following (3.6) for the univariate model, we let

$$Q = \begin{pmatrix} R_1 \\ C \end{pmatrix} \quad P = \begin{pmatrix} R_2 & A \end{pmatrix} \quad \eta = \begin{pmatrix} \Theta & K \\ \Pi & \Gamma \end{pmatrix} \quad (7.13)$$

so that

$$QBP = \begin{pmatrix} R_1BR_2 & R_1BA \\ CBR_2 & CBA \end{pmatrix} = \begin{pmatrix} \Theta & K \\ \Pi & \Gamma \end{pmatrix}. \quad (7.14)$$

Then, the restricted LS estimator of B under the hypothesis $H : CBA = \Gamma$ for the restriction $R_1BA = \Theta$ is

$$\widehat{B}_{r,\omega} = \widehat{B} - (X'X)^{-1}Q'(Q(X'X)^{-1}Q')^{-1}(Q\widehat{B}P - \eta)('P'P)^{-1}P' \quad (7.15)$$

and the error sum of products (SSP) matrix is

$$\begin{aligned}
E^*_r &= (Y - \widehat{B}_{r,\omega})'(Y - \widehat{B}_{r,\omega}) \\
&= (Y - X\widehat{B})'(Y - X\widehat{B}) + (\widehat{B}_r - \widehat{B}_{r,\omega})'(X'X)(\widehat{B}_r - \widehat{B}_{r,\omega}) \\
&= E + A(A'A)^{-1}(C\widehat{B}A - \Gamma)'(C(X'X)^{-1}C')^{-1}(C\widehat{B}A - \Gamma)'(A'A)^{-1}A'
\end{aligned}$$
$$(7.16)$$

where $E = Y'(I - X(X'X)^{-1}X')Y$, $\widehat{B} = (X'X)^{-1}X'Y$ and $\widehat{B}_r$ is the restricted LS estimator defined in (7.9) (Timm, 1980a). To test the hypothesis

$$H : CBA = \Gamma \quad (7.17)$$

where the $g = \text{rank}(C) = \nu_h$ and $u = \text{rank}(A)$, we use the following matrices

$$H_r = (C\widehat{B}_rA - \Gamma)'(CFGC')^{-1}(C\widehat{B}_rA - \Gamma) \quad (7.18)$$
$$F = I - (X'X)^{-1}R'_1(R_1(X'X)^{-1}R'_1)^{-1}R_1 \quad (7.19)$$
$$E_r = A'E^*_r = A'EA + (C\widehat{B}A - \Theta)'(C(X'X)^{-1}C')^{-1}(C\widehat{B}A - \Theta) \quad (7.20)$$

where $F = F^2$ is idempotent, $G = (X'X)^{-1}$ and H_r and E_r have independent Wishart distributions

$$E_r \sim W_u(\nu_e = n - k + r_1, A'\Sigma A) \quad (7.21)$$
$$H_r \sim W - u(\nu_h = g, A'\Sigma A, (A'\Sigma A)^{-1}\Delta) \quad (7.22)$$

with noncentrality matrix

$$\Delta = (\Gamma - \Gamma_0)'(CFG^{-1}C')^{-1}(\Gamma - \Gamma_0). \tag{7.23}$$

The test statistic assuming $A = R_2$ by (7.11) is

$$\lambda_* = \left[\frac{|(Y - X\widehat{B}_r)'(Y - X\widehat{B}_r)|}{|(Y - X\widehat{B}_{r,\omega})'(Y - X\widehat{B}_{r,\omega})|} \right]^{n/2}. \tag{7.24}$$

On simplifying (7.24) using (3.12) for partitioned matrices, the U-statistic becomes

$$\Lambda = \lambda_*^{2/n} = \frac{|E_r|}{|E_r + H_r|} \tag{7.25}$$

where $\nu_h = g = \text{rank}(C)$ and $\nu_e = n - k + r_1$. The other multivariate criteria follow by using E_r and H_r, in place of H and E in (5.55)-(5.57).

To obtain $1 - \alpha$ simultaneous confidence intervals for bilinear forms $\psi = c'Ba$, one finds that the best estimator for ψ is $\hat{\psi} = c'\widehat{B}_r a$ and $\hat{\sigma}_\psi^2 = (a'Sa)c'(FG)c$ where

$$S = \frac{E_r}{n - k + r_1} \tag{7.26}$$

so that $\mathcal{E}(S) = \Sigma$. Extended linear hypotheses may also be tested using the restricted model.

Assuming (7.1), we demonstrated how one can test the hypothesis $H : CBA = \Gamma$ when the matrix R_2 in the restriction $R_1 B R_2 = \Theta$ is chosen equal to A in the hypothesis. The problem of testing $CBA = \Gamma$ assuming $R_1 B R_2 = \Theta$ for arbitrary R_2 and A is more difficult since R_2, A and the structure of Σ are associated (Timm, 1996). This association occurs, for example, in the SUR or multiple design multivariate (MDM) model and the growth curve model. To see this, suppose we associate with $R_1 B R_2$ the more general structure $R_* B = \Theta$ and assume $\Theta = 0$. Then if we incorporate the restriction $R_* B = 0$ into (7.5) by setting appropriate parameters to zero and modifying the columns of $D = I_p \otimes X$, (7.5) becomes

$$\begin{pmatrix} y_1 \\ y_2 \\ \vdots \\ y_p \end{pmatrix} = \begin{pmatrix} X_1 & 0 & \cdots & 0 \\ 0 & X_2 & \cdots & 0 \\ \vdots & \vdots & \ddots & \vdots \\ 0 & 0 & \cdots & X_p \end{pmatrix} \begin{pmatrix} \beta_{11} & \beta_{22} & \vdots & \beta_{pp} \end{pmatrix} + \begin{pmatrix} e_1 \\ e_2 \\ \vdots \\ e_p \end{pmatrix}$$

$$y = D\beta + e \tag{7.27}$$

where $e \sim N_{np}(0, \Omega = \Sigma \otimes I_n)$. Then, the best linear unbiased estimator (BLUE) of β is

$$\hat{\beta} = (D'\Omega^{-1}D)^{-1}D'\Omega^{-1}y \tag{7.28}$$

so that Σ does not "drop out" of the equation (7.28). Model (7.27) is called the SUR model, Zellner's model or the MDM model. Using matrix notation, we may write (7.27) as

$$\Omega_0 : Y_{n \times p} = \begin{pmatrix} X_1 \beta_{11} & X_2 \beta_{22} & \cdots & X_p \beta_{pp} \end{pmatrix} + U$$

$$= \begin{pmatrix} X_1 & X_2 & \cdots & X_p \end{pmatrix} \widetilde{B} + U \tag{7.29}$$

$$\mathrm{cov}(Y) = I_n \otimes \Sigma \tag{7.30}$$

where

$$\widetilde{B} = \begin{pmatrix} \beta_{11} & 0 & \cdots & 0 \\ 0 & \beta_{22} & \cdots & 0 \\ \vdots & \vdots & \ddots & \vdots \\ 0 & 0 & \cdots & \beta_{pp} \end{pmatrix}, \tag{7.31}$$

$\beta_{ii(k_i \times 1)}$, $i = 1, \ldots, p$, and $k = \sum k_i$. When $X_1 = X_2 = \cdots = X_p = X$, (7.29) reduces to the multivariate analysis of variance (MANOVA) or multivariate regression model discussed in Chapter 5. If the design matrices are not equal, but $\Omega = \sigma^2 I$ we have the independent regression model. When the populations are independent, the problem is greatly simplified (Timm, 1975, p. 331-347). We discuss the SUR model in Chapter 8.

7.3 THE GMANOVA MODEL

Testing the restriction $R_1 B R_2 = \Theta$ when R_2 and Σ are associated is a special case of the Potthoff and Roy (1964) growth curve model also called the GMANOVA model. To see this, let the reduced form of the MANOVA model be defined as

$$\Omega_0 : Y_{n \times p} = K_{n \times q} \xi_{q \times p} + U_{n \times p} \tag{7.32}$$

subject to the joint linear restrictions

$$R_{*(q-k) \times q} \xi_{q \times p} = 0 \quad \text{and} \quad \xi_{q \times p} Q_{* p \times (p-q)} = 0 \tag{7.33}$$

where the $\mathrm{rank}(R_*) = q - k$ and the $\mathrm{rank}(Q_*) = p - q < p$. Then we may find a matrix $X'_{2(p \times q)}$ orthogonal to Q_* and a matrix $D'_{k \times q}$ orthogonal to R_* so that ξ may be represented

$$\xi = D_{q \times k} B_{k \times q} X_{w(q \times p)} \tag{7.34}$$

where $B_{k \times q}$ is a reparameterized matrix that is unrestricted. Then, (7.32) becomes

$$\Omega_0 : Y = KDBX_2 + U \tag{7.35}$$

$$Y_{n \times p} = X_{1(n \times k)} B_{k \times q} X_{2(q \times p)} + U_{n \times p} \tag{7.36}$$

$$\mathrm{cov}(Y) = I_n \otimes \Sigma \tag{7.37}$$

where $KD = X_1$ $(\text{rank}(X_1) = k)$ and X_2 $(\text{rank}(X_2) = q)$ are known and B is the matrix of parameters. This is called the (unrestricted) GMANOVA model. Assuming (7.36), we can test the linear hypothesis

$$R_1 B R_2 = \Theta \qquad (7.38)$$

which is usually written as $CBA = \Gamma$. Letting $X_2 = I$, the GMANOVA model reduces to the MANOVA model. To show that (7.36) is again a special case of the GLM defined by (1.1), we define

$$X = X_2' \otimes X_1 \quad \Omega = \Sigma \otimes I_n \quad y' = \begin{pmatrix} y_1' & y_2' & \cdots & y_p' \end{pmatrix} \qquad (7.39)$$

where $y_{i(n \times p)}$ so that $y = \text{vec}(Y)$, $\beta = \text{vec}(B)$, and $e = \text{vec}(U)$ so that $y = X\beta + e$.

The GMANOVA model is a special case of the extended GMANOVA model. The extended or restricted GMANOVA model has the general form

$$\Omega_0 : Y = X_1 B X_2 + U \quad \text{with} \quad X_3 B X_4 = \Theta \qquad (7.40)$$

where the restrictions are part of the model. Given (7.40), one may test hypotheses of the form

$$H : X_5 B X_6 = \Gamma \quad \text{or} \quad CBA = \Gamma \qquad (7.41)$$

letting $X_5 = C$ and $X_6 = A$. Setting $X_2 = I$ in (7.40), the MANOVA model with general doubly linear restrictions is a special case of the extended GMANOVA model where $X_6 \neq X_4$. Thus, we have that the MANOVA model is a special case of the GMANOVA model and that the more general doubly linear restricted MANOVA model is a special case of the extended (restricted) GMANOVA model. Before discussing hypothesis testing using the MANOVA model with doubly linear restrictions which is complicated by the association of R_2, A and Σ, we will review the GMANOVA model with no restrictions.

Given the GMANOVA model in (7.36) without restrictions, the BLUE or Gauss-Markov (GM) estimate of B is obtained by using representation (7.39) and (4.6) so that

$$\begin{aligned} \text{vec}(\widehat{B}) = \hat{\beta} &= (X'\Omega^{-1}X)^{-1}X'\Omega^{-1}y \\ &= [(X_2' \otimes X_1)'(\Sigma \otimes I_n)^{-1}(X_2' \otimes X_1)]^{-1}(X_2' \otimes X_1)'(\Sigma \otimes I_n)^{-1}y \\ &= [(X_2\Sigma^{-1}X_2')^{-1}X_2\Sigma^{-1} \otimes (X_1'X_1)^{-1}X_1']y \\ &= \text{vec}[(X_1'X_1)^{-1}X_1'Y\Sigma^{-1}X_2'(X_2\Sigma^{-1}X_2')^{-1} \qquad (7.42) \\ \widehat{B}_{GM} &= (X_1'X_1)^{-1}X_1'Y\Sigma^{-1}X_2'(X_2\Sigma^{-1}X_2')^{-1}. \qquad (7.43) \end{aligned}$$

Adding the restriction $R\beta = (R_2' \otimes R_1)\beta = \theta$ to the growth curve model and using (3.1) with β defined in (7.43), the BLUE or GM estimator of β, using matrix notation, is

$$\widehat{B}_{GM_r} = \widehat{B}_{GM} - P_1(R_1 B R_2 - \Theta)P_2 \qquad (7.44)$$

where

$$P_1 = (X_1'X_1)^{-1}R_1'[R_1(X_1'X_1)^{-1}R_1']^{-1} \tag{7.45}$$

$$P_2 = [R_2'(X_2\Sigma^{-1}X_2')^{-1}R_2]^{-1}R_2'(X_2\Sigma^{-1}X_2')^{-1}. \tag{7.46}$$

From (7.43), we see that if $X_2 = I$ that the GMANOVA estimator B reduces to the OLSE for B under the MANOVA model, $\widehat{B} = (X'X)^{-1}X'Y$. Furthermore, the BLUE estimator for B, $\widehat{B}_{GM_r}$ in (7.44), reduces to the ML estimate for B in (7.10) obtained in the MANOVA model with doubly linear restrictions and known Σ.

Potthoff and Roy (1964) recognized (7.43) and recommended reducing the GM-ANOVA model to the MANOVA model through the transformation

$$Y_0 : YG^{-1}X_2'(X_2G^{-1}X_2')^{-1} \tag{7.47}$$

where $G_{p \times p}$ is any symmetric nonsingular matrix. If we use (7.47), the GMANOVA model is reduced to the MANOVA model where $\mathcal{E}(Y_0) = X_1B$, $\mathrm{cov}(Y_0) = I_n \otimes \Sigma_0$ and Σ_0 is

$$\Sigma_{0(q \times q)} = (X_2G^{-1}X_2')^{-1}X_2G^{-1}\Sigma G^{-1}X_2'(X_2G^{-1}X_2')^{-1}. \tag{7.48}$$

Hence, the MANOVA theory discussed in Chapter 5 may be used to test

$$H : CBA = \Gamma$$

by using the transformed data matrix Y_0 given the unrestricted GMANOVA model (7.36). From (7.47), the optimal choice of G is Σ, but in practice Σ is unknown. When $p = q$, $\widehat{B}$ does not depend on Σ for either the restricted GMANOVA model or the (unrestricted) GMANOVA model and the transformation becomes

$$Y_0 = YX_2^{-1} \tag{7.49}$$

so that there is no need to choose G. If $q < p$, however, the choice of G is important since it affects the power of the MANOVA tests, the variance of the parameters and the widths of confidence intervals. Potthoff and Roy (1964) recommended selecting $G = \widehat{\Sigma}$ where $\widehat{\Sigma}$ was a consistent estimator of Σ based on a different, independent set of data.

To avoid the arbitrary choice of G, Rao (1965, 1966) and Khatri (1966) derived a likelihood ratio (LR) test for the GMANOVA model. Rao transformed the model to a multivariate analysis of covariance (MANCOVA) model and developed a conditional LR test for the GMANOVA model. Khatri used the ML method and a conditional argument to obtain the LR test criterion. Gleser and Olkin (1970) reduced the model to canonical form and using invariance arguments derived a LR test; they also showed that the Rao-Khatri test was the unconditional LR test.

To develop a LR ratio test of $H : CBA = \Gamma$ under the growth curve model defined in (7.36), Khatri (1966) derived the ML estimate of B and Σ under multivariate normality

$$\widehat{B}_{ML} = (X_1'X_1)^{-1}X_1YEX_2'(X_2E^{-1}X_2')^{-1} \tag{7.50}$$

$$n\widehat{\Sigma} = (Y - X_1\widehat{B}_{ML}X_2)'(Y - X_1\widehat{B}_{ML}X_2) \tag{7.51}$$

where $E = Y'(I - X_1(X_1'X_1)^{-1}X_1')Y$. Comparing (7.50) with (7.43), we see that the BLUE or Gauss-Markow (GM) estimate of B and the ML estimate have the same form, but that they are not equal since Σ is usually unknown and not equal to its estimate. Rao (1967) derived the covariance matrix of $C\widehat{B}_{ML}A$:

$$\text{cov}(C\widehat{B}_{ML}A) = C(X_1'X_1)^{-1}C' \otimes \left(\frac{n-k-1}{n-k-(p-q)-1}\right) A'(X_2\Sigma^{-1}X_2')^{-1}A.$$

(7.52)

The corresponding covariance matrix for the Gauss-Markov (GM) estimator of $C\widehat{B}_{GM}A$ is

$$\text{cov}(C\widehat{B}_{GM}A) = C(X_1'X_1)^{-1}C' \otimes A'(X_2\Sigma^{-1}X_2')^{-1}A \qquad (7.53)$$

which agrees with the ML estimate except for the multiplication factor. The two are equal if $p = q$, otherwise the GM estimator is best. Because Σ is unknown, an estimate of the covariance matrix is obtained by substituting an estimate $\widehat{\Sigma}$ for Σ. One can also easily establish the class of covariance matrices that ensure equality of the GM estimator $\widehat{B}_{GM}$ and the OLSE $\widehat{B}$ for the GMANOVA model by applying Theorem 4.1 to the GMANOVA case. Then, $\Sigma = X_2'\Delta X_2 + Z_2'\Gamma Z_2 + \sigma^2 I_n$ (Kariya, 1985, p. 19-31).

To test the hypothesis $H : CBA = \Gamma$ under the GMANOVA model in (7.36), Khatri (1966) using maximum likelihood methods and Rao (1965, 1966) using covariance adjustment reduced the GMANOVA model to the conditional MANCOVA model

$$\mathcal{E}(Y_0|Z) = X_1B + Z\Gamma_0 \qquad (7.54)$$

where $Y + 0(n \times q)$ is the data matrix, $X_{1(n \times k)}$ is the known design matrix, $B_{k \times q}$ is a matrix of parameters, $Z_{n \times h}$ is a matrix of covariates and $\Gamma_{0(h \times q)}$ is a matrix of unknown regression coefficients.

To reduce (7.36) to (7.54), a $(p \times p)$ nonsingular matrix $H = (H_1, H_2)$ is constructed such that $X_2H_1 = I$ and $X_2H_2 = 0$. If the rank of X_2 is q, H_1 and H_2 can be selected:

$$H_1 = G^{-1}X_2'(X_2G^{-1}X_2')^{-1} \qquad H_2 = I - H_1X_2 \qquad (7.55)$$

where G is an arbitrary nonsingular matrix. Such a matrix H is not unique; however, estimates and tests are invariant for all choices of H satisfiting the specified conditions (Khatri, 1966). Hence, G in the expression for H_1 does not affect estimates or tests under (7.54). If we equate

$$Y_0 = YH_1 = YG^{-1}X_2'(X_2G^{-1}X_2')^{-1} \qquad (7.56)$$

$$Z = YH_2 = Y(I - H_1X_2) = Y[I - G^{-1}X_2'(X_2G^{-1}X_2')^{-1}], \qquad (7.57)$$

$\mathcal{E}(Y_0) = X_1B$ and $\mathcal{E}(Z) = 0$, thus the expected value of Y given Z is seen to be of the form specified in (7.54) (Grizzle et al., 1969, p. 362). Using this Rao-Khatri conditional model, the information in the covariates Z, which is ignored in the Potthoff-Roy reduction, is utilized. To find the BLUE of B and Γ_0 under (7.54), we minimize the $\text{tr}\{[Y - \mathcal{E}(Y_0|Z)]'[Y - \mathcal{E}(Y_0|Z)]\}$ to obtain the normal equations. The normal equations are

$$\begin{pmatrix} X_1'X_1 & X_1'B \\ Z'X_1 & Z'Z \end{pmatrix} \begin{pmatrix} B \\ \Gamma_0 \end{pmatrix} = \begin{pmatrix} X_1'Y_0 \\ Z'Y_0 \end{pmatrix}. \tag{7.58}$$

Using (3.12), we have that

$$\begin{pmatrix} X_1'X_1 & X_1'Z \\ Z'X_1 & Z'Z \end{pmatrix} = \begin{pmatrix} (X_1'X_1)^{-1} & 0 \\ 0 & 0 \end{pmatrix} + \begin{pmatrix} -(X_1'X_1)^{-1}X_1'Z \\ I \end{pmatrix} (Z'\bar{X}_1 Z)^{-1} \left(-Z'X_1(X_1'X_1)^{-1} \quad I \right) \tag{7.59}$$

where $\bar{X}_1 = I - X_1(X_1'X_1)^{-1}X_1'$ so that the BLUE of B and Γ for the MANCOVA model are

$$\widehat{B} = (X_1'X_1)^{-1}X_1'Y_0 - (X_1'X_1)^{-1}X_1'Z\widehat{\Gamma}_0 \tag{7.60}$$
$$\widehat{\Gamma}_0 = (Z'\bar{X}_1 Z)^{-1}Z'\bar{X}_1 Y_0. \tag{7.61}$$

Using the reductions in (7.56) and the Theorem 7.1 regarding projection matrices,

$$\begin{aligned} \widehat{B} &= (X_1'X_1)^{-1}X_1'Y_0 - (X_1'X_1)^{-1}X_1'Z[(Z'\bar{X}_1 Z)^{-1}Z'\bar{X}_1]Y_0 \\ &= (X_1'X_1)^{-1}X_1'YH_1 - (X_1'X_1)^{-1}X_1'Y[H_2(H_2'EH_2)^{-1}H_2']EH_1 \\ &= (X_1'X_1)^{-1}X_1'YH_1 - (X_1'X_1)^{-1}X_1'Y[E^{-1} - E^{-1}X_2'(X_2E^{-1}X_2')^{-1}X_2E^{-1}]EH_1 \\ &= (X_1'X_1)^{-1}X'YE^{-1}X_2'(X_2E^{-1}X_2')^{-1}X_2H_1 \\ &= (X_1'X_1)^{-1}X_1'YS^{-1}X_2'(X_2S^{-1}X_2')^{-1}, \end{aligned} \tag{7.62}$$

the ML estimate of B under the GMANOVA model. Thus, if $q < p$, Rao's procedure using $p - q$ covariates, Khatri's ML estimate of B and Potthoff-Roy's estimate weighting by $G^{-1} = S^{-1}$ are identical. Setting $G = I$ in the Potthoff-Roy estimate is equivalent to not including any covariates in the Rao-Khatri reduction. When $p = q$, H_2 does not exist and there is no need to choose G so that $Y_0 = YX_2^{-1}$.

Theorem 7.1. *Let $E_{p \times p}$ be a nonsingular matrix, $X_{2(q \times p)}$ of rank $q \leq p$, and $H_{2(r \times q)}'$ where $r = p - q$ of rank r such that $X_2H_2 = 0$. Then,*

$$H_2(H_2'EH_2)^{-1}H_2' = E^{-1} - E^{-1}X_2'(X_2E^{-1}X_2')^{-1}X_2E^{-1}.$$

To obtain the likelihood ratio test of $H : CBA = \Gamma$ under the GMANOVA model, one may use the standard theory in Chapter 5 using the MANCOVA model where the design matrix is $X = \begin{pmatrix} X_1 \\ Z \end{pmatrix}$, the $\text{rank}(X_1) = n - k - h$, $h = p - q$, and the $\text{rank}(C) = g = \nu_h$. To find the error sum of products matrix, observe that

$$
\begin{aligned}
\widetilde{E} &= A'Y_0(I - X(X'X)^{-1}X')Y_0 A \\
&= A'[Y_0'\bar{X}_1 Y_0 - \widehat{\Gamma}_0'(Z'\bar{X}_1 Z)\widehat{\Gamma}_0]A \\
&= A'[H_1'Y'\bar{X}_1 Y H_1 - \widehat{\Gamma}_0'Z'\bar{X}Y_0]A \\
&= A'[H_1'Y'\bar{X}_1 Y H_1 - (H_1'EH_2)(H_2'EH_2)^{-1}H_2'Y'\bar{X}_1 Y H_1]A \\
&= A'[(H_1'EH_1) - H_1'EH_2(H_2'EH_2)^{-1}(H_2'EH_1)]A \\
&= A'(H_1'X_2(X_2 E^{-1}X_2')^{-1}X_2'H_1)A \\
&= A'(X_2 E^{-1}X_2')^{-1}A \qquad (7.63)
\end{aligned}
$$

where $\nu_e = n - k - h = n - k - p + q$. To find the hypothesis test matrix, let $\widetilde{C} = (C, I)$ to test the hypothesis regarding B in the MANCOVA model. The hypothesis test matrix is

$$
\begin{aligned}
\widetilde{H} &= (C\widehat{B}A - \Gamma)'[C(X_1'X_1)^{-1}C + \\
&\quad C(X_1'X_1)^{-1}X_1'Z(Z'\bar{X}_1 Z)^{-1}Z'X_1(X_1'X_1)^{-1}C]^{-1}(C\widehat{B}A - \Gamma) \\
&= (C\widehat{B}A)'\{C[(X_1'X_1)^{-1} + \\
&\quad (X_1'X_1)^{-1}X_1'Z(Z'\bar{X}_1 Z)^{-1}Z'X_1(X_1'X_1)^{-1}]C\}^{-1}(C\widehat{B}A - \Gamma) \\
&= (C\widehat{B}A - \Gamma)'\{C[(X_1'X_1)^{-1} + \\
&\quad (X_1'X_1)^{-1}X_1'Y H_2(H_2'EH_2)^{-1}H_2'Y X_1(X_1'X_1)^{-1}]C\}^{-1}(C\widehat{B}A - \Gamma) \\
&= (C\widehat{B}A - \Gamma)'(CR^{-1}C')^{-1}(C\widehat{B}A - \Gamma) \qquad (7.64)
\end{aligned}
$$

where

$$
R^{-1} = (X_1'X_1)^{-1} + \\
(X_1'X_1)^{-1}X_1'Y[E^{-1} - E^{-1}X_2'(X_2 E^{-1}X_2')^{-1}X_2 E^{-1}]Y'X_1(X_1'X_1)^{-1} \qquad (7.65)
$$

$$
\widehat{B} = (X_1'X_1)^{-1}X_1'Y S^{-1}X_2'(X_2 S^{-1}X_2')^{-1} \qquad (7.66)
$$

$$
E = Y'(I - X_1(X_1'X_1)^{-1}X_1')Y. \qquad (7.67)
$$

Thus, with $\widetilde{H}$ defined in (7.64) and $\widetilde{E}$ defined in (7.63), the likelihood ratio criterion for the hypothesis $H : CBA = \Gamma$ for the GMANOVA model is

$$
\Lambda = \frac{|\widetilde{E}|}{|\widetilde{E} + \widetilde{H}|} \sim U_{u,\nu_h,\nu_e}. \qquad (7.68)
$$

The $1 - \alpha$ simultaneous confidence intervals for parametric function $\psi = c'Ba$ are given by

$$
\hat{\psi} - c_0\hat{\sigma}_{\hat{\psi}} < \psi < \hat{\psi} + c_0\hat{\sigma}_{\hat{\psi}} \qquad (7.69)
$$

where

$$
\hat{\sigma}_{\hat{\psi}}^2 = (a'\widetilde{E}a)'(c'R^{-1}c). \qquad (7.70)
$$

and c_0 is the appropriate constant chosen to agree with the test criterion

$$c_0^2 = \frac{1 - U^{1-\alpha}}{U^{1-\alpha}} \qquad \text{Wilks} \qquad (7.71)$$

$$c_0^2 = \frac{\theta^{1-\alpha}}{1 - \theta^{1-\alpha}} \qquad \text{Roy} \qquad (7.72)$$

$$c_0^2 = U_0^{1-\alpha} \qquad \text{Bartlett-Lawley-Hotelling} \qquad (7.73)$$

$$c_0^2 = \frac{V^{1-\alpha}}{1 - V^{1-\alpha}} \qquad \text{Bartlett-Nanda-Pillai.} \qquad (7.74)$$

In testing $H : CBA = \Gamma$ under the GMANOVA model, we discussed the conditional likelihood ratio test that depends on selecting the appropriate number of $p - q$ covariates since if $p = q$, H_2 does not exist and we should use the Potthoff-Roy procedure which no longer depends on the arbitrary matrix G. To test the adequacy of the model fit to the data, we must choose the most appropriate set of covariates. Recall that the GMANOVA model has the form

$$\mathcal{E}(Y) = X_1 B X_2 = X_1 \Theta \qquad (7.75)$$

$$\mathcal{E}(Z) = X_1 B X_2 H_2 = X_1 \theta H_2 = 0 \qquad (7.76)$$

where the $\text{rank}(X_1) = k$ and $X_1 \Theta H_2 = 0$ if and only if $\Theta H_2 = 0$. Hence we may evaluate the fit of the model by testing whether $\mathcal{E}(Z) = 0$ or equivalently that $H : \Theta H_2 = 0$ for the model $\mathcal{E}(Y) = X_1 \Theta$. This is the standard MANOVA model discussed in Chapter 5 where now the hypothesis and error SSP matrices are

$$H = (\widehat{\Theta} H_2)'(X_1' X_1)(\widehat{\Theta} H_2) \qquad (7.77)$$

$$E = H_2' Y'(I - X_1(X_1' X_1)^{-1} X_1') Y H_2 \qquad (7.78)$$

with $u = \text{rank}(H_2)$, $\nu_h = \text{rank}(C) = g$, $\nu_e = n - \text{rank}(X_1) = n - k$, and where $\widehat{\Theta} = (X_1' X_1) X_1' Y$, given in (7.62). For additional detail see Grizzle and Allen (1969), and Seber (1984, Section 9.7). A discussion of growth curves may also be found in Srivastava and Carter (1983, Chapter 6), Kshirsagar and Smith (1995), and Srivastava and Rosen (1999).

Gleser and Olkin (1970) developed a canonical form for the GMANOVA model and showed that the Rao-Khatri conditional LR test is also the unconditional test. Tubbs et al. (1975) claimed that their proposed test is an unconditional test of $H : CBA = \Gamma$ for the GMANOVA model; however, as noted by Rosen (1989), the estimator of B estimated under the hypothesis $CBA = \Gamma$ with $\widehat{\Sigma}$ replacing Σ is not the constrained maximum likelihood estimator of B. The expression for B derived by Tubbs et al. (1975, Eq. 2.15) may yield a nonoptimal solution. The test procedure proposed by Tubbs et al. (1975) is not a LR test, but a test created using the substitution method as defined by Arnold (1981, p. 363) which is equivalent to setting $G = S$ in the Potthoff-Roy model. From (5.62), their statistic may be written as

$$T_0^2 = \text{tr}[(C\widehat{B}A - \Gamma)'(C(X_1' X_1)^{-1} C')^{-1}(C\widehat{B}A - \Gamma)(A'(X_1 E X_1')^{-1} A)^{-1}] \qquad (7.79)$$

where $\widehat{B}$ and E are given in (7.66)-(7.67). T_0^2 converges in distribution to a chi-square distribution with $\nu = \mathrm{rank}(C)\mathrm{rank}(A) = gu$ degrees of freedom. If E is replaced by an estimator $\widehat{\Sigma}$ of Σ so that $\mathcal{E}(\widehat{\Sigma}) = K\Sigma$, then $K^{-1}X^2 \sim \chi_{gn}^2$ as $n \to \infty$.

7.4 CANONICAL FORM OF THE GMANOVA MODEL

Gleser and Olkin (1970); Srivastava and Khatri (1979); Kariya (1985) discussed the growth curve model using an equivalent canonical form for the model, a common practice in the study of multivariate linear hypotheses; see, Lehmann, 1991, Chapter 8 or T. W. Anderson, 1984, p. 296, to reduce the GMANOVA model to canonical form, we recall a few results from matrix algebra and vector spaces (Timm, 2002, Chapter 1; Stapleton, 1995).

Theorem 7.2. (Singular-value decomposition). *Let $A_{m \times n}$ be a matrix of rank r, with $r \leq m \leq n$. Then, there exist orthogonal matrices $P_{m \times m}$ and $Q_{n \times n}$ such that $P'AQ = \begin{pmatrix} \Lambda & 0 \\ 0 & 0 \end{pmatrix}$ or $A = P\begin{pmatrix} \Lambda & 0 \\ 0 & 0 \end{pmatrix}Q'$ where $\Lambda = \mathrm{diag}(\lambda_1, \ldots, \lambda_r)$ is a diagonal matrix of nonnegative elements and 0 is a matrix of order $m \times (n - m)$. The diagonal elements of Λ^2 are the eigenvalues of AA' and $A'A$.*

Lemma 7.2.1. *For the matrix $A_{m \times n}$ in Theorem 7.2, there exists a nonsingular matrix $M_{m \times m}$ such that $A = M\begin{pmatrix} I_r & 0 \\ 0 & 0 \end{pmatrix}A'$ where A is an orthogonal matrix.*

Using Lemma 7.2.1, we may write X_1 of rank k and X_2 of rank q for the GMANOVA model following Kariya (1985)

$$X_1 = Q_1 \begin{pmatrix} I_k \\ 0 \end{pmatrix} P_1 \qquad\qquad X_2 = P_2 \begin{pmatrix} I_q & 0 \end{pmatrix} Q_2 \qquad (7.80)$$

so that $Y = Q_1\binom{I_k}{0}P_1BP_2 \begin{pmatrix} I_q & 0 \end{pmatrix} Q_2$. Since Q_1 and Q_2 are known, the matrix

$$Y^* = Q_1'YQ_2 = \begin{pmatrix} I_k \\ 0 \end{pmatrix} P_1BP_2 \begin{pmatrix} I_q & 0 \end{pmatrix} Q_2. \qquad (7.81)$$

The transformed matrix Y^* is a random matrix with a p-variate normal distribution with covariance matrix $\Sigma^* = Q_2'\Sigma Q_2$ and mean

$$\mathcal{E}(Y^*) = \begin{pmatrix} I_k \\ 0 \end{pmatrix} P_1BP_2 \begin{pmatrix} I_q & 0 \end{pmatrix} \equiv \begin{pmatrix} I_k \\ 0 \end{pmatrix} B^* \begin{pmatrix} I_p & 0 \end{pmatrix} = \begin{pmatrix} B^* & 0 \\ 0 & 0 \end{pmatrix} \qquad (7.82)$$

so that

$$Y^* \sim N_{n,p}\left[\begin{pmatrix} B^* & 0 \\ 0 & 0 \end{pmatrix}, \Sigma^*, I_p \right] \qquad (7.83)$$

where $B_{k \times q}^* = P_1BP_2$, $k = k_1 + K_2$, $q = q_1 + q_2$, $p = q_1 + q_2 + q_3$, and $n = k_1 + k_2 + k_3$.

The GMANOVA hypothesis

$$H : CBA = \Gamma = X - X_3 B X_4 \tag{7.84}$$

may be similarly reduced

$$X_3 B X_4 = X_3 P_1^{-1} P_1 B P_2 P_2^{-1} X_4 = X_3 P_1^{-1} B^* P_2^{-1} X_4. \tag{7.85}$$

Applying Lemma 7.2.1 to $X_3 P_1^{-1}$ and $P_2^{-1} X_4$ yields

$$X_3 P_1^{-1} = P_3 \begin{pmatrix} I_{k_1} & 0 \end{pmatrix} Q_3 \qquad P_2^{-1} X_4 = Q_4 \begin{pmatrix} 0 \\ I_{q_2} \end{pmatrix} P_4 \tag{7.86}$$

so that the hypothesis in (7.84) becomes

$$H : \begin{pmatrix} I_{k_1} & 0 \end{pmatrix} \Theta \begin{pmatrix} 0 \\ I_{q_2} \end{pmatrix} = P_3^{-1} X_0 P_4^{-1} \tag{7.87}$$

where $\Theta_{k \times q} = Q_3 B^* Q_4$.

Because Q_3 and Q_4 are nonsignular and known, we may use them to transform Y^* to a new random matrix Z:

$$Z_{n \times p} = \begin{pmatrix} Q_3 & 0 \\ 0 & I \end{pmatrix} Y^* \begin{pmatrix} Q_4 & 0 \\ 0 & I \end{pmatrix} = \begin{matrix} & \begin{matrix} q_1 & q_2 & q_3 \end{matrix} & \\ \begin{pmatrix} Z_{11} & Z_{12} & Z_{13} \\ Z_{21} & Z_{22} & Z_{23} \\ Z_{31} & Z_{32} & Z_{33} \end{pmatrix} & \begin{matrix} k_1 \\ k_2 \\ k_3 \end{matrix} \end{matrix} \tag{7.88}$$

where each row of Z is normally distributed with covariance matrix

$$\Sigma_{n \times p} = \begin{pmatrix} Q_4' & 0 \\ 0 & I \end{pmatrix} \Sigma^* \begin{pmatrix} Q_4 & 0 \\ 0 & I \end{pmatrix} = \begin{matrix} & \begin{matrix} q_1 & q_2 & q_3 \end{matrix} & \\ \begin{pmatrix} \Sigma_{11} & \Sigma_{12} & \Sigma_{13} \\ \Sigma_{21} & \Sigma_{22} & \Sigma_{23} \\ \Sigma_{31} & \Sigma_{32} & \Sigma_{33} \end{pmatrix} & \begin{matrix} q_1 \\ q_2 \\ q_3 \end{matrix} \end{matrix} \tag{7.89}$$

$\Sigma_i q_i = p$, $\Sigma_{ij(q_i \times q_j)}$, $k_3 = n - (k_1 + k_2)$, and $Q_3 = p - q = p - q_1 - q_2$. The matrix $Z \sim N_{n,p}(\Theta, \Sigma, I_p)$ with Σ defined in (7.89) has mean matrix

$$\tilde{\Theta} = \begin{pmatrix} \Theta_{11} & \Theta_{12} & \Theta_{13} \\ \Theta_{21} & \Theta_{22} & \Theta_{23} \\ 0 & 0 & 0 \end{pmatrix} = \begin{matrix} & \begin{matrix} q_1 & q_2 & q_3 \end{matrix} & \\ \begin{pmatrix} \Theta_{11} & \Theta_{12} & 0 \\ \Theta_{21} & \Theta_{22} & 0 \\ 0 & 0 & 0 \end{pmatrix} & \begin{matrix} k_1 \\ k_2 \\ k_3 \end{matrix} \end{matrix} \tag{7.90}$$

where $\Theta = \begin{pmatrix} \Theta_{11} & \Theta_{12} \\ \Theta_{21} & \Theta_{22} \end{pmatrix} Q_3 B^* Q_4$ and $\sum_i k_i = n$. Substituting Θ into (7.87), the canonical form of the growth curve hypothesis is equivalent to testing

$$H : \Theta_{12} = P_2^{-1} X_0 P_4^{-1} \tag{7.91}$$

or letting $X_0 = 0$, the following equivalent hypothesis is

$$H : X_3 B X_4 \equiv \Theta_{12} = 0. \tag{7.92}$$

This is the canonical form of the null hypothesis where the transformed data matrix is the canonically reduced matrix Z. For the canonical form of the GMANOVA model, we are interested in testing (7.92) given Z, the canonical data matrix. To test $H : \Theta_{12} = 0$, we further observe from the canonical form of the model that

$$\bar{Z} = \begin{pmatrix} Z_{11} & Z_{12} & Z_{13} \\ Z_{21} & Z_{22} & Z_{23} \end{pmatrix} \sim N_{(k_1+k_2),p} \left[\tilde{\Theta} = \begin{pmatrix} \Theta_{11} & \Theta_{12} & 0 \\ \Theta_{21} & \Theta_{22} & 0 \end{pmatrix}, \Sigma, I_p \right] \tag{7.93}$$

$$Z_3 = \begin{pmatrix} Z_{31} & Z_{32} & Z_{33} \end{pmatrix} \sim N_{k_3,p}(0, \Sigma, I_p) \tag{7.94}$$

$$Z_3' Z_3 = V = \begin{pmatrix} V_{11} & V_{12} & V_{13} \\ V_{21} & V_{22} & V_{23} \\ V_{31} & V_{32} & V_{33} \end{pmatrix} \sim W_p(k_3, \Sigma) \tag{7.95}$$

and that $\bar{Z}$ and V are independent. Hence $(\bar{Z}, V)$ are sufficient statistics for (Θ, Σ).

Using the canonical form of the GMANOVA model, Gleser and Olkin (1970, p. 280), Srivastava and Khatri (1979, p. 194), and Kariya (1985, p. 104) showed that the LR test statistic for testing $H : \Theta = 0$ is

$$\lambda^{2/n} = \Lambda = \frac{|I + Z_{13} V_{33}^{-1} Z_{13}'|}{\left| I + \begin{pmatrix} Z_{12} & Z_{13} \end{pmatrix} \begin{pmatrix} V_{22} & V_{23} \\ V_{32} & V_{33} \end{pmatrix}^{-1} \begin{pmatrix} Z_{12} & Z_{13} \end{pmatrix}' \right|}$$

$$= \frac{\begin{vmatrix} V_{22} & V_{23} \\ V_{32} & V_{33} \end{vmatrix} |V_{33} + Z_{13}' Z_{13}|}{|V_{33}| \left| \begin{pmatrix} V_{22} & V_{23} \\ V_{32} & V_{33} \end{pmatrix} + \begin{pmatrix} Z_{12} & Z_{13} \end{pmatrix}' \begin{pmatrix} Z_{12} & Z_{13} \end{pmatrix} \right|}$$

$$= \frac{|V_{33}||V_{22} - V_{23} V_{33}^{-1} V_{32}||V_{33} + Z_{13}' Z_{13}|}{|V_{33}| \begin{vmatrix} V_{22} + Z_{12}' Z_{12} & V_{23} + Z_{12}' Z_{13} \\ V_{32} + Z_{13}' Z_{12} & V_{33} + Z_{13}' Z_{13} \end{vmatrix}}$$

$$= \frac{|V_{22} - V_{23} V_{33}^{-1} V_{32}|}{|(V_{22} + Z_{12}' Z_{12}) - (V_{23} + Z_{12}' Z_{13})(V_{33} + Z_{13}' Z_{13})^{-1}(V_{23} + Z_{12}' Z_{13})'|}$$

$$= \frac{|V_{22\cdot3}|}{|V_{22\cdot3} + X'X|} = \frac{1}{|I + X V_{22\cdot3}^{-1} X'|} = \frac{1}{|I + T_1|} \tag{7.96}$$

by using the following

$$|V| = \begin{vmatrix} V_{11} & V_{12} \\ V_{21} & V_{22} \end{vmatrix} = |V_{22}||V_{11} - V_{12} V_{22}^{-1} V_{21}| \tag{7.97}$$

if

$$P \equiv (Y - B\xi A)(Y - B\xi A)' = S + (Y - B\xi)AA'(Y - B\xi)' \tag{7.98}$$

for $Y_{p \times n}$, $B_{q \times p}$, $\xi_{p \times k}$, and $A_{k \times n}$ then

$$|P| = |S||I_p + S^{-1}(Y - B\xi)AA'(Y - B\xi)'|$$
$$= |S||I_k + AA'(Y - B\xi)S^{-1}(Y - B\xi)'| \qquad (7.99)$$

and, defining

$$X = (I + Z_{13}V_{33}^{-1}Z_{13}')^{-1/2}(Z_{12} - Z_{13}V_{33}^{-1}V_{32}) \qquad (7.100)$$
$$T_1 = XV_{22 \cdot 3}^{-1}X' \qquad (7.101)$$
$$T_2 = Z_{13}V_{33}^{-1}Z_{13}'. \qquad (7.102)$$

The matrices X and $V_{22 \cdot 3}$ given T_2 are independent and

$$X|T_2 \sim N_{k_1,p}[(I + T_2)^{-1/2}\Theta_{12}, \Sigma_{22 \cdot 3}, I_p] \qquad (7.103)$$
$$V_{22 \cdot 3} \sim W_{q_2}(k_3 - q_3, V_{22 \cdot 3}) \qquad (7.104)$$
$$X_{13} \sim N_{k_1,p}(0, \Sigma_{33}, I_p) \qquad (7.105)$$
$$V_{33} \sim W_{q_3}(q_2, \Sigma_{22}) \qquad (7.106)$$

where $\nu_e = n - k - p + q = k_3 - q_3$.

To transform the canonical form of the test back into the original variables, we again recall some vector algebra. Using Theorem 7.3, let $X_{1(n \times k)}$ be a design matrix of rank k, then $X_1'X_1$ is symmetric and of rank k. Furthermore there exists an orthogonal matrix P such that $P'(X_1'X_1)P = \Lambda$ or $(X_1'X_1) = P\Lambda P' = P\Lambda^{1/2}P'P\Lambda^{1/2}P'$. The positive square root matrix of $(X_1'X_1)$ is defined as

$$(X_1'X_1)^{1/2} = P\Lambda^{1/2}P'. \qquad (7.107)$$

Theorem 7.3. (The spectral decomposition theorem). *Let A be a symmetric matrix of order n. Then there exists an orthogonal matrix P $(P'P = I)$, such that $P'AP = \Lambda$ or $A = P\Lambda P'$ where $\Lambda = \text{diag}(\lambda_1, \lambda_2, \ldots, \lambda_n)$ is a diagonal matrix whose elements are the eigenvalues of A and the columns of P correspond to the orthonormal eigenvectors of A.*

Theorem 7.4. *Let $A_{n \times k}$ be a matrix of rank k. Then the matrix $A(A'A)A = P$ is an orthogonal projection matrix if and only if P is symmetric and idempotent, $P^2 = P$. Furthermore, if P is a projection matrix, so is $I - P$, and $(I - P)P = 0$ where P and $I - P$ are disjoint.*

Theorem 7.5. (Gram-Schmidt). *Let $a_1, a_2, \ldots, a_n$ be a basis for a vector subspace A of R^n. Then there exists an orthonormal basis, $e_1, e_2, \ldots, e_k, e_{k+1}, e_{k+2}, \ldots, e_n$ such that $R^n = A \oplus A^\perp$ where $A^\perp$ is the orthogonal complement of A, $e_i \in A$ and $e_j \in A^\perp$ such that $e_i'e_j = 0$, and $e_{k+1}, \ldots, e_n$ is associated with $A^\perp$.*

Associating the design matrix X_1 with A and $\widetilde{X}_1$ with $A^\perp$, in Theorem 7.5, we see for the GMANOVA model that

$$\text{rank}\left(X_1 \ \ \widetilde{X}_1\right) = n \quad \text{and} \quad \widetilde{X}_1'X_1 = 0 \qquad (7.108)$$

$$\text{rank}\left(\begin{matrix} X_2 \\ \widetilde{X}_2 \end{matrix}\right) = q \quad \text{and} \quad X_2'\widetilde{X}_2 = 0 \qquad (7.109)$$

so that we may form the projection matrices

$$\widetilde{X}_1 = \widetilde{X}_1'(\widetilde{X}_1'\widetilde{X}_1)^{-1}\widetilde{X}_1 = I_n - X_1(X_1'X_1)^{-1}X_1' \tag{7.110}$$

$$\widetilde{X}_2 = \widetilde{X}_2'(\widetilde{X}_2'\widetilde{X}_2)^{-1}\widetilde{X}_2 = I_n - X_2(X_2'X_2)^{-1}X_2'. \tag{7.111}$$

Similarly, for the hypothesis $H : X_2BX_3 = 0$, we form matrices $\widetilde{X}_3$ and $\widetilde{X}_4$ such that

$$\text{rank}\begin{pmatrix} X_3 & \widetilde{X}_3 \end{pmatrix} = k_1 + k_2 \quad \text{and} \quad \widetilde{X}_3'X_3 = 0 \tag{7.112}$$

$$\text{rank}\begin{pmatrix} X_4 \\ \widetilde{X}_4 \end{pmatrix} = q_1 + q_2 \quad \text{and} \quad X_4'\widetilde{X}_4 = 0. \tag{7.113}$$

Following Kariya (1985, p. 81), we associate with the P_i's and Q_i's the following matrices

$$P_1 = (X_1'X_1)^{1/2} \tag{7.114}$$

$$P_2 = (X_2X_2')^{1/2} \tag{7.115}$$

$$P_3 = (X_3P_1^{-2}X_3)^{1/2} \tag{7.116}$$

$$P_4 = (X_4'P_2^{-2}X_4)^{1/2} \tag{7.117}$$

$$Q_1 = \left[X_1(X_1'X_1)^{-1/2} \quad \widetilde{X}_1(\widetilde{X}_1'\widetilde{X}_1)^{-1/2} \right] \tag{7.118}$$

$$Q_2 = \begin{bmatrix} (X_2'X_2)^{-1/2}X_2 \\ (\widetilde{X}_2\widetilde{X}_2')^{-1/2}\widetilde{X}_2 \end{bmatrix} \tag{7.119}$$

$$Q_3 = \begin{bmatrix} (X_3P_1^{-2}X_3')^{-1/2}X_3P_1^{-1} \\ (\widetilde{X}_3P_1^{-2}\widetilde{X}_3')^{-1/2}\widetilde{X}_3P_1 \end{bmatrix} \tag{7.120}$$

$$Q_4 = \left[P_2\widetilde{X}_4(\widetilde{X}_4'P_2^{-1}\widetilde{X}_4)^{-1/2} \quad P_2^{-1}X_4(X_4'P_2^{-2}X_4)^{-1/2} \right] \tag{7.121}$$

where $Y^* = Q_1'YQ_2$, $B^* = P_1BP_2$ and $\Sigma^* = Q_2'\Sigma Q_2$. Then the Z_{ij} in (7.88) become

$$Z_{12} = [X_3(X_1'X_1)^{-1}X_3']^{-1/2}X_3BX_4[X_4'(X_2X_2')^{-1}X_4]^{-1/2} \tag{7.122}$$

$$\widehat{B} = (X_1'X_1)^{-1}X_1'YX_2'(X_2X_2')^{-1} \tag{7.123}$$

$$Z_{13} = [X_3(X_1'X_1)X_3']^{-1/2}X_3(X_1'X_1)^{-1}X_1'YX_2'(X_2X_2')^{-1/2} \tag{7.124}$$

$$Z_{32} = (X_1'X_1)^{-1/2}X_1'YX_2'(X_2X_2')^{-1}X_4[X_4'(X_2X_2')X_4]^{-1/2} \tag{7.125}$$

$$Z_{33} = (X_1'X_1)^{-1/2}X_1YX_2'(X_2X_2')^{-1/2} \tag{7.126}$$

in terms of the original variables. To obtain the test statistic, let

$$\bar{X}_1 = I_n - X_1(X_1'X_1)^{-1}X_1' \tag{7.127}$$

$$E = Y'\bar{X}_1Y \tag{7.128}$$

$$A_1 = [X_3(X_1'X_1)^{-1}X_3']^{-1/2}X_3(X_1'X_1)^{-1}X_1' \tag{7.129}$$

$$A_2 = X_2'(X_2X_2')X_4[X_4'(X_2X_2')^{-1}X_4]^{-1/2}. \tag{7.130}$$

Then by (7.122)-(7.126),

$$V_{22\cdot3} = [X_4'(X_2X_2')^{-1}X_4]^{-1/2}X_4'(X_2E^{-1}X_2')^{-1}X_4\cdots$$
$$[X_4'(X_2X_2')^{-1}X_4]^{-1/2} \tag{7.131}$$

$$Z_{12} - Z_{13}V_{33}^{-1}Z_{32} = A_1YE^{-1}X_2'(X_2E^{-1}X_2')^{-1}X_2A_2 \tag{7.132}$$

$$T_2 = A_1Y[E^{-1}E^{-1}X_2'(X_2E^{-1}X_2')^{-1}X_2E^{-1}]Y'A_1 \tag{7.133}$$

$$T_1 = (I+T_2)^{-1/2}A_1YE^{-1}X_2'(X_2E^{-1}X_2')^{-1}X_4[X_4'\cdots$$
$$(X_2E^{-1}X_2')^{-1}X_4]^{-1}X_4'(X_2E^{-1}X_2')^{-1}X_2E^{-1}Y'A_1'(I+T_2)^{-1/2}. \tag{7.134}$$

For the MANOVA model, $X_2 = I$ and $T_2 = 0$ so that

$$T_1 = A_1YX_4(X_4'EX_4)^{-1}X_4'Y'A_1'. \tag{7.135}$$

To verify the hypothesis and error SSP matrices for the MANOVA model, we use the Bartlett-Lawley-Hotelling trace criterion. Hence,

$$\text{tr}(T_1) = \text{tr}\{[X_3(X_1'X_1)^{-1}X_3]^{-1/2}X_3(X_1'X_1)^{-1}X_1'YX_4(X_4'EX_4)^{-1}X_4'YX_1(X_1'X_1)^{-1}$$
$$X_3'[X_3(X_1'X_1)^{-1}X_3]^{-1/2}\}$$
$$= \text{tr}\{[X_3(X_1'X_1)X_3]^{-1/2}(X_3BX_4)(X_4'EX_4)^{-1}(X_3BX_4)'[X_3(X_1'X_1)^{-1}X_3]^{-1/2}\}$$
$$= \text{tr}\{(X_4'EX_4)^{-1}(X_3BX_4)'[X_3(X_1'X_1)^{-1}X_3]^{-1}(X_3BX_4)\}$$
$$= \text{tr}(E^{-1}H) \tag{7.136}$$

where

$$H = (X_3\widehat{B}X_4)'[X_3(X_1'X_1)^{-1}X_3]^{-1}(X_3\widehat{B}X_4) \tag{7.137}$$
$$E = X_4'EX_4 \tag{7.138}$$
$$\widehat{B} = (X_1'X_1)^{-1}X_1'Y. \tag{7.139}$$

H and E in (7.137)-(7.138) are equal to the expressions given in (5.39)-(5.40) for testing $H : CBA = 0$ in the MANOVA model with $X_4 = A$ and $X_3 = C$. In a similar manner, substituting T_1 in (7.134) into the test criterion for the GMANOVA model, the canonical form of the test statistic in (7.96) reduces to Λ in (7.68) using the MANCOVA model after some matrix algebra (Fujikoshi, 1974; Khatri, 1966).

7.5   RESTRICTED NONORTHOGONAL THREE-FACTOR FACTORIAL MANOVA

The standard application of the RMGLM is to formulate a full-rank model for the MANOVA and, using restrictions, to constrain the interactions to equal zero to obtain an additive model. Restrictions are imposed on the linear model $Y = XB+U$ of the form $R_1BR_2 = \Theta$ where Θ is usually zero and where $R_2 = A$ or I when testing the hypothesis $H : CBA = 0$. When $R_2 \neq A\ or\ I$, the RMGLM becomes a mixture of the GMANOVA and the MANOVA models. Applications of the more general restricted model are discussed in Chapter 8.

In Chapter 5, we illustrated how to analyze a complete three-factor factorial design which included all factors A, B, C, AB, AC, BC, and ABC in the design. This design may be analyzed using a multivariate full-rank model. The full-rank model for the three-factor design with no empty cells is

$$y_{ijkm} = \mu_{ijk} + e_{ijk} \tag{7.140}$$

$$e_{ijk} \sim IN_p(0, \Sigma) \tag{7.141}$$

$$i = 1, 2, \ldots, I; j = 1, 2, \ldots, J;$$
$$k = 1, 2, \ldots, K; m = 1, 2, \ldots, M_{ijk}.$$

For the three-factor design, either weighted or unweighted hypotheses about main effects may be tested. For a 2×3 design, for example, we may construct means as:

$$\bar{\mu}_1 = \frac{w_{11}\mu_{11} + w_{12}\mu_{12} + w_{13}\mu_{13}}{w_{11} + w_{12} + w_{13}}$$

$$\bar{\mu}_2 = \frac{w_{21}\mu_{21} + w_{22}\mu_{22} + w_{23}\mu_{23}}{w_{21} + w_{22} + w_{23}}$$

are the means for populations A_1 and A_2 collapsing across, or ignoring, factor B. Forming a weighted mean is essentially collapsing across or ignoring the factor or factors over which the mean is being found. However, other methods of weighting are available in the three-factor design. To test a hypothesis about factor A in a three-factor design, one can consider three different ways of collapsing across factor B and C: both factors B and C are collapsed or just factor B or just factor C. For a design with I by J by K levels, the mean

$$\tilde{\bar{\mu}}_{i\cdot\cdot} = \sum_i \sum_k \frac{M_{ijk}\mu_{ijk}}{M_{i++}} \quad i = 1, 2, \ldots, I, \tag{7.142}$$

where M_{i++} is the total number of observations at the ith level of factor A, may be constructed to test

$$H_{A \text{ ignoring } B\&C} : \text{all } \tilde{\bar{\mu}}_{i\cdot\cdot} \text{ are equal.} \tag{7.143}$$

Alternatively, means

$$\bar{\mu}_{i\cdot\cdot} = \frac{1}{J} \sum_j \left(\sum_k \frac{M_{ijk}\mu_{ijk}}{M_{ij+}} \right) \quad i = 1, 2, \ldots, I, \tag{7.144}$$

where M_{ij+} is the total number of observations at the ith level of factor A and the jth level of B, are formed to test

$$H_{A \text{ given } B \text{ ignoring } C} : \text{all } \bar{\mu}_{i\cdot\cdot} \text{ are equal.} \tag{7.145}$$

Finally, the mean

$$\tilde{\mu}_{i\cdot\cdot} = \frac{1}{K} \sum_k \left(\sum_j \frac{M_{ijk}\mu_{ijk}}{M_{i+k}} \right) \quad i = 1, 2, \ldots, I, \tag{7.146}$$

Table 7.1: AB Treatment Combination Means: Three-Factor Design.

	B_1	B_2	$\cdots$	B_J
A_1	$\mu_{11\cdot} = \sum_k \frac{\mu_{11k}}{K}$	$\mu_{12\cdot} = \sum_k \frac{\mu_{12k}}{K}$	$\cdots$	$\mu_{1J\cdot} = \sum_k \frac{\mu_{1Jk}}{K}$
A_2	$\mu_{21\cdot} = \sum_k \frac{\mu_{21k}}{K}$	$\mu_{22\cdot} = \sum_k \frac{\mu_{22k}}{K}$	$\cdots$	$\mu_{2J\cdot} = \sum_k \frac{\mu_{2Jk}}{K}$
$\vdots$	$\vdots$	$\vdots$	$\ddots$	$\vdots$
A_I	$\mu_{I1\cdot} = \sum_k \frac{\mu_{I1k}}{K}$	$\mu_{I2\cdot} = \sum_k \frac{\mu_{I2k}}{K}$	$\cdots$	$\mu_{IJ\cdot} = \sum_k \frac{\mu_{IJk}}{K}$

where M_{i+k} is the total number of observations at the ith level of factor A and the kth level of C, is used to test

$$H_{A \text{ given } C \text{ ignoring } B} : \text{all } \tilde{\mu}_{i\cdot\cdot} \text{ are equal.} \tag{7.147}$$

In a similar manner three different hypotheses can be tested by considering the main effects of factor B or factor C. To perform tests of either weighted or unweighted hypotheses about main effects in a three-factor design, one can construct hypothesis test matrices using the same procedures discussed in the analysis of univariate designs.

To test for two-factor interactions, we average cell means to produce a two-factor table of means and then express the hypotheses as we did for two–factor designs. For example, with an I by J by K design we form the hypothesis of no AB interaction by finding the average at each AB treatment combinations as shown in Table 7.1. A general form for the hypothesis of no AB interaction is

$$H_{AB} : \mu_{ij\cdot} - \mu_{i'j\cdot} - \mu_{ij'\cdot} + \mu_{i'j'\cdot} = 0 \text{ for all } i, i', j, j'. \tag{7.148}$$

Similarly we may form AC and BC interaction hypotheses be defining means for each AC and BC treatment combination and express the hypotheses of no AC and BC interaction as

$$H_{AC} : \mu_{i\cdot k} - \mu_{i'\cdot k} - \mu_{i\cdot k'} + \mu_{i\cdot k'} = 0 \text{ for all } i, i', k, k' \tag{7.149}$$

$$H_{BC} : \mu_{\cdot jk} - \mu_{\cdot j'k} - \mu_{\cdot jk'} + \mu_{\cdot j'k'} = 0 \text{ for all } j, j', k, k'. \tag{7.150}$$

When the cell sizes are unequal we may also use weighted averages of the cell means as the two-factor treatment combination means and test weighted hypotheses of no two-factor interactions. For the AB interaction, for example, we define the weighed means as

$$\bar{\mu}_{ij\cdot} = \sum_k \frac{M_{ijk}\mu_{ijk}}{M_{ij+}} \quad i = 1, 2, \ldots, I; j = 1, 2, \ldots, J \tag{7.151}$$

and then test the hypothesis

$$H_{AB^*} : \bar{\mu}_{ij\cdot} - \bar{\mu}_{i'j\cdot} - \bar{\mu}_{ij'\cdot} + \bar{\mu}_{i'j'\cdot} = 0 \text{ for all } i, i', j, j'. \tag{7.152}$$

Table 7.2: Data for a 3 by 3 by 2 MANOVA Design.

C_1

		B_1	B_2	B_3
A_1	$y1$	$1, 2, 3$	$3, 4$	$1, 2$
	$y2$	$22, 21, 14$	$31, 35$	$31, 41$
A_2	$y1$	$3, 5$	2	$4, 5, 6$
	$y2$	$31, 36$	45	$21, 21, 31$
A_3	$y1$	$2, 3$	$4, 5, 6$	$1, 2, 3$
	$y2$	$21, 31$	$66, 41, 51$	$61, 47, 35$

C_2

		B_1	B_2	B_3
A_1	$y1$	$3, 4$	$4, 5, 6$	$1, 2$
	$y2$	$66, 55$	$61, 11, 14$	$41, 21$
A_2	$y1$	2	$2, 3$	$4, 5$
	$y2$	41	$47, 61$	$41, 55$
A_3	$y1$	1	$2, 3, 4$	$5, 6, 7$
	$y2$	41	$18, 21, 31$	$57, 64, 77$

Similar hypotheses can be formed for the interactions AC^* and BC^*. Weighted hypotheses should not be tested in a three-factor design (or any other design, for that matter) unless the weights have meaning in the populations to which inferences are to be made.

In the three-factor design, the three-way interaction is zero if all linear combinations of cell means used in the two-factor interaction are identical at all levels of the third factor. Thus, the hypothesis of no three-factor interaction can be stated as an equality of, for example, combinations of AB means at each level of factor C. For an I by J by K design, we would have

$$H_{ABC} : (\mu_{ijk} - \mu_{i'jk} - \mu_{ij'k} + \mu_{i'j'k}) - (\mu_{ijk'} - \mu_{i'jk'} - \mu_{ij'k'} + \mu_{i'j'k'}) = 0$$
$$\text{for all } i, i', j, j', k, k'. \tag{7.153}$$

The hypothesis of no three-factor interaction can also be stated as an equality of the AC interaction combinations at different levels of B or the equality of the BC interaction combinations at different levels of A. All three forms of the hypothesis are equivalent.

To illustrate the analysis of data for a three factor design with an unequal number of cell frequencies, we use the data defined in Table 7.2, and analyzed in Section 5.11.

If the first column in the design matrix is associated with cell $A_1 B_1 C_1$, the second with $A_1 B_2 C_1$, etc., then the parameter vector is $\mu' = (\mu_{111}, \mu_{121}, \ldots, \mu_{332})$. Thus, the first six means are from level 1 of A, the second size from level 2 of A, and the last six are from level 3 of A and we can test

$$H_A : \text{all } \mu_{i..} \text{ equal}$$
$$H_B : \text{all } \mu_{.j.} \text{ equal}$$
$$H_A : \text{all } \mu_{..k} \text{ equal}$$

using the hypothesis test matrices C_A, C_B, and C_C defined below. The matrices to test the unweighted hypotheses are not unique.

If we wish to test weighted rather than unweighted hypotheses, we may form the appropriate hypothesis test matrix. We introduce some new notation here to easily differentiate among the matrices. The ordinary subscripts: A, B, C, denote an unweighted hypothesis. An asterisk on a subscript: A^*, B^*, C^*, indicates a weighted hypothesis where all other factors are ignored in the hypothesis. A letter in square brackets indicates a factor that is not ignored in the weighted hypothesis and implies that all other factors are ignored. Thus, $C_{A[B]}$ means the weighted hypothesis test matrix for testing A given B but ignoring C, C_{A^*} means the weighted hypothesis test matrix for testing A ignoring both factors B and C, and C_A means the hypothesis test matrix for testing A given B and C (an unweighted test). The hypothesis test matrices for the design example follow:

$$
\mu = \begin{pmatrix} \mu_{111} \\ \mu_{121} \\ \mu_{131} \\ \mu_{112} \\ \mu_{122} \\ \mu_{132} \\ \mu_{211} \\ \mu_{221} \\ \mu_{231} \\ \mu_{212} \\ \mu_{222} \\ \mu_{232} \\ \mu_{311} \\ \mu_{321} \\ \mu_{331} \\ \mu_{312} \\ \mu_{322} \\ \mu_{332} \end{pmatrix}
C_A' = \begin{pmatrix} 1/6 & 0 \\ 1/6 & 0 \\ 1/6 & 0 \\ 1/6 & 0 \\ 1/6 & 0 \\ 1/6 & 0 \\ 0 & 1/6 \\ 0 & 1/6 \\ 0 & 1/6 \\ 0 & 1/6 \\ 0 & 1/6 \\ 0 & 1/6 \\ -1/6 & -1/6 \\ -1/6 & -1/6 \\ -1/6 & -1/6 \\ -1/6 & -1/6 \\ -1/6 & -1/6 \\ -1/6 & -1/6 \end{pmatrix}
C_B' = \begin{pmatrix} 1/6 & 0 \\ 0 & 1/6 \\ -1/6 & -1/6 \\ 1/6 & 0 \\ 0 & 1/6 \\ -1/6 & -1/6 \\ 1/6 & 0 \\ 0 & 1/6 \\ -1/6 & -1/6 \\ 1/6 & 0 \\ 0 & 1/6 \\ -1/6 & -1/6 \\ 1/6 & 0 \\ 0 & 1/6 \\ -1/6 & -1/6 \\ 1/6 & 0 \\ 0 & 1/6 \\ -1/6 & -1/6 \end{pmatrix}
C_C' = \begin{pmatrix} 1/9 \\ 1/9 \\ 1/9 \\ -1/9 \\ -1/9 \\ -1/9 \\ 1/9 \\ 1/9 \\ 1/9 \\ -1/9 \\ -1/9 \\ -1/9 \\ 1/9 \\ 1/9 \\ 1/9 \\ -1/9 \\ -1/9 \\ -1/9 \end{pmatrix}
$$

$$\text{(7.154)}$$

$$
C'_{A[B]} =
\begin{pmatrix}
3/15 & 0 \\
2/15 & 0 \\
2/12 & 0 \\
2/15 & 0 \\
3/15 & 0 \\
2/12 & 0 \\
0 & 2/9 \\
0 & 1/9 \\
0 & 3/15 \\
0 & 1/9 \\
0 & 2/9 \\
0 & 2/15 \\
-2/9 & -2/9 \\
-1/9 & -1/9 \\
-3/15 & -3/15 \\
-1/9 & -1/9 \\
-2/9 & -2/9 \\
-2/15 & -2/15
\end{pmatrix}
\quad
C'_{A[C]} =
\begin{pmatrix}
3/14 & 0 \\
2/14 & 0 \\
2/14 & 0 \\
2/14 & 0 \\
3/14 & 0 \\
2/14 & 0 \\
0 & 3/14 \\
0 & 2/14 \\
0 & 2/14 \\
0 & 2/14 \\
0 & 3/14 \\
0 & 2/14 \\
-2/16 & -2/16 \\
-3/16 & -3/16 \\
-3/16 & -3/16 \\
-1/14 & -1/14 \\
-3/14 & -3/14 \\
-3/14 & -3/14
\end{pmatrix}
\quad
C'_{A*} =
\begin{pmatrix}
3/14 & 0 \\
2/14 & 0 \\
2/14 & 0 \\
2/14 & 0 \\
3/14 & 0 \\
2/14 & 0 \\
0 & 2/11 \\
0 & 1/11 \\
0 & 3/11 \\
0 & 1/11 \\
0 & 2/11 \\
0 & 2/11 \\
-2/15 & -2/15 \\
-3/15 & -3/15 \\
-3/15 & -3/15 \\
-1/15 & -1/15 \\
-3/15 & -3/15 \\
-3/15 & -3/15
\end{pmatrix}
\tag{7.155}
$$

$$
C'_{B[A]} =
\begin{pmatrix}
3/15 & 0 \\
0 & 2/15 \\
-2/12 & 0 \\
2/15 & 0 \\
0 & 3/15 \\
-2/12 & -2/12 \\
2/9 & 0 \\
0 & 1/9 \\
-3/15 & -3/15 \\
1/9 & 0 \\
0 & 3/18 \\
-2/15 & -3/18 \\
2/9 & 0 \\
0 & 3/18 \\
-3/18 & -3/18 \\
1/9 & 0 \\
0 & 3/18 \\
-3/18 & -3/18
\end{pmatrix}
\quad
C'_{B[C]} =
\begin{pmatrix}
3/14 & 0 \\
0 & 2/12 \\
-2/16 & -2/16 \\
2/8 & 0 \\
0 & 3/16 \\
-2/14 & -2/14 \\
2/14 & 0 \\
0 & 1/12 \\
-3/16 & -3/16 \\
1/8 & 0 \\
0 & 2/16 \\
-2/14 & -2/14 \\
2/14 & 0 \\
0 & 3/12 \\
-3/16 & -3/16 \\
1/8 & 0 \\
0 & -3/16 \\
-3/14 & -3/14
\end{pmatrix}
\quad
C'_{B*} =
\begin{pmatrix}
3/11 & 0 \\
0 & 2/14 \\
-2/15 & -2/15 \\
2/11 & 0 \\
0 & 3/14 \\
-2/15 & -2/15 \\
1/11 & 0 \\
0 & 2/14 \\
-3/15 & -3/15 \\
1/11 & 0 \\
0 & 2/14 \\
-2/15 & -2/15 \\
2/11 & 0 \\
0 & 3/14 \\
-3/15 & -3/15 \\
1/11 & 0 \\
0 & 3/14 \\
-3/15 & -3/15
\end{pmatrix}
\tag{7.156}
$$

Table 7.3: Means of AB Treatment Combinations.

	B_1	B_2	B_3
A_1	$\mu_{111}, M_{111} = 3, (\mu1)$ $\mu_{112}, M_{112} = 2, (\mu4)$	$\mu_{121}, M_{121} = 2, (\mu2)$ $\mu_{122}, M_{122} = 3, (\mu5)$	$\mu_{131}, M_{131} = 2, (\mu3)$ $\mu_{132}, M_{132} = 2, (\mu6)$
A_2	$\mu_{211}, M_{211} = 3, (\mu7)$ $\mu_{212}, M_{212} = 2, (\mu10)$	$\mu_{221}, M_{221} = 1, (\mu8)$ $\mu_{222}, M_{222} = 2, (\mu11)$	$\mu_{231}, M_{231} = 3, (\mu9)$ $\mu_{232}, M_{232} = 2, (\mu12)$
A_3	$\mu_{311}, M_{311} = 2, (\mu13)$ $\mu_{312}, M_{312} = 1, (\mu16)$	$\mu_{321}, M_{321} = 3, (\mu14)$ $\mu_{322}, M_{322} = 3, (\mu17)$	$\mu_{331}, M_{331} = 3, (\mu15)$ $\mu_{332}, M_{332} = 3, (\mu18)$

$$
C'_{C[A]} = \begin{pmatrix} 3/21 \\ 2/21 \\ 2/21 \\ -2/21 \\ -3/21 \\ -2/21 \\ 2/18 \\ 1/18 \\ 3/18 \\ -1/15 \\ -2/15 \\ -2/15 \\ 2/24 \\ 3/24 \\ 3/24 \\ -1/21 \\ -3/21 \\ -3/21 \end{pmatrix} \quad
C'_{C[B]} = \begin{pmatrix} 3/21 \\ 2/18 \\ 2/24 \\ -2/12 \\ -3/24 \\ -2/21 \\ 2/21 \\ 1/18 \\ 3/24 \\ -1/12 \\ -2/24 \\ -2/21 \\ 2/21 \\ 3/18 \\ 3/24 \\ -1/12 \\ -3/24 \\ -3/21 \end{pmatrix} \quad
C'_{C*} = \begin{pmatrix} 3/21 \\ 2/21 \\ 2/21 \\ -2/19 \\ -3/19 \\ -2/19 \\ 2/21 \\ 1/21 \\ 3/21 \\ -1/19 \\ -2/19 \\ -1/19 \\ 2/21 \\ 3/21 \\ 3/21 \\ -1/19 \\ -3/19 \\ -3/19 \end{pmatrix} \quad (7.157)
$$

If one wants to see how the two-factor interactions are tested, it is convenient to form 2 by 2 tables of means. A table of means for the AB treatment combinations is shown in Table 7.3. Each mean in Table 7.3 is accompanied by two numbers, the frequency (M_{ijk}) of the cell used in estimating the mean, and a number in parentheses indicating the position of the mean in the parameter vector. The latter number helps to place coefficients in the correct positions in hypothesis test matrices.

Using Table 7.3, we can form hypothesis test matrices for either weighted or unweighted tests of no AB interaction. One example of each is shown in (7.158). Note that each coefficient in C'_{AB} is $\frac{1}{2}$ since each unweighted AB mean is the simple average of two cell means. The coefficients in C'_{AB*} are those required to form weighted AB means. In a manner similar to the interpretation of the weighted main effect tests, the weighted interaction tests may be interpreted as test of two-factor interactions ignoring the factor not entered into the interaction.

$$
C'_{AB} = \begin{pmatrix}
\frac{1}{2} & 0 & 0 & 0 \\
-\frac{1}{2} & \frac{1}{2} & 0 & 0 \\
0 & -\frac{1}{2} & 0 & 0 \\
\frac{1}{2} & 0 & \frac{1}{2} & 0 \\
-\frac{1}{2} & \frac{1}{2} & -\frac{1}{2} & 0 \\
0 & -\frac{1}{2} & 0 & 0 \\
-\frac{1}{2} & 0 & \frac{1}{2} & 0 \\
\frac{1}{2} & -\frac{1}{2} & -\frac{1}{2} & \frac{1}{2} \\
0 & \frac{1}{2} & 0 & -\frac{1}{2} \\
-\frac{1}{2} & 0 & -\frac{1}{2} & 0 \\
\frac{1}{2} & -\frac{1}{2} & \frac{1}{2} & \frac{1}{2} \\
0 & \frac{1}{2} & 0 & -\frac{1}{2} \\
0 & 0 & -\frac{1}{2} & 0 \\
0 & 0 & \frac{1}{2} & -\frac{1}{2} \\
0 & 0 & 0 & \frac{1}{2} \\
0 & 0 & -\frac{1}{2} & 0 \\
0 & 0 & \frac{1}{2} & -\frac{1}{2} \\
0 & 0 & 0 & \frac{1}{2}
\end{pmatrix}
\qquad
C'_{AB^*} = \begin{pmatrix}
\frac{3}{5} & 0 & 0 & 0 \\
-\frac{2}{5} & \frac{2}{5} & 0 & 0 \\
0 & -\frac{1}{2} & 0 & 0 \\
\frac{2}{5} & 0 & 0 & 0 \\
-\frac{3}{5} & \frac{3}{5} & 0 & 0 \\
0 & -\frac{1}{2} & 0 & 0 \\
-\frac{2}{3} & 0 & \frac{2}{3} & 0 \\
\frac{1}{3} & -\frac{1}{3} & -\frac{1}{3} & \frac{1}{3} \\
0 & \frac{3}{5} & 0 & -\frac{3}{5} \\
-\frac{1}{3} & 0 & \frac{1}{3} & 0 \\
\frac{2}{3} & -\frac{2}{3} & -\frac{2}{3} & \frac{2}{3} \\
0 & \frac{2}{5} & 0 & -\frac{2}{5} \\
0 & 0 & -\frac{2}{3} & 0 \\
0 & 0 & \frac{1}{2} & -\frac{1}{2} \\
0 & 0 & 0 & \frac{1}{2} \\
0 & 0 & -\frac{1}{3} & 0 \\
0 & 0 & \frac{1}{2} & -\frac{1}{2} \\
0 & 0 & 0 & \frac{1}{2}
\end{pmatrix}
\tag{7.158}
$$

Using a similar procedure, we form hypothesis test matrices for weighted or unweighted tests of the AC and BC interactions, examples of which are shown in (7.159)-(7.160).

$$
C'_{AC} = \begin{pmatrix}
\frac{1}{3} & 0 \\
\frac{1}{3} & 0 \\
\frac{1}{3} & 0 \\
-\frac{1}{3} & 0 \\
-\frac{1}{3} & 0 \\
-\frac{1}{3} & 0 \\
-\frac{1}{3} & \frac{1}{3} \\
-\frac{1}{3} & \frac{1}{3} \\
-\frac{1}{3} & \frac{1}{3} \\
\frac{1}{3} & -\frac{1}{3} \\
\frac{1}{3} & -\frac{1}{3} \\
\frac{1}{3} & -\frac{1}{3} \\
0 & -\frac{1}{3} \\
0 & -\frac{1}{3} \\
0 & -\frac{1}{3} \\
0 & \frac{1}{3} \\
0 & \frac{1}{3} \\
0 & \frac{1}{3}
\end{pmatrix}
\qquad
C'_{AC^*} = \begin{pmatrix}
\frac{3}{7} & 0 \\
\frac{2}{7} & 0 \\
\frac{2}{7} & 0 \\
-\frac{2}{7} & 0 \\
-\frac{3}{7} & 0 \\
-\frac{2}{7} & 0 \\
-\frac{2}{6} & \frac{2}{6} \\
-\frac{1}{6} & \frac{1}{6} \\
-\frac{3}{6} & \frac{3}{6} \\
\frac{1}{5} & -\frac{1}{5} \\
\frac{2}{5} & -\frac{2}{5} \\
\frac{2}{5} & -\frac{2}{5} \\
0 & -\frac{2}{8} \\
0 & -\frac{3}{8} \\
0 & -\frac{3}{8} \\
0 & \frac{1}{7} \\
0 & \frac{3}{7} \\
0 & \frac{3}{7}
\end{pmatrix}
\tag{7.159}
$$

$$
C'_{BC} = \begin{pmatrix}
\frac{1}{3} & 0 \\
-\frac{1}{3} & \frac{1}{3} \\
0 & -\frac{1}{3} \\
-\frac{1}{3} & 0 \\
\frac{1}{3} & -\frac{1}{3} \\
0 & \frac{1}{3} \\
\frac{1}{3} & 0 \\
-\frac{1}{3} & \frac{1}{3} \\
0 & -\frac{1}{3} \\
-\frac{1}{3} & 0 \\
\frac{1}{3} & -\frac{1}{3} \\
0 & \frac{1}{3} \\
\frac{1}{3} & 0 \\
-\frac{1}{3} & \frac{1}{3} \\
0 & \frac{1}{3} \\
-\frac{1}{6} & 0 \\
\frac{1}{3} & -\frac{1}{3} \\
0 & \frac{1}{3}
\end{pmatrix}
\qquad
C'_{BC^*} = \begin{pmatrix}
\frac{3}{7} & 0 \\
-\frac{2}{6} & \frac{2}{6} \\
0 & -\frac{2}{8} \\
-\frac{2}{4} & 0 \\
\frac{3}{8} & -\frac{3}{8} \\
0 & \frac{2}{7} \\
\frac{2}{7} & 0 \\
-\frac{1}{6} & \frac{1}{6} \\
0 & -\frac{3}{8} \\
-\frac{1}{4} & 0 \\
\frac{1}{4} & -\frac{1}{4} \\
0 & \frac{2}{7} \\
\frac{2}{7} & 0 \\
-\frac{3}{6} & \frac{3}{6} \\
0 & -\frac{3}{8} \\
-\frac{1}{4} & 0 \\
\frac{3}{8} & -\frac{3}{8} \\
0 & \frac{3}{7}
\end{pmatrix}
\tag{7.160}
$$

To illustrate the formation of hypotheses of no three-factor interaction, we write out linear combinations of cell means like those for contrasts for any one of the two-factor interactions, and state that those combinations are identical at the different levels of the third factor. Choosing the AC interaction, for example, we write out linear combinations for each level of B and state that they are equal, as follows:

$$
(\mu_{111} - \mu_{112} - \mu_{211} + \mu_{212}) = (\mu_{121} - \mu_{122} - \mu_{221} + \mu_{222}) = (\mu_{131} - \mu_{132} - \mu_{231} + \mu_{232})
$$
$$
(\mu_{211} - \mu_{212} - \mu_{311} + \mu_{312}) = (\mu_{221} - \mu_{222} - \mu_{321} + \mu_{322}) = (\mu_{231} - \mu_{232} - \mu_{331} + \mu_{332}).
$$

From each of these two sets of equalities we then form two contrasts, for example:

$$(\mu_{111} - \mu_{112} - \mu_{211} + \mu_{212}) = (\mu_{121} - \mu_{122} - \mu_{221} + \mu_{222}) = 0$$
$$(\mu_{121} - \mu_{122} - \mu_{221} + \mu_{222}) = (\mu_{131} - \mu_{132} - \mu_{231} + \mu_{232}) = 0$$
$$(\mu_{211} - \mu_{212} - \mu_{311} + \mu_{312}) = (\mu_{221} - \mu_{222} - \mu_{321} + \mu_{322}) = 0$$
$$(\mu_{221} - \mu_{222} - \mu_{321} + \mu_{322}) = (\mu_{231} - \mu_{232} - \mu_{331} + \mu_{332}) = 0.$$

These contrasts are then used to form the hypothesis test matrices which follows:

$$
C'_{ABC} =
\begin{pmatrix}
1 & 0 & 0 & 0 \\
-1 & 1 & 0 & 0 \\
0 & -1 & 0 & 0 \\
-1 & 0 & 0 & 0 \\
1 & -1 & 0 & 0 \\
0 & 1 & 0 & 0 \\
-1 & 0 & 1 & 0 \\
1 & -1 & -1 & 1 \\
0 & 1 & 0 & -1 \\
1 & 0 & -1 & 0 \\
-1 & 1 & 1 & -1 \\
0 & -1 & 0 & 1 \\
0 & 0 & -1 & 0 \\
0 & 0 & 1 & -1 \\
0 & 0 & 0 & 1 \\
0 & 0 & 1 & 0 \\
0 & 0 & -1 & 1 \\
0 & 0 & 0 & -1
\end{pmatrix}.
\tag{7.161}
$$

The data in Table 7.2 were analyzed in Chapter 5 using PROC GLM which is equivalent to using an unrestricted full rank cell means model. The RMGLM is appropriate whenever some terms are to be excluded from the cell means model. The terms are excluded by incorporating restrictions into the full-rank means model. When restrictions are used, the estimates of model parameters are no longer the cell means.

To illustrate the use of restrictions, we will assume that all interactions involving factor C are zero. Hence, the matrix of restriction R must restrict the AC, BC, and ABC interactions to zero. To accomplish this, the matrix R is constructed from the hypothesis test matrices C_{AC}, C_{BC}, and C_{ABC}:

$$
R = \begin{pmatrix} C_{AC} \\ C_{BC} \\ C_{ABC} \end{pmatrix}
\tag{7.162}
$$

and the matrix $A = I$. Given R, we may test A, B, C, and AB using the matrices C_A, C_B, C_C, and C_{AB}, respectively, as defined for the full-rank model.

PROC GLM does not allow one to impose general restrictions on the model parameters, except to exclude the intercept from the model: the intercept equal to zero restriction. To demonstrate the use of general restrictions on model parameters, we must use PROC REG with the MTEST statement. To test H_A, H_B, H_C and H_{AB} using PROC REG, we use the RESTRICT statement to formulate the matrix R which operates on the matrix of means B. The hypotheses are tested using the MTEST statement and the hypothesis test matrices C_A, C_B, C_C and C_{AB}. The PROC REG step and the PROC GLM step are included in Program 7_5.sas to complete the comparison with the less-than full-rank model.

To analyze the three-factor design using PROC REG, the data and design matrix must be input. The data are contained in the data file 7_5.dat where the variables u1 to u28 are the design matrix. The MTEST statements are used to test the unweighted tests H_A, H_B, H_C, and H_{AB}.

7.5.1 Results and Interpretation

PROC REG outputs the matrix $\widehat{B}_r$ given in (7.9) with $R_2 = I$ and $R_1 = R$ defined in (7.158) and $\Theta = 0$, a variable at a time. An estimate for each restriction is also calculated. The hypothesis and error test matrices H_r and E_r as defined in (7.18) and (7.20) are output for each MTEST statement. Output 7.5.1-7.5.3 includes the data for the analysis and only the unweighted test of A using both PROC REG and PROC GLM. As expected, the multivariate test criteria for PROC REG and PROC GLM agree. The purpose for using PROC REG in this example was to demonstrate the equivalence between the two approaches. Often using RMGLM with PROC REG to test hypotheses is easier than trying to determine estimable functions using contrasts in PROC GLM with standard Type I, II, III, and IV sum of squares do not yield the hypothesis test desired.

7.6 RESTRICTED INTRACLASS COVARIANCE DESIGN

Other important applications of the restricted model are to formulate an intraclass regression model, which permits different regression equations for each class, and to add restriction for each class to ensure parallelism of slopes. This results in the standard MANCOVA model. The MANCOVA design, discussed in Chapter 5, requires the slopes of the regression surfaces within each population to be identical: that

$$\Gamma_1 = \Gamma_2 = \cdots = \Gamma_I = \Gamma \tag{7.163}$$

for I independent populations. More generally letting B_i and Γ_i represent the slopes and intercepts for the regression of Y on Z for the ith population, we have the intraclass covariance model

$$\underset{J_i \times p}{\mathcal{E}(Y_i)} = \underset{J_i \times k\,k \times p}{X_i\ B_i} + \underset{J_i \times h\,h \times p}{Z_i\ \Gamma_i} \tag{7.164}$$

Output 7.5.1: Restricted Nonorthogonal Three-Factor MANOVA Design.

Obs	a	b	c	y1	y2	u1	u2	u3	u4	u5	u6	u7	u8	u9	u10	u11	u12	u13	u14	u15	u16	u17	u18
1	1	1	1	1	22	1	0	0	0	0	0	0	0	0	0	0	0	0	0	0	0	0	0
2	1	1	1	2	21	1	0	0	0	0	0	0	0	0	0	0	0	0	0	0	0	0	0
3	1	1	1	3	14	1	0	0	0	0	0	0	0	0	0	0	0	0	0	0	0	0	0
4	1	2	1	3	31	0	1	0	0	0	0	0	0	0	0	0	0	0	0	0	0	0	0
5	1	2	1	4	25	0	1	0	0	0	0	0	0	0	0	0	0	0	0	0	0	0	0
6	1	3	1	1	31	0	0	1	0	0	0	0	0	0	0	0	0	0	0	0	0	0	0
7	1	3	1	2	41	0	0	1	0	0	0	0	0	0	0	0	0	0	0	0	0	0	0
8	1	1	2	3	66	0	0	0	1	0	0	0	0	0	0	0	0	0	0	0	0	0	0
9	1	1	2	4	55	0	0	0	1	0	0	0	0	0	0	0	0	0	0	0	0	0	0
10	1	2	2	4	61	0	0	0	0	1	0	0	0	0	0	0	0	0	0	0	0	0	0
11	1	2	2	5	11	0	0	0	0	1	0	0	0	0	0	0	0	0	0	0	0	0	0
12	1	2	2	6	21	0	0	0	0	1	0	0	0	0	0	0	0	0	0	0	0	0	0
13	1	3	2	1	41	0	0	0	0	0	1	0	0	0	0	0	0	0	0	0	0	0	0
14	1	3	2	2	21	0	0	0	0	0	1	0	0	0	0	0	0	0	0	0	0	0	0
15	2	1	1	3	31	0	0	0	0	0	0	1	0	0	0	0	0	0	0	0	0	0	0
16	2	1	1	5	66	0	0	0	0	0	0	1	0	0	0	0	0	0	0	0	0	0	0
17	2	2	1	2	45	0	0	0	0	0	0	0	1	0	0	0	0	0	0	0	0	0	0
18	2	3	1	4	21	0	0	0	0	0	0	0	0	1	0	0	0	0	0	0	0	0	0
19	2	3	1	5	21	0	0	0	0	0	0	0	0	1	0	0	0	0	0	0	0	0	0
20	2	3	1	6	31	0	0	0	0	0	0	0	0	1	0	0	0	0	0	0	0	0	0
21	2	1	2	2	41	0	0	0	0	0	0	0	0	0	1	0	0	0	0	0	0	0	0
22	2	2	2	2	47	0	0	0	0	0	0	0	0	0	0	1	0	0	0	0	0	0	0
23	2	2	2	3	61	0	0	0	0	0	0	0	0	0	0	1	0	0	0	0	0	0	0
24	2	3	2	4	41	0	0	0	0	0	0	0	0	0	0	0	1	0	0	0	0	0	0
25	2	3	2	5	55	0	0	0	0	0	0	0	0	0	0	0	1	0	0	0	0	0	0
26	3	1	1	2	21	0	0	0	0	0	0	0	0	0	0	0	0	1	0	0	0	0	0
27	3	1	1	3	31	0	0	0	0	0	0	0	0	0	0	0	0	1	0	0	0	0	0
28	3	2	1	4	66	0	0	0	0	0	0	0	0	0	0	0	0	0	1	0	0	0	0
29	3	2	1	5	41	0	0	0	0	0	0	0	0	0	0	0	0	0	1	0	0	0	0
30	3	2	1	6	51	0	0	0	0	0	0	0	0	0	0	0	0	0	1	0	0	0	0
31	3	3	1	1	61	0	0	0	0	0	0	0	0	0	0	0	0	0	0	1	0	0	0
32	3	3	1	2	47	0	0	0	0	0	0	0	0	0	0	0	0	0	0	1	0	0	0
33	3	3	1	3	35	0	0	0	0	0	0	0	0	0	0	0	0	0	0	1	0	0	0
34	3	1	2	1	41	0	0	0	0	0	0	0	0	0	0	0	0	0	0	0	1	0	0
35	3	2	2	2	18	0	0	0	0	0	0	0	0	0	0	0	0	0	0	0	0	1	0
36	3	2	2	3	21	0	0	0	0	0	0	0	0	0	0	0	0	0	0	0	0	1	0
37	3	2	2	4	31	0	0	0	0	0	0	0	0	0	0	0	0	0	0	0	0	1	0
38	3	3	2	5	57	0	0	0	0	0	0	0	0	0	0	0	0	0	0	0	0	0	1
39	3	3	2	6	64	0	0	0	0	0	0	0	0	0	0	0	0	0	0	0	0	0	1
40	3	3	2	7	77	0	0	0	0	0	0	0	0	0	0	0	0	0	0	0	0	0	1

Output 7.5.2: Restricted Nonorthogonal Three-Factor MANOVA Design: Analysis Using PROC REG.

L Ginv(X'X) L' LB-cj			
5.2782407407	2.6712962963	-3.138888889	-56.34236111
2.6712962963	6.1351851852	0.8777777778	8.9430555556

Inv(L Ginv(X'X) L') Inv()(LB-cj)			
0.2430049584	-0.105805811	-0.855639554	-14.63770036
-0.105805811	0.2090627476	0.5156233191	7.8310089877

Error Matrix (E)	
57.866666667	226.69166667
226.69166667	7927.0822917

Hypothesis Matrix (H)	
3.138360179	52.820000701
52.820000701	894.75574822

Hypothesis + Error Matrix (T)	
61.005026846	279.51166737
279.51166737	8821.8380399

Eigenvectors	
0.046549	0.008552
0.130419	-0.007711

Eigenvalues
0.114302
0.000344

Multivariate Statistics and F Approximations					
S=2 M=-0.5 N=13.5					
Statistic	Value	F Value	Num DF	Den DF	Pr > F
Wilks' Lambda	0.88539287	0.91	4	58	0.4643
Pillai's Trace	0.11464651	0.91	4	60	0.4628
Hotelling-Lawley Trace	0.12939766	0.93	4	33.787	0.4591
Roy's Greatest Root	0.12905308	1.94	2	30	0.1619
NOTE: F Statistic for Roy's Greatest Root is an upper bound.					
NOTE: F Statistic for Wilks' Lambda is exact.					

Output 7.5.3: Restricted Nonorthogonal Three-Factor MANOVA Design: Analysis Using PROC GLM.

H = Type III SSCP Matrix for a		
	y1	y2
y1	3.138360179	52.820000701
y2	52.820000701	894.75574822

Characteristic Roots and Vectors of: E Inverse * H, where H = Type III SSCP Matrix for a E = Error SSCP Matrix			
		Characteristic Vector V'EV=1	
Characteristic Root	Percent	y1	y2
0.12905308	99.73	0.04946173	0.00908752
0.00034458	0.27	0.13044106	-0.00771247

MANOVA Test Criteria and F Approximations for the Hypothesis of No Overall a Effect H = Type III SSCP Matrix for a E = Error SSCP Matrix					
S=2 M=-0.5 N=13.5					
Statistic	Value	F Value	Num DF	Den DF	Pr > F
Wilks' Lambda	0.88539287	0.91	4	58	0.4643
Pillai's Trace	0.11464651	0.91	4	60	0.4628
Hotelling-Lawley Trace	0.12939766	0.93	4	33.787	0.4591
Roy's Greatest Root	0.12905308	1.94	2	30	0.1619
NOTE: F Statistic for Roy's Greatest Root is an upper bound.					
NOTE: F Statistic for Wilks' Lambda is exact.					

Table 7.4: Two-Factor Intraclass Design.

		B_1	B_2	B_3
	$y1$	$2, 2, 3, 3, 4$	$3, 4, 5, 5$	$5, 6, 8$
A_1	$y2$	$22, 21, 14, 25, 31$	$31, 25, 30, 31$	$31, 41, 50$
	z	$1, 2, 4, 4, 5$	$3, 5, 6, 5$	$6, 5, 6$
	$y1$	$3, 5, 5$	$5, 6, 6, 7$	$3, 4, 4$
A_2	$y2$	$31, 66, 45$	$45, 34, 31, 30$	$21, 21, 31$
	z	$2, 4, 3$	$5, 8, 4, 6$	$3, 3, 4$

where the $\sum_i J_i = N$, also called the heterogeneous data model. Given the model, three natural tests are evident. The test of coincidence

$$H_c : B_1 = B_2 = \cdots = B_I = B \tag{7.165}$$
$$\Gamma_1 = \Gamma_2 = \cdots = \Gamma_I = \Gamma \tag{7.166}$$

the test of parallelism,

$$H_p : \Gamma_1 = \Gamma_2 = \cdots = \Gamma_I = \Gamma \tag{7.167}$$

and given parallelism , the tests of equal intercepts

$$H_g : B_1 = B_2 = \cdots = B_I = B, \tag{7.168}$$

or the test of group differences given the MANCOVA model.

In Chapter 3, a single-factor nonorthogonal intraclass covariance model was discussed. In this chapter we illustrate the analysis of a multivariate two-factor nonorthogonal intraclass design and show how restrictions are used to produce a nonorthogonal MANCOVA design.

The data given in Table 7.4 are used for the analysis. For this example the combined design matrix has the general form:

$$X = \begin{pmatrix} X_1 & 0 & \cdots & 0 & Z_1 & 0 & \cdots & 0 \\ 0 & X_2 & \cdots & 0 & 0 & Z_2 & \cdots & 0 \\ \vdots & \vdots & \ddots & \vdots & \vdots & \vdots & \ddots & \vdots \\ 0 & 0 & \cdots & X_{IJ} & 0 & 0 & \cdots & Z_{IJ} \end{pmatrix}. \tag{7.169}$$

The design matrix X for the example is

$$X = \begin{pmatrix}
1 & 0 & 0 & 0 & 0 & 0 & 1 & 0 & 0 & 0 & 0 & 0 \\
1 & 0 & 0 & 0 & 0 & 0 & 2 & 0 & 0 & 0 & 0 & 0 \\
1 & 0 & 0 & 0 & 0 & 0 & 4 & 0 & 0 & 0 & 0 & 0 \\
1 & 0 & 0 & 0 & 0 & 0 & 4 & 0 & 0 & 0 & 0 & 0 \\
1 & 0 & 0 & 0 & 0 & 0 & 5 & 0 & 0 & 0 & 0 & 0 \\
0 & 1 & 0 & 0 & 0 & 0 & 0 & 3 & 0 & 0 & 0 & 0 \\
0 & 1 & 0 & 0 & 0 & 0 & 0 & 5 & 0 & 0 & 0 & 0 \\
0 & 1 & 0 & 0 & 0 & 0 & 0 & 6 & 0 & 0 & 0 & 0 \\
0 & 1 & 0 & 0 & 0 & 0 & 0 & 5 & 0 & 0 & 0 & 0 \\
0 & 0 & 1 & 0 & 0 & 0 & 0 & 0 & 6 & 0 & 0 & 0 \\
0 & 0 & 1 & 0 & 0 & 0 & 0 & 0 & 5 & 0 & 0 & 0 \\
0 & 0 & 1 & 0 & 0 & 0 & 0 & 0 & 6 & 0 & 0 & 0 \\
0 & 0 & 0 & 1 & 0 & 0 & 0 & 0 & 0 & 2 & 0 & 0 \\
0 & 0 & 0 & 1 & 0 & 0 & 0 & 0 & 0 & 4 & 0 & 0 \\
0 & 0 & 0 & 1 & 0 & 0 & 0 & 0 & 0 & 3 & 0 & 0 \\
0 & 0 & 0 & 0 & 1 & 0 & 0 & 0 & 0 & 0 & 5 & 0 \\
0 & 0 & 0 & 0 & 1 & 0 & 0 & 0 & 0 & 0 & 8 & 0 \\
0 & 0 & 0 & 0 & 1 & 0 & 0 & 0 & 0 & 0 & 4 & 0 \\
0 & 0 & 0 & 0 & 1 & 0 & 0 & 0 & 0 & 0 & 6 & 0 \\
0 & 0 & 0 & 0 & 0 & 1 & 0 & 0 & 0 & 0 & 0 & 3 \\
0 & 0 & 0 & 0 & 0 & 1 & 0 & 0 & 0 & 0 & 0 & 3 \\
0 & 0 & 0 & 0 & 0 & 1 & 0 & 0 & 0 & 0 & 0 & 4
\end{pmatrix} \tag{7.170}$$

and the parameter matrix B is

$$B = \begin{pmatrix} B_1 \\ B_2 \\ B_3 \\ B_4 \\ B_5 \\ B_6 \\ \Gamma_1 \\ \Gamma_2 \\ \Gamma_3 \\ \Gamma_4 \\ \Gamma_5 \\ \Gamma_6 \end{pmatrix} = \begin{pmatrix} \mu_{111} & \mu_{112} \\ \mu_{121} & \mu_{122} \\ \mu_{131} & \mu_{132} \\ \mu_{211} & \mu_{212} \\ \mu_{221} & \mu_{222} \\ \mu_{231} & \mu_{232} \\ \gamma_{111} & \gamma_{112} \\ \gamma_{121} & \gamma_{122} \\ \gamma_{131} & \gamma_{132} \\ \gamma_{211} & \gamma_{212} \\ \gamma_{221} & \gamma_{222} \\ \gamma_{231} & \gamma_{232} \end{pmatrix}. \tag{7.171}$$

The parameters in the model are associated with the cells in the design as shown in Table 7.5. Given the multivariate intraclass model, the first hypothesis of interest is the test of parallelism. Given parallelism, the intraclass model reduces to the MANCOVA model or from the original specification a restricted intraclass model. For our example, the test of parallelism is

$$H_p : \Gamma_1 = \Gamma_2 = \cdots = \Gamma_6 = \Gamma. \tag{7.172}$$

Table 7.5: Model Parameters.

	B_1	B_2	B_3
A_1	B_1, Γ_1	B_2, Γ_2	B_3, Γ_3
A_2	B_4, Γ_4	B_5, Γ_5	B_6, Γ_6

To test for parallelism, the hypothesis test matrix for the example is

$$C_{5\times 6} = \begin{pmatrix} 1 & 0 & 0 & 0 & 0 & -1 \\ 0 & 1 & 0 & 0 & 0 & -1 \\ 0 & 0 & 1 & 0 & 0 & -1 \\ 0 & 0 & 0 & 1 & 0 & -1 \\ 0 & 0 & 0 & 0 & 1 & -1 \end{pmatrix}. \tag{7.173}$$

The SAS program to perform the analysis is provided in Program 7_6.sas.

7.6.1 Results and Interpretation

The SAS code to perform the test of parallelism and coincidence is given in the first part of the program. PROC REG, used with MTEST statements, are used for the tests. The results of these tests are found in Output 7.6.1 which follows.

Because we did not reject the hypothesis of parallelism, we would not perform the test of coincidence. Instead, the matrix used to test for parallelism becomes the matrix of restrictions for the intraclass covariance model to create the nonorthogonal MANCOVA model. The SAS code to perform the MANCOVA analysis using the RMGLM to test the weighted and unweighted tests of main effects and interaction is found in Program 7_6.sas. Also included in the code is the test $H : \Gamma = 0$, labeled REG in the output. If $\Gamma = 0$, the MANCOVA model again becomes a MANOVA model. The output for the RMGLM (MANCOVA) model is displayed in Output 7.6.2-7.6.3.

While PROC GLM may be used to obtain the correct unweighted multivariate tests for the nonorthogonal MANCOVA design using the HTYPE=3 and ETYPE=3 option, the correct multivariate weighted tests are not obtained using HTYPE=1 and ETYPE=1 options. This is because the Type 1 sum of squares does not adjust for the covariate. One sees this from the univariate analysis in the output where we have displayed all sum of squares. Thus, one must be very careful when using SAS to analyze nonorthogonal designs that involve a covariate.

It is also possible to specify a model with pooled estimates for each level of a factor, called intrarow and intracolumn covariance models (Searle, 1987, p. 451). For example, if $\Gamma_1 = \Gamma_2 = \Gamma_3 = \Gamma_{1A}$ and $\Gamma_4 = \Gamma_5 = \Gamma_6 = \Gamma_{2A}$ we would have an intrarow covariance model. Of course, if $\Gamma_{1A} = \Gamma_{2A} = \Gamma$, we again have the standard MANCOVA model with restrictions. Similar models may be hypothesized for columns, or we may formulate an intrarow plus intracolumn model (Searle, 1987, p. 451).

Output 7.6.1: Restricted Intraclass Covariance Design: Testing Equality of Intraclass Regression Coefficients.

The REG Procedure
Model: MODEL1
Multivariate Test: Parallel

Multivariate Statistics and F Approximations					
S=2 M=1 N=3.5					
Statistic	Value	F Value	Num DF	Den DF	Pr > F
Wilks' Lambda	0.38546201	1.10	10	18	0.4126
Pillai's Trace	0.68602300	1.04	10	20	0.4446
Hotelling-Lawley Trace	1.40883655	1.20	10	11.059	0.3804
Roy's Greatest Root	1.26186988	2.52	5	10	0.0998
NOTE: F Statistic for Roy's Greatest Root is an upper bound.					
NOTE: F Statistic for Wilks' Lambda is exact.					

The REG Procedure
Model: MODEL1
Multivariate Test: Coin

Multivariate Statistics and F Approximations					
S=2 M=3.5 N=3.5					
Statistic	Value	F Value	Num DF	Den DF	Pr > F
Wilks' Lambda	0.07952611	2.29	20	18	0.0412
Pillai's Trace	1.33467229	2.01	20	20	0.0640
Hotelling-Lawley Trace	6.36615446	2.68	20	11.788	0.0426
Roy's Greatest Root	5.40200678	5.40	10	10	0.0067
NOTE: F Statistic for Roy's Greatest Root is an upper bound.					
NOTE: F Statistic for Wilks' Lambda is exact.					

Output 7.6.2: Restricted Intraclass Covariance Design: MANCOVA Analysis Using
PROC REG.

The REG Procedure
Model: MODEL1
Multivariate Test: A

Multivariate Statistics and Exact F Statistics					
S=1 M=0 N=6					
Statistic	Value	F Value	Num DF	Den DF	Pr > F
Wilks' Lambda	0.85235220	1.21	2	14	0.3268
Pillai's Trace	0.14764780	1.21	2	14	0.3268
Hotelling-Lawley Trace	0.17322393	1.21	2	14	0.3268
Roy's Greatest Root	0.17322393	1.21	2	14	0.3268

The REG Procedure
Model: MODEL1
Multivariate Test: B

Multivariate Statistics and F Approximations					
S=2 M=-0.5 N=6					
Statistic	Value	F Value	Num DF	Den DF	Pr > F
Wilks' Lambda	0.70266524	1.35	4	28	0.2761
Pillai's Trace	0.30143190	1.33	4	30	0.2813
Hotelling-Lawley Trace	0.41732193	1.43	4	15.818	0.2689
Roy's Greatest Root	0.40284784	3.02	2	15	0.0790
NOTE: F Statistic for Roy's Greatest Root is an upper bound.					
NOTE: F Statistic for Wilks' Lambda is exact.					

The REG Procedure
Model: MODEL1
Multivariate Test: AB

Multivariate Statistics and F Approximations					
S=2 M=-0.5 N=6					
Statistic	Value	F Value	Num DF	Den DF	Pr > F
Wilks' Lambda	0.36903633	4.52	4	28	0.0061
Pillai's Trace	0.72734922	4.29	4	30	0.0073
Hotelling-Lawley Trace	1.44857857	4.97	4	15.818	0.0086
Roy's Greatest Root	1.23752726	9.28	2	15	0.0024
NOTE: F Statistic for Roy's Greatest Root is an upper bound.					
NOTE: F Statistic for Wilks' Lambda is exact.					

The REG Procedure
Model: MODEL1
Multivariate Test: Reg

Multivariate Statistics and Exact F Statistics					
S=1 M=0 N=6					
Statistic	Value	F Value	Num DF	Den DF	Pr > F
Wilks' Lambda	0.65831474	3.63	2	14	0.0536
Pillai's Trace	0.34168526	3.63	2	14	0.0536
Hotelling-Lawley Trace	0.51903024	3.63	2	14	0.0536
Roy's Greatest Root	0.51903024	3.63	2	14	0.0536

Output 7.6.3: Restricted Intraclass Covariance Design: MANCOVA Analysis Using PROC REG — Weighted.

The REG Procedure
Model: MODEL1
Multivariate Test: Awt

Multivariate Statistics and Exact F Statistics					
S=1 M=0 N=6					
Statistic	Value	F Value	Num DF	Den DF	Pr > F
Wilks' Lambda	0.75008682	2.33	2	14	0.1336
Pillai's Trace	0.24991318	2.33	2	14	0.1336
Hotelling-Lawley Trace	0.33317901	2.33	2	14	0.1336
Roy's Greatest Root	0.33317901	2.33	2	14	0.1336

The REG Procedure
Model: MODEL1
Multivariate Test: Bwt

Multivariate Statistics and F Approximations					
S=2 M=-0.5 N=6					
Statistic	Value	F Value	Num DF	Den DF	Pr > F
Wilks' Lambda	0.70044260	1.36	4	28	0.2716
Pillai's Trace	0.30279252	1.34	4	30	0.2788
Hotelling-Lawley Trace	0.42305005	1.45	4	15.818	0.2631
Roy's Greatest Root	0.41183518	3.09	2	15	0.0753
NOTE: F Statistic for Roy's Greatest Root is an upper bound.					
NOTE: F Statistic for Wilks' Lambda is exact.					

Because the test of interaction is significant, both the weighted and unweighted tests of A and B are confounded. Some may consider the test REG ($\Gamma = 0$) to be nonsignificant ($p = 0.0536$). If $\Gamma = 0$, then the MANCOVA model is appropriate for the analysis. This may be accomplished by using the intraclass model and imposing the restriction that all $\Gamma_i = 0$. The SAS code to perform this restricted MANOVA analysis is included in Program 7.6.sas. Unweighted tests of A, B, and AB are performed using PROC REG and MTEST statements using PROC REG. The results are displayed in Output 7.6.4 and are seen to agree with the tests using PROC GLM.

7.7   GROWTH CURVE ANALYSIS

While the theory associated with the growth curve model is complicated, application of the model in practice is straightforward. A convenient form of the growth curve model given in (7.36) is

$$\mathcal{E}(Y_{n\times p}) = X_{n\times k}B_{k\times q}P_{q\times p} \tag{7.174}$$

$$\text{cov}(Y) = I_n \otimes \Sigma. \tag{7.175}$$

More specifically, suppose we have k groups with n_i subjects per group and suppose that the growth curve for the ith group is

$$\beta_{i0} + \beta_{i1}t + \beta_{i2}t^2 + \cdots + \beta_{i,q-1}t^{q-1}. \tag{7.176}$$

Letting $Y_{n\times p}$ represent the data matrix, we define $n = \sum_i n_i$, X, B, and P:

$$X_{n\times k} = \begin{pmatrix} 1_{n_1} & 0 & \cdots & 0 \\ 0 & 1_{n_2} & \cdots & 0 \\ \vdots & \vdots & \ddots & \vdots \\ 0 & 0 & \cdots & 1_{n_k} \end{pmatrix} \tag{7.177}$$

$$B_{k\times q} = \begin{pmatrix} \beta_{10} & \beta_{11} & \cdots & \beta_{1,q-1} \\ \beta_{20} & \beta_{21} & \cdots & \beta_{2,q-1} \\ \vdots & \vdots & \ddots & \vdots \\ \beta_{k0} & \beta_{k1} & \cdots & \beta_{k,q-1} \end{pmatrix} \tag{7.178}$$

$$P_{q\times p} = \begin{pmatrix} 1 & 1 & \cdots & 1 \\ t_1 & t_2 & \cdots & t_p \\ t_1^2 & t_2^2 & \cdots & t_p^2 \\ \vdots & \vdots & \ddots & \vdots \\ t_1^{q-1} & t_2^{q-1} & \cdots & t_p^{q-1} \end{pmatrix} \tag{7.179}$$

where 1_{n_i} is a vector of n_i $1's$, a unit vector.

Here we have assumed that the repeated observations follow a polynomial growth curve and that we have only one dependent variable observed over time. For simplicity, suppose we have two measurements at each time point where fore each variable

Output 7.6.4: Restricted Intraclass Covariance Design: Restricted Unweighted MANOVA Tests Using PROC REG.

The REG Procedure
Model: MODEL1
Multivariate Test: A

Multivariate Statistics and Exact F Statistics					
S=1 M=0 N=6.5					
Statistic	Value	F Value	Num DF	Den DF	Pr > F
Wilks' Lambda	0.90935857	0.75	2	15	0.4904
Pillai's Trace	0.09064143	0.75	2	15	0.4904
Hotelling-Lawley Trace	0.09967622	0.75	2	15	0.4904
Roy's Greatest Root	0.09967622	0.75	2	15	0.4904

The REG Procedure
Model: MODEL1
Multivariate Test: B

Multivariate Statistics and F Approximations					
S=2 M=-0.5 N=6.5					
Statistic	Value	F Value	Num DF	Den DF	Pr > F
Wilks' Lambda	0.47779057	3.35	4	30	0.0221
Pillai's Trace	0.52223783	2.83	4	32	0.0409
Hotelling-Lawley Trace	1.09290775	4.03	4	17.014	0.0178
Roy's Greatest Root	1.09285335	8.74	2	16	0.0027
NOTE: F Statistic for Roy's Greatest Root is an upper bound.					
NOTE: F Statistic for Wilks' Lambda is exact.					

The REG Procedure
Model: MODEL1
Multivariate Test: AB

Multivariate Statistics and F Approximations					
S=2 M=-0.5 N=6.5					
Statistic	Value	F Value	Num DF	Den DF	Pr > F
Wilks' Lambda	0.28662870	6.51	4	30	0.0007
Pillai's Trace	0.87267678	6.19	4	32	0.0008
Hotelling-Lawley Trace	1.93304375	7.12	4	17.014	0.0015
Roy's Greatest Root	1.58164312	12.65	2	16	0.0005
NOTE: F Statistic for Roy's Greatest Root is an upper bound.					
NOTE: F Statistic for Wilks' Lambda is exact.					

we have the growth curves:

$$\beta_{i0} + \beta_{i1}t + \beta_{i2}t^2 + \cdots + \beta_{i,q_1-1}t^{q_1-1} \tag{7.180}$$

$$\theta_{i0} + \theta_{i1}t + \theta_{i2}t^2 + \cdots + \theta_{i,q_2-1}t^{q_2-1}. \tag{7.181}$$

Then the parameter matrices are defined:

$$B_{k\times q} = \begin{pmatrix} \beta_{10} & \beta_{11} & \cdots & \beta_{1,q_1-1} & \theta_{10} & \theta_{11} & \cdots & \theta_{1,q_2-1} \\ \beta_{20} & \beta_{21} & \cdots & \beta_{2,q_1-1} & \theta_{20} & \theta_{21} & \cdots & \theta_{2,q_2-1} \\ \vdots & \vdots & \ddots & \vdots & \vdots & \vdots & \ddots & \vdots \\ \beta_{k0} & \beta_{k1} & \cdots & \beta_{k,q_1-1} & \theta_{k0} & \theta_{k1} & \cdots & \theta_{k,q_2-1} \end{pmatrix} \tag{7.182}$$

$$P_{(q_1+q_2)\times 2p} = \begin{pmatrix} 1 & 1 & \cdots & 1 & 0 & 0 & \cdots & 0 \\ t_1 & t_2 & \cdots & t_p & 0 & 0 & \cdots & 0 \\ t_1^2 & t_2^2 & \cdots & t_p^2 & 0 & 0 & \cdots & 0 \\ \vdots & \vdots & \ddots & \vdots & \vdots & \vdots & \ddots & \vdots \\ t_1^{q_1-1} & t_2^{q_1-1} & \cdots & t_p^{q_1-1} & 0 & 0 & \cdots & 0 \\ 0 & 0 & \cdots & 0 & 1 & 1 & \cdots & 1 \\ 0 & 0 & \cdots & 0 & t_1 & t_2 & \cdots & t_p \\ 0 & 0 & \cdots & 0 & t_1^2 & t_2^2 & \cdots & t_p^2 \\ \vdots & \vdots & \ddots & \vdots & \vdots & \vdots & \ddots & \vdots \\ 0 & 0 & \cdots & 0 & t_1^{q_2-1} & t_2^{q_2-1} & \cdots & t_p^{q_2-1} \end{pmatrix}$$

$$\tag{7.183}$$

where X remains defined as above, but the data matrix becomes $Y_{n\times 2p}$ where the first p columns are associated with the first variable and the second p columns are associated with the second variable.

For a growth curve model with only a single variable observed over time, we might be interested in testing hypotheses of the form $H : CBA = 0$. Some specific hypotheses of interest include:

1. the regression equations are coincident (equal) across groups or

2. the regression equations are parallel.

To test (1), the matrices C and A are:

$$C_{(k-1)\times k} = \begin{pmatrix} 1 & 0 & \cdots & -1 \\ 0 & 1 & \cdots & -1 \\ \vdots & \vdots & \ddots & \vdots \\ 0 & 0 & \cdots & -1 \end{pmatrix} \quad \text{and } A = I_q. \tag{7.184}$$

To test (2), the matrices are:

$$C_{(k-1)\times k} = \begin{pmatrix} 1 & 0 & \cdots & -1 \\ 0 & 1 & \cdots & -1 \\ \vdots & \vdots & \ddots & \vdots \\ 0 & 0 & \cdots & -1 \end{pmatrix} \quad \text{and } A_{q\times(q-1)} = \begin{pmatrix} 0 & 0 & \cdots & 0 \\ 1 & 0 & \cdots & 0 \\ \vdots & \vdots & \ddots & \vdots \\ 0 & 0 & \cdots & 1 \end{pmatrix}. \tag{7.185}$$

We may also evaluate the degree of the polynomial. For example, to test that the growth curves are of degree $q - 2$ or less, we select the matrices:

$$C = I_k \text{ and } A_{q \times 1} = \begin{pmatrix} 0 \\ \vdots \\ 0 \\ 1 \end{pmatrix}. \tag{7.186}$$

To test hypotheses $H : CBA = 0$ using the growth curve model: $\mathcal{E}(Y_{n \times p}) = X_{n \times k} B_{k \times q} P_{q \times p}$, Potthoff and Roy (1964) suggested the data matrix transformation

$$Y_0 = Y G^{-1} P' (P G^{-1} P)^{-1} \tag{7.187}$$

where $G_{p \times p}$ is any symmetric, positive definite matrix. Using the transformation, the GMANOVA (growth curve) model reduces to the MANOVA model. However, the analysis depends on the unknown matrix G which is critical when $q < p$, which occurs frequently in practice.

To avoid the choice of the matrix G, one may use the Rao-Khatri transformation which reduces the GMANOVA model to a MANCOVA model. To apply the Rao-Khatri method, a matrix $H_{p \times p} = (H_1, H_2)$ is created where $Y_0 = Y H_1$ and $Z = Y H_2$ and the matrix H is nonsingular. The matrix $H_1 (p \times q)$ is any matrix that forms a basis for the vector space generated by the rows of P, $\text{rank}(H_1) = q$ and $P H_1 = I_q$, and where the matrix $H_2 [p \times (p - q)]$ is created such that the $\text{rank}(H_2) = p - q$ and $P H_2 = 0_{p-q}$. Using the transformation, the GMANOVA model is reduced to the MANCOVA model

$$\mathcal{E}(Y_0 | Z) = X_{n \times k} B_{k \times q} + Z_{n \times h} \Gamma_{h \times q}. \tag{7.188}$$

To form the matrices, we set $G = I$ in the Potthoff and Roy transformation so that $H_1 = P'(PP')^{-1}$ and let H_2 be any $p - q$ linearly independent columns of the matrix $I - P'(PP')^{-1}P$. More conveniently, suppose P is a matrix of orthogonal polynomials; then $PP' = I$ so that $H_1 = P'$ and H_2 can be formed such that $H = (H_1, H_2)$ is an orthogonal matrix. That is select H_1 to be a matrix of orthogonalized polynomials of degree 0 to $q - 1$ and H_2 to be the similar matrix of the higher polynomials of order q to $p - 1$. For example, if

$$P = \begin{pmatrix} 0.577350 & 0.577350 & 0.577350 \\ -0.707107 & -0.707107 & -0.707107 \\ \hline 0.408248 & -0.816497 & 0.408248 \end{pmatrix} = \begin{pmatrix} P_1(q \times p) \\ \hline P_2[(p - q) \times p] \end{pmatrix}$$

so that $H_1 = P_1'$ and $H_2 = P_2'$. And, $Y_0 = Y H_1$ and $Z = Y H_2$, so that $\mathcal{E}(Y_0) = X B P H_1 = X B I_q$ and $\mathcal{E}(Z) = \mathcal{E}(Y H_2) = X B = 0$ as required.

A limitation of the application of the Rao-Khatri reduction lies in the determination of the number of (not necessarily all) higher order terms to include in the model as covariates. To test that the higher order polynomials are zero, the test of model adequacy using (7.77) and (7.78) may be employed using several combinations of the higher order terms while still trying to control the size of the multiple dependent tests.

For our first example of the growth curve model, we analyze the dog data of Grizzle and Allen (1969). The data consist of four groups of dogs (control and three treatments) with responses of coronary sinus potassium (MIL equivalents per liter) at times 1-13 minutes after coronary occlusion every 2 minutes. The SAS code to perform the analysis is provided in Program 7_7.sas.

7.7.1 Results and Interpretation

The data for this example are in the data set 7_7.dat and listed in Output 7.7.1. The data include corrections of errors in the published data set. For dog 2, the value 3.7 has be changed to 4.7. For dog 14, the value 2.9 has been changed to 3.9 and for dog 19, the value of 5.3 has been changed to 4.3. For dog 20, the value 3.7 has been changed to 4.0.

When performing a growth curve analysis, one must determine the degree of the polynomial to be fit to the means. To do this, one can plot the data and perform tests of fit. Using SAS code suggested by Freund and Littell (1991, p. 281) included in Program 7_7.sas, we summarize the data for this example using PROC SUMMARY and PROC GPLOT. The plot is displayed in Graph 7.7.1.

The plot suggests that a third degree polynomial should fit the data. To try to determine the order of the polynomial to fit to the data, we evaluate polynomial contrast fit to the seven time points. This is done in Program 7_7.sas using PROC GLM and the POLYNOMIAL option in the REPEATED statement. The results are displayed in Output 7.7.2.

The significance of the variable time_1, time_2, and time_3 suggests that a third degree polynomial is adequate for the data. Assuming a third degree polynomial is adequate for the data, the GMANOVA model for the example can be written as:

$$\mathcal{E}(Y_{36 \times 7}) = X_{36 \times 4} B_{4 \times 4} P_{4 \times 7} \qquad (7.189)$$

$$\text{cov}(Y) = I_{36} \otimes \Sigma, \qquad (7.190)$$

where the design matrix X is

$$X_{36 \times 4} = \begin{pmatrix} 1_9 & 0 & 0 & 0 \\ 0 & 1_{10} & 0 & 0 \\ 0 & 0 & 1_8 & 0 \\ 0 & 0 & 0 & 1_7 \end{pmatrix} \qquad (7.191)$$

and $P' = H_1$ consists of the first four columns of the 7×7 matrix of normalized orthogonal polynomials (constant, linear, quadratic, cubic, $\ldots, 6^{th}$). In Program 7_7.sas we generate $H = (H_1, H_2)$ using PROC IML code with the ORPOL function. The transformed data matrix $Y_0 = YH$ is created and named TRANS. It is this data set that is used throughout the rest of the analysis to perform the Rao-Khatri solution.

Output 7.7.1: Growth Curve Analysis — Grizzle and Allen Data.

Obs	group	y1	y2	y3	y4	y5	y6	y7
1	1	4.0	4.0	4.1	3.6	3.6	3.8	3.1
2	1	4.2	4.3	4.7	4.7	4.8	5.0	5.2
3	1	4.3	4.2	4.3	4.3	4.5	5.8	5.4
4	1	4.2	4.4	4.6	4.9	5.3	5.6	4.9
5	1	4.6	4.4	5.3	5.6	5.9	5.9	5.3
6	1	3.1	3.6	4.9	5.2	5.3	4.2	4.1
7	1	3.7	3.9	3.9	4.8	5.2	5.4	4.2
8	1	4.3	4.2	4.4	5.2	5.6	5.4	4.7
9	1	4.6	4.6	4.4	4.6	5.4	5.9	5.6
10	2	3.4	3.4	3.5	3.1	3.1	3.7	3.3
11	2	3.0	3.2	3.0	3.0	3.1	3.2	3.1
12	2	3.0	3.1	3.2	3.0	3.3	3.0	3.0
13	2	3.1	3.2	3.2	3.2	3.3	3.1	3.1
14	2	3.8	3.9	4.0	3.9	3.5	3.5	3.4
15	2	3.0	3.6	3.2	3.1	3.0	3.0	3.0
16	2	3.3	3.3	3.3	3.4	3.6	3.1	3.1
17	2	4.2	4.0	4.2	4.1	4.2	4.0	4.0
18	2	4.1	4.2	4.3	4.3	4.2	4.0	4.2
19	2	4.5	4.4	4.3	4.5	4.3	4.4	4.4
20	3	3.2	3.3	3.8	3.8	4.4	4.2	4.0
21	3	3.3	3.4	3.4	3.7	3.7	3.6	3.7
22	3	3.1	3.3	3.2	3.1	3.2	3.1	3.1
23	3	3.6	3.4	3.5	4.6	4.9	5.2	4.4
24	3	4.5	4.5	5.4	5.7	4.9	4.0	4.0
25	3	3.7	4.0	4.4	4.2	4.6	4.8	5.4
26	3	3.5	3.9	5.8	5.4	4.9	5.3	5.6
27	3	3.9	4.0	4.1	5.0	5.4	4.4	3.9
28	4	3.1	3.5	3.5	3.2	3.0	3.0	3.2
29	4	3.3	3.2	3.6	3.7	3.7	4.2	4.4
30	4	3.5	3.9	4.7	4.3	3.9	3.4	3.5
31	4	3.4	3.4	3.5	3.3	3.4	3.2	3.4
32	4	3.7	3.8	4.2	4.3	3.6	3.8	3.7
33	4	4.0	4.6	4.8	4.9	5.4	5.6	4.8
34	4	4.2	3.9	4.5	4.7	3.9	3.8	3.7
35	4	4.1	4.1	3.7	4.0	4.1	4.6	4.7
36	4	3.5	3.6	3.6	4.2	4.8	4.9	5.0

Graph 7.7.1: Growth Curve Plots of Group Means.

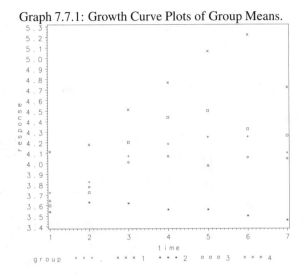

Using the Rao-Khatri model to perform the overall test of fit for the third degree polynomial model we have that

$$Y_0 = YH = (YH_1, YH_2) = (yt1, ty2, yt3, yt4, yt5, yt6, yt7);$$

the Rao-Khatri reduction becomes:

$$\mathcal{E}(Y_0|Z = YH_2) = XB + Z\Gamma;$$

using (7.77) and (7.78), we test $H : BH_2 = 0$ where $H_2(7 \times 3)$ corresponds to the normalized polynomials in H of 4th, 5th, and 6th degrees. This analysis is performed in Program7_7.sas using PROC GLM with the transformed data and the MODEL statement

```
model yt5-yt7/nouni; .
```

The adequacy of a second degree polynomial is also tested. The results of these test are included in the Output 7.7.3-7.7.4. Wilks' Λ criterion for the test of cubic fit is $\Lambda = 0.95296544$ with P-value 0.6563. Thus, we do not reject the hypothesis that the third degree polynomial adequately fits the data. Testing the adequacy of the second degree polynomial, the P-value is 0.1001. Thus, the overall tests of fit suggest a second degree polynomial is adequate for the data. However, this was not the case for the trend analysis using contrasts. Because it is usually better to overfit by one variable than to underfit, we have chosen to fit a third degree polynomial to these data.

Having found an adequate model to account for mean growth, one may next be interested in testing hypotheses regarding the structure of the curves across groups. For example, one may test that the curves are identical across groups: the hypothesis

Output 7.7.2: Growth Curve Analysis — Grizzle and Allen Data: Transformed Data Polynomials.

Contrast Variable: time_1

Source	DF	Type III SS	Mean Square	F Value	Pr > F
Mean	1	6.65307281	6.65307281	18.95	0.0001
group	3	4.93401885	1.64467295	4.69	0.0080
Error	32	11.23299504	0.35103109		

Contrast Variable: time_2

Source	DF	Type III SS	Mean Square	F Value	Pr > F
Mean	1	3.07031855	3.07031855	13.97	0.0007
group	3	0.93658036	0.31219345	1.42	0.2549
Error	32	7.03361475	0.21980046		

Contrast Variable: time_3

Source	DF	Type III SS	Mean Square	F Value	Pr > F
Mean	1	0.34404995	0.34404995	2.73	0.1080
group	3	1.27958796	0.42652932	3.39	0.0298
Error	32	4.02647685	0.12582740		

Contrast Variable: time_4

Source	DF	Type III SS	Mean Square	F Value	Pr > F
Mean	1	0.04159233	0.04159233	0.47	0.4982
group	3	0.41893921	0.13964640	1.58	0.2144
Error	32	2.83604455	0.08862639		

Contrast Variable: time_5

Source	DF	Type III SS	Mean Square	F Value	Pr > F
Mean	1	0.08284696	0.08284696	1.30	0.2622
group	3	0.07749041	0.02583014	0.41	0.7496
Error	32	2.03484755	0.06358899		

Contrast Variable: time_6

Source	DF	Type III SS	Mean Square	F Value	Pr > F
Mean	1	0.00956449	0.00956449	0.36	0.5531
group	3	0.02366098	0.00788699	0.30	0.8277
Error	32	0.85161649	0.02661302		

Output 7.7.3: Growth Curve Analysis — Grizzle and Allen Data: Test of Cubic Fit.

MANOVA Test Criteria and Exact F Statistics for the Hypothesis of No Overall Intercept Effect H = Type III SSCP Matrix for Intercept E = Error SSCP Matrix					
S=1 M=0.5 N=15.5					
Statistic	Value	F Value	Num DF	Den DF	Pr > F
Wilks' Lambda	0.95296544	0.54	3	33	0.6563
Pillai's Trace	0.04703456	0.54	3	33	0.6563
Hotelling-Lawley Trace	0.04935599	0.54	3	33	0.6563
Roy's Greatest Root	0.04935599	0.54	3	33	0.6563

Output 7.7.4: Growth Curve Analysis — Grizzle and Allen Data: Test of Quadratic Fit.

MANOVA Test Criteria and Exact F Statistics for the Hypothesis of No Overall Intercept Effect H = Type III SSCP Matrix for Intercept E = Error SSCP Matrix					
S=1 M=1 N=15					
Statistic	Value	F Value	Num DF	Den DF	Pr > F
Wilks' Lambda	0.78985294	2.13	4	32	0.1001
Pillai's Trace	0.21014706	2.13	4	32	0.1001
Hotelling-Lawley Trace	0.26605847	2.13	4	32	0.1001
Roy's Greatest Root	0.26605847	2.13	4	32	0.1001

of coincidence. The hypothesis test matrices for the example are

$$C_{3\times4} = \begin{pmatrix} 1 & -1 & 0 & 0 \\ 1 & 0 & -1 & 0 \\ 1 & 0 & 0 & -1 \end{pmatrix} \text{ and } A = I_4. \tag{7.192}$$

Another hypothesis is that the growth curves are parallel, that all polynomial components are coincident except for the intercept. The hypothesis test matrices for the example are:

$$C_{3\times4} = \begin{pmatrix} 1 & -1 & 0 & 0 \\ 1 & 0 & -1 & 0 \\ 1 & 0 & 0 & -1 \end{pmatrix} \text{ and } A = \begin{pmatrix} 0 & 0 & 0 \\ 1 & 0 & 0 \\ 0 & 1 & 0 \\ 0 & 0 & 1 \end{pmatrix}. \tag{7.193}$$

The results of these tests are displayed in Output 7.7.5-7.7.6. Since the P-values for both tests are small, we reject both the hypothesis of coincidence and the test of parallelism for this example using any of the multivariate criteria.

Finally, the code in Program 7_7.sas also outputs the columns of the parameter matrix B and differences in trends comparing group 1 with 2, 3, and 4 for which Grizzle and Allen (1969) establish Bonferroni type confidence sets.

Output 7.7.5: Growth Curve Analysis — Grizzle and Allen Data: Test of Group Differences.

MANOVA Test Criteria and F Approximations for the Hypothesis of No Overall GROUP Effect H = Type III SSCP Matrix for GROUP E = Error SSCP Matrix					
S=3 M=0 N=12					
Statistic	Value	F Value	Num DF	Den DF	Pr > F
Wilks' Lambda	0.37304366	2.60	12	69.081	0.0065
Pillai's Trace	0.73453561	2.27	12	84	0.0151
Hotelling-Lawley Trace	1.39788294	2.94	12	41.379	0.0049
Roy's Greatest Root	1.16390208	8.15	4	28	0.0002
NOTE: F Statistic for Roy's Greatest Root is an upper bound.					

Output 7.7.6: Growth Curve Analysis — Grizzle and Allen Data: Test of Parallelism.

MANOVA Test Criteria and F Approximations for the Hypothesis of No Overall parallel Effect on the Variables Defined by the M Matrix Transformation H = Contrast SSCP Matrix for parallel E = Error SSCP Matrix					
S=3 M=-0.5 N=12.5					
Statistic	Value	F Value	Num DF	Den DF	Pr > F
Wilks' Lambda	0.44098587	2.93	9	65.862	0.0056
Pillai's Trace	0.63307895	2.59	9	87	0.0109
Hotelling-Lawley Trace	1.10030957	3.22	9	39.366	0.0051
Roy's Greatest Root	0.91977535	8.89	3	29	0.0002
NOTE: F Statistic for Roy's Greatest Root is an upper bound.					

From (7.62),

$$\widehat{B} = (X'X)^{-1}X'YS^{-1}P'(PS^{-1}P')^{-1} \tag{7.194}$$

$$= \begin{pmatrix} 12.36 & 0.78 & -0.51 & -0.47 \\ 9.48 & -0.13 & -0.11 & 0.04 \\ 10.80 & 0.78 & -0.26 & -0.20 \\ 10.25 & 0.36 & -0.17 & 0.00 \end{pmatrix}. \tag{7.195}$$

This matrix does not agree with the matrix reported by Grizzle and Allen (1969). Their estimate of regression is

$$\beta = \begin{pmatrix} 4.671 & 0.148 & -0.056 & -0.190 \\ 3.582 & -0.025 & -0.012 & 0.018 \\ 4.084 & 0.146 & -0.029 & -0.080 \\ 3.874 & 0.068 & -0.018 & 0.001 \end{pmatrix} \text{ or} \tag{7.196}$$

$$\Xi = \begin{pmatrix} 32.701 & 4.135 & -4.680 & -1.142 \\ 25.071 & -0.687 & -0.995 & 0.110 \\ 28.585 & 4.101 & -2.402 & -0.480 \\ 27.116 & 1.908 & -1.549 & 0.004 \end{pmatrix} = \beta(Q_1'Q_1). \tag{7.197}$$

This difference is due to the fact that the matrix $P = Q_1$ of orthogonal polynomials was not normalized to unity in the Grizzle and Allen analysis where

$$Q' = \left(\begin{array}{cccc|ccc} 1 & -3 & 5 & -1 & 3 & -1 & 1 \\ 1 & -2 & 0 & 1 & -7 & 4 & -6 \\ 1 & -1 & -3 & 1 & 1 & -5 & 15 \\ 1 & 0 & -4 & 0 & 6 & 0 & -20 \\ 1 & 1 & -3 & 1 & 1 & 5 & 15 \\ 1 & 2 & 0 & -1 & -7 & -4 & 6 \\ 1 & 3 & 5 & 1 & 3 & 1 & 1 \end{array} \right) = \left(\begin{array}{c|c} Q_1 & Q_2 \end{array} \right). \tag{7.198}$$

In the SAS analysis each column of Q' is normalized. For additional detail, see Seber (1984, p. 491). Seber does not use the modified data set, but rather the published data which contain printing errors already noted.

7.8 MULTIPLE RESPONSE GROWTH CURVES

For our second application of the growth curve model, the dental data of Dr. Thomas Zullo, analyzed in Chapter 6, are again used. The data were partially discussed by Timm (1980b). The design consists of three dependent variables obtained over three time points. Our goal is to analyze the trend for the two groups over the three dependent variables simultaneously and to determine the best model for the study data. The SAS code for the multiple response growth curve analysis is provided in Program 7_8.sas.

Graph 7.8.1: Multiple Response Growth Curve Analysis — Zullo Data: Plot of Means.

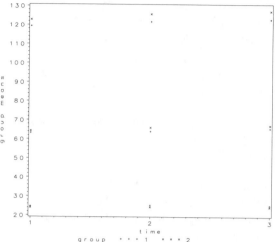

7.8.1   Results and Interpretation

The data file mmm.dat, which was used for the mixed model analysis, must first be reorganized to nest the time dimension within each variable. This was accomplished in Program 7_8.sas using a DO loop in the DATA step. Next, using PROC SUMMARY and GPLOT, the means are plotted for each variable against time. The plot is displayed in Graph 7.8.1 suggesting a first degree (linear) polynomial should be fit to the mean data over time.

Assuming that $p = q$, the parameter matrix has the form

$$B = \begin{pmatrix} \beta_{10} & \beta_{11} & \beta_{12} & \theta_{10} & \theta_{11} & \theta_{12} & \xi_{10} & \xi_{11} & \xi_{12} \\ \beta_{20} & \beta_{21} & \beta_{22} & \theta_{20} & \theta_{21} & \theta_{22} & \xi_{20} & \xi_{21} & \xi_{22} \end{pmatrix} \qquad (7.199)$$

where B is partitioned for variables 1, 2, and 3 represented by β, θ, and ξ, respectively. If $q < p$ for all variables, and if we assume a linear model as suggested by the plot, then the parameter matrix becomes:

$$B = \begin{pmatrix} \beta_{10} & \beta_{11} & \theta_{10} & \theta_{11} & \xi_{10} & \xi_{11} \\ \beta_{20} & \beta_{21} & \theta_{20} & \theta_{21} & \xi_{20} & \xi_{21} \end{pmatrix}. \qquad (7.200)$$

For the GMANOVA model we may fit different curves to each variable. For example, assume variable one is fit best by a quadratic polynomial and that the second and third variables follow a linear model. Then the parameter matrix would have the form:

$$B = \begin{pmatrix} \beta_{10} & \beta_{11} & \beta_{12} & \theta_{10} & \theta_{11} & \xi_{10} & \xi_{11} \\ \beta_{20} & \beta_{21} & \beta_{22} & \theta_{20} & \theta_{21} & \xi_{20} & \xi_{21} \end{pmatrix}. \qquad (7.201)$$

If we fit the full model, $p = q$, quadratic trend model for each variable, the normalized matrix of orthogonal polynomials has the structure:

$$P' = \begin{pmatrix} P_1' & 0 & 0 \\ 0 & P_2' & 0 \\ 0 & 0 & P_3' \end{pmatrix} \text{ where } P_1' = P_2' = P_3' = \begin{pmatrix} 0.577 & -0.707 & 0.408 \\ 0.577 & 0.000 & -0.816 \\ 0.577 & 0.707 & 0.408 \end{pmatrix}.$$

$$(7.202)$$

The matrix P_i' is constructed by normalizing the columns of the orthogonal polynomial matrix

$$Q' = \begin{pmatrix} 1 & -1 & 1 \\ 1 & 0 & 2 \\ 1 & 1 & 1 \end{pmatrix}.$$

The matrix Q' is generated by the linear and quadratic polynomials $(x - 2)$ and $x^2 - 12x + 10$. For Q' the polynomial has the form

$$y = \alpha_0 + \alpha_1(x - 2) + \alpha_3(3x^2 - 12x + 10)$$

for each variable. Normalizing the polynomial, we have that

$$y = \frac{\alpha_0}{\sqrt{3}} + \frac{\alpha_1}{\sqrt{2}}(x - 2) + \frac{\alpha_3}{\sqrt{6}}(3x^2 - 12x + 10)$$

$$= \alpha_0' + \alpha_1'(x - 2) + \alpha_3'(3x^2 - 12x + 10).$$

In Program 7_8.sas, PROC IML is used to create a new transformed data set using normalized orthogonal polynomials. The transformed data $Y_0 = YP$. It is this transformed data set that is used for the analysis in the remainder of Program 7_8.sas. The test of model adequacy or fit is next performed. PROC GLM is used to test that a linear trend is appropriate for each of the three variables by testing that the coefficients associated with yt3, yt6, and yt9 are equal to zero.

The result of this test is found in Output 7.8.1. We see from the output labeled No Overall Intercept Effect that we reject the null hypothesis of linear fit ($p = 0.0052$). Upon investigating the results of the univariate test of fit of linear trend for each variable found in the output, we see that the linear trend is adequate for variables two ($p = 0.8263$) and three ($p = 0.3239$), but not for variable one ($p = 0.0011$). Next PROC GLM is used to test the overall fit of a model assuming a linear trend for variables two and three and a quadratic trend for variable one. This is done with PROC GLM by testing that yt6 and yt9 are equal to zero. The results are displayed in Output 7.8.2. Since the P-value is 0.6205, this model appears to fit the data best.

Assuming a quadratic model for each variable, Program 7_8.sas also contains the SAS code to a test that the quadratic beta weights are zero. The results are displayed in Output 7.8.3. The joint test is not conclusive regarding model fit since the P-values range from 0.0052 to 0.0526. However, they suggest that the quadratic terms should be retained for all variables.

Output 7.8.1: Multiple Response Growth Curve Analysis — Zullo Data: Test of Fit — Linear.

Dependent Variable: YT3

Source	DF	Sum of Squares	Mean Square	F Value	Pr > F
Model	1	4.89814815	4.89814815	15.32	0.0011
Error	17	5.43518519	0.31971678		
Uncorrected Total	18	10.33333333			

Dependent Variable: YT6

Source	DF	Sum of Squares	Mean Square	F Value	Pr > F
Model	1	0.05787037	0.05787037	0.05	0.8263
Error	17	19.81712963	1.16571351		
Uncorrected Total	18	19.87500000			

Dependent Variable: YT9

Source	DF	Sum of Squares	Mean Square	F Value	Pr > F
Model	1	0.37925926	0.37925926	1.03	0.3239
Error	17	6.24740741	0.36749455		
Uncorrected Total	18	6.62666667			

MANOVA Test Criteria and Exact F Statistics for the Hypothesis of No Overall Intercept Effect H = Type III SSCP Matrix for Intercept E = Error SSCP Matrix					
S=1 M=0.5 N=6.5					
Statistic	Value	F Value	Num DF	Den DF	Pr > F
Wilks' Lambda	0.43811339	6.41	3	15	0.0052
Pillai's Trace	0.56188661	6.41	3	15	0.0052
Hotelling-Lawley Trace	1.28251413	6.41	3	15	0.0052
Roy's Greatest Root	1.28251413	6.41	3	15	0.0052

Output 7.8.2: Multiple Response Growth Curve Analysis — Zullo data: Test of Fit — Quadratic (1)/Linear (2).

MANOVA Test Criteria and Exact F Statistics for the Hypothesis of No Overall Intercept Effect H = Type III SSCP Matrix for Intercept E = Error SSCP Matrix					
S=1 M=0 N=7					
Statistic	Value	F Value	Num DF	Den DF	Pr > F
Wilks' Lambda	0.94208276	0.49	2	16	0.6205
Pillai's Trace	0.05791724	0.49	2	16	0.6205
Hotelling-Lawley Trace	0.06147786	0.49	2	16	0.6205
Roy's Greatest Root	0.06147786	0.49	2	16	0.6205

Output 7.8.3: Multiple Response Growth Curve Analysis — Zullo Data: Test of Quadratic Beta Weights.

MANOVA Test Criteria and F Approximations for the Hypothesis of No Overall order Effect on the Variables Defined by the M Matrix Transformation H = Contrast SSCP Matrix for order E = Error SSCP Matrix					
S=2 M=0 N=6					
Statistic	Value	F Value	Num DF	Den DF	Pr > F
Wilks' Lambda	0.40079253	2.70	6	28	0.0337
Pillai's Trace	0.64651410	2.39	6	30	0.0526
Hotelling-Lawley Trace	1.37702373	3.12	6	17	0.0300
Roy's Greatest Root	1.28518252	6.43	3	15	0.0052
NOTE: F Statistic for Roy's Greatest Root is an upper bound.					
NOTE: F Statistic for Wilks' Lambda is exact.					

Since our analysis is inconclusive regarding the degree of the polynomial to use, we analyze the data assuming $p = q$ for all variables and also assuming $p < q$ for the second and third variables. Program 7_8.sas contains the SAS code to perform the test of coincidence, the test of parallelism, and to estimate the model parameter matrix B for both models.

With $p = q = 3$, the parameter matrix $\widehat{B}$ is

$$\widehat{B} = \begin{pmatrix} 210.155 & 2.553 & -0.476 & 111.332 & 1.689 & 0.340 & 43.269 & 0.314 & -0.227 \\ 216.955 & 3.182 & -0.567 & 113.994 & 2.043 & -0.227 & 41.864 & 0.204 & -0.064 \end{pmatrix}$$

using normalized orthogonal polynomials. For test of parallelism:

$$H_p : \begin{pmatrix} \beta_{11} & \beta_{12} & \theta_{11} & \theta_{12} & \xi_{11} & \xi_{12} \end{pmatrix} = \begin{pmatrix} \beta_{21} & \beta_{22} & \theta_{21} & \theta_{22} & \xi_{21} & \xi_{22} \end{pmatrix},$$
(7.203)

the hypothesis test matrices are

$$C = \begin{pmatrix} 1 & -1 \end{pmatrix} \text{ and } A = I_3 \otimes A_1 \text{ where } A_1 = \begin{pmatrix} 0 & 0 \\ 1 & 0 \\ 0 & 1 \end{pmatrix}.$$

Using Wilks Λ criterion, $\Lambda = 0.583$ with P-value 0.3292. Given parallelism, we next test for coincidence. For this test, $C = (1, -1)$ and $A = I_3$. For Wilks' Λ criterion, $\Lambda = 0.422$ with P-value 0.3965. Hence, the regression curves appear parallel.

For $q < p$, the parameter matrix estimate for

$$B = \begin{pmatrix} \beta_{10} & \beta_{11} & \beta_{12} & \theta_{10} & \theta_{11} & \xi_{10} & \xi_{11} \\ \beta_{20} & \beta_{21} & \beta_{22} & \theta_{20} & \theta_{21} & \xi_{20} & \xi_{21} \end{pmatrix}$$
(7.204)

is the matrix

$$\widehat{B} = \begin{pmatrix} 210.430 & 2.695 & -0.565 & 112.328 & 1.846 & 45.505 & -0.054 \\ 217.738 & 3.218 & -0.571 & 114.353 & 2.036 & 42.009 & 0.094 \end{pmatrix}.$$
(7.205)

Using Wilks' Λ criterion, we do not reject the null hypothesis of coincidence since from the output we see that $\Lambda = 0.461$ with P-value 0.3456. Likewise, we fail to reject the null hypothesis of parallelism, $\Lambda = 0.637$ with a P-value of 0.2513. These results are consistent with the $p = q$ model analysis. However, we have not conclusively determined whether the full $p = q$ model or the $p < q$ model best fits the data. To determine which model is best, one may replicate the study and use each model to obtain predicted values using the estimates from the first study. The model with the highest correlation between the vector of predicted values and the observation vector of the replicated data matrix is usually considered "best".

Alternatively, the best model should have the smallest residuals and standard error for the estimates of the coefficients in the parameter matrix B. Recall from matrix algebra that the Euclidean vector norm squared of a matrix R is defined as

$$\|R\|^2 = \mathrm{tr}(R'R).$$

If the matrix R is a matrix of residuals, one may compare the norm squared values for two regression models. The model with the smaller value may be considered better in that its index of fit is smaller. For our example, the index for the model with $p = q$, $\|R\|^2 = 5075$; however, the index for the $p < q$ model is 4304. Thus, based upon the residual Euclidean vector norm squared index of fit, we conclude that the model $p < q$ is marginally better.

7.9   SINGLE GROWTH CURVE

In the previous example, we had difficulty establishing the best covariates for the analysis. For our last example, we consider the Zullo dental data as if it were obtained from a single population and we show how the REG procedure may be used to help determine the best set of covariates. Program 7_9.sas contains the code for the analysis.

In program 7_9.sas, the first several sections of the code is the same as in Program 7_8.sas. Recall from the prior example that both variables yt6 and yt9 were candidates for covariates and that the model with $p < q$ was marginally better than the model with $p = q$ using the Euclidean vector norm as a fit index. We now address the problem of covariate selection. Again, the best model should have the smallest residual and standard errors for the estimates of the coefficients of the parameter matrix B. In addition, as in the prior example, it should have the smallest index of fit as estimated using the Euclidean vector norm using the matrix of residuals.

7.9.1   Results and Interpretation

From the output, we summarize the estimates $\widehat{B}$ and standard errors (s.e) for several models:

$$\begin{pmatrix} \widehat{B} \\ (s.e.) \end{pmatrix} = \begin{pmatrix} 213.555 & 2.868 & -0.522 & 112.663 & 1.866 & 0.057 \\ (2.892) & (0.197) & (0.133) & (1.903) & (0.176) & (0.254) \end{pmatrix}$$

$$\begin{pmatrix} 42.567 & 0.259 & -0.145 \\ (2.297) & (0.247) & (0.142) \end{pmatrix}.$$

Using the variable yt6 as a covariate, we have that:

$$\begin{pmatrix} \widehat{B} \\ (s.e.) \end{pmatrix} = \begin{pmatrix} 213.500 & 2.874 & -0.526 & | & 112.680 & 1.877 & | & 42.651 & 0.256 \\ (2.975) & (0.197) & (0.083) & | & (1.962) & (0.173) & | & (2.338) & (0.255) \end{pmatrix}.$$

Using the variable yt9 as a covariate, we have that:

$$\begin{pmatrix} \widehat{B} \\ (s.e.) \end{pmatrix} = \begin{pmatrix} 214.273 & 2.965 & -0.566 & | & 113.382 & 1.940 & | & 43.654 & 0.219 \\ (2.977) & (0.179) & (0.134) & | & (1.874) & (0.170) & | & (2.157) & (0.085) \end{pmatrix}.$$

Finally, using both variable (yt6,yt9) as covariates:

$$\begin{pmatrix} \widehat{B} \\ (s.e.) \end{pmatrix} = \begin{pmatrix} 214.236 & 2.967 & -0.568 & | & 113.382 & 1.945 & | & 43.685 & 0.023 \\ (3.056) & (0.183) & (0.137) & | & (1.937) & (0.168) & | & (2.209) & (0.087) \end{pmatrix}.$$

Reviewing the standard errors, we see that the model with only one covariate (yt9) appears to be better than the model using other subsets. In addition, comparing the Euclidean vector norm squares for the model with only one covariate, the model with covariate yt9 versus yt6 has a smaller value: 5649 versus 5260. Thus, we conclude that the mode using one covariate is best. Rao (1987) suggests other procedures for selecting the best subset of covariates in the development of a GMANOVA model.

While the Rao-Khatri approach to fitting polynomial growth curves is straight forward, there are serious limitations. These include:

1. The covariates are not independent of the treatment.

2. Different covariates may not be applied to different treatments.

3. The set of covariates may vary from sample to sample.

4. The observation vector may not contain missing values.

More importantly, this approach does not model individual effects, but rather group effects; it does not work well with long series, and does not provide a mechanism for modeling the structure of Σ (Jöreskog, 1973). The mixed linear model discussed in Chapter 11 provides a mechanism for modeling Σ using PROC MIXED. Furthermore, Galecki (1994) discusses a general model that permits on to specify structured covariance matrices in the analysis of double multivariate linear models. A comprehensive review of growth curve modeling using standard statistical model is presented by Rosen (1991). We discuss some new approach using structural equation modeling in Chapter 13.

CHAPTER 8

SUR Model and Restricted GMANOVA Model

8.1 INTRODUCTION

In this chapter, we show that the restricted multivariate general linear model (RMGLM) is really a mixture of the multivariate analysis of variance (MANOVA) and generalized MANOVA (GMANOVA) models. For this model, approximate likelihood ratio tests are discussed. More generally, the MANOVA-GMANOVA model is shown to be a sum-profiles model which may be analyzed using the completely GMANOVA (CGMANOVA) model. Because likelihood ratio tests for these models are difficult to construct, we show how the seemingly unrelated regression (SUR) model may be used to estimate model parameters and to test hypotheses. We also show how the SUR model may be used to analyze the restricted GMANOVA model.

8.2 MANOVA–GMANOVA MODEL

Our discussion in Chapter 7 began with the analysis of the multivariate general linear model (GLM) with restrictions where we were interested in testing the general hypothesis $H : CBA = \Gamma$ using the model

$$\Omega_0 : Y_{n \times p} = X_{n \times k} B_{k \times p} + U_{n \times p} \qquad R_{1(r_1 \times k)} B_{k \times p} R_{2(p \times r_2)} = \Theta_0 \qquad (8.1)$$

where R_1, R_2, C and A are known, the $\mathrm{rank}(R_1) = r_1 \leq k$, $\mathrm{rank}(R_2) = r_2 \leq p$, C is $(g \times k)$ with $\mathrm{rank}(C) = g \leq k$, and A is $(p \times u)$ with $\mathrm{rank}(A) = u \leq p$. To test $H : CBA = \Gamma$ we assumed in Chapter 7 that $A = R_2$ since this eliminated the association of R_2 with Σ. Testing the restriction when Σ and R_2 are associated led to the GMANOVA model. Using (8.1), we may incorporate the restrictions into the model. To do this, we expand the matrices R_1 and R_2 to full rank using Theorem 7.5 and construct new matrices R and P such that

$$R_{k \times k} = \begin{pmatrix} R_1 \\ \widetilde{R}_1 \end{pmatrix} \qquad\qquad \mathrm{rank}(R) = k \qquad (8.2)$$

$$P_{p \times p} = \begin{pmatrix} R_2 & \widetilde{R}_2 \end{pmatrix} \qquad\qquad \mathrm{rank}(P) = p \qquad (8.3)$$

Since R^{-1} and P^{-1} exist, XB may be written as

$$XB = XR^{-1}RBPP^{-1}$$

$$= XR^{-1}\begin{pmatrix} R_1 BR_2 & R_1 B\widetilde{R}_2 \\ \widetilde{R}_1 BR_2 & \widetilde{R}_1 B\widetilde{R}_2 \end{pmatrix} P^{-1}. \tag{8.4}$$

Letting $X^* = XR^{-1} = \begin{pmatrix} X_1^* & X_2^* \end{pmatrix}$ and $P^{-1} = \begin{pmatrix} P^{11} \\ P^{21} \end{pmatrix}$, (8.4) becomes

$$XB = \begin{pmatrix} X_1^* & X_2^* \end{pmatrix} \begin{pmatrix} 0 & \beta_{12} \\ \beta_{21} & \beta_{22} \end{pmatrix} \begin{pmatrix} P^{11} \\ P^{21} \end{pmatrix}$$

$$= X_1^* \beta_{12} P^{21} + X_2^* \begin{pmatrix} \beta_{21} & \beta_{22} \end{pmatrix} P^{-1}. \tag{8.5}$$

Finally, if we let $X_1 = X_1^*$, $\Theta = \beta_{12} = R_1 B\widetilde{R}_2$, $X_2 = P^{21}$, $Z = X_2^*$, and $\Gamma_0 = \begin{pmatrix} \beta_{21} & \beta_{22} \end{pmatrix} P^{-1} = \begin{pmatrix} \widetilde{R}_1 BR_2 & \widetilde{R}_1 B\widetilde{R}_2 \end{pmatrix} P^{-1}$, the restricted multivariate GLM has the form

$$\mathcal{E}(Y|Z) = X_1\Theta X_2 + Z\Gamma_0. \tag{8.6}$$

This is a mixture of the GMANOVA and the MANOVA models, Chinchilli and El-swick (1985). The matrix $X_1(n \times m_1)$ is a known matrix of rank $m_1 \leq n$, $X_2(q \times p)$ is a matrix of rank $q \leq p$, $\Theta(m_1 \times q)$ is a matrix of unknown parameters, $Z(n \times m_2)$ is a known matrix of rank $m_2 \leq n$ and $\Gamma_0(m_2 \times p)$ is a matrix of unknown parameters, where $q = p - m_2$ and $n > m_1 + m_2 + p$.

From Chapter 7, we saw how to test hypotheses of the form $H : CBA = \Gamma = 0$ for the GMANOVA model. Recall that the maximum likelihood (ML) estimate of B and Σ are

$$\widehat{B} = (X_1'X_1)^{-1}X_1'YE^{-1}X_2'(X_2E^{-1}X_2')^{-1} \tag{8.7}$$

$$n\widehat{\Sigma} = (Y - X_1\widehat{B}X_2)'(Y - X_1\widehat{B}X_2)$$

$$= (Y - X_1\widehat{B}X_2)'(I_n - X_1(X_1'X_1)^{-1}X_1')(Y - X_1\widehat{B}X_2) +$$

$$(Y - X_1\widehat{B}X_2')(X_1(X_1'X_1)^{-1}X_1')(Y - X_1\widehat{B}X_2)$$

$$= E + W'Y'X_1(X_1'X_1)^{-1}X_1'YW \tag{8.8}$$

where $E = Y'(I_n - X_1(X_1'X_1)^{-1}X_1')Y$ and $W = I_p - E^{-1}X_2'(X_2E^{-1}X_2')^{-1}X_2$. To obtain the likelihood ratio (LR) test, we employed Rao's conditional model which reduced the GMANOVA model $\mathcal{E}(Y) = X_1BX_2$ to a multivariate analysis of co-variance (MANCOVA) model using the transformation

$$Y_0 = YH_1 = YG^{-1}X_2'(X_2G^{-1}X_2')^{-1}$$

and incorporating a set of covariates $Z = YH_2 = Y(I - H_1X_2) = Y(I - G^{-1}X_2'(X_2G^{-1}X_2')^{-1}X_2)$ into the model. Khatri using the MLEs in (8.8) and a conditional argument derived a LR test of $H : CBA = \Gamma$ (Srivastava & Khatri, 1979, p. 193). In both cases the LR test for the growth curve hypothesis is to reject H if

$$\Lambda = \frac{|\tilde{E}|}{|\tilde{E} + \tilde{H}|} < U_{u,\nu_h,\nu_e}^{1-\alpha} \tag{8.9}$$

where

$$\tilde{E} = A'(X_2 E^{-1} X_2')A \tag{8.10}$$
$$\tilde{H} = (C\hat{B}A)'(CR^{-1}C')^{-1}(C\hat{B}A) \tag{8.11}$$

with E and $\hat{B}$ defined in (8.8), and R^{-1} defined as

$$R^{-1} = (X_1'X_1)^{-1} + (X_1'X_1)^{-1}X_1'Y[E^{-1} - E^{-1}X_2'(X_2 E^{-1} X_2')^{-1}]$$
$$Y'X_1(X_1'X_1)^{-1}. \tag{8.12}$$

For this LR test, $u = \text{rank}(A)$, $\nu_h = \text{rank}(C)$, and $\nu_e = n - \text{rank}(X_1) - p + \text{rank}(X_2)$.

To obtain LR tests of the parameters in the mixed MANOVA-GMANOVA model (8.6), Chinchilli and Elswick (1985) derive the ML estimate of Θ, Γ_0, and Σ. Given model (8.5), they form the augmented matrix $X = (X_1, Z)$ and associate

$$(X'X) = \begin{pmatrix} X_1'X_1 & X_1'Z \\ Z'X_1 & Z'Z \end{pmatrix} = \begin{pmatrix} X_{11} & X_{12} \\ X_{21} & X_{22} \end{pmatrix} \tag{8.13}$$

with

$$(X'X)^{-1} = \begin{pmatrix} X^{11} & X^{12} \\ X^{21} & X^{22} \end{pmatrix} \tag{8.14}$$

where the partitioned matrices X^{ij} are defined in (3.12). Then, the ML estimate of the model parameters are

$$\hat{\Theta} = (X^{11}X_1' + X^{12}Z')YE^{-1}X_2'(X_2 E^{-1} X_2')^{-1} \tag{8.15}$$
$$\hat{\Gamma}_0 = (X^{21}X_1' + X^{22}Z')YE^{-1}X_2'(X_2 E^{-1} X_2')^{-1}X_2 + (Z'Z)^{-1}Z'YW \tag{8.16}$$
$$n\hat{\Sigma} = E + W'Y'(X(X'X)^{-1}X' - Z(Z'Z)^{-1}Z')YW \tag{8.17}$$

where $E = Y'(I_n - X(X'X)^{-1}X')Y$ and $W = I_p - E^{-1}X_2'(X_2 E^{-1} X_2')^{-1}X_2$.

For the MANOVA-GMANOVA model (8.6), three hypotheses are of interest

$$H_1 : C_1 \Theta A_1 = 0 \tag{8.18}$$
$$H_2 : C_2 \Gamma_0 = 0 \tag{8.19}$$
$$H_3 : H_1 \cap H_2 = 0 \tag{8.20}$$

where $C_{1(g_1 \times m_1)}$ of rank $g_1 \leq m_1$, $\Theta_{m_1 \times g}$, $A_{1(g \times u)}$ of rank $u \leq g$, $C_{2(g_2 \times m_2)}$ of rank $g_2 \leq m_2$ and $\Gamma_{0(m_2 \times p)}$.

To obtain the LR test for H_1 under the mixed MANOVA-GMANOVA model, one may employ the direct approach due to Khatri or the conditional approach due to Rao. Using the conditional approach, the data matrix Y is transformed to the matrix

$$Y_0 = \begin{pmatrix} Y_1 & Y_2 \end{pmatrix} = \begin{pmatrix} YH_1' & YH_2' \end{pmatrix} \tag{8.21}$$

(Chinchilli & Elswick, 1985) where

$$\underset{p x q}{H_1'} = X_2'(X_2 X_2')^{-1} \qquad \underset{p \times (p-q)}{H_2'} = I_p - X_2'(X_2 X_2')^{-1} X_2 \tag{8.22}$$

so that $H_1 X_2' = I$, $H_2 X_2' = 0$, and H_2' is any $(p-q)$ columns of $I_p - H_1' X_2$. Assuming Y has a matrix multivariate normal distribution, by Definition 5.1 we may write the density for the MANOVA-GMANOVA model as

$$(2\pi)^{np/2} |\Sigma|^{-n/2} \mathrm{etr} \left[-\frac{l}{2}(Y - X_1 \Theta X_2 - Z\Gamma_0)\Sigma^{-1}(Y - X_1 \Theta X_2 - Z\Gamma_0)' \right].$$
$$\tag{8.23}$$

Using Theorem 7.1 with $H_2 = H_2'$, observe that

$$\Sigma = X_2'(X_2 \Sigma^{-1} X_2')^{-1} X_2 + \Sigma H_2'(H_2 \Sigma H_2')^{-1} H_2 \Sigma. \tag{8.24}$$

On post multiplying (8.24) by $X_2'(X_2 X_2')^{-1}$ and pre-multiplying by Σ^{-1}, we have that

$$\Sigma^{-1} X_2'(X_2 \Sigma^{-1} X_2')^{-1} = X_2'(X_2 X_2')^{-1} - H_2'(H_2 \Sigma H_2')^{-1} H_2 \Sigma X_2'(X_2 X_2')^{-1}. \tag{8.25}$$

Using the relations (8.24) and (8.25), and results of conditional normal distributions, the joint density of $Y_0 = (Y_1, Y_2)$ in (8.21) may be expressed as a product of Y_2 and the conditional density of Y_1 given Y_2 where

$$Y_1 | Y_2 \sim N_{n,p}(X_1 \Theta + Z\xi + Y_2 \gamma, (X_2 \Sigma^{-1} X_2')^{-1}, I_n) \tag{8.26}$$
$$Y_2 \sim N_{n,p-q}(Z\Gamma_0 H_2', H_2 \Sigma H_2', I_n) \tag{8.27}$$

and

$$\xi = \Gamma_0 \Sigma^{-1} X_2'(X_2 \Sigma^{-1} X_2')^{-1} \tag{8.28}$$
$$\gamma = (H_2 \Sigma H_2')^{-1} H_2 \Sigma X_2'. \tag{8.29}$$

Now the density of Y_2 does not involve Θ; thus to test $H_1 : C_1 \Theta A_1 = 0$ we use the MANOVA model and the conditional distribution of $Y_1 | Y_2$. The LR test of H_1 under the MANOVA-GMANOVA model is to reject $H_1 : C_1 \Theta A_1 = 0$ if

$$\Lambda = \frac{|\widetilde{E}|}{|\widetilde{E} + \widetilde{H}|} < U_{u,\nu_h,\nu_e}^{1-\alpha}$$

where the maximum likelihood estimate of $\widehat{\Theta}$ is given in (8.17). Using identities (8.24) and (8.25), $\widehat{\Theta}$ becomes

$$\widehat{\Theta} = (X^{11}X_1' + X^{12}Z')[I_n - Y_2(Y_2'Y_2 - Y_2'X(X'X)^{-1}X'Y_2)^{-1}Y_2'\widetilde{X}]Y_1 \quad (8.30)$$

where $X = (X_1, Z)$, $Y_2 = YH_2'$, $Y_1 = YH_1' = YX_2'(X_2X_2')^{-1}$, and $\widetilde{X} = I_n = X(X'X)^{-1}X'$. The hypothesis test matrices are

$$\widetilde{E} = A_1'(X_2E^{-1}X_2')^{-1}A_1 \quad (8.31)$$

$$\widetilde{H} = (C_1\widehat{\Theta}A_1)'(C_1R_1^{-1}C_1')^{-1}(C_1\widehat{\Theta}A_1) \quad (8.32)$$

where R_1^{-1} is

$$R_1^{-1} = X^{11} + (X^{11}X_1' + X^{12}Z')Y_2$$
$$[Y_2'Y_2 - Y_2'X(X'X)^{-1}X'Y_2]^{-1}Y_2'(X_1X^{11} + ZX^{21}) \quad (8.33)$$

and $E = Y(I - X(X'X)^{-1}X')Y$. The matrices $\widetilde{E}$ and $\widetilde{H}$ have independent Wishart distributions

$$\widetilde{E} \sim W_u(\nu_e = n - m_1 - m_2 - p + q, A_1'(X_2\Sigma^{-1}X_2')^{-1}A_1, \Delta = 0) \quad (8.34)$$

$$\widetilde{H} \sim W_u(\nu_h = g_1, A_1'(X_2\Sigma^{-1}X_2)^{-1}A_1, [A_1'(X_2\Sigma^{-1}X_2')^{-1}A_1]^{-1}\Delta) \quad (8.35)$$

where the noncentrality matrix is

$$\Delta = (C_1\widehat{\Theta}A_1)'(C_1R_1^{-1}C_1')^{-1}(C_1\widehat{\Theta}A_1). \quad (8.36)$$

For $Z = 0$, the LR test $H : C_1\widehat{\Theta}A_1 = 0$ reduces to the GMANOVA model.

If one wants to test the hypothesis H_2 under the mixed MANOVA-GMANOVA model, the LR test of $H_2 : C_2\Gamma_0 = 0$ is to reject for small values of

$$\Lambda = \left(\frac{|\widetilde{E}_1|}{|\widetilde{E}_1 + \widetilde{H}_1|}\right)\left(\frac{|\widetilde{E}_2|}{|\widetilde{E}_2 + \widetilde{H}_2|}\right) = \lambda_1\lambda_2 \quad (8.37)$$

where

$$X = \begin{pmatrix} X_1 & Z \end{pmatrix} \quad (8.38)$$

$$E = Y'(I_n - X(X'X)^{-1}X')Y \quad (8.39)$$

$$\widetilde{E}_1 = (X_2E^{-1}X_2')^{-1} \quad (8.40)$$

$$\widetilde{H}_1 = (C_2\widehat{\xi})'(C_2R_2C_2')^{-1}(C_2\widehat{\xi}) \quad (8.41)$$

$$\widehat{\xi} = \widehat{\Gamma}_0E^{-1}X_2'(X_2E^{-1}X_2')^{-1} \quad (8.42)$$

$$\widetilde{E}_2 = H_2E_2H_2' \quad (8.43)$$

$$\widetilde{H}_2 = (C_2\widehat{\Gamma}_0H_2')'(C_2(Z'Z)^{-1}C_2')^{-1}(C_2\widehat{\Gamma}_0H_2') \quad (8.44)$$

$$R_2 = (X^{22} + X^{21}X_1' + X^{22}Z')Y$$
$$[E^{-1} - E^{-1}X_2'(X_2E^{-1}X_2')^{-1}X_2E^{-1}]Y'(X_1X^{12} + ZX^{22}) \quad (8.45)$$

$$E_2 = Y'[I_n - Z(Z'Z)^{-1}Z']Y \quad (8.46)$$

$$\widehat{\Gamma}_0H_2' = (Z'Z)^{-1}Z'YH_2' \qquad \text{since } X_2H_2' = 0 \quad (8.47)$$

and $\widehat{\Gamma}_0$ is defined in (8.17). The matrices $\widetilde{E}_1$, $\widetilde{H}_1$, $\widetilde{E}_2$, and $\widetilde{H}_2$ have independent Wishart distributions

$$\widetilde{E}_1 \sim W_q(\nu_e = n - m_1 - m_2 - p + q, (X_2 \Sigma^{-1} X_2')^{-1}, \Delta = 0) \qquad (8.48)$$

$$\widetilde{H}_1 \sim W_q(\nu_h = g_1, X_2 \Sigma^{-1} X_2', (X_2 \Sigma^{-1} X_2')^{-1} \Delta) \qquad (8.49)$$

$$\widetilde{E}_2 \sim W_{p-q}(\nu_e = n - m_2, H_2 \Sigma H_2', \Delta = 0) \qquad (8.50)$$

$$\widetilde{H}_2 \sim W_{p-q}(\nu_h = g_2, H_2 \Sigma H_2', (H_2 \Sigma H_2')^{-1} \Delta). \qquad (8.51)$$

The distribution of Λ in (8.37) follows a matrix inverted Dirichlet distribution (a generalization of the multivariate beta II density also, called the inverted multivariate beta density) which arises when one tests the equality of means and covariances of several multivariate normal distributions (T. W. Anderson, 1984, p. 409; Johnson & Kotz, 1972, p. 237).

Since λ_1 and λ_2 are independent likelihood ratio statistics, we may use Theorem 5.3 with $\nu_e \equiv \nu_1 = n - m_1 - m_2 - p + q$ and $\nu_e \equiv \nu_2 = n - m_2$ to obtain an approximate critical value for the test of H_2 in that

$$X^2 = [\nu_1 - (q - g_1 + 1)/2] \log \lambda_1 + [\nu_2 - (p - q - g_2 + 1)/2] \log \lambda_2$$
$$\sim \chi^{\alpha}_{(qg_1)} + \chi^{\alpha}_{[(p-q)g_2]}. \qquad (8.52)$$

Alternatively to obtain an approximate p-value, we employ Theorem 5.4 using the two independent chi-square distributions where for λ_1, $k = m_1 + m_2 + p - q$, $g \equiv g_1$ and $u \equiv q$ and for λ_2, $k = m_2$, $g \equiv g_2$ and $u \equiv p - q$ for the MANOVA-GMANOVA model.

The LR test of H_3 ($C_1 \Theta A_1 = 0$, $C_2 \Gamma_0 = 0$) rejects H_3 for small values of Λ with

$$\Lambda = \left(\frac{|E_1^*|}{|E_1^* + H_1^*|} \right) \left(\frac{|\widetilde{E}_2|}{|\widetilde{E}_2 + \widetilde{H}_2|} \right) = \lambda_1 \lambda_2 \qquad (8.53)$$

where

$$E_1^* = (X_2 E^{-1} X_2')^{-1} \qquad (8.54)$$

$$H_1^* = (C^* \Theta^*)'(C^* R^{-1} C^{*'})^{-1}(C^* \Theta^*) \qquad (8.55)$$

and $\widetilde{E}_2$ and $\widetilde{H}_2$ are given in (8.47), R^{-1} is defined in (8.12) with X_1 replaced by $X = (X_1, Z)$

$$C^* = \begin{pmatrix} C_1 & 0 \\ 0 & C_2 \end{pmatrix} \qquad\qquad \widehat{\Theta}^* = \begin{pmatrix} \widehat{\Theta} \\ \widehat{\xi} \end{pmatrix} \qquad (8.56)$$

with $\widehat{\Theta}$ defined in (8.15), $\widehat{\xi}$ defined in (8.47), and $\widehat{\Gamma}_0$ given in (8.16), and $E = Y'(I_n - X(X'X)^{-1}X')Y$. The distribution of the test matrices are identical to those given in (8.51) with $\widetilde{E}_1$ and $\widetilde{H}_1$ replaced by E_1^* and H_1^*. The degrees of freedom for

H_1^* is $g_1 + g_2$, not g_2. To evaluate the test statistic in (8.53), we may alternatively use a chi-square distribution to approximate the test of $H_3(H_1 \cap H_2)$ as given in (8.53) with g_1 replaced by $g_1 + g_2$ in (8.52). Theorem 5.4 is used to obtain an approximate p-value for the test. This is illustrated later in this chapter when applications of the MANOVA-GMANOVA model are discussed.

8.3 TESTS OF FIT

In our discussion of the GMANOVA model in Chapter 7, we presented a test (7.76) for the inclusion of the higher order polynomials as covariates in the MAN-COVA model. The test is evaluating whether the best model for the study is a MANOVA or GMANOVA model. To see this, recall that a MANOVA model $\mathcal{E}(Y) = X_1 B$ may be represented as a GMANOVA model if B can be rewritten as $B = \Theta X_2$. If $H_{2[(p-q) \times p]}$ is a matrix of full rank such that $BH_2' = 0$, the condition $B = \Theta X_2$ is equivalent to the requirement that $BH_2' = \Theta X_2 H_2' = 0$. Thus to evaluate the adequacy of the GMANOVA model versus the MANOVA model, we test the hypotheses that $BH_2' = 0$ under the MANOVA model. Failing to reject the null hypothesis supports the adequacy of the GMANOVA model. For example, one may evaluate whether a quadratic model is adequate for a set of data which would suggest that a subset of higher order polynomials should be included as covariates in the GMANOVA model. If no covariates are included in the model, then one has a MANOVA model. Thus, one needs to test the null hypothesis that the model is GMANOVA versus the alternative that the model is MANOVA. For the MANOVA-GMANOVA models we may evaluate the adequacy of the GMANOVA model versus the MANOVA-GMANOVA model and the adequacy of the MANOVA-GMANOVA model versus the MANOVA model. These goodness-of-fit tests were derived by Chinchilli and Elswick (1985).

An alternative approach to the evaluation of the adequacy of the GMANOVA model versus the MANOVA model is to test

$$\Omega_0 : \mathcal{E}(Y) = X_1 \Theta X_2 \qquad (8.57)$$
$$\omega : \mathcal{E}(Y) = X_1 \Theta^* \qquad (8.58)$$

using the likelihood ratio test directly. We compare the ML estimate of Σ under Ω_0, $\hat{\Sigma}_{\Omega_0}$, with the corresponding estimate under ω, $\hat{\Sigma}_\omega$. From (8.8),

$$n\hat{\Sigma}_{\Omega_0} = E + W'Y'X_1(X_1'X_1)^{-1}X_1'YW. \qquad (8.59)$$

Under ω, $\Theta^* = \Theta X_2$. Hence,

$$n\hat{\Sigma}_\omega = E \qquad (8.60)$$

so that

$$\Lambda = \lambda^{2/n} = \frac{|E|}{|E + W'Y'X_1(X_1'X_1)^{-1}X_1'YW|}. \qquad (8.61)$$

If we let $M_{p \times p} = (H_2', E^{-1} X_2' (X_2 X_2')^{-1})$ be a matrix of full rank, and H_2 is any set of $p - q$ independent rows of $I_p - X_2'(X_2 X_2')^{-1} X_2$ so that $X_2 H_2' = 0$, we may reduce Λ in (8.61) to Λ in (7.78). That is,

$$
\begin{aligned}
\Lambda &= \frac{|E|}{|E + W'Y'X_1(X_1'X_1)^{-1}X_1'YW|} \\[4pt]
&= \frac{|M'||E||M|}{|M'||E + W'Y'X_1(X_1'X_1)^{-1}X_1'YW||M|} \\[4pt]
&= \frac{|M'EM|}{|M'EM + M'W'Y'X_1(X_1'X_1)^{-1}X_1'YWM|} \\[4pt]
&= \frac{|\mathrm{diag}(H_2 E H_2')(X_2 X_2')^{-1} X_2 E^{-1} X_2'(X_2 X_2')^{-1}|}{|\mathrm{diag}(H_2 E H_2' + H_2 Y'X_1(X_1'X_1)^{-1}X_1'YH_2', (X_2 X_2')^{-1}X_2 E^{-1}X_2'(X_2 X_2')^{-1})|} \\[4pt]
&= \frac{|H_2 E H_2'|}{|H_2 E H_2' + H_2 Y'X_1(X_1'X_1)^{-1}X_1'YH_2'|} \\[4pt]
&= \frac{|H_2 Y'X_1(X_1'X_1)^{-1}X_1'YH_2'|}{|H_2 Y'(I - X_1(X_1'X_1)^{-1}X_1')YH_2'|} \quad\quad (8.62)
\end{aligned}
$$

which is the same statistic as given in (7.78) since H_2 in (8.62) is H_2' in (7.78). This is the approach employed by Chinchilli and Elswick (1985) to test model adequacy for the MANOVA-GMANOVA model.

To test

$$\Omega_0 : \mathcal{E}(Y) = X_1 \Theta X_2 + Z\Gamma_0 \quad\quad (8.63)$$
$$\omega : \mathcal{E}(Y) = X_1 \Theta X_2 + Z\Gamma_0^* X_2 \quad\quad (8.64)$$

the null hypothesis that the model is GMANOVA versus the alternative that the model is MANOVA-GMANOVA, we derive the LR test by using $\hat{\Sigma}_{\Omega_1}$ in (8.17) and $\hat{\Sigma}_\omega$ in (8.8) with $X = (X_1, Z)$. The statistic is

$$
\begin{aligned}
\Lambda &= \frac{|E + W'Y'(X(X'X)^{-1}X' - Z(Z'Z)^{-1}Z')YW|}{|E + W'Y'X(X'X)^{-1}X'YW|} \\[4pt]
&= \frac{|\widetilde{E}|}{|\widetilde{E} + \widetilde{H}|} \quad\quad (8.65)
\end{aligned}
$$

where

$$\widetilde{H} = H_2 Y' Z(Z'Z)^{-1} Z' Y H_2' \quad\quad (8.66)$$
$$\widetilde{E} = H_2 Y' (I - Z(Z'Z)^{-1}Z') Y H_2' \quad\quad (8.67)$$

have independent Wishart distributions with m_2 and $n - m_2$ degrees of freedom, respectively, and parameter matrix is $H_2 \Sigma H_2'$ of rank $p - q$. Hence Λ may be compared to a U-distribution with $u = p - q$, $\nu_h = m_2$, and $\nu_e = n - m_2$ degrees of freedom.

To test

$$\Omega_0 : \mathcal{E}(Y) = X_1 \Theta^* + Z\Gamma_0 \qu\quad (8.68)$$
$$\omega : \mathcal{E}(Y) = X_1 \Theta X_2 + Z\Gamma_0 \quad\quad (8.69)$$

that the model is mixed MANOVA-GMANOVA versus the alternative that the model is MANOVA, the LR test is

$$\Lambda = \frac{|\widetilde{H}|}{|\widetilde{H} + \widetilde{E}|} \tag{8.70}$$

where

$$\widetilde{H} = H_2 Y' (X(X'X)^{-1}X' - Z(Z'Z)^{-1}Z')Y H_2' \tag{8.71}$$

$$\widetilde{E} = H_2 E H_2' \tag{8.72}$$

have independent Wishart distribution with degrees of freedom m_1 and $n - m_1 - m_2$, respectively, with parameter matrix $H_2 \Sigma H_2'$ of rank $p - q$. For this test $u = p - q$, $\nu_h = m_1$ and $\nu_e = n - m_1 - m_2$ for the U-distribution.

The restricted multivariate general liner model (RMGLM) may be used to test hypotheses on the between groups design of a repeated measurement experiment with covariates, subject to the restriction of equality of slopes over groups. One application of the MANOVA-GMANOVA model is in experiments where data are collected on a response made before the application of repeated treatments and the initial response is then used as a covariate. Letting the data matrix Y contain differences between consecutive time points and $Z_{n \times 1}$ vector of initial response data, we have a simple application of the model.

8.4 SUM OF PROFILES AND CGMANOVA MODELS

An extension of the mixed MANOVA-GMANOVA which allows for both between design and within design covariates was proposed by Patel (1986). The model is written

$$\mathcal{E}(Y_{n \times p}) = XB + \sum_{j=1}^{r} Z_j \Gamma_j \tag{8.73}$$

where $Y_{n \times p}$ is a data matrix on p occasions for n subjects, $X_{n \times k}$ is known, $B_{k \times p}$ is a matrix of parameters, $Z_{j(n \times p)}$ is a within design matrix of covariates of full rank p that change with time and $\Gamma_{j(p \times p)}$ are diagonal matrices with diagonal elements γ_{jk} for $k = 1, 2, \ldots, p$ and $j = 1, 2, \ldots, r$. Letting $XB = X_1 B_1 + X_2 B_2$, we see that we have both between and within covariates in that X_1 is the between MANOVA model with covariates X_2. To estimate the parameters in (8.73) is complicated using ML methods. Verbyla (1988) relates (8.73) to a SUR model. To establish the relationship, Verbyla lets

$$B = \begin{pmatrix} \beta_1 & \beta_2 & \ldots & \beta_0 \end{pmatrix} \qquad Z_j = \begin{pmatrix} Z_{j1} & Z_{j2} & \ldots & Z_{jp} \end{pmatrix}. \tag{8.74}$$

Then (8.73) is written as

$$
\begin{aligned}
\mathcal{E}(Y) &= X \begin{pmatrix} \beta_1 & \beta_2 & \dots & \beta_p \end{pmatrix} + \sum_{j=1}^{r} \begin{pmatrix} Z_{j1}\gamma_{j1} & \dots & Z_{jp}\gamma_{jp} \end{pmatrix} \\
&= \begin{pmatrix} X\beta_1 + Z_{.1}\gamma_{.1} & \dots & X\beta_p + Z_{.p}\gamma_{.p} \end{pmatrix} \\
&= \begin{pmatrix} X_1\theta_{11} & X_2\theta_{22} & \dots & X_p\theta_{pp} \end{pmatrix} \\
&= \begin{pmatrix} X_1 & X_2 & \dots & X_p \end{pmatrix} \widetilde{\Theta}
\end{aligned} \tag{8.75}
$$

where

$$
\widetilde{\Theta} = \begin{pmatrix}
\theta_{11} & 0 & \dots & 0 \\
0 & \theta_{22} & \dots & 0 \\
\vdots & \vdots & \ddots & \vdots \\
0 & 0 & \dots & \theta_{pp}
\end{pmatrix} \tag{8.76}
$$

$$
Z_{.k} = \begin{pmatrix} Z_{1k} & Z_{2k} & \dots & Z_{rk} \end{pmatrix} \tag{8.77}
$$

$$
\gamma'_{.k} = \begin{pmatrix} \gamma_{1k} & \gamma_{2k} & \dots & \gamma_{rk} \end{pmatrix} \tag{8.78}
$$

$$
X_k = \begin{pmatrix} X & Z_{.k} \end{pmatrix} \tag{8.79}
$$

$$
\theta'_{kk} = \begin{pmatrix} \beta'_k & \gamma'_{.k} \end{pmatrix} \tag{8.80}
$$

which is a SUR model.

Verbyla and Venables (1988) extended the MANOVA-GMANOVA model further as a sum of GMANOVA profiles given by

$$
\mathcal{E}(Y_{n \times p}) = \sum_{i=1}^{r} X_i B_i M_i \tag{8.81}
$$

where $X_{i(n \times k_i)}$ and $M_{i(q_i \times p)}$ are known matrices, and $B_{i(k_i \times q_i)}$ is a matrix of unknown parameters. Reducing (8.81) to a canonical form, Verbyla and Venables show that (8.81) is a multivariate version of the SUR model calling their model the extended GMANOVA model or sum of profiles model. More recently, Rosen (1989) has found closed form maximum likelihood estimates for the parameters B_i in (8.81) under the restrictions: 1) rank$(X_1) = p \leq n$ and 2) $V(X_r) \subseteq V(X_{r-1}) \subseteq, \dots, \subseteq V(X_1)$ where $V(X_i)$ represents the column space of X_i. Using von Rosen's expressions for the maximum likelihood estimate $\widehat{B}_i$ and the unique maximum likelihood estimate of Σ, one may obtain likelihood ratio tests of hypotheses of the form $H : A_i B_i C_i = 0$ for $i = 1, 2, \dots, r$ when the nested condition $V(A'_r) \subseteq V(A'_{r-1}) \subseteq, \dots, \subseteq V(A'_1)$ is satisfied. However, the distribution of the likelihood statistics was not studied.

A further generalization of (8.81) was considered by Hecker (1987) as discussed in Chapter 6. To see this generalization, we roll out the matrices M_i, X_i, and B_i as vectors, columnwise. Then (8.81) becomes

$$\mathcal{E}[\text{vec}(Y')] = \left[\sum_{i=1}^{r}(X_i \otimes M_i')\right]\text{vec}(B_i')$$

$$= \left[(X_1 \otimes M_1') \quad (X_2 \otimes M_2') \quad \ldots \quad (X_r \otimes M_r')\right]\begin{pmatrix}\text{vec}(B_1')\\\text{vec}(B_2')\\\vdots\\\text{vec}(B_r')\end{pmatrix} \qquad (8.82)$$

so that Hecker's model generalizes (8.81). Hecker calls his model the CGMANOVA.

ML estimators of model parameters for general sum of profile models of the form proposed by Verbyla and Venables (1988) involve solving likelihood equations of the form

$$n\Sigma = (Y - XBM)'(Y - XBM) \qquad (8.83)$$

$$M\Sigma^{-1}(Y - XBM)X' = 0. \qquad (8.84)$$

These were solved by Chinchilli and Elswick (1985) for the mixed MANOVA-GMANOVA model. Rosen (1989) showed that closed form maximum likelihood solutions to the likelihood equations for models

$$\mathcal{E}(Y) = \sum_{i=1}^{r}X_iB_iM_i + X_{r+1}B_{r+1} \qquad (8.85)$$

have closed form if and only if the X_i for $i = 1, 2, \ldots, r$ are nested. Srivastava (2001) developed the LR test for several special cases of nested models. To test hypotheses, one has to use large sample chi-square approximations to the LR statistics. An alternative approach to the estimation problem is to utilize the SUR model, which only involve regression calculations.

8.5 SUR MODEL

Before applying the procedure to the general sum of profile models, we first investigate the SUR model introduced in Chapter 7. For convenience, we repeat the model here. The SUR or multiple design multivariate (MDM) model is:

$$\mathcal{E}(Y_{n \times p}) = \begin{pmatrix}X_1\theta_{11} & X_2\theta_{22} & \ldots & X_p\theta_{pp}\end{pmatrix} \qquad (8.86)$$

$$\text{cov}(Y) = I_n \otimes \Sigma \qquad (8.87)$$

$$\mathcal{E}(y_j) = X_j\theta_{jj} \qquad (8.88)$$

$$\text{cov}(y_i, y_j) = \sigma_{ij}I_n. \qquad (8.89)$$

Letting $y = \text{vec}(Y)$ and $\theta' = (\theta'_{11}, \theta'_{22}, \ldots, \theta'_{pp})$, (8.86) becomes

$$\mathcal{E}(y) = D\theta \tag{8.90}$$

$$\text{cov}(y) = \Sigma \otimes I_n \tag{8.91}$$

$$\underset{np \times \sum_{j=1}^{p} k_j}{D} = \begin{pmatrix} X_1 & 0 & \ldots & 0 \\ 0 & X_2 & \ldots & 0 \\ \vdots & \vdots & \ddots & \vdots \\ 0 & 0 & \ldots & X_p \end{pmatrix} \tag{8.92}$$

where each $X_{i(n \times k_j)}$ has full rank k_j, and $\theta_{jj(k_j \times 1)}$ are fixed parameters to be estimated.

Several BAN estimators for θ have been suggested in the literature. Kmenta and Gilbert (1968) compare their ML estimate with Telser's iterative estimator (TIE) (Telser, 1964), and Zellner's two-stage Aitken estimator (ZEF), also called the feasible generalized least squares (FGLS) estimator, and Zellner's iterative Aitken estimator (IZEF) proposed by Zellner (1962, 1963).

From the vector version of the GLM, the best linear unbiased estimator (BLUE) of θ by (4.6) is

$$\hat{\theta} = [D'(\Sigma \otimes I_n)^{-1}D]^{-1}D'(\Sigma \otimes I_n)^{-1}y \tag{8.93}$$

if Σ is known. Zellner proposed replacing Σ by a consistent estimator $\widehat{\Sigma} = (\widehat{\Sigma}_{ij})$ where

$$\widehat{\Sigma}_{ij} = \frac{1}{n-m} y'_i (I_n - X_i(X'_iX_i)^{-1}X'_i)(I_n - X_j(X'_jX_j)^{-1}X'_j)y_j \tag{8.94}$$

where $m = \text{rank}(X_i)$ so that the number of variables in each regression is the same. Relaxing this assumption we may let $m = 0$ or let

$$m_{ij} = \text{tr}[(I_n - X_i(X'_iX_i)^{-1}X'_i)(I_n - X_j(X'_jX_j)^{-1}X'_j)]. \tag{8.95}$$

Substituting $\widehat{\Sigma}$ for Σ in (8.93), the ZEF or FGLS estimator of θ is $\breve{\theta}$ defined as

$$\breve{\theta} = \begin{pmatrix} \breve{\theta}_{11} \\ \breve{\theta}_{22} \\ \vdots \\ \breve{\theta}_{pp} \end{pmatrix} = \begin{pmatrix} \hat{\sigma}^{11}X'_1X_1 & \hat{\sigma}^{12}X'_1X_2 & \ldots & \hat{\sigma}^{1p}X'_1X_p \\ \hat{\sigma}^{21}X'_2X_1 & \hat{\sigma}^{22}X'_2X_2 & \ldots & \hat{\sigma}^{2p}X'_2X_p \\ \vdots & \vdots & \ddots & \vdots \\ \hat{\sigma}^{p1}X'_pX_1 & \hat{\sigma}^{p2}X'_pX_2 & \ldots & \hat{\sigma}^{pp}X'_pX_p \end{pmatrix}^{-1} \begin{pmatrix} X'_1\sum_{j=1}^{p}\hat{\sigma}^{1j}y_j \\ X'_2\sum_{j=1}^{p}\hat{\sigma}^{2j}y_j \\ \vdots \\ X'_p\sum_{j=1}^{p}\hat{\sigma}^{pj}y_j \end{pmatrix} \tag{8.96}$$

where $\widehat{\Sigma}^{-1} = (\widehat{\Sigma}^{ij})$. The asymptotic covariance matrix of $\breve{\theta}$ is

$$[D'(\Sigma^{-1} \otimes I_n)D]^{-1} = \begin{pmatrix} \sigma^{11}X'_1X_1 & \sigma^{12}X'_1X_2 & \ldots & \sigma^{1p}X'_1X_p \\ \sigma^{21}X'_2X_1 & \sigma^{22}X'_2X_2 & \ldots & \sigma^{2p}X'_2X_p \\ \vdots & \vdots & \ddots & \vdots \\ \sigma^{p1}X'_pX_1 & \sigma^{p2}X'_pX_2 & \ldots & \sigma^{pp}X'_pX_p \end{pmatrix}^{-1}. \tag{8.97}$$

To estimate θ we completed the step of estimating Σ and with $\widehat{\Sigma}$ and then estimated $\breve{\theta}$, a two-stage process. However, the two-stage ZEF estimator may be used to calculate an estimate of $\Omega = \Sigma \otimes I_n$ as

$$\widehat{\Omega}_2 = (y - D\breve{\theta}_1)(y - D\breve{\theta}_1)' \tag{8.98}$$

where $\breve{\theta}_1$ is the ZEF estimator in (8.96). Then, a new estimate of $\breve{\theta}$ is provided

$$\breve{\theta} = (D'\widehat{\Omega}_2^{-1}D)^{-1}D'\widehat{\Omega}_2^{-1}y. \tag{8.99}$$

Continuing in this manner, the ith iteration is

$$\widehat{\Omega}_i = (y - D\breve{\theta}_{i-1})(y - D\breve{\theta}_{i-1})' \tag{8.100}$$

$$\breve{\theta}_i = (D'\widehat{\Omega}_i^{-1}D)^{-1}D'\widehat{\Omega}_i^{-1}y. \tag{8.101}$$

The process continues until $\breve{\theta}_i$ converges in that $\|\breve{\theta}_i - \breve{\theta}_{i-1}\|^2 < \varepsilon$. This process results in the IZEF estimator of θ, also called the iterative FGLS estimate.

(Telser, 1964) proposed an alternative method for estimating θ_{jj} in (8.86). He introduces into the system a set of new errors v_j, disturbances of all the other equations in the system. Letting $y_j = X_j\theta_{jj} + e_j$ for $j = 1, 2, \ldots, p$, Telser defines

$$e_1 = a_{12}e_2 + a_{13}e_3 + \cdots + a_{1p}e_p + v_1$$
$$e_2 = a_{21}e_1 + a_{23}e_3 + \cdots + a_{2p}e_p + v_2$$

$$\vdots$$

$$e_p = a_{p1}e_1 + a_{p2}e_2 + \cdots + a_{p-1,p-1}e_{p-1} + v_p \tag{8.102}$$

where the a_{ij} are scalars and v_j are random errors such that $\text{cov}(e_i, v_j) = 0$ and Telser shows that $a_{jp} = -\sigma^{jp}/\sigma^{jj}$. Telser substitutes e_j into the equation $y_j = X_j e_{jj} + e_j$ for $j = 1, 2, \ldots, p$. The iterative process yields estimates of a_{ij} and θ_{jj}, and continues until convergence in the θ_{jj}. Telser showed that his estimator has the same asymptotic properties as the IZEF estimator.

To obtain ML estimators of θ and Σ, Kmenta and Gilbert (1968) employed an iterative procedure to solve the likelihood equations for the likelihood

$$L(\theta, \Sigma|y) = (2\pi)^{-np/2}|\Sigma|^{-n/2}\text{etr}\left[-\frac{1}{2}(\Sigma^{-1} \otimes I)(y - D\theta)(y - D\theta)'\right]. \tag{8.103}$$

In their study, they found that all estimators appeared to give the same results with the ZEF estimator favored in small samples. Because the estimator is always unbiased, and two-stage required the least calculation, they recommended the ZEF estimator for the SUR model. Park and Woolson (1992) showed that the ML and IZEF estimators are mathematically equivalent.

To test hypotheses of the form $H : C_*\theta = \xi$ given the SUR model in (8.86), we utilize Wald's statistic defined in (4.15). We use $\widehat{\Sigma} = (\widehat{\Sigma}_{ij})$ defined in (8.94), $\breve{\theta}$ given in (8.96) and the design matrix D specified in (8.86). Then

$$W = (C_*\breve{\theta} - \xi)'\{C_*[D'(\widehat{\Sigma} \otimes I_n)^{-1}D]^{-1}C_*'\}^{-1}(C_*\breve{\theta} - \xi). \tag{8.104}$$

Under the null hypothesis, the statistic W has an asymptotic chi-square distribution with degrees of freedom $\nu_h = \text{rank}(C_*)$. Alternatively, we may write W as T_0^2 so that

$$T_0^2 = \text{tr}\{[C_*(D'(\hat{\Sigma} \otimes I)^{-1}D)^{-1}C_*']^{-1}(C_*\breve{\theta} - \xi)(C_*\breve{\theta} - \xi)'\}. \qquad (8.105)$$

For small sample sizes it is more conservative to use W/ν_h which is approximately distributed as a F-distribution with $\nu_h = \text{rank}(C_*)$ and $\nu_e = n - \sum_j k_j/p$ where $\nu_e = n - m$ if the $\text{rank}(X_i) = m$ for each of the p equations (Theil, 1971, p. 402-403).

Comparing (8.86) with (8.75), we may use the SUR model to estimate the parameters in the model proposed by Patel (1986) which provides for the analysis of repeated measures designs with changing covariates. The estimators using the SUR model are always unbiased, generally have smaller variances than the ML estimator, and are easy to compute since they only involve regression calculations (Kmenta & Gilbert, 1968). Patel (1986) derived LR tests of

$$H_{01} : CBA = 0 \qquad (8.106)$$

$$H_{02} : \Gamma_j = 0 \qquad \text{for } j > r_2 < r \qquad (8.107)$$

and other tests for two-period and three-period crossover designs. For higher-order designs he used a Wald statistic using the asymptotic covariance matrix of the ML estimators. While both approaches are asymptotically equivalent, it is unknown what happens in small samples.

The GMANOVA model and the mixed MANOVA-GMANOVA models may also be represented as SUR models (Stanek & Koch, 1985; Verbyla, 1988). To establish the equivalence, recall that the GMANOVA model is

$$Y_{n \times p} = X_{n \times k} B_{k \times q} P_{q \times p} + U_{n \times p}. \qquad (8.108)$$

Next, consider the SUR model

$$\underset{n \times 1}{y_j} = \underset{n \times k}{X} \underset{k \times 1}{\beta_j} + \underset{n \times 1}{u_j} \qquad \text{for } j = 1, \ldots, q \qquad (8.109)$$

$$\underset{n \times 1}{y_j} = \underset{n \times 1}{u_j} \qquad \text{for } j = q + 1, \ldots, p \qquad (8.110)$$

where the $\text{rank}(X) = k$, $\mathcal{E}(u_j) = 0$, $\text{cov}(u_j, u_j') = \sigma_{jj}I_n$. Expressing (8.110) as a univariate model by stacking the y_j vectors, we have that

$$y_{np \times 1} = (P' \otimes X)_{np \times qk} \beta_{qk \times 1} + u_{np \times 1} \qquad (8.111)$$

where

$$P_{q \times p} = (I_q, 0_{q \times (p \times q)}) \qquad (8.112)$$

and the $\text{cov}(y) = \Sigma \otimes I_n$. Applying (4.6) to (8.111),

$$\begin{aligned}
\hat{\beta} &= [(P' \otimes X)'(\Sigma \otimes I_n)^{-1}(P' \otimes X)]^{-1}(P' \otimes X)'(\Sigma \otimes I)^{-1}y \\
&= [(P\Sigma^{-1}P')^{-1} \otimes (X'X)^{-1}](P\Sigma^{-1} \otimes X')y \\
&= [(P\Sigma^{-1}P')^{-1}P\Sigma^{-1} \otimes (X'X)^{-1}X']y.
\end{aligned} \tag{8.113}$$

Substituting S for Σ in (8.113) with $S = Y'(I - X(X'X)^{-1}X')Y/_{(n-k)}$, the two-stage ZEF estimator is equal to the ML estimate for B in (8.113) since

$$\hat{\beta} = \text{vec}(\hat{B}) = \text{vec}[(X'X)^{-1}X'YS^{-1}P'(PS^{-1}P')^{-1}] \tag{8.114}$$

for the GMANOVA model (8.108) as shown in (7.62).

To establish the equivalence of the GMANOVA and SUR models for arbitrary P, let

$$H = \begin{pmatrix} H_1 & H_2 \end{pmatrix} = \begin{pmatrix} S^{-1}P'(PS^{-1}P')^{-1} & I_p - H_1 P \end{pmatrix} \tag{8.115}$$

and $Y_0 = YH$, $X_2 = PH$ and $U_0 = UH$ in (8.108),

$$Y_{0(n \times p)} = X_{n \times k} B_{k \times p} X_{2(q \times p)} + U_{0(n \times p)} \tag{8.116}$$

with P defined in (8.112). Then, the ML estimate of B is

$$\begin{aligned}
\hat{B} &= (X'X)^{-1}X'Y_0(H'SH)^{-1}X_2'(X_2(H'SH)^{-1}X_2')^{-1} \\
&= (X'X)^{-1}X'YS^{-1}P'(PS^{-1}P')^{-1}.
\end{aligned} \tag{8.117}$$

Hence, model (8.116) is equivalent to the SUR model given in (8.111) using the transformed data matrices $Y_0 = YH$ and $X_2 = PH$.

The MANOVA-GMANOVA model or more generally the sum of profiles model may be represented as a multivariate SUR (MSUR) model (Verbyla & Venables, 1988). With the MSUR model, as with the SUR model, we may estimate parameters using regression equations rather than solving likelihood equations. The MSUR model has the following general form

$$\Omega_0 : \mathcal{E} \begin{pmatrix} Y_1 & Y_2 & \dots & Y_q \end{pmatrix} = \begin{pmatrix} X_1\Theta_{11} & X_2\Theta_{22} & \dots & X_q\Theta_{qq} \end{pmatrix} \tag{8.118}$$

$$Y_{i(n \times p_i)} = X_{i(n \times k_i)}\Theta_{i(k_i \times p_i)} + U_{i(n \times p_i)} \qquad i = 1, 2, \dots, q \tag{8.119}$$

where the $\text{cov}(Y_i, Y_j) = \Sigma_{ij}$ is of order $(p_i \times p_j)$, $\Sigma = (\Sigma_{ij})$ of order $(p \times p)$ and $p = \sum_{i=1}^q p_i$. If $p_1 = p_2 = \dots = p_q = p_0$, then (8.73) becomes

$$\begin{pmatrix} Y_1 \\ Y_2 \\ \vdots \\ Y_q \end{pmatrix} = \begin{pmatrix} X_1 & 0 & \dots & 0 \\ 0 & X_2 & \dots & 0 \\ \vdots & \vdots & \ddots & \vdots \\ 0 & 0 & \dots & X_q \end{pmatrix} \begin{pmatrix} \Theta_{11} \\ \Theta_{22} \\ \vdots \\ \Theta_{qq} \end{pmatrix} + \begin{pmatrix} U_1 \\ U_2 \\ \vdots \\ U_q \end{pmatrix}, \tag{8.120}$$

a matrix generalization of (7.27).

Representing (8.19) as a GLM, let

$$U = \begin{pmatrix} E_1 & E_2 & \ldots & E_q \end{pmatrix} \tag{8.121}$$

$$U_i = \begin{pmatrix} e_{i1} & e_{i2} & \ldots & e_{ip_i} \end{pmatrix} \tag{8.122}$$

$$Y = \begin{pmatrix} Y_1 & Y_2 & \ldots & Y_q \end{pmatrix} \tag{8.123}$$

$$Y_i = \begin{pmatrix} y_{i1} & y_{i2} & \ldots & y_{ip_i} \end{pmatrix} \tag{8.124}$$

$$\Theta = \begin{pmatrix} \Theta_{11} & \Theta_{22} & \ldots & \Theta_{qq} \end{pmatrix} \tag{8.125}$$

$$\Theta_{jj} = \begin{pmatrix} \Theta_{jj_1} & \Theta_{jj_2} & \ldots & \Theta_{jj_{p_1}} \end{pmatrix} \tag{8.126}$$

where

$$e = \mathrm{vec}(U) = \begin{bmatrix} \mathrm{vec}(U_1) & \mathrm{vec}(U_2) & \ldots & \mathrm{vec}(U_q) \end{bmatrix} \tag{8.127}$$

$$y = \mathrm{vec}(Y) = \begin{bmatrix} \mathrm{vec}(Y_1) & \mathrm{vec}(Y_2) & \ldots & \mathrm{vec}(Y_q) \end{bmatrix} \tag{8.128}$$

$$\theta = \mathrm{vec}(\Theta) = \begin{bmatrix} \mathrm{vec}(\Theta_{11}) & \mathrm{vec}(\Theta_{22}) & \ldots & \mathrm{vec}(\Theta_{qq}) \end{bmatrix} \tag{8.129}$$

so that y_{ij} is the jth column of Y_i and θ_{jj_t} is the tth column of Θ_{jj}, then (8.119) becomes

$$y = D\theta + e \tag{8.130}$$

where

$$D = \mathrm{diag} \begin{pmatrix} I_{p_1} \otimes X_1 & I_{p_2} \otimes X_2 & \ldots & I_{p_q} \otimes X_q \end{pmatrix} \tag{8.131}$$

and $\mathrm{diag}(A_1, A_2, \ldots, A_q)$ is a block diagonal matrix with $A_i = I_{p_i} \otimes X_i$. Applying (4.6) to the GLM, we find that the BLUE of θ is

$$\hat{\theta} = [D'(\Sigma^{\otimes} I_n)^{-1} D]^{-1} D'(\Sigma \otimes I_n)^{-1} y \tag{8.132}$$

so that the multivariate FGLS estimator for θ is

$$\breve{\theta} = [D'(\hat{\Sigma}^{-1} \otimes I_n) D]^{-1} D'(\hat{\Sigma}^{-1} \otimes I_n) y \tag{8.133}$$

where $\widehat{\Sigma}$ is any consistent estimator of Σ. Following (8.94), we may set

$$\widehat{\Sigma}_{ij} = \frac{1}{n - m_{ij}} [Y_i'(I_n - X_i(X_i'X_i)^{-1}X_i')(I_n - X_j(X_j'X_j)^{-1}X_j')Y_j] \tag{8.134}$$

where $m_{ij} = 0$ or

$$m_{ij} = \mathrm{tr}[(I_n - X_i(X_i'X_i)^{-1}X_i')(I_n - X_j(X_j'X_j)^{-1}X_j')]. \tag{8.135}$$

Since (8.130) is a GLM, the Wald statistic

$$X^2 = (C_*\breve{\theta} - \xi)'\{C_*[D'(\widehat{\Sigma} \otimes I_n)^{-1}D]^{-1}C_*\}^{-1}(C_*\breve{\theta} - \xi) \tag{8.136}$$

may be used to test hypotheses $H : C_*\theta = \xi$ where the $\text{rank}(C_*) = \nu_h$ and D is defined in (8.130).

That the parameters in (8.81) may be estimated using the MSUR model was shown by Verbyla and Venables (1988) employing a canonical reduction of (8.81). Using a conditional argument to partition the data matrix, they relate the MANOVA-GMANOVA model to an MSUR analysis as suggested by Stanek and Koch (1985). For the sum of profiles model,

$$\mathcal{E}(Y_{n\times p}) = \sum_{i=1}^{r} X_{i(n\times k_i)} B_{i(k_i \times q_i)} M_{i(q_i \times p)} \tag{8.137}$$

where the $\text{rank}(X_i) = k_i$, $\text{rank}(M_i) = q_i$ where $q_i \leq p$ so that X_i and M_i' have full column rank. Letting $r = 2$, we may consider the vector space $M_1' + M_2'$. An orthogonal decomposition of $M_1' + M_2'$ where $A|B$ represents the orthocomplement of A relative to B is

$$M_1' + M_2' = M_1' \oplus (M_1' + M_2')|M_1' \tag{8.138}$$
$$M_1' + M_2' = M_2' \oplus (M_1' + M_2')|M_2' \tag{8.139}$$

which is not unique since $M_1' \cap M_2' \neq \{0\}$. Using this fact, we can obtain a basis for $M_1' \cap M_2' = N_{12}$. Similarly, we can construct a basis of M_1', M_2', N_1, and N_2, respectively. Hence, there exist nonsingular matrices A_1' and A_2' such that $M_1' = (N_1', N_{12}')A_1'$ and $M_2' = (N_{12}', N_1')A_2'$. The jth column of M_i' is a linear combination of the rows of A_i' by construction. Then (8.137) becomes

$$\mathcal{E}(Y) = X_1 B_1 M_1 + X_2 B_2 M_2 = X_1 B_1 A_1 \begin{pmatrix} N_1 \\ N_{12} \end{pmatrix} + X_2 B_2 A_2 \begin{pmatrix} N_{12} \\ N_2 \end{pmatrix}$$

$$= (X_1 X_2) \begin{pmatrix} B_{11}^* & B_{12}^* & 0 \\ 0 & B_{21}^* & B_{22}^* \end{pmatrix} \begin{pmatrix} N_1 \\ N_{12} \\ N_2 \end{pmatrix}$$

$$= X B^* Q \tag{8.140}$$

where $\binom{B_{11}^*}{B_{12}^*} = B_1 A_1$ and $\binom{B_{21}^*}{B_{22}^*} = B_2 A_2$. The design matrix $M' = M_1' + M_2'$ is of full rank since N_1', N_2', and N_{12}' form a basis for the sum. If Q is less than full rank, $q < p$, by the GMANOVA model, there exists a matrix $H = (H_1, H_2)$ such that

$$Y_0 = (Y_1 \quad Y_2) = [Y H_1 \quad Y(I - H_1 Q)] = (Y H_1 \quad Y H_2) \tag{8.141}$$

so that (8.140) becomes

$$\mathcal{E}(Y_0) = \mathcal{E}(Y_1 \quad Y_2) = \left[X_1 B_{11}^* \quad (X_1 \quad X_2) \begin{pmatrix} B_{12}^* \\ B_{21} \end{pmatrix} \quad X_2 B_{22}^* \quad 0 \right]$$

$$= (X_1^* \Theta_{11} \quad X_2^* \Theta_{22} \quad X_3^* \Theta_{33} \quad 0) \tag{8.142}$$

which has the form of (8.119), a MSUR model. To estimate Θ_{11}, Θ_{22}, and Θ_{33} we would use (8.133) to determine the multivariate FGLS estimator (Park & Woolson, 1992).

8.6 RESTRICTED GMANOVA MODEL

We saw in Chapter 7 that the MANOVA model is a special case of the GM-
ANOVA model and that estimation for the GMANOVA model may be accomplished
using the SUR model. We also showed that the restricted GLM is a special case of the
mixed MANOVA-GMANOVA model which is a special case of the sum of profiles
model whose parameters may be estimated using the MSUR model. However, we
also noted in Chapter 7 that the restricted multivariate GLM is a special case of
the GMANOVA model with double restrictions termed the extended GMANOVA
model by Kariya (1985). To develop the restricted GMANOVA model, recall that
the hypothesis for the extended GMANOVA model was written as

$$H : X_5 B X_6 = 0 \qquad (8.143)$$

where the $\text{rank}(X_5) = r_5 \leq k$ and the $\text{rank}(X_6) = r_6 \leq q$. The restricted
GMANOVA model is defined

$$Y_{n \times p} = X_{1(n \times k)} B_{k \times q} X_{2(q \times p)} + U_{n \times p} \qquad (8.144)$$

$$X_{3(r_3 \times k)} B_{k \times q} X_{4(q \times r_4)} = 0 \qquad (8.145)$$

where the $\text{rank}(X_1) = k$, $\text{rank}(X_2) = q$ and the $\text{rank}(X_i) = r_i$ $(i = 3, 4, 5, 6)$.
When $X_2 = I$, (8.145) reduces to (8.136). To test (8.143), we allow X_2 to be
arbitrary and $X_4 \neq X_6$. We may not allow the matrices X_3, X_4, and X_5 to be
entirely arbitrary but have to impose restrictions on the matrices to ensure that the
parameters are identifiable and testable. This was also the case for the RMGLM
discussed in Chapter 7 with $X_4 = X_6$ and $X_2 = I$. Recall that we required the
matrices X_3 and X_5 to be disjoint (linearly independent or that the intersection of
the column space of X_3' and X_5' must be null, $V(X_3') \cap V(X_5') = 0$). Furthermore,
the parameter space Ω_0 was restricted by X_3, the restricted parameter space $\widetilde{\Omega}$ is
the intersection of the column space of X_1 and the null space of $X_3(X_1'X_1)^{-1}X_1'$,
$[N(X_3(X_1'X_1)^{-1}X_1')]$, and the hypothesis test space $\omega = N(X_5'(X_1'X_1)^{-1}X_1') \cap \Omega_0$.
Thus, for RMGLM, X_5 is a subspace of X_3 relative to $(X_1'X_1)^{-1}$ so that the column
space of $(X_1'X_1)^{-1}X_5'$ is included in the column space of $(X_1'X_1)^{-1}X_3'$.

Testing (8.143) given (8.145) is complicated. The first solution appears to be that
of Gleser and Olkin (1970, p. 288) which was developed using the canonical form of
the model as an "aside". Kabe (1975, p. 36 eq. 7) claims to have solved the problem
using the original variables, but his results do not appear to agree with those of Gleser
and Olkin. The test developed by Kabe is not an LR test. Kabe (1981, p. 2549) result
for the restricted MANOVA model also seems to be incorrect. A complete analysis
of the model in canonical form is provided by Kariya (1985, Chapter 4). Testing
hypothesis for the extended GMANOVA model is complicated since it is really a
sum of profiles model. Following (8.3), observe that by incorporating the restrictions

into the model that

$$
\begin{aligned}
X_1 B X_2 &= \begin{pmatrix} X_{11}^* & X_{12}^* \end{pmatrix} \begin{pmatrix} 0 & \beta_{12} \\ \beta_{21} & \beta_{22} \end{pmatrix} \begin{pmatrix} Q^{11} \\ Q^{21} \end{pmatrix} \begin{pmatrix} X_{21}^* \\ X_{22}^* \end{pmatrix} \\
&= X_{11}^* \beta_{12} Q^{21} X_{22}^* + X_{12}^* \begin{pmatrix} \beta_{12} & \beta_{22} \end{pmatrix} Q^{-1} X_2 \\
&= A_1 B_1 C_1 + A_2 B_2 C_2,
\end{aligned} \tag{8.146}
$$

a sum of profiles model which can be expressed as an MSUR model (Timm, 1996).

To ensure identifiability and hence testability of (8.143) given (8.145), Kariya (1985) uses the following result from matrix algebra to reduce the restricted GMANOVA model to canonical form.

Theorem 8.1. *Given two symmetric matrices A and B, the matrices A and B may be simultaneously diagonalized by an orthogonal matrix P such that $P'AP = D_1$ and $P'AP = D_2$ if and only if A and B commute, $AB = BA$.*

Following Kariya (1985, p. 143), we have the following result.

Theorem 8.2. *The restricted GMANOVA model may be reduced to canonical form if the matrices X_3, X_4, X_5, and X_6 satisfy the condition that*

$$
M_3 M_5 = M_5 M_3 \qquad\qquad M_4 M_6 = M_6 M_4 \tag{8.147}
$$

where

$$
M_i = (X_1'X_1)^{-1/2} X_i' (X_i (X_1'X_1)^{-1} X_i')^{-1} X_i (X_1'X_1)^{-1/2} \quad i = 3,5 \tag{8.148}
$$

$$
M_j = (X_2 X_2')^{-1/2} X_j (X_j' (X_2 X_2')^{-1} X_j)^{-1} X_j' (X_2 X_2')^{-1/2} \quad j = 4,6. \tag{8.149}
$$

Using Theorem 8.2, a canonical form for the restricted GMANOVA problem leads one to consider four cases: 1) $M_3 M_5 = M_5$ and $M_4 M_6 = M_4$, 2) $M_3 M_5 = M_5$ and $M_4 M_6 = 0$, 3) $M_3 M_5 = M_3$ and $M_4 M_6 = M_6$, and 4) $M_3 M_5 = 0$ and $M_4 M_6 = M_6$. When $X_2 = I$, we have as a special case the MANOVA model with double linear restrictions. For this problem,

$$
M_{4(q \times q)} = X_4 (X_4'X_4)^{-1} X_4' \tag{8.150}
$$

$$
M_{6(q \times q)} = X_6 (X_6'X_6)^{-1} X_6' \tag{8.151}
$$

$$
M_{3(k \times k)} = (X_1'X_1)^{-1/2} X_3' (X_3 (X_1'X_1)^{-1} X_3')^{-1} X_3 (X_1'X_1)^{-1/2} \tag{8.152}
$$

$$
M_{5(k \times k)} = (X_1'X_1)^{-1/2} X_5' (X_5 (X_1'x_1)^{-1} X_5')^{-1} X_5 (X_1'X_1)^{-1/2} \tag{8.153}
$$

and by Theorem 8.2, we must have $M_3 M_5 = M_5 M_3$ and $M_4 M_6 = M_6 M_4$. If we further restrict the matrices to the situation discussed in Chapter 7 where $X_4 = X_6 = I$ or $X_4 = X_6 = A$ (say), we observe that $M_4 = M_6 = I$ so that $M_4 M_6 = M_6 M_4$. In addition, we require $M_3 M_5 = M_5 M_3$. This condition implies that the row space of $X_5 (X_1'X_1)^{-1/2}$ is a subspace of the row space of $X_3 (X_1'X_1)^{-1/2}$ or that X_5 is a subspace of X_3 relative to $(X_1'X_1)^{-1}$ as required in the development of this special case.

Kariya (1985) has been able to reduce the restricted GMANOVA model to canonical form, using Lemma 7.2.1 and Theorem 8.2 and develop LR tests, the distribution of the LR criterion is unknown. An alternative analysis is to utilize the SUR model to estimate parameters and use the large sample Wald statistic to test hypotheses. The more general approach is to use the CGMANOVA model (8.82) proposed by Hecker (1987). Since the model is linear, we may estimate parameters of the sum of profiles model and test hypotheses using the Wald statistic. Timm (1996) developed goodness-of-fit tests.

Returning to the SUR model (8.86), recall that the FGLS estimator for θ,

$$\check{\theta} = [D'(\hat{\Sigma}^{-1} \otimes I_n)D]^{-1}D'(\hat{\Sigma}^{-1} \otimes I_n)y, \tag{8.154}$$

was obtained with no restrictions on θ. We can also test hypotheses of the form $H : C_*\theta = \theta_0$ with the restriction

$$R\theta = \xi \tag{8.155}$$

added to the SUR model. Again the rank of R is s where the rows of R are linearly independent, the intersection of R and C_*. Following the arguments for the restricted GLM (RGLM), if $\hat{\Sigma}$ is a consistent estimator of Σ, the restricted FGLS estimate of θ is

$$\check{\theta}_r = \check{\theta} - \widehat{W}^{-1}R(R\widehat{W}^{-1}R')^{-1}(R\check{\theta} - \xi) \tag{8.156}$$

where

$$\hat{W}^{-1} = [D'(\hat{\Sigma}^{-1} \otimes I_n)D]^{-1}. \tag{8.157}$$

Furthermore, the covariance matrix for $\check{\theta}_r$ is

$$\begin{aligned} \text{cov}(\check{\theta}_r) &= \Omega^{-1} - \Omega^{-1}R'(R\Omega^{-1}R')^{-1}R\Omega^{-1} \\ &= [I - \Omega^{-1}R'(R\Omega^{-1}R')^{-1}R]\Omega^{-1} \\ &= F\Omega^{-1} \end{aligned} \tag{8.158}$$

where $\Omega^{-1} = [D'(\Sigma^{-1} \otimes^{-1} I_n)D]^{-1}$.

To test the hypotheses $H : C_*\theta = \theta_0$, assuming multivariate normality, we again may use the Wald statistic

$$X^2 = (C_*\check{\theta}_r - \theta_0)'[C_*(\widehat{F}\widehat{W}^{-1}\widehat{F}')C_*']^{-1}(C_*\check{\theta}_r - \theta_0) \tag{8.159}$$

where $\widehat{F} = I - \widehat{W}^{-1}R'(R\widehat{W}^{-1}R')R$. The asymptotic null distribution of X^2 is chi-square with $\nu_h = g = \text{rank}(C)$ degrees of freedom. If $H : C_*\theta = \theta_0$ is rejected, approximate $1 - \alpha$ simultaneous intervals for parametric functions $\psi = c'\theta$ are obtained using the formula:

$$\hat{\psi} - c_0\hat{\sigma}_{\hat{\psi}} \leq \psi \leq \hat{\psi} + c_0\hat{\sigma}_{\hat{\psi}} \tag{8.160}$$

where

$$\hat{\psi} = c'\breve{\theta}_r \quad \text{and} \quad \hat{\sigma}^2_{\hat{\psi}} = c'\widehat{FW}^{-1}c \qquad (8.161)$$

and c_0^2 is the $\chi_\alpha^2(\nu_h)$ critical value for level α.

Restrictions may also be added to the MSUR model (8.119). Because of the complications involved with double restrictions for the MANOVA and GMANOVA model, it is convenient to utilize the vector form (8.130) of the model and modify the Wald statistic in (8.136) as shown in (8.159) to test hypotheses of the form H : $C_*\theta = \theta_0$ where $\breve{\theta}$ is defined in (8.133) and the restricted estimator has the form given in (8.156).

8.7 GMANOVA–SUR: ONE POPULATION

In Section 7.7 we illustrated how to estimate the parameter matrix B of the GMANOVA model using the Rao-Khatri reduction which utilized the MANCOVA model, following a transformation of the data matrix Y. Since the GMANOVA model may be represented as a SUR model as in (8.111), we have shown in (8.116) that the estimate of B may be obtained from the SUR model for general P. To illustrate the equivalence between the two models we use the Elston and Grizzle (1962) ramus height data found in the data set 8_7.dat. The data consist of ramus heights, measured in mm, of a single cohort of boys aged 8, 8.5, 9, and 9.5 years of age. The study was conducted to establish a "normal" growth curve for boys to be used by orthodontists. The SAS code to perform the analysis of the data utilizing the two models is found in Program 8_7.sas. These data were also examined by Lindsey (1993, p. 90).

8.7.1 Results and Interpretation

To estimate the parameter matrix B using the SUR model, we calculate the FGLS estimator for $\beta = \text{vec}(B)$. To accomplish this task, we use PROC IML to transform the data matrix Y to a vector, $\text{vec}(Y)$, and calculate the covariance matrix S. A matrix of orthogonal polynomial P is created and the design matrix X is a vector of 1's, $1_{20\times1}$. The estimate for β is calculated as

$$\hat{\beta}\left[(PS^{-1}P')^{-1}PS^{-1} \otimes (X'X)^{-1}X'\right] \text{vec}(Y) = \begin{pmatrix} 50.0496 \\ 0.4654 \end{pmatrix}. \qquad (8.162)$$

The results are given in Output 8.7.1.

The SAS code to perform the GMANOVA analysis is similar to that used in Section 7.7. From the results of the test of linear fit, we see that a linear model fits the data. The estimate of the parameter matrix B using the GMANOVA model and PROC GLM is found in Output 8.7.2.

As expected the estimate of B from the GMANOVA model and the estimate of β from the SUR model agree. We next estimate B using PROC SYSLIN which estimates parameters for systems of equations using OLS, 2SLS, 3SLS, and SUR,

Output 8.7.1: GMANOVA-SUR One Population Ramus Height Data: Plot of Ramus
Heights.

S			
6.3299737	6.1890789	5.777	5.5481579
6.1890789	6.4493421	6.1534211	5.9234211
5.777	6.1534211	6.918	6.9463158
5.5481579	5.9234211	6.9463158	7.4647368

Estimate of B using SUR estimate and PROC IML

BETA
50.049582
0.4654022

P prime matrix

P_PRIME			
1	-3	1	-1
1	-1	-1	3
1	1	-1	-3
1	3	1	1

Output 8.7.2: GMANOVA-SUR One Population Ramus Height Data.

Test of Linear Fit

MANOVA Test Criteria and Exact F Statistics for the Hypothesis of No Overall Intercept Effect
H = Type III SSCP Matrix for Intercept
E = Error SSCP Matrix

S=1   M=0   N=8

Statistic	Value	F Value	Num DF	Den DF	Pr > F
Wilks' Lambda	0.98952499	0.10	2	18	0.9096
Pillai's Trace	0.01047501	0.10	2	18	0.9096
Hotelling-Lawley Trace	0.01058590	0.10	2	18	0.9096
Roy's Greatest Root	0.01058590	0.10	2	18	0.9096

Estimate of B Using PROC GLM
Dependent Variable: YT1

Parameter	Estimate	Standard Error	t Value	Pr > |t|
beta	50.0495824	0.59120793	84.66	<.0001

Dependent Variable: YT2

Parameter	Estimate	Standard Error	t Value	Pr > |t|
beta	0.46540223	0.05551038	8.38	<.0001

among other estimation methods. To estimate B for the GMANOVA model using the SUR model we use the transformed data matrix $Y_0 = YH$ where H is defined in (8.115). For PROC SYSLIN, the parameter estimates are associated with the independent (regression) variable x. As expected the PROC SYSLIN output agrees with the IML and GLM procedures. For an alternative approach, see Verbyla (1986). The results are displayed in Output 8.7.3.

8.8 GMANOVA–SUR: SEVERAL POPULATIONS

As a second illustration of the equivalence of the GMANOVA and the SUR models, we reanalyze the Grizzle and Allen (1969) dog data analyzed in Section 7.7 using the GMANOVA model. We now use PROC SYSLIN to analyze the example which assumes a SUR model. The SAS code is provided in Program 8_8.sas.

8.8.1 Results and Interpretation

Recall from Section 7.7 that the application involves fitting a third-degree polynomial to the four groups, a control group and three treatments. For convenience we have included the code to perform the GMANOVA analysis in Program 8_8.sas; however, only the SUR analysis using PROC SYSLIN results are provided in Output 8.8.1-8.8.2. From Output 8.8.1-8.8.2, we observe that PROC SYSLIN yields the same estimate for the matrix B as the GMANOVA model, again fitting a third-degree polynomial.

While the estimate of the parameter matrix B is identical for the two models, this is not the case for tests of hypotheses. For the SUR model, we usually employ the Wald statistic W defined in (8.104) which has an asymptotic chi-square distribution. For the GMANOVA model, we have an exact test. PROC SYSLIN employs a F-statistic:

$$F = \frac{(L\tilde{\beta} - c)'[L(s^2(X'X)^{-1})L'](L\tilde{\beta} - c)}{\text{rank}(L)} \qquad (8.163)$$

which is valid only for OLS. For our example, the test of coincidence is tested using PROC SYSLIN with the statement

```
TEST x1=x2=x3=x4/print;
```

The results for the test are displayed in Output 8.8.3. The P-value for the F-test is 0.0014 which compares favorably with the GMANOVA exact test of coincidence of profiles which (from Section 7.7) was $p = 0.0065$.

8.9 SUR MODEL

Sections 8.7 and 8.8 illustrated the equivalence of the estimates of the parameter matrix B from the GMANOVA and the SUR model in the analysis of growth curve data. The SUR model may also be used to estimate model parameters for multiple

Output 8.7.3: GMANOVA-SUR One Population Ramus Height Data: Estimate of B Using PROC SYSLIN.

Model	YT1
Dependent Variable	YT1

Analysis of Variance					
Source	DF	Sum of Squares	Mean Square	F Value	Pr > F
Model	3	50151.55	16717.18	2416.72	<.0001
Error	17	117.5943	6.917310		
Uncorrected Total	20	50269.14			

Root MSE	2.63008	R-Square	0.99766
Dependent Mean	50.07500	Adj R-Sq	0.99725
Coeff Var	5.25228		

Note: The NOINT option changes the definition of the R-Square statistic to:
1 - (Residual Sum of Squares/Uncorrected Total Sum of Squares).

Parameter Estimates					
Variable	DF	Parameter Estimate	Standard Error	t Value	Pr > |t|
X	1	50.04958	0.591208	84.66	<.0001
YT3	1	-0.95993	2.568346	-0.37	0.7132
YT4	1	-1.90961	6.368749	-0.30	0.7679

Model	YT2
Dependent Variable	YT2

Analysis of Variance					
Source	DF	Sum of Squares	Mean Square	F Value	Pr > F
Model	3	4.928648	1.642883	26.94	<.0001
Error	17	1.036702	0.060982		
Uncorrected Total	20	5.965350			

Root MSE	0.24695	R-Square	0.82621
Dependent Mean	0.46650	Adj R-Sq	0.79554
Coeff Var	52.93598		

Note: The NOINT option changes the definition of the R-Square statistic to:
1 - (Residual Sum of Squares/Uncorrected Total Sum of Squares).

Parameter Estimates					
Variable	DF	Parameter Estimate	Standard Error	t Value	Pr > |t|
X	1	0.465402	0.055510	8.38	<.0001
YT3	1	0.109250	0.241150	0.45	0.6563
YT4	1	-1.77795	0.597982	-2.97	0.0085

Output 8.8.1: GMANOVA–SUR Several Population Dog Data: Estimate of B Using PROC SYSLIN — YT1 and YT2.

Model	YT1
Dependent Variable	YT1

Analysis of Variance					
Source	DF	Sum of Squares	Mean Square	F Value	Pr > F
Model	7	4172.931	596.1331	303.37	<.0001
Error	29	56.98567	1.965023		
Uncorrected Total	36	4229.917			

Root MSE	1.40179	R-Square	0.98653
Dependent Mean	10.71109	Adj R-Sq	0.98328
Coeff Var	13.08730		

Note: The NOINT option changes the definition of the R-Square statistic to:
1 - (Residual Sum of Squares/Uncorrected Total Sum of Squares).

Parameter Estimates					
Variable	DF	Parameter Estimate	Standard Error	t Value	Pr > |t|
X1	1	12.36734	0.488032	25.34	<.0001
X2	1	9.448517	0.488482	19.34	<.0001
X3	1	10.74846	0.522161	20.58	<.0001
X4	1	10.35984	0.477636	21.69	<.0001
YT5	1	1.012136	0.795901	1.27	0.2136
YT6	1	-0.20819	0.894399	-0.23	0.8176
YT7	1	0.266191	1.338977	0.20	0.8438

Model	YT2
Dependent Variable	YT2

Analysis of Variance					
Source	DF	Sum of Squares	Mean Square	F Value	Pr > F
Model	7	11.78700	1.683856	4.64	0.0014
Error	29	10.52300	0.362862		
Uncorrected Total	36	22.31000			

Root MSE	0.60238	R-Square	0.52833
Dependent Mean	0.41576	Adj R-Sq	0.41448
Coeff Var	144.88627		

Note: The NOINT option changes the definition of the R-Square statistic to:
1 - (Residual Sum of Squares/Uncorrected Total Sum of Squares).

Parameter Estimates					
Variable	DF	Parameter Estimate	Standard Error	t Value	Pr > |t|
X1	1	0.800823	0.209718	3.82	0.0007
X2	1	-0.11175	0.209911	-0.53	0.5985
X3	1	0.736464	0.224384	3.28	0.0027
X4	1	0.352830	0.205250	1.72	0.0963
YT5	1	-0.47040	0.342016	-1.38	0.1795
YT6	1	0.064871	0.384343	0.17	0.8671
YT7	1	-0.09138	0.575387	-0.16	0.8749

Output 8.8.2: GMANOVA–SUR Several Population Dog Data: Estimate of B Using
PROC SYSLIN — YT3 and YT4.

Model	YT3
Dependent Variable	YT3

Analysis of Variance					
Source	DF	Sum of Squares	Mean Square	F Value	Pr > F
Model	7	7.816146	1.116592	8.63	<.0001
Error	29	3.751473	0.129361		
Uncorrected Total	36	11.56762			

Root MSE	0.35967	R-Square	0.67569
Dependent Mean	-0.26186	Adj R-Sq	0.59741
Coeff Var	-137.35051		

Note: The NOINT option changes the definition of the R-Square statistic to:
1 - (Residual Sum of Squares/Uncorrected Total Sum of Squares).

Parameter Estimates					
Variable	DF	Parameter Estimate	Standard Error	t Value	Pr > |t|
X1	1	-0.48903	0.125218	-3.91	0.0005
X2	1	-0.13986	0.125333	-1.12	0.2736
X3	1	-0.27789	0.133975	-2.07	0.0470
X4	1	-0.14167	0.122550	-1.16	0.2571
YT5	1	-1.18513	0.204210	-5.80	<.0001
YT6	1	-0.03110	0.229482	-0.14	0.8932
YT7	1	-0.10981	0.343551	-0.32	0.7515

Model	YT4
Dependent Variable	YT4

Analysis of Variance					
Source	DF	Sum of Squares	Mean Square	F Value	Pr > F
Model	7	3.830252	0.547179	7.28	<.0001
Error	29	2.179748	0.075164		
Uncorrected Total	36	6.010000			

Root MSE	0.27416	R-Square	0.63731
Dependent Mean	-0.12021	Adj R-Sq	0.54977
Coeff Var	-228.07433		

Note: The NOINT option changes the definition of the R-Square statistic to:
1 - (Residual Sum of Squares/Uncorrected Total Sum of Squares).

Parameter Estimates					
Variable	DF	Parameter Estimate	Standard Error	t Value	Pr > |t|
X1	1	-0.46662	0.095448	-4.89	<.0001
X2	1	-0.02437	0.095536	-0.26	0.8004
X3	1	-0.19709	0.102123	-1.93	0.0635
X4	1	0.004771	0.093415	0.05	0.9596
YT5	1	0.185747	0.155661	1.19	0.2424
YT6	1	-0.69693	0.174925	-3.98	0.0004
YT7	1	0.649324	0.261875	2.48	0.0192

Output 8.8.3: GMANOVA–SUR Several Population Dog Data.

Test Results for Variable COIN			
Num DF	Den DF	F Value	Pr > F
3	116	5.55	0.0014

design multivariate (MDM) regression models. PROC SYSLIN performs the analysis of systems of equations with one dependent (endogenous) variable and several regressor (instrument) variables. The SAS procedure provides numerous methods for estimating model parameters for a system of multiple regression equations. For a detailed discussion of several of the estimation methods, see Greene (1992). PROC SYSLIN provides the SUR estimator of β which substitutes for the unknown covariance matrix Σ the unbiased estimate S. Replacing S with the ML estimate $\widehat{\Sigma}$, the Feasible Generalized Least Squares (FGLS) estimator iterates to the ML estimate of β, the FIML option in PROC SYSLIN. For the SUR model, the 2SLS option in PROC SYSLIN is equal to the OLS estimator, and the 3SLS estimator is equal to the SUR estimator. The IT3SLS estimator does not converge to the ML estimator since S, and not the ML estimate of Σ is used at each step of the iteration process. For additional detail, one should consult the SAS/ETS (time series) on-line documentation.

To illustrate the analysis of the SUR model using PROC SYSLIN, we use the Grunfeld investment data given in Greene (1992, p. 445). The data consist of 20 yearly observations of gross investment, firm value, and stock value of plant and equipment for five firms. The model to be estimated is

$$I_{it} = \beta_0 + \beta_1 F_{it} + \beta_2 C_{it} + e_{it} \text{ for } i = 1, 2, \ldots, 5 \qquad (8.164)$$

companies. The companies are General Motors (GM), Chrysler (Ch), General Electric (GE), Westinghouse Electric (WE), and U. S. Steel (US). Program 8_9.sas contains the SAS code for the example. The data are in the data set 8_9.sas.

8.9.1 Results and Interpretation

When using the SYSLIN procedure, one must specify the method of estimation on the PROC SYSLIN statement. The default option is the OLS estimation method which ignores the dependency across model. The parameter estimates for the OLS (Table 8.1) and SUR (Table 8.2) solutions follow. Also provided in the Output 8.9.1-8.9.2 is the FIML solution.

Because the SUR estimates take into account the covariance matrix across equations, ignored by the OLSE, there is a small gain in efficiency (smaller standard errors) for the SUR solution. The gain in efficiency can also be measured by investigating the diagonal elements of the Cross Model Inverse Correlation matrix, the model inflation factor (MIF). For the example, the values are between 1.23 and 3.36. The ML estimate of β is found in Output 8.9 under the FIML results. The difference between the SUR and FIML estimates is small. While both methods are asymptotically efficient, both methods are sensitive to specification error and sample size.

Output 8.9.1: SUR — Greene–Grunfeld Investment Data: OLS Estimation.

Parameter Estimates						
Variable	DF	Parameter Estimate	Standard Error	t Value	Pr > |t|	Variable Label
Intercept	1	-149.782	105.8421	-1.42	0.1751	Intercept
gm_f	1	0.119281	0.025834	4.62	0.0002	Market Value GM prior yr
gm_c	1	0.371445	0.037073	10.02	<.0001	Stock Value GM prior yr

Parameter Estimates						
Variable	DF	Parameter Estimate	Standard Error	t Value	Pr > |t|	Variable Label
Intercept	1	-6.18996	13.50648	-0.46	0.6525	Intercept
ch_f	1	0.077948	0.019973	3.90	0.0011	Market Value CH prior yr
ch_c	1	0.315718	0.028813	10.96	<.0001	Stock Value CH prior yr

Parameter Estimates						
Variable	DF	Parameter Estimate	Standard Error	t Value	Pr > |t|	Variable Label
Intercept	1	-9.95631	31.37425	-0.32	0.7548	Intercept
ge_f	1	0.026551	0.015566	1.71	0.1063	Market Value GE prior yr
ge_c	1	0.151694	0.025704	5.90	<.0001	Stock Value GE prior yr

Parameter Estimates						
Variable	DF	Parameter Estimate	Standard Error	t Value	Pr > |t|	Variable Label
Intercept	1	-0.50939	8.015289	-0.06	0.9501	Intercept
we_f	1	0.052894	0.015707	3.37	0.0037	Market Value WE prior yr
we_c	1	0.092406	0.056099	1.65	0.1179	Stock Value WE prior yr

Parameter Estimates						
Variable	DF	Parameter Estimate	Standard Error	t Value	Pr > |t|	Variable Label
Intercept	1	-30.3685	157.0477	-0.19	0.8490	Intercept
us_f	1	0.156571	0.078886	1.98	0.0635	Market Value US prior yr
us_c	1	0.423866	0.155216	2.73	0.0142	Stock Value US prior yr

Output 8.9.2: SUR — Greene–Grunfeld Investment Data: FIML Estimation.

						Parameter Estimates
Variable	DF	Parameter Estimate	Standard Error	t Value	Pr > |t|	Variable Label
Intercept	1	-162.364	97.03216	-1.67	0.1126	Intercept
gm_f	1	0.120493	0.023460	5.14	<.0001	Market Value GM prior yr
gm_c	1	0.382746	0.035542	10.77	<.0001	Stock Value GM prior yr

						Parameter Estimates
Variable	DF	Parameter Estimate	Standard Error	t Value	Pr > |t|	Variable Label
Intercept	1	0.504304	12.48742	0.04	0.9683	Intercept
ch_f	1	0.069546	0.018328	3.79	0.0014	Market Value CH prior yr
ch_c	1	0.308545	0.028053	11.00	<.0001	Stock Value CH prior yr

						Parameter Estimates
Variable	DF	Parameter Estimate	Standard Error	t Value	Pr > |t|	Variable Label
Intercept	1	-22.4389	27.67879	-0.81	0.4287	Intercept
ge_f	1	0.037291	0.013301	2.80	0.0122	Market Value GE prior yr
ge_c	1	0.130783	0.023916	5.47	<.0001	Stock Value GE prior yr

						Parameter Estimates
Variable	DF	Parameter Estimate	Standard Error	t Value	Pr > |t|	Variable Label
Intercept	1	1.088877	6.788627	0.16	0.8745	Intercept
we_f	1	0.057009	0.012324	4.63	0.0002	Market Value WE prior yr
we_c	1	0.041506	0.044689	0.93	0.3660	Stock Value WE prior yr

						Parameter Estimates
Variable	DF	Parameter Estimate	Standard Error	t Value	Pr > |t|	Variable Label
Intercept	1	85.42325	121.3481	0.70	0.4910	Intercept
us_f	1	0.101478	0.059421	1.71	0.1059	Market Value US prior yr
us_c	1	0.399991	0.138613	2.89	0.0103	Stock Value US prior yr

Table 8.1: SUR — Greene–Grunfeld Investment Data: OLS Estimates (Standard Errors).

	β_0		β_1		β_3	
GM	−149.782	(105.84)	0.11928	(0.0258)	0.37145	(0.0371)
CH	−6.18996	(13.51)	0.77950	(0.01997)	0.31572	(0.02881)
GE	−9.95630	(31.37)	0.02655	(0.01557)	0.15169	(0.02570)
WE	−0.50939	(8.015)	0.05289	(0.01571)	0.09241	(0.05610)
US	−30.36850	(157.05)	0.15657	(0.07889)	0.42387	(0.15522)

Table 8.2: SUR — Greene–Grunfeld Investment Data: FIML Estimates (Standard Errors).

	β_0		β_1		β_3	
GM	−162.3641	(97.03)	0.12049	(0.0235)	0.38275	(0.0355)
CH	0.5043	(12.49)	0.06955	(0.01833)	0.30850	(0.02805)
GE	−22.4390	(27.68)	0.03729	(0.01330)	0.13078	(0.02392)
WE	1.0889	(6.788)	0.05701	(0.01232)	0.04150	(0.04469)
US	85.4230	(121.35)	0.10145	(0.05942)	0.39990	(0.13861)

With serious misspecification and small sample, system methods are to be avoided; instead, one should employ OLS or 2SLS methods.

For the investment data, we may also test the equivalence of the regression equations, the test of coincidence:

$$H : \beta_1 = \beta_2 = \cdots = \beta_5 \qquad (8.165)$$

for the five firms. The test is performed using the STEST statement. The approximate F statistic is 116.7857 with P-value 0.0001. Thus, the hypothesis of coincidence is rejected.

In our analysis of the investment data, we fit to the data a model that allowed for variability of the β_i across firms by estimating the fixed parameters β_i for each firm. Another approach to modeling parameter variation across firms as examined by Greene (1992, p. 459) represents parameter heterogeneity as stochastic or random variation. Thus, the parameters β_i are represented as a fixed effect plus a vector for random variation

$$\beta_i = \beta + v_i \quad i = 1, 2, \ldots, n. \qquad (8.166)$$

Then for each firm we have the linear model:

$$y_i = X_i\beta_i + e_i = X_i\beta + (X_iv_i + e_i) = X_i\beta + u_i. \qquad (8.167)$$

More generally, the linear model is a special case of the more general random coefficient regression model (Swamy, 1971)

$$y_i = X_i \beta_i + Z_i b_i + e_i. \tag{8.168}$$

In the random coefficient regression model, $y_i (n_i \times 1)$ is a vector of repeated measurements at n_i time points for the ith firm (or subject), so the number of points may be different for each firm: $X_i(n_i \times p)$ and $Z_i(n_i \times k)$ are known design matrices; $\beta_{p \times 1}$ is a fixed vector of parameters common for all firms; $b_i(k \times 1)$ is a random component representing stochastic variation across firms; and $e_i(n_i \times 1)$ is a vector of random errors. For the random coefficient model, we further assume that the random vector b_i have mean zero and common covariance structure D and that the b_i are uncorrelated, $\text{cov}(b_i, b_{i'}) = 0$, $\text{cov}(e_i, e_{i'}) = 0$, and the $\text{cov}(b_i, e_{i'}) = 0$ for $i \neq i'$. Thus, the covariance structure for the vector y_i is

$$\text{cov}(y_i) = \Omega_i = Z_i D Z_i' + \Psi_i. \tag{8.169}$$

Stacking the vector one upon another into a single vector, the general linear model becomes

$$\begin{pmatrix} y_1 \\ y_2 \\ \vdots \\ y_n \end{pmatrix} = \begin{pmatrix} X_1 \\ X_2 \\ \vdots \\ X_n \end{pmatrix} \beta + \begin{pmatrix} Z_1 & 0 & \cdots & 0 \\ 0 & Z_2 & \cdots & 0 \\ \vdots & \vdots & \ddots & \vdots \\ 0 & 0 & \cdots & Z_n \end{pmatrix} \begin{pmatrix} b_1 \\ b_2 \\ \vdots \\ b_n \end{pmatrix} + \begin{pmatrix} e_1 \\ e_2 \\ \cdots \\ e_n \end{pmatrix} \tag{8.170}$$

$$y_{N \times 1} = X_{N \times p} \beta_{p \times 1} + Z_{N \times N} b_{N \times 1} + e_{N \times 1} \tag{8.171}$$

where $N = \sum_i n_i$. Employing the vector form of the model, we have that the

$$\text{cov}(b) = \begin{pmatrix} D & 0 & \cdots & 0 \\ 0 & D & \cdots & 0 \\ \vdots & \vdots & \ddots & \vdots \\ 0 & 0 & \cdots & D \end{pmatrix} = I_n \otimes D = V \tag{8.172}$$

and the

$$\text{cov}(e) = \begin{pmatrix} \Psi_1 & 0 & \cdots & 0 \\ 0 & \Psi_2 & \cdots & 0 \\ \vdots & \vdots & \ddots & \vdots \\ 0 & 0 & \cdots & \Psi_n \end{pmatrix} = \Psi. \tag{8.173}$$

Hence, the

$$\text{cov}(y) = \begin{pmatrix} \Omega_1 & 0 & \cdots & 0 \\ 0 & \Omega_2 & \cdots & 0 \\ \vdots & \vdots & \ddots & \vdots \\ 0 & 0 & \cdots & \Omega_n \end{pmatrix} = \Omega. \tag{8.174}$$

Under multivariate normality of y,

$$y \sim N(X\beta, \Omega), \tag{8.175}$$

the ML estimate of the parameter vector β, the mean in the random process $\beta_i + v_i$ is

$$\hat{\beta} = [X'(ZVZ' + \Psi)^{-1}X'(ZVZ' + \Psi)^{-1}]y$$

$$= \left(\sum_{i=1}^{n} X_i'\Omega_i^{-1}X_i\right)^{-1} \left(\sum_{i=1}^{n} X_i'\Omega_i^{-1}y_i\right) \tag{8.176}$$

where $\Omega_i = Z_i DZ_i' + \Psi_i$ for $i = 1, 2, \ldots, n$. Because D and Ψ_i are usually unknown in practice, we cannot apply the model directly, but must estimate D and Ψ_i in order to estimate β. More will be said about this model in Chapter 11 since the random coefficient regress model is equivalent to a two-level hierarchical linear model. In SAS, PROC MIXED is used to analyze these designs.

8.10   TWO-PERIOD CROSSOVER DESIGN WITH CHANGING COVARIATES

We have illustrated how the SUR model may be used to estimate the parameters of the growth curve model and how to analyze data containing both cross-sectional and time series data. We now show how Patel (1986) MANOVA-GMANOVA model may be formulated as SUR model to analyze repeated measures data with changing covariates. This topic is addressed in more detail in Chapter 13 when we discuss Structural Equation Modeling. To illustrate the theory, we consider a simple two-group, cross-sectional example with $p = 3$ repeated measures and $k = 2$ covariates.

From (8.73), the model for our example, considered by Patel (1986) using likelihood methods, has the form

$$\mathcal{E}(Y_{n\times 3}) = X_{n\times 2}B_{2\times 3} + Z_{1(n\times 3)}\Gamma_{1(3\times 3)} + Z_{2(n\times 3)}\Gamma_{2(3\times 3)}. \tag{8.177}$$

Expanding the model, we write

$$\mathcal{E}\begin{pmatrix} Y_1 \\ Y_2 \end{pmatrix} = \begin{pmatrix} 1 & 0 \\ 0 & 1 \end{pmatrix}\begin{pmatrix} \mu_{11} & \mu_{12} & \mu_{13} \\ \mu_{21} & \mu_{22} & \mu_{23} \end{pmatrix} + \begin{pmatrix} z_{11} & z_{12} & z_{13} \end{pmatrix}\begin{pmatrix} \gamma_{11} & 0 & 0 \\ 0 & \gamma_{12} & 0 \\ 0 & 0 & \gamma_{13} \end{pmatrix}$$

$$= \begin{pmatrix} z_{21} & z_{22} & z_{23} \end{pmatrix}\begin{pmatrix} \gamma_{21} & 0 & 0 \\ 0 & \gamma_{22} & 0 \\ 0 & 0 & \gamma_{23} \end{pmatrix}$$

$$= X \begin{pmatrix} \beta_1 & \beta_2 & \beta_3 \end{pmatrix} + \sum_{j=1}^{2} \begin{pmatrix} z_{j1}\gamma_{j1} & z_{j2}\gamma_{j2} & z_{j3}\gamma_{j3} \end{pmatrix}$$

$$= \begin{pmatrix} X\beta_1 + Z_{.1}\gamma_{.1} & X\beta_2 + Z_{.2}\gamma_{.2} & X\beta_3 + Z_{.3}\gamma_{.3} \end{pmatrix}$$

$$= \begin{pmatrix} X_1\theta_{11} & X_2\theta_{22} & X_3\theta_{33}\theta_{33} \end{pmatrix}, \tag{8.178}$$

where

$$Z_{\cdot i} = \begin{pmatrix} z_{1i} & z_{2i} \end{pmatrix},$$
$$\gamma'_{\cdot i} = \begin{pmatrix} \gamma_{1i} & \gamma_{2i} \end{pmatrix},$$
$$X_i = \begin{pmatrix} X & Z_{\cdot i} \end{pmatrix} = \begin{pmatrix} X & z_{1i} & z_{2i} \end{pmatrix},$$
$$\theta'_{ii} = \begin{pmatrix} \beta'_i & \gamma'_{\cdot i} \end{pmatrix} = \begin{pmatrix} \mu_{1i} & \mu_{2i} & \gamma_{i1} & \gamma_{i2} \end{pmatrix}.$$

Thus, letting y_i represent the ith column of the data matrix Y, we have the SUR model in (8.86) with

$$y_i = X_i \theta_{ii} + e_{ii} \quad i = 1, 2, 3. \tag{8.179}$$

To illustrate the application of the SUR model for a design with changing covariates, we consider the two-period crossover design analyzed by Patel (1986) who used likelihood ratio methods. Table 8.3 contains the data on the FEV_1 (forced expired volume in one second) obtained from a cross over trial involving 17 patients with mild-to-moderate bronchial asthma. The data are provided in Patel (1986). Patients were randomly assigned to two treatment sequences in which acute effects of single oral doses of two active drugs are compared. The variable z_i contains baseline values for Period i, and the variable y_i contains the average of the responses obtained two and three hours after the initiation of a dose for Period i ($i = 1, 2$). Thus, we have a repeated measures design with $p = 2$ variables and one covariate. The model for the study is

$$\mathcal{E}(Y) = \begin{pmatrix} 1 & 0 \\ 0 & 1 \end{pmatrix} \begin{pmatrix} \mu_{11} & \mu_{12} \\ \mu_{21} & \mu_{22} \end{pmatrix} + \begin{pmatrix} z_{11} & z_{12} \end{pmatrix} \begin{pmatrix} \gamma_{11} & 0 \\ 0 & \gamma_{12} \end{pmatrix}. \tag{8.180}$$

The data are analyzed using the SAS code in Program 8_10.sas.

8.10.1   Results and Interpretation

To analyze the data in Table 8.3 using PROC SYSLIN, we arrange the data as shown in Output 8.10.1. Because PROC SYSLIN fits an intercept, the design matrix must be reparameterized to a constant term plus the treatment effect. Hence, the parameter matrix is defined:

$$\begin{pmatrix} \mu_{11} & \mu_{12} \\ \mu_{21} & \mu_{22} \end{pmatrix} \equiv \begin{pmatrix} \mu_1 & \alpha_{11} \\ \mu_2 & \alpha_{21} \end{pmatrix}$$

where $\delta = \alpha_{11} - \alpha_{21}$ is the difference in the treatment effects.

Using PROC SYSLIN with the SUR option, the unconstrained design parameters for the model are estimated:

$$\begin{pmatrix} \hat{\mu}_1 & \hat{\alpha}_{11} \\ \hat{\mu}_2 & \hat{\alpha}_{21} \end{pmatrix} = \begin{pmatrix} 0.628801 & -0.186543 \\ 0.022622 & 0.07496 \end{pmatrix}$$

Table 8.3: FEV Measurements in Cross Over Trial of Two Active Drugs in Asthmatic Patients.

| | Treatment Sequence 1 | | | |
Patient	Period 1	Drug A	Period 2	Drug B
	z_1	y_1	z_2	y_2
1	1.09	1.28	1.24	1.33
2	1.38	1.60	1.90	2.21
3	2.27	2.46	2.19	2.43
4	1.34	1.41	1.47	1.81
5	1.31	1.40	0.85	0.85
6	0.96	1.12	1.12	1.20
7	0.66	0.90	0.78	0.90
8	1.69	2.41	1.90	2.79
Mean	1.34	1.57	1.43	1.69
SD	0.49	0.47	0.52	0.73

| | Treatment Sequence 2 | | | |
Patient	Period 1	Drug A	Period 2	Drug B
	z_1	y_1	z_2	y_2
9	1.74	3.06	1.54	1.38
10	2.41	2.68	2.13	2.10
11	3.05	2.60	2.18	2.32
12	1.20	1.48	1.41	1.30
13	1.70	2.08	2.21	2.34
14	1.89	2.72	2.05	2.48
15	0.89	1.94	0.72	1.11
16	2.41	3.35	2.83	3.23
17	0.96	1.16	1.01	1.25
Mean	1.81	2.34	1.79	1.95
SD	0.73	0.73	0.67	0.72

z = baseline measurement

y = average of measurement made two and three hours after administration of the drug

Output 8.10.1: SUR — 2-Period Crossover Design with Changing Covariate.

Obs	group	x	y1	z1	y2	z2	x1	x2
1	1	1	1.28	1.09	1.33	1.24	1	0
2	1	1	1.60	1.38	2.21	1.90	1	0
3	1	1	2.46	2.27	2.43	2.19	1	0
4	1	1	1.41	1.34	1.81	1.47	1	0
5	1	1	1.40	1.31	0.85	0.85	1	0
6	1	1	1.12	0.96	1.20	1.12	1	0
7	1	1	0.90	0.66	0.90	0.78	1	0
8	1	1	2.41	1.69	2.79	1.90	1	0
9	1	-1	3.06	1.74	1.38	1.54	0	1
10	1	-1	2.68	2.41	2.10	2.13	0	1
11	1	-1	2.60	3.05	2.32	2.18	0	1
12	1	-1	1.48	1.20	1.30	1.41	0	1
13	1	-1	2.08	1.70	2.34	2.21	0	1
14	1	-1	2.72	1.89	2.48	2.05	0	1
15	1	-1	1.94	0.89	1.11	0.72	0	1
16	1	-1	3.35	2.41	3.23	2.83	0	1
17	1	-1	1.16	0.96	1.25	1.01	0	1

and the covariate parameters are estimated as

$$\hat{\gamma}_{11} = 0.84504 \quad \hat{\gamma}_{12} = 1.115726.$$

To estimate the difference between the two drugs, we have that the treatment effect is:

$$\hat{\delta} = \hat{\alpha}_{11} - \hat{\alpha}_{21} = -0.257039.$$

Patel (1986) proposed an iterative scheme to obtain ML estimates for the model parameters where the initial estimates are obtained from the sum of squares and products (SSP) matrix of y_1, z_1, y_2, z_2. Using his method, he obtained estimates of μ_{ij} where i = sequence and j = period:

$$\begin{pmatrix} \hat{\mu}_{11} & \hat{\mu}_{12} \\ \hat{\mu}_{21} & \hat{\mu}_{22} \end{pmatrix} = \begin{pmatrix} 0.422 & 0.093 \\ 0.815 & -0.048 \end{pmatrix}, \hat{\gamma}_{11} = 0.845, \text{ and } \hat{\gamma}_{12} = 1.15$$

so that the treatment effect $\hat{\delta} = \frac{(\hat{\mu}_{11} + \hat{\mu}_{22} - \hat{\mu}_{12} - \hat{\mu}_{21})}{2} = -0.257$, which is the same as obtained using the SUR model. The corresponding estimates for the means using the SUR model are:

$$\hat{\mu}_{11} = 0.628801 - 0.186543 = 0.442258$$
$$\hat{\mu}_{12} = 0.628801 + 0.186543 = 0.815343$$
$$\hat{\mu}_{21} = 0.022622 + 0.074960 = 0.097582$$
$$\hat{\mu}_{22} = 0.022622 - 0.074960 = -0.052338.$$

For this example there is close agreement between Patel's iterative ML estimates and the SUR estimate (these data are reanalyzed in Chapter 11 using PROC MIXED).

To obtain an estimate of the covariance matrix Σ, Patel (1986) suggests dividing the residual covariate matrix

$$\widehat{V} = (Y - X\widehat{B} - Z\widehat{\Gamma})'(Y - X\widehat{B} - Z\widehat{\Gamma})$$

by n so that $\widehat{\Sigma} = \frac{\widehat{V}}{n}$. The degrees of freedom associated with this estimator is $n - m = n - \text{rank}(X) - r$. Program 8_10.sas contains the PROC IML code to perform this analysis. The results are found in Output 8.10.2. From the PROC IML output, we see that the estimates of V and Σ using the SUR estimates for $\widehat{B}$ and $\widehat{\Gamma}$ are

$$\widehat{V} = \begin{pmatrix} 2.494 & 0.498 \\ 0.498 & 0.858 \end{pmatrix} \text{ and } \widehat{\Sigma} = \begin{pmatrix} 0.1467 & 0.0293 \\ 0.0293 & 0.0505 \end{pmatrix}$$

which are in agreement with Patel's estimates to three significant figures. The estimate of the covariance matrix Σ is computed by PROC SYSLIN, defined in (8.94), and is labeled the Cross Model Covariance matrix in the output. It is approximately equal to the unbiased estimator

$$S = \frac{\widehat{V}}{(n - m)}.$$

The two estimators $\widehat{\Sigma}$ and S are asymptotically equivalent.

To obtain the asymptotic covariance matrix of $\theta' = (\hat{\theta}_{11}, \hat{\theta}_{12}, \ldots, \hat{\theta}_{pp})$, using the SUR model, we use expression (8.97). The covariance matrix is

$$\widehat{\Sigma}^{-1} \otimes X_i'X_j)^{-1}$$

where $\widehat{\Sigma}$ is any consistent estimator of Σ. For our example, $X_1 = (X, z_{11})$ and $X_2 = (X, z_{12})$. The asymptotic covariance matrix of the parameter vector

$$\hat{\theta}' = \begin{pmatrix} \hat{\mu}_{11} & \hat{\mu}_{21} & \hat{\gamma}_{11} & \hat{\mu}_{12} & \hat{\mu}_{22} & \hat{\gamma}_{12} \end{pmatrix}$$

is given in the output using $\widehat{\Sigma} = \frac{\widehat{V}}{n}$. The asymptotic variance of the treatment effect $\hat{\delta} = \frac{(\hat{\mu}_{11} + \hat{\mu}_{22} - \hat{\mu}_{12} - \hat{\mu}_{21})}{2}$ is 0.00736 which compares favorably with the ML estimate of 0.0094 calculated by Patel. The ease of the estimation procedure for the SUR model supports its use over the iterative ML approach proposed by Patel.

To test the hypotheses given in (8.106) and (8.107), Patel derived likelihood ratio tests. In particular he considered the hypotheses

$$H_{01} : CBA = 0$$
$$H_{02} : \Gamma_i = 0 \text{ for } i > r_1 < r$$

and H_{03}, the equality of equal carryover effects as measured in terms of the baseline values. Because the exact distribution of the likelihood ratio statistic is unknown, he

Output 8.10.2: SUR — 2-Period Crossover Design with Changing Covariate: Estimates using PROC IML.

SS and SP matrix of (y1,z1,y2,z2)			
S			
6.5142389	4.8637944	5.4266444	4.5598083
4.8637944	5.8809722	5.0529222	4.6975917
5.4266444	5.0529222	7.9154222	6.2118667
4.5598083	4.6975917	6.2118667	5.4654875

Residual SS matrix	
VHAT	
2.4935959	0.4982712
0.4982712	0.8579608

Estimate of SUR Covariance Matrix using vhat/n	
SIGMA	
0.1466821	0.0293101
0.0293101	0.0504683

Asymptotic Covariance Matrix of Parameters					
COVEHAT					
0.0781091	0.0806916	-0.044691	0.0156078	0.0161238	-0.00893
0.0746172	0.1170274	-0.055789	0.01491	0.0233845	-0.011148
-0.041763	-0.056378	0.031225	-0.008345	-0.011266	0.0062394
0.0156078	0.0161238	-0.00893	0.0268747	0.0277632	-0.015377
0.01491	0.0233845	-0.011148	0.0256732	0.0402651	-0.019195
-0.008345	-0.011266	0.0062394	-0.014369	-0.019398	0.0107434

CHI_T	DFT	P_T
7.6545847	2	0.0217685

CHI_G	DFGAMMA	P_G
70.611637	1	0

CHI_TP	DFTP	P_TP
0.7420086	1	0.3890182

used the asymptotic chi-square distribution to obtain approximate critical values for the statistics. Patel also developed the approximate test based upon Wilks' Λ criterion. Using the latter approach, Verbyla (1988) suggests obtaining SUR estimates with constraints imposed by the hypothesis to obtain approximate test based upon Wilks' Λ criterion. For example, for the hypothesis H_{01},

$$E = A'(Y - X\widehat{B}_{\Omega_0} - Z\widehat{\Gamma}_{\Omega_0})'(Y - X\widehat{B}_{\Omega_0} - Z\widehat{\Gamma}_{\Omega_0})A$$
$$H = A'(Y - X\widehat{B}_\omega - Z\widehat{\Gamma}_\omega)'(Y - X\widehat{B}_\omega - Z\widehat{\Gamma}_\omega)A$$
$$U = \frac{|E|}{|H|} \simeq U(u, \nu_h, \nu_e)$$

where $u = \operatorname{rank}(A)$, $\nu_h = \operatorname{rank}(C)$, $k = \operatorname{rank}(X)$, and $\nu_e = n - k - r$, and $\widehat{B}_\omega$ and $\widehat{\Gamma}_\omega$ are SUR estimates subject to the restriction of the hypothesis. However, this process is also complicated since it involves the sum of profiles model.

Instead of either of these approaches, we propose using the Wald statistic given in (8.136)

$$X^2 = (C_*\tilde{\theta} - \xi)\{C_*[D'(\widehat{\Sigma} \otimes I_n)^{-1}D]C_*\}^{-1}(C_*\tilde{\theta} - \xi),$$

which has an asymptotic chi-square distribution with degrees of freedom $\nu = \operatorname{rank}(C_*)$. To test for differences in treatments, no covariates, and no treatment by period interaction, the hypothesis test matrices are:

$$H_{0t} : C_* = \begin{pmatrix} 1 & -1 & 0 & 0 & 0 & 0 \\ 0 & 0 & 0 & -1 & 1 & 0 \end{pmatrix}$$
$$H_\Gamma : C_* = \begin{pmatrix} 0 & 0 & 1 & 0 & 0 & 1 \end{pmatrix}$$
$$H_{t\times p} : C_* = \begin{pmatrix} 1 & -1 & 0 & 1 & -1 & 0 \end{pmatrix}$$

where the Wald statistics have asymptotic chi-square distributions with 2, 1, and 1 degrees of freedom, respectively. The PROC IML code to perform these tests is included in Program 8_10.sas. From the output, the P-value for the test of interaction is 0.389. The corresponding P-value obtained by Pates was 0.351 using his chi-square LR approximation and 0.404 using his approximate Wilks' Λ criterion. For the test of treatment the P-value for the Wald statistic is 0.022. Patel obtained P-values of 0.001 and 0.020 for his two methods. While all three procedures are asymptotically equivalent, it is unknown how each perform for small samples.

8.11 REPEATED MEASUREMENTS WITH CHANGING COVARIATES

For our next example, we analyze repeated measures with time varying covariates using the SUR model and the Wald statistic. The structure of the covariance matrix can be arbitrary, and we do not require equality of the covariates across variables. To illustrate the approach, data from Winer (1971, p. 806) are used. The design involves a pretest, posttest with varying covariates. An exact solution exists assuming a common regression equation. And, the circularity condition is met since the covariance of the difference is a constant; hence all we require is homogeneity of variances across groups. We use the data to show how one may test for equal slopes and intercepts using the SUR model. The SAS code to perform the calculations is in Program 8_11.sas.

8.11.1 Results and Interpretation

The data for this example are in the data set 8_11.dat and are listed in the Output. Using PROC SYSLIN, the estimates of the model parameters are:

$$\begin{pmatrix} \hat{\mu}_1 & \hat{\mu}_2 \\ \hat{\alpha}_{11} & \hat{\alpha}_{21} \\ \hat{\alpha}_{12} & \hat{\alpha}_{22} \end{pmatrix} = \begin{pmatrix} 4.991191 & 8.677820 \\ 1.037876 & 2.084727 \\ -1.925506 & -3.104175 \end{pmatrix}.$$

Transforming the estimates to cell means, we have that

$$\widehat{B} = \begin{pmatrix} \hat{\mu}_{11} & \hat{\mu}_{12} \\ \hat{\mu}_{21} & \hat{\mu}_{22} \\ \hat{\mu}_{32} & \hat{\mu}_{32} \end{pmatrix} = \begin{pmatrix} 6.325 & 11.102 \\ 3.362 & 5.913 \\ 5.287 & 9.018 \end{pmatrix}, \text{ and}$$

$$\widehat{\Gamma} = \begin{pmatrix} \hat{\gamma}_{11} & 0 \\ 0 & \hat{\gamma}_{22} \end{pmatrix} = \begin{pmatrix} 0.890147 & 0 \\ 0 & 0.804161 \end{pmatrix}.$$

The estimates of the parameter matrices are FGLS estimates and not ML estimates. It is a more complicated process to obtain ML estimates. This will be discussed in Chapter 11. To test the equality of intercepts and slopes, we use the Wald statistic. The hypothesis test matrices in Program 8_11.sas to test $H : CBA = 0$ are obtained be recalling that the test of H is equivalent to the test of $C_* \text{vec}(B) = 0$ where $C_* = A' \otimes C$. The PROC IML output containing the results are displayed in Output 8.11.1. The Wald statistic for the equality of intercepts is $X^2 = 17.33$ with P-value 0.0017. The test of equal slopes is not rejected (P-Value=0.2900)); hence, a common slope model is appropriate for the study. Using the matrix $\widehat{\Gamma}$, the between and within regression slopes given by Winer (1971, p. 806) are easily obtained

$$\hat{\gamma}_b = \frac{(\hat{\gamma}_{11} + \hat{\gamma}_{12})}{2} = 0.847$$

$$\hat{\gamma}_w = \frac{(8\hat{\gamma}_{11} + 9\hat{\gamma}_{12})}{17} = 0.845.$$

A reanalysis of the data with equal slopes is illustrated in Chapter 11 using a mixed linear model, PROC MIXED, and PROC GLM. An analysis of these data using the ANCOVA model is provided in BMDP Statistical Software Inc. (1992, p. 1289). The SUR model provides a methodology for the analysis of complex multivariate designs that do not require iterative methods. How well the procedure performs for small samples is, in general, unknown.

Output 8.11.1: Repeated Measures with Changing Covariates: Using Proc IML.

Residual SS matrix

VHAT	
22.474349	23.51342
23.51342	29.623092

Estimate of SUR Covariance Matrix using vhat/n

SIGMA	
2.4971499	2.6126022
2.6126022	3.2914546

Asymptotic Covariance Matrix of Parameters

COVEHAT							
2.5004014	1.7558086	2.1069703	-0.263371	2.616004	1.8369859	2.2043831	-0.275548
1.1737905	2.0679523	1.4826828	-0.185335	1.2280591	2.1635612	1.5512325	-0.193904
1.6062397	1.6907786	2.8613176	-0.253617	1.6805019	1.7689494	2.9936067	-0.265342
-0.185335	-0.19509	-0.234108	0.0292635	-0.193904	-0.20411	-0.244931	0.0306164
2.616004	1.8369859	2.2043831	-0.275548	3.2957404	2.314304	2.7771648	-0.347146
1.2280591	2.1635612	1.5512325	-0.193904	1.5471551	2.7257359	1.9543012	-0.244288
1.6805019	1.7689494	2.9936067	-0.265342	2.1171596	2.2285891	3.7714584	-0.334288
-0.193904	-0.20411	-0.244931	0.0306164	-0.244288	-0.257145	-0.308574	0.0385717

THETA							
6.324944	3.361562	5.287068	0.890147	11.102363	5.913461	9.017636	0.804161

Multivariate test of intercepts

MT							
1	0	-1	0	0	0	0	0
0	1	-1	0	0	0	0	0
0	0	0	0	1	-1	0	0
0	0	0	0	0	1	-1	0

CHI_MT	DFMT	P_MT
17.331075	4	0.0016666

Equality of Covariates

CGAMMA							
0	0	0	1	0	0	0	-1

CHI_G	DFGAMMA	P_G
1.1198432	1	0.2899522

8.12   MANOVA–GMANOVA MODEL

We saw in Chapter 7 that testing the hypothesis $H : CBA = \Gamma$ for the restricted MANOVA model $\mathcal{E}(Y) = XB$ subject to the restriction $R_1 BR_2 = 0$ is complicated since parameter estimates depend on the unknown covariance. Verbyla (1988) following (8.4) showed that the MANOVA model $\mathcal{E}(Y) = XB$ subject to the restriction $R_1 BR_2 = 0$ reduces to a mixture of the GMANOVA and MANOVA models as given in (8.5). In particular

$$XB = X_1^* \beta_{12} P^{21} + X_2^* (\beta_{21}, \beta_{22}) P^{-1} = X_1^* \beta_{12} P^{21} + X_2^* \beta_2 \qquad (8.181)$$

where $\beta_2 = (\beta_{21}, \beta_{22}) P^{-1}$, $X^* R^{-1} = (X_1^*, X_2^*) < P^{-1} = \begin{pmatrix} P^{11} \\ P^{21} \end{pmatrix}$, and P and R are defined in (8.2). Adding covariates to the model, XB in Patel's model defined in (8.73) can be replaced by XB defined above with $X_2^* \beta_2$ and the covariates $\sum_{j=1}^r Z_j \Gamma_j$ incorporated into a SUR model as in (8.75). Thus, Patel's model with restrictions has the general form under the hypothesis:

$$\mathcal{E}(XB) = X_1^* \beta_{12} P^{21} + (Z_1^* \gamma_1^*, \dots, Z_p^* \gamma_p^*) \qquad (8.182)$$

where $Z_j^* = (X_j, X_2^*)$, $\gamma_j^{*\prime} = (\gamma_j', \beta_{2 \cdot j}')$ and $\beta_{2 \cdot j}$ is the jth column of β_2., a mixture of the GMANOVA and the MANOVA models.

Verbyla (1988) recommends writing the GMANOVA-MANOVA and sum of profiles model as a MSUR model and estimating the parameters using an iterative least squares procedure following Verbyla and Venables (1988). We will not pursue this approach here since an alternative approach, which seems to work reasonably well, is to use the CGMANOVA model developed by Hecker (1987). Before discussing this approach, we turn to the analysis of the MANOVA-GMANOVA model considered by Chinchilli and Elswick (1985), a special case of the Patel model with closed form ML estimates and LR tests.

A simple application of the GMANOVA-MANOVA model is a design with m_1 groups, m_2 covariates, and p repeated measures. The MANOVA model contains the covariates and the GMANOVA model the growth curves. This design was considered by Chinchilli and Elswick (1985) using data from Danford, Hughes, and McNee (1960). The Danford et al. data set (8_12.dat) consists of 45 patients with cancerous lesions who were subjected to whole-body x-radiation. The radiation dosage was at four levels, a control and three treatment levels. A baseline measurement was taken at day 0 (pre-treatment), and the measurements were taken daily for ten consecutive days (post-treatment). For our example, only the first five measurements are used. Program 8_12 contains the SAS code to analyze both the GMANOVA model

$$\mathcal{E}(Y_{45 \times 5}) = X_{1(45 \times 4)} \Theta_{4 \times 2} X_{2(2 \times 5)} \qquad (8.183)$$

and the MANOVA-GMANOVA model

$$\mathcal{E}(Y_{45 \times 5}) = X_{1(45 \times 4)} \Theta_{4 \times 2} X_{2(2 \times 5)} + Z_{45 \times 1} \Gamma_{0(1 \times 5)} \qquad (8.184)$$

where the matrix X_2 is a matrix of orthogonal polynomials, and Z contains the base line covariate measurements.

8.12.1   Results and Interpretation

Plotting the data, we see that it appears to be linear. Ignoring the baseline data, this is confirmed by the sequential polynomial tests and the test of model adequacy (P-value $= 0.9162$). The matrix of parameters, ignoring the covariates, and assuming a linear model is

$$\widehat{\Theta}' = \left(\begin{array}{cccc} 399.67 & 312.96 & 418.73 & 453.73 \\ 49.06 & 31.35 & 34.01 & 26.90 \end{array} \right).$$

The question that now arises is whether we can improve upon these estimated by reducing their variability by including a baseline covariate in the model. This would may also improve upon the power of tests regarding model parameters. First, we must evaluate whether the MANOVA-GMANOVA model fits the data. To evaluate the adequacy of the model, we perform the goodness-of-fit test given in (8.57)-(8.58) and (8.68)-(8.69). In (8.58), we state that the model associated with the covariates is the same as the model for the repeated measures, linear for our example. Letting $Y_2 = Y H_2'$ ignoring the growth curve model, we see that $\mathcal{E}(Y_2) = Z\Gamma^*$ where $\Gamma^* = \Gamma_0 H_2'$ is the $m_2 \times p$ matrix of parameters. To test $H : \Gamma^* = 0$, from (8.65), we have that

$$\widetilde{H} = Y_2' Z (Z'Z)^{-1} Z' Y_2$$
$$\widetilde{E} = Y_2' Y_2 - \widetilde{H}$$

where $\Lambda = \frac{|\widetilde{E}|}{|\widetilde{H}+\widetilde{E}|} \sim U_{p-q}(\nu_h = m_2, \nu_e = n - m_2)$. In Program 8_12.sas, we perform these calculations using PROC IML and PROC GLM using the NOINT option in the MODEL statement and the MANOVA statement. The results are displayed in Output 8.12.1. The null hypothesis that the covariate for the model is also linear is rejected ($p = 0.0120$).

We next want to evaluate whether the growth curve portion of the model is a linear function of time with the covariate term in the model. This is not the same as the test of model adequacy ignoring the covariate, but is the test given in (8.69). The null hypothesis ω is that we have a MANOVA-GMANOVA model versus the alternative that the model is MANOVA. Again letting $Y_2 = Y H_2'$, we have that

$$\widetilde{H} = Y_2' (X(X'X)^{-1}X - Z(Z'Z)^{-1}Z') Y_2$$
$$\widetilde{E} = H_2 E H_2' = Y_2' (I - X(X'X)^{-1}X') Y_2$$

where $X = (X_1, Z)$ and $E = Y'(I - X(X'X)^{-1}X')Y$. Alternatively, we may employ PROC GLM with the NOINT option in the MODEL statement. For this test, $\Lambda = \frac{|\widetilde{E}|}{|\widetilde{H}+\widetilde{E}|} \sim U_{p-q}(\nu_h = m_1, \nu_e = n - m_1 - m_2)$. The results for the test are displayed in Output 8.12.2. Because the P-value for the test is nonsignificant ($p = 0.0766$), the MANOVA-GMANOVA model appears reasonable for the experiment.

To obtain the ML estimates of the model parameter matrices Θ, Γ_0 and Σ we use PROC IML and the expression given in (8.15)-(8.17). The results are displayed in Output 8.12.3. Thus, we have that

Output 8.12.1: MANOVA–GMANOVA Model: Test of Fit of GMANOVA versus GMANOVA–MANOVA, Using PROC IML.

Gamma Star

GAM_STR		
-0.030038	0.0412179	0.0086033

Hypothesis Matrix

H		
792.41047	-1087.35	-226.9609
-1087.35	1492.0675	311.43704
-226.9609	311.43704	65.005793

Error Matrix

E		
20686.875	4854.0334	1398.7245
4854.0334	10126.233	-144.6413
1398.7245	-144.6413	5541.2085

H + E Matrix

DEN		
21479.286	3766.6835	1171.7635
3766.6835	11618.3	166.79572
1171.7635	166.79572	5606.2143

Eigenvalues

VALUES
0.2940672
9.5286E-7
-8.187E-6

Wilks Lamda

LAMDA
0.772756

MANOVA Test Criteria and Exact F Statistics for the Hypothesis of No Overall Z Effect H = Type III SSCP Matrix for Z E = Error SSCP Matrix					
S=1 M=0.5 N=20					
Statistic	Value	F Value	Num DF	Den DF	Pr > F
Wilks' Lambda	0.77275595	4.12	3	42	0.0120
Pillai's Trace	0.22724405	4.12	3	42	0.0120
Hotelling-Lawley Trace	0.29406962	4.12	3	42	0.0120
Roy's Greatest Root	0.29406962	4.12	3	42	0.0120

Output 8.12.2: MANOVA–GMANOVA Model: Test of GMANOVA–MANOVA versus MANOVA, Using PROC IML.

Hypothesis Matrix

H		
788.50214	392.37063	-22.72803
392.37063	2522.6501	-577.5972
-22.72803	-577.5972	731.82554

Error Matrix

E		
19898.373	4461.6628	1421.4525
4461.6628	7603.5825	432.95587
1421.4525	432.95587	4809.383

H + E Matrix

DEN		
20686.875	4854.0334	1398.7245
4854.0334	10126.233	-144.6413
1398.7245	-144.6413	5541.2085

Eigenvalues

VALUES
0.4133794
0.1106642
0.0384564

Wilks Lambda

LAMBDA
0.613405

MANOVA Test Criteria and F Approximations for the Hypothesis of No Overall GROUP Effect
H = Type III SSCP Matrix for GROUP
E = Error SSCP Matrix

S=3   M=0   N=18

Statistic	Value	F Value	Num DF	Den DF	Pr > F
Wilks' Lambda	0.61340500	1.70	12	100.83	0.0766
Pillai's Trace	0.42919846	1.67	12	120	0.0819
Hotelling-Lawley Trace	0.56255143	1.74	12	62.318	0.0787
Roy's Greatest Root	0.41335519	4.13	4	40	0.0068

NOTE: F Statistic for Roy's Greatest Root is an upper bound.

Output 8.12.3: MANOVA–GMANOVA Model: GMANOVA–MANOVA Parameter
Estimates and Tests.

N	M1	M2	P	Q
45	4	1	5	2

ML estimate of sigma Gmanova Model

SIGMA1				
4485.756	4171.786	4767.3581	4927.917	4439.5311
4171.786	4374.0923	4866.1494	4977.0937	4503.038
4767.3581	4866.1494	5704.589	5753.2162	5222.0549
4927.917	4977.0937	5753.2162	6112.5601	5548.1936
4439.5311	4503.038	5222.0549	5548.1936	6001.389

ML estimate of sigma Gmanova-Manova Model

SIGMA				
4411.6058	4170.1745	4759.1262	4869.0095	4407.5881
4170.1745	4374.0573	4865.9705	4975.8135	4502.3439
4759.1262	4865.9705	5703.6751	5746.6766	5218.5087
4869.0095	4975.8135	5746.6766	6065.7619	5522.8169
4407.5881	4502.3439	5218.5087	5522.8169	5987.6283

Theta Hat Matrix

THETA	
150.83889	32.387865
102.79981	17.265428
139.86553	15.326229
132.18866	5.3569949

Gamma Zero Hat Matrix

GAMMA0				
0.7825576	0.8870342	0.9257084	0.9277608	0.9943531

Output 8.12.4: MANOVA–GMANOVA Model: For test of gamma0.

LAMDA1	CHI1	DF1
0.0524856	109.04703	6

LAMDA2	CHI2	DF2
0.772756	10.95616	3

RHO1	RHO2	GAMMA1	GAMMA2
0.8888889	0.9666667	0.0038542	0.0024222

	PVAL1	PVAL2	PVALUE
p-value of test H2: gamma0=zero	0	0.0122767	0.0122767

$$\widehat{\Theta}' = \begin{pmatrix} 150.84 & 102.80 & 139.86 & 132.19 \\ 32.39 & 17.27 & 15.33 & 5.36 \end{pmatrix}$$

$$\widehat{\Gamma}_0' = \begin{pmatrix} 0.783 & 0.887 & 0.926 & 0.928 & 0.994 \end{pmatrix}.$$

The inclusion of the covariate clearly changes the estimate of the parameter matrix Θ; more importantly inclusion of the covariates reduces the standard errors which increases the precision of the estimates.

Given the MANOVA-GMANOVA model for the data, we next show how to test hypotheses. Likelihood ratio tests for Θ and Γ_0 are given in (8.30) and (8.37), respectively. A joint LR test of Θ and Γ_0 is provided in (8.53). To perform these tests in SAS, we utilize PROC IML since standard routines are not yet available. For our example, the primary tests of interest are:

$$H_1 : C_1 \Theta A_1 = 0$$
$$H_2 : C_2 \Gamma_0 = 0.$$

If we select $C_2 = I$, the test of H_2 evaluates the significance of the covariates in the mixed GMANOVA-MANOVA model. To perform the test of $H_2 : \Gamma_0 = 0$, we use the expressions given in (8.38)-(8.47) to obtain λ_1 and λ_2, as shown in Program 8.12.sas. The results are found in Output 8.12.4. From the output, we have that $\lambda_1 = 0.0524$ and $\lambda_2 = 0.7728$. Using Theorem 5.3, we see that the chi-square statistics are

$$
\begin{aligned}
X_1^2 &= -[\nu_e - (u - \nu_h + 1)/2] \log \lambda_1 \\
&= [(n - m_1 - m_2 - p + q) - (u_1 - g_1 + 1)/2] \log \lambda_1 \\
&= -37 \log \lambda_1 \\
&= 109.05 \\
X_2^2 &= -[\nu_e - (u - \nu_h + 1)/2] \log \lambda_2 \\
&= [(n - m_2) - (u_2 - g_2 + 1)/2] \log \lambda_2 \\
&- [44.5 - 1.5] \log \lambda_2 \\
&- 42.5 \log \lambda_2 \\
&= 10.96
\end{aligned}
$$

since for our example $n = 45$, $m_1 = 4$, $m_2 = 1$, $p = 5$, and $q = 2$. Thus, for λ_1, $\nu_e = 37 = \nu_1$, $u = \mathrm{rank}(X_2) = 2 = q = u_1$, and $\nu_h = \mathrm{rank}(C_1) = g_1$. For λ_2, $\nu_e = 44 = \nu_2$, $u = p - q = 3 = u_2$, and $\nu_h = \mathrm{rank}(C_2) = 1 = g_2$. If we invoke formula (8.52),

$$
\begin{aligned}
X_1^2 &= 109.05 > \chi_{(qg_1)}^2 = \chi_6^2 = 14.4494, \\
X_2^2 &= 10.96 > \chi_{[(p-q)g_2]}^2 = \chi_3^2 = 9.3484.
\end{aligned}
$$

The P-value for the test of H_2 is approximated using Theorem 5.4. That is the sum of the probabilities

$$
P(\rho_1 n \log \lambda_1 > 109.5 \simeq (1 - \gamma_1) P(X_{\nu_1}^2 > 109.05 + \gamma_1 P(X_{\nu_1+4}^2 > 109.05 \simeq 0
$$

where $\nu_1 = u_1 g_1 = 6$, $\gamma_1 = \nu_1(u_1^2 + g_1^2 - 5)/48(\rho_1 n)^2 = 0.0039$ since $k = m_1 + m_2 + p - q = 8$, $g_1 = 3$, $u_1 = 2$, $\rho_1 = 1 - n^{-1}[k - g_1 + (u_1 + g_1 + 1)/2] = 0.8889$ and the

$$
P(\rho_2 n \log \lambda_1 > 10.96 \simeq (1 - \gamma_1) P(X_{\nu_2}^2 > 10.96 + \gamma_1 P(X_{\nu_2+4}^2 > 10.96 \simeq 0.0123
$$

where $\nu_2 = u_2 g_2 = (p - q) g_2 = 3$, $\gamma_2 = \nu_2(u_2^2 + g_2^2 - 5)/48(\rho_2 n)^2 = 0.0024$ since $k = m_2 = 1$, $g_2 = 1$, and $\rho_2 = 0.9667$. Hence, the approximate P-value for the test of H_2 is 0.0122767 as seen in the output. The test of H_2 is rejected. The covariates should remain in the model.

We next test the hypothesis H_1. Since we have fit linear polynomials to each group, the test of group differences is equivalent to testing the polynomials are coincident. For the test, the hypothesis test matrices are

$$
C_1 = \begin{pmatrix} 1 & -1 & 0 & 0 \\ 1 & 0 & -1 & 0 \\ 1 & 0 & 0 & -1 \end{pmatrix} \text{ and } A_1 = I_2
$$

which compares the control to each treatment. To perform the test of coincidence, $H_1 : C_1 \Theta A_1 = 0$, we evaluate the expressions given in (8.32)-(8.33) to obtain the Λ

Output 8.12.5: MANOVA–GMANOVA Model: Test of Coincidence.

	LAMBDA	DFE	DFH	CHI_2
Test of Coincidence Wilks and Chi-square	0.8321483	37	3	6.7985487

DF
6

RHO	OMEGA
0.8222222	0.0007305

	PVAL
p-value of test H1: Coincidence	0.3401753

criterion in (8.30) as shown in Program 8_12.sas. The results are displayed in Output 8.12.5. Using Theorem 5.3, the P-value for the test is 0.3402 so that the test is not significant.

To test for parallelism of profiles, one would select the matrices

$$C_1 = \begin{pmatrix} 1 & -1 & 0 & 0 \\ 1 & 0 & -1 & 0 \\ 1 & 0 & 0 & -1 \end{pmatrix} \text{ and } A_1 = \begin{pmatrix} 0 \\ 1 \end{pmatrix},$$

with the test following that outlined for the test of coincidence.

Because the GMANOVA-MANOVA model allows one to incorporate a covariate into the model, the mixed GMANOVA-MANOVA model is usually more resolute than the GMANOVA model when the covariate is significant. For our example, this was not the case because of the large variability within patients.

8.13 CGMANOVA MODEL

The SUR model may be used to represent numerous multivariate linear models as special cases. We have already demonstrated how the model may be used to analyze the GMANOVA model and Patel's model which allows for changing covariates. The SUR model may be used to analyze the MANOVA model with double restriction, the extended GMANOVA model

$$\mathcal{E}(Y_{n\times p}) = \sum_{i=1}^{r}(X_i)_{n\times k_i}(B_i)_{k_i\times q_i}(M_i)_{q_i\times p} \tag{8.185}$$

proposed by Verbyla and Venables (1988), and the mixed GMANOVA-MANOVA model. To see this, we let the SUR model in (8.86) be written in its most general form

$$(y_i)_{(p\times 1)} = (D_i)_{p\times m}\theta_{m\times 1} + (e_i)_{(p\times 1)} \tag{8.186}$$

where $i = 1, 2, \ldots, n$ and the $\text{var}(e_i) = \Sigma$, the structure of the p variates. Letting

$$D_i = \text{diag}(x'_{ij}),$$

where x'_{ij} $j = 1, \ldots, p$ is a $k_j \times 1$ vector of variables for the jth variable and the ith subject and $m = \sum_{j=1}^{p} k_j$, we have the MDM model in another form. When $x_{ij} = x_i$ so that $D_i = I_p \otimes x'_i$, the MDM model reduces to the standard multivariate GLM. The most general SUR model may be represented as:

$$y_{np \times 1} = D_{np \times m} \theta_{m \times 1} + e_{np \times 1} \qquad (8.187)$$

where $y = (y'_1, y'_2, \ldots, y'_n)'$, $D = (D'_1, D'_2, \ldots, D'_n)'$ and $e = (e'_1, e'_2, \ldots, e'_n)'$ and the $\text{var}(y) = I_n \otimes \Sigma$. This is the exact form of the CGMANOVA model proposed by Hecker (1987). To see this, we let the ith row of the data matrix $Y_{n \times p}$ be y'_i in y, the design matrix $D = [(X_1 \otimes M'_1), \ldots, (X_r \otimes M'_r)]$ and $\theta = (\theta'_1, \theta'_2, \ldots, \theta'_r)'$ where $\theta_i = \text{vec}(B_i)$. Hence, to estimate θ we may use the most general SUR model instead of the MSUR model. The estimate of θ is

$$\tilde{\theta} = [D'(I_n \otimes \widehat{\Sigma}^{-1})D]^{-1} D'(I_n \otimes \widehat{\Sigma})^{-1} y \qquad (8.188)$$

where $\widehat{\Sigma}$ is a consistent estimator of Σ. A convenient candidate is the estimate

$$\widehat{\Sigma} = \frac{Y'(I - 1(1'1)^{-1}1'Y}{n} \qquad (8.189)$$

which uses just the within subject variation.

To test hypotheses, we may use the Wald statistic given in (8.104), instead of a complicated LR test. Thus, we have that

$$W = (C\tilde{\theta} - \xi)'[C(D'(I_n \otimes \widehat{\Sigma}^{-1})D)^{-1}C']^{-1}(C\tilde{\theta} - \xi). \qquad (8.190)$$

We may analyze the data previously analyzed in Section 8.12 now using Hecker's CGMANOVA model. The SAS code to do the analysis is found in Program 8_13.sas. To represent the MANOVA-GMANOVA model as a SUR model, we use the generalized vector form of the SUR model. Thus for the Danford data, the model is:

$$\begin{pmatrix} y_1 \\ y_2 \\ \vdots \\ y_{45} \end{pmatrix} = [X_1 \otimes X'_2, Z \otimes I_5] \begin{pmatrix} \theta_{11} \\ \vdots \\ \theta_{42} \\ \gamma_1 \\ \vdots \\ \gamma_4 \end{pmatrix} + \begin{pmatrix} e_1 \\ e_2 \\ \cdots \\ e_{45} \end{pmatrix}. \qquad (8.191)$$

Alternatively, we have that

$$y_i = D_i \theta + e_i \text{ for } i = 1, 2, \ldots, 45 \qquad (8.192)$$

where $D_i = (I_p \otimes x'_{ip})$ are sets of five rows of the design matrix D, and x_{ij} is a vector of independent variables for response j on subject i, for the most general SUR model.

Output 8.13.1: CGMANOVA Model.

Consistent Estimator of Covariance Matrix

SIGMA				
4877.4588	4691.8854	5197.364	5306.9077	4903.0178
4691.8854	4897.6277	5337.644	5469.7165	5013.28
5197.364	5337.644	6109.998	6166.5551	5671.2533
5306.9077	5469.7165	6166.5551	6516.3388	6028.1244
4903.0178	5013.28	5671.2533	6028.1244	6520.4711

GLSE of Parameter Matrix

THETA
147.99995
32.04349
107.71277
18.078296
141.46387
15.817764
132.79786
5.5771335
0.7795546
0.8832298
0.9211025
0.9223535
0.9881443

Wald Statistic for Coincidence

W	DF	PVALUE
4.7744378	6	0.5730516

Wald Statistic for Gamma0=0

W	DF	PVALUE
42.329283	5	5.0524E-8

8.13.1   Results and Interpretation

The results of the analysis of the Danford data using the CGMANOVA model are found in Output 8.13.1. In Program 8_13.sas, we estimate both Θ and Γ_0 using the SUR estimates using the estimate Σ based upon only within subject variation. For the output, the estimates are:

$$\widehat{\Theta}' = \begin{pmatrix} 148.00 & 107.71 & 141.46 & 132.80 \\ 32.04 & 18.07 & 15.82 & 5.57 \end{pmatrix}$$

$$\widehat{\Gamma}_0' = \begin{pmatrix} 0.780 & 0.883 & 0.921 & 0.922 & 0.988 \end{pmatrix}$$

which are nearly identical to the ML estimates using the MANOVA-GMANOVA model in Section 8.12.

To test hypotheses using the CGMANOVA model we use the Wald statistic. To illustrate, we perform the test of coincidence. The hypothesis test matrix is

$$C = \begin{pmatrix} 1 & 0 & 0 & 0 & 0 & 0 & -1 & 0 & 0 & 0 & 0 & 0 & 0 \\ 0 & 1 & 0 & 0 & 0 & 0 & 0 & -1 & 0 & 0 & 0 & 0 & 0 \\ 0 & 0 & 1 & 0 & 0 & 0 & -1 & 0 & 0 & 0 & 0 & 0 & 0 \\ 0 & 0 & 0 & 1 & 0 & 0 & 0 & -1 & 0 & 0 & 0 & 0 & 0 \\ 0 & 0 & 0 & 0 & 1 & 0 & -1 & 0 & 0 & 0 & 0 & 0 & 0 \\ 0 & 0 & 0 & 0 & 0 & 1 & 0 & -1 & 0 & 0 & 0 & 0 & 0 \end{pmatrix}$$

where the $\text{rank}(C) = 6$. From Output 8.13.1, the Wald statistic is 4.774, with P-value 0.5732. Because the Wald statistic is a large sample result its P-value is larger than the corresponding LR test P-value of 0.34 obtained using the MANOVA-GMANOVA model. It is currently unknown under what conditions the CGMANO-VA model is optimal. Hecker (1987) obtained approximate bounds for a LR test as discussed in Section 5.5.

Finally, we illustrate the test that $\Gamma_0 = 0$ using the CGMANOVA model and the Wald statistic. For this hypothesis, the hypothesis test matrix is

$$C = (0_{5 \times 8}, I_5)$$

where the $\text{rank}(C) = 5$. As seen in the output, $W = 42.33$ with P-value < 0.0001. This result is consistent with the LR test.

In conclusion, we see that the SUR model is a very flexible model for the analysis of multivariate data; however, the precision of the results is unknown for small samples when compared to the LR test. Finally to incorporate restriction into the SUR model, one may transform each observation vector y_i by $\Sigma^{-1/2}$. Then $z = (I_n \otimes \Sigma^{-1/2})y$ has a multivariate normal distribution with mean $\mu = (I_n \otimes \Sigma^{-1/2})\theta$ and covariance matrix $\Omega = I_n \otimes \Sigma^{-1/2}\Sigma\Sigma^{-1/2} = I_n \otimes I_p$ so that the restricted linear model in Chapter 3 may be applied to the transformed model. Thus, depending on Σ,

$$Q_h(\Sigma) = (C\tilde{\theta} - \xi)'[C(D'(I_n \otimes \widehat{\Sigma}^{-1})D)^{-1}C']^{-1}(C\tilde{\theta} - \xi) \qquad (8.193)$$

may be used to test $H : C\theta = \xi$. Thus, the theory in Chapter 3 may be followed exactly to obtain approximate tests for a restricted SUR model.

CHAPTER 9

Simultaneous Inference Using Finite Intersection Tests

9.1 INTRODUCTION

In conducting an experiment to compare the equality of a finite number of means from several univariate or multivariate normal populations (assuming a linear model for the means), we have discussed how one may perform an overall univariate or multivariate test of significance. Following the significant overall test one evaluates the significance of an infinite number of post hoc contrasts to isolate specific population mean differences. This approach to the analysis of means is often called exploratory data analysis. Alternatively, one may be interested in investigating a finite number of planned simultaneous contrasts in population means. In this case one formulates a finite number of comparisons, contrasts in the population means. Significance of these planned hypotheses (comparisons) does not involve an overall test. Instead, one needs a method to study the significance of the family of planned comparisons. The simultaneous testing of a finite family of hypotheses is called a simultaneous test procedure (STP). The primary objective of a STP is to both test hypotheses and generate simultaneous confidence sets. Analysis of mean differences involving STPs is often called confirmatory data analysis since the hypotheses are specified a priori.

Fisher (1935) least significant difference (LSD) test and his other proposed method known today as the Bonferroni procedure set the direction for the development of STPs in the analysis of mean differences. The early works of Tukey (1949, 1953); Dunnett (1955); Krishnaiah (1964, 1965a, 1965c, 1965b, 1979); Gabriel (1969); Shaffer (1977); Scheffé (1947), among others, set the modern day direction for STPs. As shown by Cox, Krishnaiah, Lee, Reising, and Schuurmann (1980) and Hochberg and Tamhane (1987), many STPs are special applications of Krishnaiah's finite intersection test (FIT) procedure.

9.2 FINITE INTERSECTION TESTS

To construct a FIT, Krishnaiah (1965a) utilized Roy (1953) union-intersection (UI) principle introduced in Chapter 1. To review, one formulates the null hypotheses for a finite family of hypotheses (not necessarily contrasts) and corresponding alternatives. To construct a Roy UI test, one formulates an overall null hypothesis H_0 as

349

the intersection of an infinite or finite family of elementary or composite hypotheses so that the alternative H_A is the union of corresponding alternative hypotheses:

$$H_0 = \bigcap_{m \in \Gamma} H_m \qquad\qquad H_A = \bigcup_{m \in \Gamma} A_m \qquad (9.1)$$

where Γ is an arbitrary index set, finite or infinite. Further, suppose a suitable test of H_m versus A_m is available; then according to Roy's UI method, the critical region for the rejection of H_0 is the union of the rejection regions for each H_m, hence the term union-intersection. The overall null hypothesis H_0 is rejected if and only if at least one elementary hypothesis H_m is rejected.

To construct a FIT, one must construct a test statistic for testing each elementary member H_m of the finite null hypothesis against the corresponding alternative A_m and find the joint distribution of the test statistics so that the familywise (FW) error rate is controlled at the nominal level α. Thus, if H_0 and H_A are represented:

$$H_0 = \bigcap_{i=1}^{m} H_i \qquad\qquad H_A = \bigcup_{i=1}^{m} A_i \qquad (9.2)$$

and F_i represents a test statistic for testing H_i versus A_i, we then accept or reject H_i versus A_i using the FIT criterion when

$$F_i \gtrless F_{i\alpha}^* \qquad (9.3)$$

and the critical values $F_{i\alpha}^*$ are selected such that the joint probability

$$P\left(F_i \le F_{i\alpha}^*; i = 1, 2, \ldots, m | H_0\right) = 1 - \alpha. \qquad (9.4)$$

The critical values are usually selected such $F_{i\alpha}^* = F_\alpha$ for $i = 1, 2, \ldots, m$ to simplify the task of computing the critical constants and to treat each hypothesis equally. The overall null hypothesis is rejected if at least one subhypothesis H_i is found to be significant.

9.3   FINITE INTERSECTION TESTS OF UNIVARIATE MEANS

To apply the FIT procedure to compare the means of k normal populations with means μ_i and homogeneous variances σ^2, we again formulate a linear model:

$$y_{ij} = \mu_i + \varepsilon_{ij} \qquad\qquad i = 1, 2, \ldots, k \quad j = 1, 2, \ldots, n_i \qquad (9.5)$$

where the y_{ij} are independently distributed normal random variables with means μ_i and common variance σ^2, $y_{ij} \sim IN(\mu_i, \sigma^2)$. For this model the least squares or maximum likelihood estimate of the population means μ_i are the sample means $\hat{\mu}_i = \bar{y}_{i\cdot} = \sum_{j=1}^{n_i} y_{ij}/n_i$ and the unbiased estimate of the common variance is the sample variance $s^2 = \sum_{i=1}^{k} \sum_{j=1}^{n_i} (y_{ij} - \bar{y}_{i\cdot})^2/\nu_e$ where $\nu_e = N - k$ and $N = \sum_{i=1}^{k} n_i$. Now we are interested in testing the hypotheses $H_1, H_2, \ldots, H_m$ simultaneously where $H_i : \psi_i = 0$, and the overall null hypothesis is

$$H_0 = \bigcap_{i=1}^{m} H_i. \tag{9.6}$$

The alternative hypotheses are $A_i : \psi_i \neq 0$, where the overall alternative hypothesis is

$$H_A = \bigcap_{i=1}^{m} A_i. \tag{9.7}$$

When comparing k means, the contrasts ψ_i with unbiased estimates $\hat{\psi}_i$ have the form:

$$\psi_i = c_{i1}\mu_1 + c_{i2}\mu_2 + \cdots + c_{ik}\mu_k$$
$$\hat{\psi}_i = c_{i1}\hat{\mu}_1 + c_{i2}\hat{\mu}_2 + \cdots + c_{ik}\hat{\mu}_k \tag{9.8}$$

where the sum, $c_{i1} + c_{i2} + \cdots + c_{ik} = 0$ for $i = 1, 2, \ldots, m$. Letting the test statistic for testing H_i versus A_i be

$$F_i = T_i^2 \tag{9.9}$$

where

$$T_i = \frac{(\hat{\psi}_i - \psi_i)}{\hat{\sigma}_{\hat{\psi}_i}} = \frac{(\hat{\psi}_i - \psi_i)}{s\sqrt{d_i}}$$

the estimated sample variance of the estimated contrast $\hat{\psi}_i$ is

$$\hat{\sigma}_{\hat{\psi}_i}^2 = s^2 \sum_{j=1}^{k} \frac{c_{ij}^2}{n_j} = s^2 d_i. \tag{9.10}$$

The FIT procedure accepts or rejects H_i depending on whether

$$F_i \gtrless F_\alpha \tag{9.11}$$

where F_α is selected such that the

$$P(F_i \leq F_\alpha; i = 1, 2, \ldots, m | H_0) = 1 - \alpha. \tag{9.12}$$

The overall null hypothesis H_0 is rejected if at least one subhypothesis H_i is rejected or the

$$\max_{1 \leq i \leq m} F_i \gtrless F_\alpha. \tag{9.13}$$

To use the FIT procedure to test H_0, the joint distribution of the test statistics F_i must be determined when the null hypothesis is true. Before stating the result, we define the m-variate multivariate F-distribution.

Definition 9.1. Let the $m \times 1$ random vector $\hat{\psi}$ be distributed as a multivariate normal distribution with mean ψ and covariance matrix $\Sigma = \sigma^2 \Omega$ where $\Omega = (\rho_{ij})$ is a known correlation matrix. Also, letting $u = {\nu_e s^2}/{\sigma^2}$ be distributed independent of $\hat{\psi}$ as a χ^2 random variable with ν_e degrees of freedom; and defining

$$F_i = \frac{(\hat{\psi}_i - \psi_i)/\hat{\sigma}_{\hat{\psi}_i}}{u/\nu_e} = \frac{(\hat{\psi}_i - \psi_i)^2}{\hat{\sigma}^2_{\hat{\psi}_i}} \qquad \text{for } i = 1, 2, \ldots, m, \qquad (9.14)$$

then the joint distribution of $F_1, F_2, \ldots, F_m$ is known (see Krishnaiah, 1964; Johnson & Kotz, 1972, p.241) to be a central or non-central m-variate F-distribution with $(1, \nu_e)$ degrees of freedom where Ω is the correlation matrix of the accompanying multivariate normal distribution.

For the test of univariate means, $\nu_e = N - k$. A special case of the multivariate F-distribution is the multivariate t-distribution. When $\Omega = I$, the joint multivariate t-distribution is known as the Studentized Maximum Modulus distribution. Hochberg and Tamhane (1987) have collected tables for the multivariate t-distribution for the equicorrelated case ($\rho_{ij} = \rho$ for $i \neq j$). Tables for the multivariate F-distribution are given in Krishnaiah and Armitage (1970); Schuurmann, Krishnaiah, and Chattogpadhyay (1975), again for the equicorrelated case. In other situations when the probability integral cannot be determined exactly, one can establish bounds on F_α as described by Krishnaiah (1979) given below.

Using Poincarè's formula (see, e.g., Feller, 1967, p. 109) it is known that the

$$P(E_1, E_2, \ldots, E_m) = 1 - \sum_{i=1}^{m} P(E_i^c) + \sum_{i<j} P(E_i^c E_j^c) + \cdots + (-1)^m P(E_1^c, \ldots, E_m^c)$$

$$(9.15)$$

where E_i^c is the complement of the event E_i. Letting

$$S_r = 1 - \sum_{i=1}^{m} P(E_i^c) + \cdots + (-1)^r \sum_{i_1 < \cdots < i_r} P(E_{i_1}^c, \ldots, E_{i_r}^c), \qquad (9.16)$$

we may construct monotonic upper and lower bounds $S_2 \leq S_4 \leq S_6 \leq \cdots \leq S_m \leq S_1 \leq S_3 \leq \ldots$. Associating $P(E_1, E_2, \ldots, E_m)$ with $P(F_i \leq c_j; i = 1, 2, \ldots, m)$, we have for the multivariate F-distribution that $S_1 \leq P(F_i \leq c_j; i = 1, 2, \ldots, m) \leq S_2$ so that

$$1 - \sum_{i=1}^{m} P(F_i > c) \leq P(F_i \leq c_j; i = 1, 2, \ldots, m)$$

$$\leq 1 - \sum_{i=1}^{m} P(F_i > c) + \sum_{i<j} P(F_i > c, F_j > c). \quad (9.17)$$

The upper (1st order) bound S_1 is the famous Bonferroni inequality while S_2 is the (2nd order) bound. Šidák (1967) showed that the

$$P(F_i \leq c; i = 1, 2, \ldots, m) \geq P(F_{io} \leq c; i = 1, 2, \ldots, m) \qquad (9.18)$$

where $F_{io} = {}^{w_i}/_{s^2}$, $w_i = y_i^2$ and $y' = (y_1, y_2, \ldots, y_m)$ is distributed independent of s^2 as a multivariate normal distribution with mean vector 0 and covariance matrix $\Sigma = \sigma^2 I_m$. Kimball (1951) showed that the

$$P\left(F_{io} \leq c; i = 1, 2, \ldots, m\right) \geq \prod_{i=1}^{m} P\left(F_{io} \leq c\right). \tag{9.19}$$

Letting $c_1, c_2, \ldots, c_5$ represent critical values, we have the following relationships for the bound on the joint probability of the multivariate F-distribution:

Bonferroni (UB) $1 - \displaystyle\sum_{i=1}^{m} P\left(F_i > c_1\right) \leq 1 - \alpha$

Kimball (UB) $\displaystyle\prod_{i=1}^{m} P\left(F_i \leq c_2\right) \leq 1 - \alpha$

Šidák (UB) $\displaystyle\prod_{i=1}^{m} P\left(F_i \leq c_3\right) \leq 1 - \alpha$

$$\tag{9.20}$$

Bonferroni (LB) $1 - \displaystyle\sum_{i=1}^{m} P\left(F_i > c_4\right) + \sum_{i<j} P\left(F_i > c_4, F_j > c_4\right) \geq 1 - \alpha$

Exact $P\left(F_i \leq c_5; i = 1, 2, \ldots, m\right) = 1 - \alpha.$

Furthermore letting $F_{\alpha L}$ and $F_{\alpha U}$ be the calculated lower and upper bounds of F_α, we accept or reject H_i according to the rule $F_i \leq F_{\alpha L}$ or $F_i \geq F_{\alpha L}$. If F_i lies between the two bounds, we are unable to reach a decision. For the equicorrelated case, $\rho_{ij} = \rho \ (i \neq j)$, we have that $c_4 \leq c_5 \leq c_3 \leq c_2 \leq c_1$. A Vax-alpha Fortran 66 computer program written by Cox, Fang, Boudreau, and Timm (1994) and modified for the PC to run under Visual Fortran by Timm and Lin (2005) may be used to compute the bounds for the multivariate F-distribution. To construct simultaneous confidence intervals for the contrasts ψ_i, we know that the joint distribution of the estimated contrasts $\hat{\psi}_i$ under the null hypothesis H_0 is the same as the joint distribution of $\hat{\psi}_i - \psi_i$ under H_i. Hence, the set of all ψ_i that will not be rejected by the FIT procedure constitutes a $(1 - \alpha)$-level confidence set given by

$$P\left(\hat{\psi}_i - \hat{\sigma}_{\hat{\psi}_i} \sqrt{F_\alpha} \leq \psi_i \leq \hat{\psi}_i - \hat{\sigma}_{\hat{\psi}_i} \sqrt{F_\alpha}; i = 1, 2, \ldots, m | H_0\right) = 1 - \alpha \quad (9.21)$$

where F_α is the critical value of the m-variate multivariate F-distribution with 1 and $\nu_e = N - k$ degrees of freedom. By construction, the confidence sets become shorter, are more resolute, as the number of hypotheses m decrease.

In our discussion of the FIT to evaluate differences in univariate means, we have assumed that the alternatives A_i are to be two-sided. The FIT procedure also allows one to investigate one-sided tests. For example, consider the problem of testing the hypotheses $H_1, H_2, \ldots, H_m$ simultaneously where $H_i : \psi_i = 0$, and the alternatives are $A_i^* : \psi_i > 0$. To test this hypothesis using the FIT procedure, we accept or reject H_i depending upon whether

$$T_i \gtrless T_{\alpha 1} \tag{9.22}$$

where the

$$P\left(T_i \le T_{\alpha 1}; i = 1, 2, \ldots, m | H_0\right) = 1 - \alpha. \tag{9.23}$$

As before, we can obtain simultaneous confidence sets:

$$P\left(\hat{\psi}_i - \hat{\sigma}_{\hat{\psi}_i} T_{\alpha 1} \le \psi_i; i = 1, 2, \ldots, m | H_0\right) = 1 - \alpha. \tag{9.24}$$

If the alternatives are $A_i^{**} : \psi_i < 0$, we accept or reject H_i depending upon whether

$$T_i \gtrless T_{\alpha 2} \tag{9.25}$$

where the

$$P\left(T_i \ge T_{\alpha 2}; i = 1, 2, \ldots, m | H_0\right) = 1 - \alpha. \tag{9.26}$$

And, the confidence sets become:

$$P\left(\hat{\psi}_i - \hat{\sigma}_{\hat{\psi}_i} T_{\alpha 2} \ge \psi_i; i = 1, 2, \ldots, m | H_0\right) = 1 - \alpha. \tag{9.27}$$

When H_0 is true, the joint distribution of the statistics $T_1, T_2, \ldots, T_m$ follows a multivariate t-distribution and $T_{1\alpha}$ and $T_{2\alpha}$ are critical values for the tests from the distribution.

9.4   FINITE INTERSECTION TESTS FOR LINEAR MODELS

We have presented the FIT procedure to evaluate univariate means; more generally the procedure may be used to test hypotheses using the linear model

$$y = X\beta + e \tag{9.28}$$

where y is a random $(N \times 1)$ vector of observations distributed multivariate normal with mean $X\beta$ and covariance matrix $\Sigma = \sigma^2 I$, $X = (x_{ij})$ is an $N \times (m + 1)$ known design matrix and $\beta' = (\beta_0, \beta_1, \ldots, \beta_m)$ is the parameter vector of fixed unknown constants. For the linear model, recall that the least squares or maximum likelihood estimate of β is $\hat{\beta} = (X'X)^{-1}X'y$ where the variance of the estimate is $\sigma_{\hat{\beta}} = \sigma^2(X'X)^{-1} = \sigma^2(w_{ij})$. The unbiased estimate of the variance σ^2 is the sample estimate $s^2 = \{y'[I - X(X'X)^{-1}X']y\}/(N - m - 1)$.

The problem of selecting variables may be formulated as a simultaneous testing procedure. Clearly, the independent variables x_i will be called significant or nonsignificant depending upon whether $\beta_i \ne 0$ or $\beta_i = 0$. To test the hypothesis

$H_0 : \beta = 0$ using the FIT, the null hypothesis is represented as the intersection of elementary hypotheses:

$$H_0 = \bigcap_{i=0}^{m} H_i : \beta_i = 0. \tag{9.29}$$

Then, we accept or reject H_i and hence H_0 depending upon whether

$$F_i \gtrless F_\alpha \tag{9.30}$$

where F_α is selected such that the

$$P\left(F_i \leq F_\alpha; i = 1, 2, \ldots, m | H_0\right) = 1 - \alpha \tag{9.31}$$

and

$$F_i = \frac{\hat{\beta}_i^2}{s^2 w_{ii}}. \tag{9.32}$$

The joint distribution of $F_0, F_1, \ldots, F_m$ is an $(m+1)$-variate multivariate F-distribution with 1 and $N - m - 1$ degrees of freedom.

Using the overall F test to test H_0, we reject the null hypothesis if

$$F = \frac{\hat{\beta}'(X'X)^{-1}\hat{\beta}/(m+1)}{s^2} > F_\alpha^* \tag{9.33}$$

where F_α^* is the upper α-level critical value of the F-distribution with $m + 1$ and $N - m + 1$ degrees of freedom. Comparing the critical values of the two procedures, the length of the confidence set for β_i for the FIT procedure is always less than or equal to the corresponding value obtained using the F-distribution since $F_\alpha \leq F_\alpha^*(m+1)$.

In general, one may be interested in some subset $r \leq (m + 1)$ of the β_i's. The relative efficiency of the two procedures may be evaluated using the ratio $R^2 = F_\alpha/_{r}F_\alpha^*$. For $\alpha = .05$ and $\nu_e = 10$, values of R^2 are shown in Table 9.1. Similar tables for the Analysis of Variance (ANOVA) were developed by Cox et al. (1980). As the number of tests increase the advantage of the FIT over the F-test decreases, but the relationship remains stable as ν_e increases.

9.5   COMPARISON OF SOME TESTS OF UNIVARIATE MEANS WITH THE FIT PROCEDURE

9.5.1   Single-Step Methods

In their study of Krishnaiah's FIT, Cox et al. (1980) showed that the FIT procedure is best in the Neyman sense (has shortest confidence sets) when compared to a wide class of classical single-step STPs. A STP is called a single-step procedure by Gabriel (1969) if the collection of tests in the family depends upon a common critical value (see Hochberg & Tamhane, 1987, p. 345). The FIT will always generate

Table 9.1: Relative Efficiency of the F-test to the FIT for $\alpha = .05$ and $\nu_e = 10$.

r	ρ		
	.10	.50	.90
1	1.00	1.00	1.00
2	.83	.80	.71
3	.72	.68	.57
4	.64	.60	.48
5	.58	.54	.42
6	.52	.49	.37
7	.49	.43	.33
8	.45	.42	.30
9	.43	.39	.28
10	.40	.36	.26

shorter confidence sets for a finite set of contrasts defined in the family when compared to Scheffé (1953) S-method. If one is only interested in pairwise comparison of means, the FIT is equivalent to Tukey (1953) test based on the studentized range distribution, sometimes called the honestly significant difference (HSD) or wholly significant difference (WSD) test. If one is only interested in a subset of all pairwise comparisons, the FIT test will yield shorter confidence intervals. In addition, the FIT does not require equal sample sizes and is the optimal method to be used when making pairwise comparisons (see Hochberg & Tamhane, 1987, p. 86) when sample sizes are unequal. The FIT procedure is better than the Tukey-Kramer procedure (see Hochberg & Tamhane, 1987, p. 91). Finally, the FIT does better than the Bonferroni-Dunn, Dunn-Šidák, or Hochberg's GT2 procedure if one is interested in comparisons of cell means in a two-way ANOVA (see Hochberg & Tamhane, 1987, p. 94-95).

When comparing several means to a control, Dunnett (1955) method is equivalent to the FIT procedure (see Hochberg & Tamhane, 1987, p. 140). Shaffer (1977) extended Dunnett's test to all contrasts (see Hochberg & Tamhane, 1987, p. 147). Her extended test is not equivalent to the FIT of Krishnaiah. While the FIT is not better than Shaffer's extended test, it does better if one is interested in the subset of contrasts for only adjacent mean differences. While Tukey's test may be used for all contrasts in mean differences (see Scheffé, 1959, p. 74; Hochberg & Tamhane, 1987, p. 81), the FIT is uniformly better than Tukey's extended procedure.

The FIT procedure is a very general method for investigating simultaneous contrasts of means in several univariate populations and does as well as or better than many classical single-step procedures. Because one may model means using a general linear model, the procedure may be used to analyze complicated ANOVA designs. For two way designs, it outperforms the maximal F-statistic using the Studentized Maximum Root (SMR) distribution developed by Boik (1993).

9.5.2 Stepdown Methods

If one is primarily interested in test of significance and not in the construction of simultaneous confidence intervals when investigating contrasts in means from univariate normal populations, one can construct a stepdown FIT that controls the FW error rate at any nominal level α that is uniformly more powerful than the single-step FIT procedure. A general method for constructing stepdown test procedures was developed by Marcus, Peritz, and Gabriel (1976) called the closed testing procedure for finite families. Letting $H_i(i = 1, 2, \ldots, m)$ be a finite family of hypotheses, the closed family of tests is formed by taking all possible intersections of the hypotheses H_i over a finite index set $\Gamma : \{1, 2, \ldots, m\}$ of elementary hypotheses $H_i, i \in \Gamma$ so that

$$H_\Gamma = \bigcap_{i=\Gamma} H_i \qquad (9.34)$$

for $\Gamma \subseteq \{1, 2, \ldots, m\}$. Given a test of size α for each hypothesis H_Γ, the closed testing procedure rejects H_Γ if and only if every H_R belonging to Γ is rejected by its size α test for all $R \supseteq \Gamma$. Associating H_Γ with the FIT null hypothesis

$$H_0 = \bigcap_{i=1}^{m} H_i \qquad (9.35)$$

and recalling that whenever the overall test is rejected, at least on of the H_i's implied by H_0 is rejected, we observe that to reject any H_i it is not necessary to test all intersection of H_0 containing H_i; we need only test the latter. Holm (1979) showed that this is accomplished by ordering the H_i to ensure coherence and to test the H_i sequentially in a stepdown manner. Because the FIT is a UI test of the form

$$\max_{1 \leq i \leq m} F_i \gtrless F_\alpha, \qquad (9.36)$$

Holm's condition is satisfied if the subhypotheses H_i are tested in the order of magnitude of the F_i from largest to smallest. Thus, the hypothesis H_i corresponding to the largest F_i is tested first. If it is rejected, then any intersection hypothesis containing it will be rejected so that it may be set aside without further testing. Next the hypothesis associated with the second largest F_i is tested and again set aside if found to be significant. This process continues until some F_i is found to be nonsignificant. At this point those H_i remaining are nonsignificant by implication.

While the stepdown FIT test procedure is more powerful for testing a finite hypothesis than single-step procedure, one may not construct simultaneous confidence intervals. Furthermore, as illustrated by Timm (1995), Ryan-Einot-Gabriel-Tukey-Welch's and Peritz's stepdown procedures based upon the studentized range distribution out perform the stepdown FIT procedure for a subset of pairwise comparisons. However, the FIT procedure is not limited to pairwise comparison since the hypotheses H_i are contrasts. And, while Hayter and Hsu (1994) show that it is difficult to construct two-sided confidence sets for the stepdown FIT procedure, using the method of Stefánsson, Kim, and Hsu (1988) one can easily construct $(1 - \alpha)$-level simultaneous confidence sets for one-sided stepdown FITs, Timm (1995).

With the ability to use the stepdown FIT procedure with any finite set of contrasts, not just pairwise comparisons, and the ability to easily construct one-sided confidence sets, the procedure cannot be ignored by researchers interested in investigating differences in univariate means. If one is willing to adopt a STP that only permits testing and not the creation of confidence sets, the sequential stepup procedures of Hochberg (1988) is an attractive option since it is uniformly more powerful than Holm (1979) sequentially rejective stepdown procedure. Hochberg's stepup procedure is however not uniformly better than the stepdown FIT procedure.

To improve upon Hochberg (1988) stepup method, Dunnett and Tamhane (1992) (abbreviated as DT) proposed a stepup STP that uses the multivariate t-distribution with equicorrelated structure. However, unlike the stepdown FIT procedure which allows one to compute critical values at each step of the process independently of each other, the critical values of the DT stepup STP procedure must be calculated recursively and the method is not uniformly more powerful than the stepdown FIT procedure. A robust competitor to the stepdown FIT is a stepdown resampling method proposed by Troendle (1995). Simulations performed by Troendle under multivariate normality show that the stepdown FIT procedure is generally more powerful than the stepup procedures of Hochberg or DT.

9.6   ANALYSIS OF MEANS ANALYSIS

The analysis of means analysis (ANOMA) is a procedure introduced by Scheffé (1947) to investigate whether k treatment means are significantly different from an overall population mean

$$H_0 = \mu_1 = \mu_2 = \cdots = \mu_k = \mu. \tag{9.37}$$

While the assumptions for test H_0 are identical to that for the ANOVA, and with only two groups the two procedures are equivalent, more generally the ANOMA is a FIT procedure for testing H_0. To see this, we again assume that the observations $y_{ij} \sim IN(\mu_i, \sigma^2)$ for $= 1, \ldots, k$ and $j = 1, 2, \ldots, n_i$. Letting $\bar{y}_{i\cdot}$ represent the sample mean for the ith group and $\bar{y}_{\cdot\cdot}$ the overall sample mean so that

$$n_i \bar{y}_{i\cdot} = \sum_{j=1}^{n_i} y_{ij} \qquad \text{for } i = 1, 2, \ldots, k,$$

$$N \bar{y}_{\cdot\cdot} = \sum_{i=1}^{k} \sum_{j=1}^{n_i} y_{ij}. \tag{9.38}$$

The least squares (maximum likelihood) estimates of μ_i and μ are the sample mean for the ith group and the overall sample mean, respectively. If the null hypothesis

H_0 is true, the $\mathcal{E}(\bar{y}_{i.} - \bar{y}_{..}) = 0$ and the variance of the difference is

$$\text{var}(\bar{y}_{i.} - \bar{y}_{..}) = \text{var}(\bar{y}_{i.}) + \text{var}(\bar{y}_{..}) - 2\text{cov}(\bar{y}_{i.}, \bar{y}_{..})$$

$$= \frac{\sigma^2}{n_i} + \frac{\sigma^2}{N} - 2\text{cov}\left(\bar{y}_{i.}, \sum_{i=1}^{k} \frac{n_i \bar{y}_{i.}}{N}\right)$$

$$= \frac{\sigma^2}{n_i} + \frac{\sigma^2}{N} - \frac{2n_i}{N\sigma^2/n_i}$$

$$= \sigma^2 \left[\frac{N - n_i}{Nn_i}\right] \tag{9.39}$$

so that

$$(\bar{y}_{i.} - \bar{y}_{..}) \sim N\left\{0, \sigma^2\left[(N - n_i)/Nn_i\right]\right\}. \tag{9.40}$$

Letting $s^2 = \sum_{i=1}^{k} \sum_{j=1}^{n_i} (y_{ij} - \bar{y}_{i.})^2/\nu(N - k)$ be an unbiased estimator of σ^2, when H_0 is true the statistic

$$T_i = \frac{(y_{ij} - \bar{y}_{i.})}{s\sqrt{(k-1)/nk}} \sim t(\nu_e) \tag{9.41}$$

has a central t-distribution. Testing that the means are equal to an unknown parameter μ is equivalent to test

$$H_0 = \bigcap_{i=m}^{m} H_i = 0 \tag{9.42}$$

where $H_i : \psi_i = \mu_i - \mu$, that the k treatment effects are equal to zero. The FIT for testing H_0 is to accept or reject H_I depending upon whether

$$|T_i| \gtrless T_\alpha \tag{9.43}$$

where T_α is selected such that the

$$P\left(|T_i| \le T_\alpha; i = 1, 2, \ldots, k | H_0\right) = 1 - \alpha. \tag{9.44}$$

Clearly the joint distribution of $(T_1, T_2, \ldots, T_k)$ has a central multivariate t-distribution with $\nu = N - k$ degrees of freedom when H_0 is true. The FIT is optimal if one is interested in investigating only fixed main effect $\psi_i = \mu_i - \mu = 0$. When all the sample sizes $n_i = n$, the associated normal distribution has equicorrelated structure $\Omega = (\rho_{ij} = \rho$ for $i \ne j$ where $\rho = -1/(k-1))$ since the

$$\text{corr}(\bar{y}_{i.} - \bar{y}_{..}, \bar{y}_{i.} - \bar{y}_{..}) = \frac{\text{cov}(\bar{y}_{i.} - \bar{y}_{..}, \bar{y}_{i.} - \bar{y}_{..})}{\sqrt{V(\bar{y}_{i.} - \bar{y}_{.})}\sqrt{V(\bar{y}_{i.} - \bar{y}_{.})}}$$

$$= \frac{0 - \sigma^2/nk - \sigma^2/nk + \sigma^2/nk}{\sigma^2(k-1)/nk}$$

$$= \frac{-\sigma^2/nk}{\sigma^2(k-1)/nk}$$

$$= -\frac{1}{k-1} \tag{9.45}$$

so that published multivariate t-distribution tables may be used to test for treatment difference. This is also the case when investigating treatment effects in all balanced complete designs and certain incomplete designs such as Latin squares, Graeco-Latin squares, balanced incomplete block designs, Youden squares, and axial mixture designs, Nelson (1993).

9.7 SIMULTANEOUS TEST PROCEDURES FOR MEAN VECTORS

To test the equality of mean vectors from k multivariate normal populations with means μ_i and common unknown covariance matrix Σ, one may use the overall tests proposed by Roy, Wilks, Bartlett-Lawley-Hotelling, or Bartlett-Nanda-Pillai discussed in Chapter 5 using the multivariate linear model. To review the approach using means, we let y_{ijp} represent the jth observation on the pth variate from the ith population. Letting $y'_{ij} = (y_{ij1}, y_{ij2}, \ldots, y_{ijp})$ represent the observation vector, a linear model for the population means similar to (9.5) follows:

$$y_{ij} = \mu_i + \epsilon_{ij} \qquad\qquad i = 1, 2, \ldots, k \quad j = 1, 2, \ldots, n_i \qquad (9.46)$$

where the y_{ij} are jointly independent multivariate normal, $y_{ij} \sim IN(\mu_i, \Sigma)$. For this model, extending (9.6), we form the contrasts and estimates of the contrasts vectors:

$$\psi_i = c_{i1}\mu_1 + c_{i2}\mu_2 + \cdots + c_{ik}\mu_k$$
$$\hat{\psi}_i = c_{i1}\hat{\mu}_{i1} + c_{i2}\hat{\mu}_{i2} + \cdots + c_{ik}\hat{\mu}_k \qquad (9.47)$$

where the least square (maximum likelihood) estimate of μ_i is the vector of sample means $\hat{\mu}_i = \bar{y}_i$. and the $\sum_{j=1}^{k} c_{ij} = 0$ for $i = 1, 2, \ldots, m$. Given the linear model in (9.34), the hypothesis and error sum of squares for testing the equality of the k population mean vectors: $H_0 : \mu_1 = \mu_2 = \cdots = \mu_k$ are

$$H = \sum_{i=1}^{k} n_i (\bar{y}_i. - \bar{y}..)(\bar{y}_i. - \bar{y}..)'$$

$$E = \sum_{i=1}^{k} \sum_{j=1}^{n_i} (\bar{y}_{ij} - \bar{y}_i.)(\bar{y}_{ij} - \bar{y}_i.)' \qquad (9.48)$$

where $n_i \bar{y}_i. = \sum_{j=1}^{n_i} y_{ij}$, $N\bar{y}.. = \sum_{i=1}^{k} \sum_{j=1}^{n_i} y_{ij}$, $N = \sum_{i=1}^{k} n_i$, and $E/(n-k) = S$ is an unbiased estimate of Σ. To test the equality of the mean vector using Roy (1953) largest root test, we accept or reject the null hypothesis depending upon whether

$$(N - k)C_L(HE^{-1}) \gtrless (k - 1)c_\alpha \qquad (9.49)$$

where the critical constant c_α is chosen such that the

$$P\left[(N - k)C_L(HE^{-1}) \le (k - 1)c_\alpha | H_0\right] = 1 - \alpha \qquad (9.50)$$

and $C_L(HE^{-1})$ is the largest characteristic root of (HE^{-1}). When the overall test is rejected, one may evaluate an infinite number of contrasts to determine which mean difference and associated variable led to the rejection of the null hypothesis. However, for a finite number of contrasts, the procedure is very conservative.

As a STP, Roy's test is equivalent to accepting or rejecting $H_i : \psi_i = 0$ for $i = 1, 2, \ldots, m$ simultaneous contrasts depending upon whether

$$T_i^2 \gtrless (k-1)c_\alpha \tag{9.51}$$

where $T_i^2 = \hat{\psi}_i' S^{-1} \hat{\psi}_i / d_i$ where $d_i = \sum_{i=1}^{k} c_{ij}^2 / n_j$ such that the

$$P\left[T_i^2(k-1)c_\alpha | H_0\right] = 1 - \alpha. \tag{9.52}$$

Because of the conservative nature of Roy's test, Roy and Bose (1953) proposed a test based upon correlated Hotelling T^2 statistics designed to investigate pairwise comparisons of mean vectors, called the $T_{\max}^2$ test. To construct their test, observe that H_0, the equality of mean vectors, can be written as

$$H_0 = \bigcap_{a \neq 0} H_a \tag{9.53}$$

where $H_a : a'\mu_1 = \cdots = a'\mu_k$ for nonnull vectors a. Alternatively, because

$$H_i = \bigcap_{a \neq 0} H_{ia} \tag{9.54}$$

where $H_{ia} : a'\psi_i = 0$, we can test hypotheses $H_1, H_2, \ldots, H_m$ simultaneously by testing the hypotheses H_{ia} simultaneously using the UI method as follows: accept the hypothesis $H_i : \psi_i = c_{i1}\mu_1 + c_{i2}\mu_2 + \cdots + c_{ik}\mu_k = 0$ if

$$F_{ia} = \frac{(a'\hat{\psi}_i - a'\psi_i)^2}{(a'Sa)d_i} = \frac{(a'\hat{\psi}_i - a'\psi_i)^2}{\hat{\sigma}_{\hat{\psi}_i}^2} \leq T_\alpha^2 \tag{9.55}$$

for all nonnull vectors a and reject H_i otherwise, where T_α^2 is chosen such that the

$$P\left[F_{ia} \leq T_\alpha^2; i = 1, 2, \ldots, m \text{ and nonnull } a | H_0\right] = 1 - \alpha. \tag{9.56}$$

But the

$$\bigcap_{a \neq 0}(F_{ia} \leq T_\alpha^2) \text{ if and only if the } \max_{1 \leq i \leq m} T_i^2 \leq T_\alpha^2. \tag{9.57}$$

Hence, we accept or reject H_i depending upon whether

$$T_i^2 \gtrless T_\alpha^2 \tag{9.58}$$

where

$$T_i^2 = \frac{\hat{\psi}_i' S^{-1} \hat{\psi}_i}{\sum_{i=1}^{k} c_{ij}^2 / n_j} \qquad i = 1, 2, \ldots, m \tag{9.59}$$

such that the

$$P(T_{\max}^2 \le T_\alpha^2) = 1 - \alpha \qquad (9.60)$$

and $T_{\max}^2 = \max(T_1^2, T_2^2, \ldots, T_m^2)$. However, the distribution of the $T_{\max}^2$ is unknown. When the sample size N is large it may be approximated by a multivariate χ^2 distribution. Alternatively, using Poincarè's formula (see, e.g., Feller, 1967, p. 109), the upper and lower bounds are:

$$1 - P_0 \le P(T_{\max}^2 \le T_\alpha^2 | H_0) \le 1 - P_0 + P_1 \qquad (9.61)$$

for

$$P_0 = \sum_{i=1}^m P(T_i^2 \ge T_\alpha^2 | H_0)$$

$$P_1 = \sum_{i=1}^m P\left[(T_i^2 \ge T_\alpha^2 \text{ and } \ge T_j^2 \ge T_\alpha^2) | H_0\right]. \qquad (9.62)$$

When H_0 is true, $T_i^2 (N - k - p + 1)/p(N - k)$ has a central F-distribution with degrees of freedom p and $N - k - p + 1$. Hence P_0 may be evaluated to obtain a Bonferroni upper bound for the test. Finally, a $(1 - \alpha)$-level simultaneous confidence set for $a'\psi_i$ are:

$$a'\hat{\psi}_i - \hat{\sigma}_{\hat{\psi}_i}^2 \sqrt{T_\alpha^2} \le T_\alpha^2 \le a'\psi_i \le a'\hat{\psi}_i - \hat{\sigma}_{\hat{\psi}_i}^2 \sqrt{T_\alpha^2}. \qquad (9.63)$$

Because the STP to test the equality of mean vectors depends upon the joint distribution of the $T_{\max}^2$ which is unknown, a FIT to test the equality of mean vectors is developed.

9.8   FINITE INTERSECTION TEST OF MEAN VECTORS

When investigating the equality of mean vectors across k independent populations, the overall null hypothesis

$$H_0 : \mu_1 = \mu_2 = \cdots = \mu_k \qquad (9.64)$$

is equivalent to testing the m simultaneous hypotheses

$$H_0 = \bigcap_{i=1}^m H_i : \psi_i = 0 \qquad (9.65)$$

where $\psi_i = c_{i1}\mu_1 + c_{i2}\mu_2 + \cdots + c_{ik}\mu_k$ and the $\sum_{j=1}^k c_{ij} = 0$ for $i = 1, 2, \ldots, m$. However, when investigating contrasts across population means many researchers are only interested in hypotheses of the form $H_{ia_j} : a'\psi_i = 0$ where the vectors a_j are known. Often the known coefficients are selected to construct contrasts a variable at a time so that the vectors a_j have a one in the jth position for $j = 1, 2, \ldots, p$ and zero's otherwise. For this case, (9.65) becomes

$$H_0 = \bigcap_{i=1}^{m} H_i : \psi_i = 0 = \bigcap_{i=1}^{m} \bigcap_{j=1}^{p} H_{ij} : a_j' \psi_j = 0. \tag{9.66}$$

To test the hypotheses H_{ij} depending upon using the FIT procedure, we accept or reject H_{ij} depending upon whether

$$F_{ia} \gtrless F_\alpha \tag{9.67}$$

where F_α is chosen such that the

$$P\left[F_{ia_j} \le F_\alpha; i = 1, 2, \ldots, m, \text{ and } j = 1, 2, \ldots, p | H_0\right] = 1 - \alpha \tag{9.68}$$

where F_{ia_j} is defined in (9.46). However, the joint distribution of these statistics is not multivariate F as in the univariate case since they involve nuisance parameters. Thus, one may only test the hypothesis H_{ij} simultaneously using for example the Bonferroni-Dunn or Dunn-Šidák inequality (Dunn, 1958; Šidák, 1967).

To remove the dependence of the ratio on the nuisance parameters a stepdown FIT procedure based upon the conditional distributions of the p variates, similar to Roy's stepdown F test (Roy, 1958), was developed by Krishnaiah (1965a, 1965c, 1969) which he called a stepdown FIT. The two procedures are very different because Roy's stepdown procedure depends upon the F-distribution and Krishnaiah's uses the multivariate F-distribution; however, they are also similar because both require one to arrange the p-response variates in their order of importance. For both procedures, the ordered, dependent response variates enter the analysis as fixed covariates.

To develop the stepdown FIT procedure, it is convenient to utilize the multivariate linear model. The multivariate linear model is given by

$$Y = XU + E_0 \tag{9.69}$$

where $Y_{(N \times p)}$ is a data matrix of p variables $y_1, y_2, \ldots, y_p$ and whose rows are distributed independently as p-variate multivariate normal random variables with common covariance matrix Σ, $X_{(N \times k)}$ is a known design matrix of full rank $k \le N - p$, $E_{0(N \times p)}$ is a matrix of random error, and $U_{(k \times p)}$ is a matrix of unknown parameters:

$$U = \begin{pmatrix} \mu_{11} & \mu_{12} & \cdots & \mu_{1p} \\ \mu_{21} & \mu_{22} & \cdots & \mu_{2p} \\ \vdots & \vdots & \ddots & \vdots \\ \mu_{k1} & \mu_{k2} & \cdots & \mu_{kp} \end{pmatrix} = \begin{pmatrix} \mu_1' \\ \mu_2' \\ \vdots \\ \mu_k' \end{pmatrix} \tag{9.70}$$

where $\mathcal{E}(Y) = XU$. For the linear model, the standard multivariate hypothesis is represented as

$$H : CU = 0 \tag{9.71}$$

where $C_{(\nu_h \times k)}$ is the hypothesis test matrix of rank(ν_h).

Next, suppose the p variables are ordered from most important to least important and let the order matrix $Y = (y_1, y_2, \ldots, y_p)$, $U = (u_1, u_2, \ldots, u_p)$, $Y_j = (y_1, y_2, \ldots, y_j)$, $U_j = (u_1, u_2, \ldots, u_j)$ for $j = 1, 2, \ldots, p$, then with Y_j fixed, the elements of y_{j+1} are distributed univariate normal with common variance $\sigma^2_{j+1} = |\Sigma_{j+1}|/|\Sigma_j|$ for $j = 0, 1, 2, \ldots, p-1$ where $|\Sigma_0| = 1$ and $|\Sigma_j|$ is the first principal minor of order j containing the first j rows and j columns of $\Sigma = (\sigma_{ij})$, and the conditional means are

$$\mathcal{E}(y_{j+1}|Y_j) = X\eta_{j+1} + y_j\beta_j = (X \quad Y_j)\begin{pmatrix}\eta_{j+1}\\\beta_j\end{pmatrix} \tag{9.72}$$

where $\eta_{j+1} = \mu_{j+1} - U_j\beta'_j$, $\beta'_j = (\sigma_{1,j+1}, \ldots, \sigma_{j,j+1})'\Sigma_j^{-1}$, and $\beta_0 = 0$.

With this reparameterization, the multivariate hypothesis in (9.65) becomes

$$H_0 = \bigcap_{i=1}^{m} H_i : \psi_i = 0 = \bigcap_{i=1}^{m}\bigcap_{j=1}^{p} H_{ij} : c'_i\eta_j = 0 \tag{9.73}$$

so that the null hypothesis H_0 is equivalent to testing for $j = 1, 2, \ldots, p$, the hypotheses

$$H_{ij} : c_{i1}\eta_{1j} + c_{i2}\eta_{2j} + \cdots + c_{ik}\eta_{kj} = \psi_{ij} = 0 \tag{9.74}$$

simultaneously or sequentially, where the vector $c'_j = (c_{i1}, c_{i2}, \ldots, c_{ik})$ and the $\sum_{j=1}^{k} c_{ij} = 0$ for $i = 1, 2, \ldots, m$. To test the hypotheses H_{ij} simultaneously, one recognizes the reparameterized model as the standard multivariate analysis of covariance (MANCOVA) linear model.

Assuming the multivariate linear model in (9.68), the least squares (maximum likelihood) estimate of U is

$$\hat{U} = (\hat{\mu}_1, \hat{\mu}_2, \ldots, \hat{\mu}_p) = (X'X)^{-1}X'Y. \tag{9.75}$$

Letting $E = Y'(I - (X'X)^{-1}X')Y$, the least squares estimate of η_{j+1} is

$$\hat{\eta}_{j+1} = \hat{\mu}_{j+1} - \hat{U}_j\hat{\beta}_j \tag{9.76}$$

where $\hat{\beta}_j = (s_{1,j+1}, \ldots, s_{j,j+1})'E_j^{-1}$, $\hat{\beta}_0 = 0$ and $E_j = Y'_j(I - (X'X)^{-1}X')Y_j$. Furthermore, the variance of the estimate $\hat{\psi}_{ij}$ is

$$V\left(\hat{\psi}_{ij}\right) = V(c'_i\hat{\eta}_{j+1}) = c'_i\left[(X'X)^{-1} + \hat{U}_jE_j^{-1}\hat{U}'_j\right]c_i\sigma^2_{j+1}. \tag{9.77}$$

Thus, for $\hat{\psi}_{ij} = c'_i\hat{\eta}_{j+1}$, the estimated variance of the contrast has the general structure $\hat{\sigma}^2_{\hat{\psi}_{ij}} = d_{ij}s^2_j/(N-k-j+1)$ where $s^2_j/(N-k-j+1)$ is an unbiased estimate of the jth variance σ^2_j; we accept or reject H_0 defined in (9.72) using the stepdown FIT procedure if

$$F_{ij} \gtreqless F_{j\alpha} \qquad\qquad j = 1, 2, \ldots, p \tag{9.78}$$

where

$$F_{ij} = \frac{\hat{\psi}_{ij}^2(N-k-j+1)}{\left(c_i'\left[(X'X)^{-1}\right]c_i + \sum_{k=1}^{j-1}\hat{\psi}_{ij}^2/s_k\right)s_j^2} = \frac{\hat{\psi}_{ij}^2(N-k-j+1)}{d_{ij}s_j^2}, \quad (9.79)$$

where $s_j^2 = |E_j|/|E_{j-1}|$, $|E_0| = 1$ and the constants $F_{j\alpha}$ are chosen such that the

$$P(F_{ij} \le F_{j\alpha}; j = 1, 2, \ldots, p \text{ and } i = 1, 2, \ldots, m | H_0)$$

$$= \prod_{j=1}^{p} P(F_{ij} \le F_{j\alpha}; i = 1, 2, \ldots, m | H_0) = 1 - \alpha. \quad (9.80)$$

For given j, the joint distribution of $F_{1j}, F_{2j}, \ldots, F_{mj}$ is a central m-variate multi-variate F-distribution with $(1, N - k - j + 1)$ degrees of freedom, when the null hypothesis defined by (9.72) is true.

Comparing the stepdown FIT of Krishnaiah to Roy's stepdown F test, Mud-holkar and Subbaiah (1980a) derived $(1-\alpha)$-level simultaneous confidence intervals for the original population mean μ_{ij} and showed that the confidence intervals for the stepdown FIT procedure are uniformly shorter than the corresponding intervals obtained using Roy's stepdown F test, if one is only interested in contrasting a variable at a time. For arbitrary contrasts across variables the FIT procedure is not uniformly better. In their study of Krishnaiah's FIT, Cox et al. (1980) showed that the stepdown FIT, in the Neyman sense, is uniformly better than Roy's largest root test or the $T_{\max}^2$ test. Mudholkar and Subbaiah (1980b) and Subbaiah and Mudholkar (1978) showed that the stepdown FIT is unbiased, invariant, has a monotonic increasing power function, and is Bahadar optimal. The $(1-\alpha)$-level simultaneous confidence intervals for contrasts in the original means, $\theta_{ij} = c_i'\mu_j$ where $i = 1, 2, \ldots, m$ and $j = 1, 2, \ldots, p$ are:

$$\hat{\theta}_{ij} - c_0\sqrt{c_i'(X'X)^{-1}c_i} \le \theta_{ij} \le \hat{\theta}_{ij} + c_0\sqrt{c_i'(X'X)^{-1}c_i} \quad (9.81)$$

where

$$c_0 = \sum_{q=1}^{j} \frac{|t_{qi}|}{\sqrt{c_q^*}}$$

$$c_1^* = c_1$$

$$c_j = \frac{F_{j\alpha}}{(N-k-j+1)}$$

$$c_j^* = c_1^* + \cdots + c_{j-1}^* \quad (9.82)$$

where the t_{qi} are the elements of the upper triangular matrix T for the Cholesky decomposition of the sum of squares and cross product matrix $E = T'T$.

Replacing θ_{ij} with $\gamma_{ij} = \sum_{i=1}^{m} a_i \theta_{ij}$ for arbitrary coefficients a_i where the $\sum_{i=1}^{m} a_i = 0$, the $(1 - \alpha)$-level simultaneous confidence intervals for the γ_{ij} are

$$\gamma_{ij} \in \hat{\gamma}_{ij} \pm \sum_{i=1}^{m} |a_i| \sqrt{c_i'(X'X)^{-1}c_i} \left(\sum_{q=1}^{j} |t_{qi}| \sqrt{c_q^*} \right). \tag{9.83}$$

Finally, using the multivariate t-distribution, one may also test one-sided hypotheses H_{ij} simultaneously and construct simultaneous confidence sets for directional alternatives.

9.9   FINITE INTERSECTION TEST OF MEAN VECTORS WITH COVARIATES

To develop the stepdown FIT procedure to test the equality of k population means with covariates, we utilize the multivariate linear model

$$Y = XU + Z\Gamma + E_0 \tag{9.84}$$

where $Y_{(N \times p)}$ is a data matrix of p variables $y_1, y_2, \ldots, y_p$ with rows distributed independently as p-variate multivariate normal random variables with common covariance matrix Σ, $X_{(N \times k)}$ is a known design matrix of full rank $k \leq N - p$, $E_{0(N \times p)}$ is a matrix of random error, $Z_{(N \times h)}$ is a matrix of known covariates, $\Gamma_{(h \times p)} = (\gamma_{ij})$ is a matrix of unknown regression coefficients, and $U_{(k \times p)} = (\mu_{ij})$ is a matrix of unknown means, where $\mathcal{E}(Y) = XU + Z\Gamma$. For the linear model, the standard multivariate hypothesis is represented as

$$H_0 : CU = 0 \tag{9.85}$$

where $C_{(\nu_h \times k)}$ is the hypothesis test matrix of $\text{rank}(\nu_h)$.

Next, suppose the p variables are ordered from most important to least important and let the order matrix be $Y_j = (y_1, y_2, \ldots, y_j)$, $U_j = (U_1, U_2, \ldots, U_j)$ and $\Gamma_j = (\gamma_1, \gamma_2, \ldots, \gamma_j)$ where y_j, μ_j, and γ_j are the jth columns of Y, U, and Γ, respectively, for $j = 1, 2, \ldots, p$. Then with Y_j fixed, the elements of y_{j+1} are distributed univariate normal with common variance $\sigma_{j+1}^2 = |\Sigma_{j+1}|/|\Sigma_j|$ for $j = 0, 1, 2, \ldots, p-1$ where $|\Sigma_0| = 1$ and $|\Sigma_j|$ is the first principal minor of order j containing the first j rows and j columns of $\Sigma = (\sigma_{ij})$, and the conditional means are

$$\mathcal{E}(y_{j+1}|Y_j, Z) = \begin{pmatrix} X & Z & Y_j \end{pmatrix} \begin{pmatrix} \eta_{j+1} \\ \gamma_{j+1}^* \\ \beta_j \end{pmatrix} \tag{9.86}$$

where $\eta_{j+1} = \mu_{j+1} - U_j \beta_j'$, $\beta_j' = (\sigma_{1,j+1}, \ldots, \sigma_{j,j+1})' \Sigma_j^{-1}$, $\beta_0 = 0$ and $\gamma_{j+1}^* = \gamma_{j+1} - Z_j \beta_j$ so that

$$\eta_{j+1} = \mu_{j+1} - U_j \left[(Z_j'Z)^{-1}(\gamma_{j+1} - \gamma_{j+1}^*) \right] \tag{9.87}$$

for $j = 0, 1, 2, \ldots, p - 1$. For fixed Z, the elements of y_1 are independent normally distributed with variance σ_1^2 and mean

$$\mathcal{E}(y_1|Z) = \begin{pmatrix} X & Z \end{pmatrix} \begin{pmatrix} \eta_1 \\ \gamma_1^* \end{pmatrix} \tag{9.88}$$

where $\eta_1 = \mu_1$ and $\gamma_1^* = \gamma_1$.

Letting $\hat{\eta}_{j+1}$ represent the least squares estimate of η_{j+1} where $\hat{\eta}_1 = \hat{\mu}_1$ and the vector of estimates have the general form:

$$\begin{pmatrix} \hat{\eta}_{j+1} \\ \hat{\gamma}_{j+1}^* \\ \hat{\beta}_j \end{pmatrix} = \left[\begin{pmatrix} X' \\ Z' \\ Y_j' \end{pmatrix} \begin{pmatrix} X & Z & Y_j \end{pmatrix} \right]^{-1} \begin{pmatrix} X' \\ Z' \\ Y_j' \end{pmatrix} y_{j+1}$$

$$= \begin{pmatrix} X'X & X'Z & X'Y_j \\ Z'X & Z'Z & Z'Y_j \\ Y_j'X & Y_j'Z & Y_j'Y_j \end{pmatrix}^{-1} \begin{pmatrix} X'y_{j+1} \\ Z'y_{j+1} \\ Y_j'y_{j+1} \end{pmatrix} \tag{9.89}$$

where $\hat{\gamma}_{j+1}^*$ and $\hat{\beta}_j$ are least squares estimates of γ_{j+1}^* and β_j, respectively. Thus, each $\hat{\eta}_{j+1}$ is an adjusted estimator in the ANCOVA model, as one enters each y_j, sequentially, into the analysis.

At each step, let $s_j^2/(N - k + p - j + 1)$ be the unbiased estimator of the variance σ_j^2 and $\hat{\sigma}_{\hat{\psi}_{ij}}^2$ be the variance estimate of $\hat{\psi}_{ij} = c_i'\hat{\eta}_j$ which is the sample estimator for ψ_{ij} with corresponding variance estimate $\hat{\sigma}_{\hat{\psi}_{ij}}^2 = V_{ij}s_j^2/(N - k - j + 1)$ where the covariance matrix of $\hat{\psi}_j = C\hat{\eta}_j = \{V_{ij}\}s_j^2/(N - k - j + 1)$.

Letting the overall hypothesis be represented by the subhypotheses

$$H_0 = CU = \bigcap_{i=1}^{m} H_i : \psi_i = 0 = \bigcap_{i=1}^{m} \bigcap_{j=1}^{p} H_{ij} : c_i'\eta_j = 0 \tag{9.90}$$

we accept or reject H_0 if

$$F_{ij} = \frac{\hat{\psi}_{ij}^2 (N - k - j + 1)}{V_{ij}s_j^2} \gtrless F_{j\alpha} \qquad j = 1, 2, \ldots, p, \tag{9.91}$$

where the constants $F_{j\alpha}$ are chosen such that the

$$P(F_{ij} \leq F_{j\alpha}; j = 1, 2, \ldots, p \text{ and } i = 1, 2, \ldots, m | H_0)$$

$$= \prod_{j=1}^{p} P(F_{ij} \leq F_{j\alpha}; i = 1, 2, \ldots, m | H_0) = 1 - \alpha. \tag{9.92}$$

When H_0 is true, for given j the joint distribution of $F_{1j}, F_{2j}, \ldots, F_{mj}$ follows a m-variate multivariate F-distribution with $(1, N - k + p - j + 1)$ degrees of freedom.

Following Mudholkar and Subbaiah (1980a), one may construct approximate confidence intervals for the elements of U. The $(1 - \alpha)$ simultaneous confidence intervals for contrasts $\theta_{ij} = c'_j \mu_j$ and $i = 1, 2, \ldots, m; j = 1, 2, \ldots, p$:

$$\theta_{ij} \in c'_i \hat{\mu}_j \pm c_0 \sqrt{c'_i \left[(X'X)^{-1} + (X'X)^{-1} X Z (Z'PZ)^{-1} Z' X (X'X)^{-1} \right] c_i}$$

$$(9.93)$$

and

$$P = I - X(X'X)^{-1}X'$$
$$\widehat{U} = (X'X)^{-1}X'Y - (X'X)^{-1}X'Z\widehat{\Gamma} = \begin{pmatrix} \hat{\mu}_1 & \hat{\mu}_2 & \cdots & \hat{\mu}_p \end{pmatrix}$$
$$\widehat{\Gamma} = (Z'PZ)^{-1}Z'PY$$
$$c_0 = \sum_{q=1}^{j} |t_{qi}| \sqrt{c_q^*}$$
$$c_j = \frac{F_{j\alpha}}{(N - k + p - j + 1)}$$
$$c_1^* = c_1$$
$$c_j^* = c_1^* + \cdots + c_{j-1}^*$$

and t_{qj} are the elements of the upper triangular matrix T for the Cholesky decomposition of the sum of squares and cross product matrix

$$E = Y'PY - Y'ZP(Z'PZ)^{-1}Z'PY = T'T.$$

We have shown how the stepdown FIT may be used to analyze the MANCOVA model. Using the Rao-Khatri reduction for the growth curve model (GCM) discussed in Chapter 5, the FIT procedure may be easily adapted to test hypotheses for the GCM, Timm (1995).

9.10 SUMMARY

Researchers performing tests of significance in an analysis of variance fixed effects experiments are often only interested in a finite number of simultaneous comparisons among means or effects. For this situation, Krishnaiah (1964, 1965a, 1965c) developed the FIT procedure. For univariate comparisons among means or effects, the procedure yields confidence intervals that are shorter than Scheffé's method for a finite number of contrasts. When the sample sizes are equal, the FIT is equivalent to Tukey's method for all pairwise comparisons. For unequal sample sizes, the procedure is optimal. The procedure is not restricted to pairwise comparison, but may be used to test (a) $\mu_i = \mu$ $i = (1, 2, \ldots, k)$, (b) $\mu_i - \mu_j = 0$ $(i - 1, 2, \ldots, k - 1)$, (c) $\mu_i = \mu_{i+1}$ $i = 1, 2, \ldots, k - 1$, (d) $\mu_i - \mu_i = 0 (i \neq j)$, and $\mu_i = 0$ $i = 1, 2, \ldots, k$, among others. The procedure is optimal for evaluating the significance of regression coefficients in multiple linear regression. Finally, one may use the stepdown FIT procedure to test for mean or effect differences, a procedure that is uniformly more powerful than the single-step FIT procedure.

When performing a multivariate analysis using several responses, the measurements obtained are usually of unequal value to the researcher. In this situation, most multivariate tests do not take into account the hierarchy, treating each variate equally. This is not the case for stepdown procedures which require one to order the importance of the dependent response variables. For this situation, the stepdown FIT procedure generally yields shorter confidence interval for variables high in priority than any of the standard multivariate test criteria when investigating a finite number of comparisons a variable at a time. Furthermore, for these types of comparisons Krishnaiah's procedure yields shorter confidence interval than Roy's stepdown F-tests. The procedure may be used with any fixed effects MANOVA design including growth curve models. The procedure is asymptotically equivalent to likelihood ratio tests which are known to be optimal in terms of exact slopes, Bahadur or B-optimal. The FIT procedure is a STP that should not be ignored by applied researchers when analyzing univariate and multivariate means differences for significance.

9.11   UNIVARIATE: ONE-WAY ANOVA

The FIT procedure is not available in SAS. One must use the Visual Fortran program called Manova.for. To access the program and all other programs used in this text, go to the Neil Timm's Web site: http://www.pitt.edu/~timm. Click on *Current Projects*, then click on *SAS programs and software*. Within the library is the folder LMbook which contains all programs illustrated in the Second Edition of the book. Using Visual Fortran, compile and execute the program. The program is interactive and will prompt you for information about the data set under analysis such as the number of populations, sample sizes, whether or not the covariance matrix is known, whether the data contains covariates, etc. The input data set for the program must be in the data file named fort.1. After execution, output for the program is stored in the file fort.36.

Data from R. G. J. Miller (1981, p. 82) are used to compare the means of five populations and to illustrate the FIT procedure. The data are provided in Table 9.2 and are available in the file Miller.1. The file must be renamed to fort.1 when used as input to the FIT program: Manova.for. To illustrate the effectiveness of the FIT procedure, we compare the method with several single-step and stepdown methods illustrated and discussed by Toothaker (1991).

We first employ the FIT to evaluate the significance of all ten pairwise comparisons. The critical bounds for the example are shown in Table 9.3 for $\alpha = 0.05$.

As expected the exact value for the FIT is identical to Tukey's procedure. For ten comparisons, critical values for some other well-known classical methods are as follows: Bonferroni-Dunn (3.160), Dunn-Šidák (3.143), Scheffé (3.338), and Hochberg-GT2 (3.110).

If one reduces the number of comparisons from ten to six, the Bonferroni-Dunn

Table 9.2: Miller's Data with Means, Variances, and Pairwise Differences.

Groups	1	2	3	4	5
	18.61	18.86	18.22	22.43	26.32
	13.54	19.17	19.42	17.22	27.01
	16.08	13.69	20.25	22.31	27.08
	18.98	14.47	25.25	19.58	22.32
	13.31	18.81	20.26	23.96	29.77
Means	16.10	17.00	20.68	21.10	26.50
Variances	7.2044	7.2044	7.2023	7.1943	7.1985

Pairwise Group Differences

Groups	1	2	3	4	5
1	—	0.90	4.58	5.00	10.40
2		—	3.68	4.10	9.50
3			—	0.42	5.82
4				—	5.40
5					—

Table 9.3: FIT Critical Values for Ten Comparisons.

ν	α	c_1	c_2	c_3	c_4	c_5 (Exact)	
20	0.05	9.444	9.880	9.697	8.385	8.950	F_α
20	0.05	3.073	3.143	3.114	2.896	2.991	$T_\alpha = \sqrt{F_\alpha}$

Table 9.4: Stepdown FIT Result for the Miller Data.

Step	$F_\alpha = c_3$	$\sqrt{F_\alpha}$	Sig. Diff
1	9.697	3.114	(1 vs 5)[a]
2	9.424	3.069	(2 vs 5)[a]
3	9.121	3.020	(3 vs 5)[a]
4	8.782	2.963	(4 vs 5)[a]
5	8.395	2.879	(1 vs 4)
6	7.946	2.819	N.S. (STOP)

[a] Significant Using the single-step FIT for pairwise comparisons.

and Dunn-Šidák critical values change to 2.930 and 2.916, respectively. The corresponding value for the FIT procedure using Šidák's bound is $c_3 = \sqrt{8.395} = 2.897$. Hence, the FIT out performs two of the better known classical methods. The advantage of the FIT procedure is that it may be used to test the significance of any finite set of contrasts, not just pairwise differences in means. We next illustrate the stepdown FIT procedure.

To perform stepdown FITs for the example, we perform overall tests, removing in a sequential manner the significant comparisons associated with the largest F_i. The results using Šidák's c_3 bound are shown in Table 9.4. As expected, the stepdown FIT procedure is better than the single-step procedure in that the pairwise comparison 1 versus 4 is found to be significant using the stepdown method while it was not found to be significant using the single-step procedure. A limitation of all stepdown procedures is that the construction of two-sided confidence intervals that control the Type I error rate at some nominal level is complicated (Hayter & Hsu, 1994).

For the Miller data, the stepdown FIT procedure for pairwise comparisons using Šidák's c_3 bound performed as well as the Ryan-Einot-Gabriel-Tukey-Welch stepdown procedure (illustrated by Toothaker, 1991, p. 47, 70 with theory provided by Hochberg & Tamhane, 1987, p. 69) and the Peritz stepdown procedure (illustrated by Toothaker, 1991, p. 61 with theory provided by Hochberg & Tamhane, 1987, p. 121-124) based upon the studentized range distribution. However, comparing critical values, the probability of finding more comparisons not significant is higher for the stepdown FIT, even if one could calculate the critical values exactly. In general, the univariate stepdown FIT is to be avoided if one is only interested in making pairwise comparisons. The advantage of the FIT method is that it is not limited to pairwise comparisons since the hypotheses in the finite set may involve a number of finite contrasts. In addition, by using the method of Stefánsson et al. (1988) one can construct $(1 - \alpha)$-level simultaneous confidence sets for one-sided stepdown FITs for general contrasts ψ_i (see, Hochberg & Tamhane, 1987, p. 60). Briefly, recall that if we are testing $H_i{:}\psi_i \leq 0$ versus $A_i^* : \psi_i > 0$ (say) for $i = 1, 2, \ldots, m$, one calculates corresponding to the test H_i a statistic T_i and one orders the statistics T_i such that $T_1 \leq T_2 \leq \cdots \leq T_m$ for tests H_i. At each step one discards the most significant ψ_i corresponding to the test statistic. The process terminates with rejecting

$H_m, \ldots, H_{p+1}$, and retaining $H_p, \ldots, H_1$. Corresponding to the first nonsignificant hypothesis is a corresponding critical value for the multivariate t-distribution, $T_{\alpha p}$. Using the critical value, the $(1 - \alpha)$-level simultaneous confidence set for each significant ψ_i is:

$$\psi_i \geq [\hat{\psi}_i - T_{\alpha p} \hat{\sigma}_{\hat{\psi}_i}]^- \qquad\qquad i = 1, 2, \ldots, p - 1 \qquad (9.94)$$

where $[x]^- = \min(x, 0)$.

With the ability to use the stepdown FIT procedure with any finite set of contrasts, not just pairwise comparisons, and the ability to construct one-sided confidence sets, the STP cannot be ignored by researchers. If one is willing to adopt a STP that only permits testing and not the creation of confidence sets, the sequential stepup procedures of Hochberg (1988) is an attractive option since it is uniformly more powerful than Holm (1979) sequentially rejective stepdown procedure. This is not the case if one employs Shaffer (1986) modification. Hochberg's stepup procedure is not uniformly better than the stepdown FIT.

To improve upon Hochberg (1988) stepup method, Dunnett and Tamhane (1992) (abbreviated as DT) proposed a stepup STP that uses the multivariate t-distribution with equicorrelated structure. However, unlike the stepdown FIT procedure which allows one to compute critical values at each step of the process independently of each other, the critical values of the DT stepup procedure must be calculated recursively; to determine the critical value of the ith value c_i, one must know all prior values. And, the DT stepup procedure is not uniformly more powerful than Hochberg's stepup procedure which ignores the correlation structure of the test statistics. A robust competitor to the stepdown FIT is the stepdown resampling method proposed by Troendle (1995). Simulations performed by Troendle under multivariate normality show that the stepdown FIT is generally more powerful than the stepup procedures of Hochberg or Dunnett and Tamhane.

9.12 MULTIVARIATE: ONE-WAY MANOVA

Following Mudholkar and Subbaiah (1980a), the Iris data made famous by Fisher (1936) are used to evaluate the FIT procedure to compare mean vectors in multivariate normal populations. The Iris data file, iris.dat, is in the folder LMbook. To use the FIT procedure, the file iris.dat with the FIT program, it must be renamed to fort.1.

The data consist of four measurements: Sepal Length (SL), Sepal Width (SW), Petal Length (PL), and Petal Width (PW) for 50 independent observations on three Iris species: Setosa(1), Versicolor(2), and Virginia(3). While Fisher was interested in developing a discriminant function to categorize species into three groups based upon the four measurements, we use the data to investigate group separation based upon the four measurements. Because the analysis of a one-way MANOVA design was not included in Chapter 5, we also include an exploratory analysis of the data at this time and compare some standard STPs with the FIT procedure. The SAS code for the exploratory analysis is provided in Program 9_1.sas and the output for the FIT procedure is included in the file iris.36.

Output 9.12.1: Iris Data Sample Means (in cm) and $SSCP$ Matrix E.

Level of Group	N	SL Mean	SL Std Dev	SW Mean	SW Std Dev	PL Mean	PL Std Dev	PW Mean	PW Std Dev
Setosa	50	5.00600000	0.35248969	3.42800000	0.37906437	1.47200000	0.17847426	0.24400000	0.10720950
Versicolor	50	5.93600000	0.51617115	2.77000000	0.31379832	4.26000000	0.46991098	1.32600000	0.19775268
Virginia	50	6.58800000	0.63587959	2.97400000	0.32249664	5.55400000	0.55703662	2.02600000	0.27465006

E = Error SSCP Matrix	SL	SW	PL	PW
SL	38.9562	13.63	24.9828	5.5956
SW	13.63	16.962	8.1194	4.8012
PL	24.9828	8.1194	27.585	6.2434
PW	5.5956	4.8012	6.2434	6.1756

For an exploratory analysis, one is interested in testing that the means for the three species are equal for all four measurements:

$$H_0 : \mu_1 = \mu_2 = \mu_3. \tag{9.95}$$

Following the overall test, simultaneous confidence intervals are created to investigate significance.

To test H_0 using the GLM procedure, all the criteria depend on the roots of the eigen equation $|H - \lambda E| = 0$. Using any of the multivariate criteria, all indicate that the test for means differences is significant. Thus, there is a significant difference among the three species for the four measurements obtained on each flower. The sample means and sample error sum of squares and cross product matrix E for the sample data in the file iris.dat are provided in Output 9.12.1. The data have been transformed to have the unit of measurement in centimeters.

While one may investigate arbitrary contrasts across the species and measurements following a multivariate test, the most natural comparison are the pairwise means comparisons: Setosa vs. Versicolor, Setosa vs. Virginia, and Versicolor vs. Virginia for each variable. The least squares estimates for these contrasts are shown in Output 9.12.2.

To evaluate the significance of the contrasts following the overall test, one may calculate the confidence intervals for the elements of the contrasts in Output 9.12.2 which have the general form:

$$\hat{\psi}_{ij} - c_0 \hat{\sigma}_{\hat{\psi}_{ij}} \leq \psi_{ij} \leq \hat{\psi}_{ij} + c_0 \hat{\sigma}_{\hat{\psi}_{ij}} \tag{9.96}$$

where c_0 depends on the overall multivariate test criterion. If the interval does not contain zero, the comparison is significant. Alternatively, one may calculate the half-width intervals $c_0 \hat{\sigma}_{\hat{\psi}_{ij}}$ and compare the $|\hat{\psi}_{ij}|$ to the half-width interval; a contrast whose absolute value is greater than its half-width is significant. We use the half-width intervals to compare several post hoc and planned STPs used in practice.

Output 9.12.2: Estimates of Contrasts for Pairwise Comparisons on Each Variable with Tukey Adjustment $\alpha = .0125$.

SL

		Least Squares Means for Effect Group	
i	j	Difference Between Means	Simultaneous 98.75% Confidence Limits for LSMean(i)-LSMean(j)	
1	2	-0.930000	-1.226888	-0.633112
1	3	-1.582000	-1.878888	-1.285112
2	3	-0.652000	-0.948888	-0.355112

SW

		Least Squares Means for Effect Group	
i	j	Difference Between Means	Simultaneous 98.75% Confidence Limits for LSMean(i)-LSMean(j)	
1	2	0.658000	0.462096	0.853904
1	3	0.454000	0.258096	0.649904
2	3	-0.204000	-0.399904	-0.008096

PL

		Least Squares Means for Effect Group	
i	j	Difference Between Means	Simultaneous 98.75% Confidence Limits for LSMean(i)-LSMean(j)	
1	2	-2.788000	-3.037828	-2.538172
1	3	-4.082000	-4.331828	-3.832172
2	3	-1.294000	-1.543828	-1.044172

PW

		Least Squares Means for Effect Group	
i	j	Difference Between Means	Simultaneous 98.75% Confidence Limits for LSMean(i)-LSMean(j)	
1	2	-1.082000	-1.200207	-0.963793
1	3	-1.782000	-1.900207	-1.663793
2	3	-0.700000	-0.818207	-0.581793

1 = Setosa, 2 = Versicolor, 3 = Virginia

Table 9.5: Half-Intervals for Multivariate Criteria and Each Variable.

	SL	SW	PL	PW
Roy (Exact)	0.3851	0.2541	0.3241	0.1533
Roy	0.3235	0.2134	0.2722	0.1287
BLH	0.4545	0.2735	0.3488	0.1650
BNP	0.0331	0.0218	0.0278	0.0131
Wilks	0.0925	0.0611	0.0779	0.0368

For $\alpha = 0.05$, approximate half-width intervals for each of the variables are calculated in Program 9_1.sas using the multivariate criteria due to Roy, Bartlett-Lawley-Hotelling (BLH), Bartlett-Nanda-Pillai (BNP), and Wilks by employing F-distribution approximations (Timm, 2002, p. 102-104). The half-interval widths are displayed in Table 9.5.

For these data and pairwise comparisons, the BNP criterion yields the shortest approximate half-intervals. Wilks' (exact) half-intervals are a close second. The approximate half-intervals for the Roy and BLH criteria are wider for these data. In general, there is no uniformly best criterion. The exact half-intervals for the Roy criterion are also given in Table 9.5. While all pairwise comparison for all variables is significant for the criteria, Wilks and BNP, this is not the case for the other criteria, even though all criteria led to significant differences in the population means.

Instead of performing overall tests followed by the construction of simultaneous confidence sets (exploratory analysis), one may use a STP to investigate planned

Table 9.6: Half-Intervals for $T^2_{\max}$ and Each Variable.

	SL	SW	PL	PW
$T^2_{\max}$	0.3682	0.2430	0.3099	0.1466

comparisons (confirmatory analysis). One STP option is to use the $T^2_{\max}$ criterion and the Bonferroni inequality for the three pairwise comparisons. For this approach, we sent the test size at $\alpha^* = \frac{\alpha}{C} = \frac{0.05}{3} = 0.0167$ for $C = 3$ comparisons. The formula for the half-intervals for the $T^2_{\max}$ criterion is:

$$
\begin{aligned}
c_0 \hat{\sigma}_{\hat{\psi}_{ij}} &= \sqrt{\frac{p \nu_e F^{\alpha^*}_{(p, \nu_e - p - 1)}}{\nu_e - p + 1}} \sqrt{\left(\frac{E_{ij}}{\nu_e}\right)\left(\frac{1}{50} + \frac{1}{50}\right)} \\
&= \sqrt{\left[\frac{(4)\nu_e F^{(0.0167)}_{(4,144)}}{144}\right][E_{ij}(.04)]} \\
&= \sqrt{\left[\frac{4(3.1304)(0.04)}{144}\right][E_{ij}]} \\
&= 0.0590\sqrt{E_{ij}}
\end{aligned}
\tag{9.97}
$$

where p is the number variables and ν_e is the degrees of freedom for error. The half-intervals for the $T^2_{\max}$ criterion are displayed in Table 9.6.

Comparing the exact critical values for the largest root test and the $T^2_{\max}$ test, the size of the interval for $T^2_{\max}$ is shorter than the largest root test. This is true in general (Krishnaiah, 1969, Theorem 3.4, p. 131).

Instead of using the $T^2_{\max}$ test, one might perform 12 univariate t-tests by using the Dunn-Šidák inequality so that $\alpha^* = 1 - (1 - \alpha)^{1/C}$ for C equal to 12 comparisons. Then $\alpha^* = 0.0043$ and $t^{\alpha^*/2}_{\nu_e} = 2.9079$ for $\nu_e = 144$. The formula for the half-interval t-test is:

$$
\begin{aligned}
c_0 \hat{\sigma}_{\hat{\psi}_{ij}} &= t^{\alpha^*}_{\nu_e} \sqrt{\left(\frac{E_{ij}}{\nu_e}\right)\left(\frac{1}{50} + \frac{1}{50}\right)} \\
&= \sqrt{\left(\frac{2.9079^2}{144}\right)\left(\frac{1}{50} + \frac{1}{50}\right)(E_{ij})} \\
&= 0.0485\sqrt{E_{ij}}.
\end{aligned}
\tag{9.98}
$$

Comparing the standard errors, the Dunn-Šidák STP is better than the $T^2_{\max}$ criterion for all pairwise comparisons.

Finally, we may use Tukey (1953) method. For this procedure, we perform univariate studentized range tests a variable at a time using $\alpha^* = \frac{\alpha}{p} = \frac{0.05}{4} = 0.0125$

to control the family-wise error rate at the nominal $\alpha = 0.05$ level. For Tukey's method, the half-intervals are:

$$c_0 \hat{\sigma}_{\hat{\psi}_{ij}} = q^{\alpha^*}_{(k,\nu_e)} \sqrt{\frac{2MS_e}{n}}$$

$$= \sqrt{\left(\frac{\left[q^{\alpha^*}_{(k,\nu_e)} \right]^2}{2\nu_e} \right) \left(\frac{1}{n} + \frac{1}{n} \right) (E_{ij})}$$

$$= \sqrt{\frac{4.12^2}{144(50)} E_{ij}}$$

$$= 0.0486 \sqrt{E_{ij}} \tag{9.99}$$

where k is the number of groups. Tukey's method and the multiple t-tests are essentially equivalent.

To complete our evaluation of STPs, we now calculate the half-intervals for the contrasts using Roy (1958) stepdown F test procedure and Krishnaiah's FIT. For the comparison, we consider only the order (1): SL,SW,PL,PW. Comparisons for the order (2): SW,SL,PW,PL are discussed by Mudholkar and Subbaiah (1980a). To calculate the size of the half-intervals, (9.81) is used. The half-intervals for the FIT procedure are:

$$c_0 \hat{\sigma}_{\hat{\psi}_{ij}} = c_0 \sqrt{c_i'(X'X)^{-1}c_i}$$

$$= \left(\sum_{q=1}^{j} |t_{qi}| \sqrt{c_q^*} \right) \sqrt{\frac{1}{50} + \frac{1}{50}} \qquad j = 1, 2, \ldots, p \tag{9.100}$$

where

$$(t_{qi}) = T$$
$$T'T = E$$
$$c_j = \frac{f_{j\alpha}}{N - k - j + 1}$$
$$c_1^* = c_1$$
$$c_j^* = (c_1^* + \cdots + c_{j-1}^*)$$

and $f_{j\alpha}$ are the critical values of the multivariate F-distribution such that the

$$P(F_{ij} \leq f_{j\alpha} \text{ for all } i \text{ and } j | H_0) = \prod_{j=1}^{p} P(F_{ij} \leq f_{j\alpha}; i = 1, \ldots, m | H_0) = 1 - \alpha$$

$$\tag{9.101}$$

where m is the number of contrasts and p is the number of variables. For the Iris data, $m = 3$, $k = 3$, $p = 4$, and $\nu_e = N - k = 147$. Using the FIT program and

Table 9.7: Critical Values Multivariate F-distribution.

Step/Variable	$f_{j\alpha}$	ν_e
$j = 1$	8.428	147
$j = 2$	8.430	146
$j = 3$	8.433	145
$j = 4$	8.433	144

Output 9.12.3: The Cholesky Triangular Matrix T for the Matrix E.

T			
6.2414902	2.1837734	4.0026979	0.8965167
0	3.4918668	-0.178009	0.8142953
0	0	3.3958389	0.8244993
0	0	0	2.0072324

Bonferroni's upper bound with $\alpha_1 = \alpha_2 = \alpha_3 = \alpha_4 = 0.012$, the critical values for the multivariate F-distribution with $(1, \nu_e - j + 1)$ degrees of freedom are shown in Table 9.7.

Using the SAS/IML command $t=root(e)$ the Cholesky triangular matrix T for the matrix E is displayed in Output 9.12.3 for the variable order (1): SL,SW, PL, and PW. Using (9.100), for $j = 1$,

$$c_0 \hat{\sigma}_{\hat{\psi}_{ij}} = t_{11} \sqrt{\left(\frac{f_{1\alpha}}{144}\right) \left(\frac{1}{50} + \frac{1}{50}\right)}$$

$$= 6.2414902 \sqrt{\left(\frac{8.428}{144}\right) \left(\frac{1}{50} + \frac{1}{50}\right)}$$

$$= 6.2414902(0.0484)$$

$$= 0.3020.$$

For $j = 2$; $c_1^* = c_1 = \frac{8.428}{147} = 0.0573$, $c_2^* = (1 + c_1^*)c_2 = (1 + c_1)c_2 = (1 + 0.0573) \left(\frac{8.430}{146}\right) = (1.0573)(0.0577) = 0.0611$ so that

$$c_0 \hat{\sigma}_{\hat{\psi}_{ij}} = \left(|t_{12}| \sqrt{c_1^*} + |t_{22}| \sqrt{c_2^*}\right) \sqrt{\frac{1}{50} + \frac{1}{50}}$$

$$= (2.1837734 \sqrt{0.0573} + 3.4918668 \sqrt{0.0611}) \sqrt{\frac{1}{50} + \frac{1}{50}}$$

$$= (0.5227 + 0.8631)(.20)$$

$$= 0.2772.$$

Table 9.8: Critical Values Multivariate F-distribution and Stepdown F.

Step/Variable	$f_{j\alpha}$	ν_e	$F^{\alpha^*}_{(2,\nu_e)}$	$2F^{\alpha^*}_{(2,\nu_e)}$
$j = 1$	8.428	147	4.4951	8.9902
$j = 2$	8.430	146	4.4960	8.9920
$j = 3$	8.433	145	4.4970	8.8940
$j = 4$	8.433	144	4.4979	8.9958

For $j = 3$; $c_3^* = (1 + c_1^* + c_2^*)c_3 = (0.0573 + 0.0611)\left(\frac{8.433}{145}\right) = 0.0650$ so that

$$c_0\hat{\sigma}_{\hat{\psi}_{ij}} = \left(|t_{13}|\sqrt{c_1^*} + |t_{23}|\sqrt{c_2^*} + |t_{33}|\sqrt{c_3^*}\right)\sqrt{\frac{1}{50} + \frac{1}{50}}$$

$$= \left(4.0026979\sqrt{0.0573} + 0.178009\sqrt{0.0611} + 3.3958389\sqrt{0.0650}\right)\sqrt{\frac{1}{50} + \frac{1}{50}}$$

$$= (0.9581 + 0.0440 + 0.8658)(.20)$$

$$= 0.3736.$$

Finally, for $j = 4$,

$$c_0\hat{\sigma}_{\hat{\psi}_{ij}} = \left(|t_{14}|\sqrt{c_1^*} + |t_{24}|\sqrt{c_2^*} + |t_{34}|\sqrt{c_3^*} + |t_{44}|\sqrt{c_4^*}\right)\sqrt{\frac{1}{50} + \frac{1}{50}}$$

$$= 0.1982$$

where $c_4^* = (1 + c_1^* + c_2^* + c_3^*)c_4$ and $c_4 = \frac{8.443}{144} = 0.0586$ and $c_4^* = 0.0693$.

To perform a similar calculation for Roy's stepdown F test, one replaces the critical values $f_{j\alpha}$ from the multivariate F-distribution with $\nu_h F_{\alpha j}$ where $F_{\alpha j}$ is the upper $100\alpha_j$ percentage point of the F-distribution with $(\nu_h = k - 1, \nu_e = N - k - j + 1)$ degrees of freedom at each step j. Because the stepdown F's are independent,

$$1 - \alpha = \prod_{j=1}^{p}(1 - \alpha_j) = (1 - \alpha^*)^p. \tag{9.102}$$

If all $\alpha_j = \alpha^*$, then $\alpha^* = 1 - (1 - \alpha)^{1/p}$. For $\alpha = 0.05$, we set $\alpha^* = 1 - (1 - .05)^{1/4} = 0.01274$ to control the FW error rate at the nominal level $\alpha = 0.05$. The critical values for the stepdown F tests are shown in Table 9.8 where they are compared with the stepdown FIT procedure. Since the critical value for the FIT procedure is always smaller than the corresponding Roy stepdown procedure, the FIT procedure will lead to shorter half-intervals. This was proven in general by Mudholkar and Subbaiah (1980a).

Finally, half-intervals for the stepdown FIT test are shown in Table 9.9 for the order (1): SL, SW, PL, and PW and for order (2): SW, SL, PW, PL and compared

Table 9.9: Half-Intervals for FIT and Some Others for Each Variable.

	SL	SW	PL	PW
FIT (order 1)	0.3021	0.2772	0.3736	0.1982
FIT (order 2)	0.4342	0.1974	0.4961	0.1830
Roy (exact)	0.3851	0.2541	0.3241	0.1533
BLH	0.4545	0.2735	0.3488	0.1650
T^2_{max}	0.3682	0.2430	0.3099	0.1466
Dunn-Šidák/Tukey	0.3027	0.1997	0.2547	0.1205

with some common single step (exploratory) methods and the confirmatory method of Tukey and Dunn-Šidák.

Depending on the order of the variables, the FIT procedure compares favorably with several of the more traditional STP methods when one is interested in comparisons a variable at a time. And, while several exploratory methods may out perform the procedure depending on the structure of the population eigenvalues, it is a viable procedure that cannot be ignored.

9.13   MULTIVARIATE: ONE-WAY MANCOVA

In Chapter 5 we illustrated the analysis of a multivariate MANCOVA design with two dependent variables and one covariate. In MANCOVA designs with several variables and several covariates, a researcher is often interested in the sequential effect of a set of variables rather than the joint effect. Thus, one has predetermined knowledge about the order of importance of the set of dependent variables. When this is the case the FIT for differences in means vectors is often preferred to a joint analysis of the dependent variables. We also investigated in Chapter 5 data collected by Dr. William Rohwer for 32 students in kindergarten from a High-SES-class area where Rohwer was interested in predicting the performance on three standardized tests: Peabody Picture Vocabulary test (PPVT), the Raven Progressive Matrices test (RPMT), Student Achievement test (SAT), and using five paired-associate, learning-proficiency tasks: Named (N:x1), Still (S:x2), Named Still (NS:x3), Named Action (NA:x4), and Sentence Still (SS:x5) for 32 school children in an upper-class, white residential school; data set Rohwer.dat. Dr. Rohwer also collected similar data for 37 students in a Low-SES class area. While this data was analyzed by Timm and Lin (2005, p. 239) using a MANCOVA model, we analyze the same data using the FIT procedure. The SAS code for the MANCOVA analysis is provided in Program 9_14.sas. This analysis showed that a difference existed between the two groups retaining only the covariates NS and NA. And an analysis of additional information indicated that the dependent variables RMMT and SAT were significant.

We now analyze the same data using the FIT procedure. For this analysis we assume the order PPVT, RPMT, and SAT and evaluate the effect of each dependent variable in the set in a stepwise manner, fitting the dependent variables sequentially.

For this analysis we again use program `mancova.for`. The input for this example is in the file fort.1 and the output is in the file fort.36. Investigating the data sequentially, we see that a difference exists for the variable PPVT, given the two covariates since the confidence set for the difference is [11.7332, 24.0904], which does not contain zero and is shorter than the corresponding Roy confidence interval (10.343006, 25.480608) using the MANCOVA approach. However, the associated confidence sets for the variables RPMT given PPVT and SAT given both RPMT and PPVT are found to be nonsignificant using the FIT procedure, but shorter than the corresponding interval for the unconditional analysis.

CHAPTER 10

Computing Power for Univariate and Multivariate GLM

10.1 INTRODUCTION

Assuming a univariate GLM with normal errors with a fixed or random design matrix X and using the likelihood ratio or UI principle for hypothesis testing with normal errors, the F test is the uniformly most powerful, invariant, unbiased (UMPIU) test for testing for significant differences in model parameters, Scheffé (1959, Section 2.10). Under the null hypothesis, the distribution of the parameter estimates (under normality) and the test statistic do not differ; both follow a central F distribution. This is not the case for the alternative hypothesis. Study of the alternative distribution by varying the tuple (β, X, N, σ^2) for fixed Type I error α is known as power analysis. For fixed X the parameter estimates follow a noncentral F distribution, Sampson (1974). The overall goal of a univariate power analysis is to plan a study to ensure that the sample size N is not too large so as to waste resources and perhaps claim that meaningless differences in population parameters are significant or too small so as to limit one's ability to find significance. In this chapter we discuss the calculation of power and sample size determination for univariate normal GLMs with fixed design matrices and use standard SAS procedures to illustrate the methodology. When the design matrix is random and not fixed, the situation is more complicated since random predictors add an additional source of randomness to the design, Glueck and Muller (2003). One may consult Gatsonis and Sampson (1989) for a general discussion and tables for computing power with random design matrices.

Using the univariate hypothesis testing paradigm for fixed effect GLMs, a sample size is selected so as to ensure a certain level of statistical power, the probability of rejecting the null hypothesis at some desired level given that the null hypothesis is false. Because the power analysis is performed for some Type I level α, the confidence set for the population parameter has confidence $1 - \alpha$. In the long run, the confidence set is expected to contain the true parameter or the expected validity of coverage is $1 - \alpha$. However, the probability of the interval containing the true parameter is zero or one. And, because the endpoints of the confidence set are estimated from the data, the width of the interval is not controlled by the power of the test. The

381

univariate power paradigm selects a sample size that provides sufficient statistical power for a test. It does not directly control the width of the interval or the validity of the interval. To offset these limitations, Jiroutek et al. (2003) show how to select a sample that controls the power of the test (rejection), the expected length of the confidence set (width), and its associated coverage probability (validity) for GLMs with fixed design matrix X under normality. We review their approach and illustrate the calculation of power using SAS software developed by Michael R. Jiroutek and Keith E. Muller.

In many applications of the GLM, the design matrix is fixed and the parameter vector is random or mixed, a GLMM. For a fixed effect design, the number of fixed populations (J) is known. However, if the J levels are selected from a population of effects, they are random and not fixed. Then, one has the additional cost of sampling the random effects as well as sampling the observations. Letting C_r represent the additional cost of sampling a random effect from among the population of effects, the total cost of the study for J selected random effects may be represented by the linear cost function: $T = C_I + J(nC_o + C_r)$ where C_I is some initial overall start-up cost. This assumes that the cost for each observation or a random effect may be unequal but are constant, do not vary from random effect to random effect or observation to observation. In many applications, the cost of sampling a treatment level is larger than the cost of sampling an observation. To determine the optimal sample size n_0 for the study now requires trade-offs between the number of levels for the random effect and the number of observations in each cell. Removing the initial cost from the linear cost function, we see that J is related to the total cost T by the inequality $J \leq \frac{T}{nC_o + C_r}$. The strategy for planning such designs is to choose the optimal sample size that attains some acceptable level of power for tests of significance and the classical A-optimal (as opposed to D- and E-optimal) criterion of minimizing the asymptotic errors of estimates for the model parameters. By varying study costs and population values of model parameters, one may evaluate the power of the statistical tests and standard errors of estimates for the estimated model parameters. One may also consider the effect of adding one or more covariates to the design. However, adding even a single additional random factor with say K levels complicates the power analysis since the cost function must take into account the additional random cost component. A simple revised linear cost function may take the form: $T = C_I + JKnC_o + JC_{r_1} + KC_{r_2}$. The single additional random effect complicates the power analysis since along with an additional unknown variance component, the new variable cost component must be factored into the analysis. We address some of these complications in estimating power and sample size for some mixed models.

In univariate fixed effect GLMs with nonrandom design matrices, power analysis is not complicated since the F test is UMPIU. And power analysis is related to a single distribution; however, random design matrices and mixed models introduce some complications. For MGLMs the situation becomes more complicated since even for designs with fixed design matrices UMPIU tests are difficult to find since they depend on the dimension of the subspace occupied by the mean vector. For the dimension $s = 1$, Hotelling's T^2 test is UMPIU, T. W. Anderson (1984). For $s > 1$, there is no uniformly most powerful test. However, one may still want to evaluate power for various multivariate test criteria and estimate sample sizes for multivariate designs with a fixed design matrix. Muller and Peterson (1984) show how to approximate the noncentral distributions of several multivariate criteria and Muller et al. (1992) developed SAS software to perform multivariate power calculations. We conclude this chapter with a brief overview of power analysis for the MGLM with fixed design matrices and illustrate how one may estimate power and determine sample sizes for several multivariate designs. The examples are illustrated with a new program called PowerLib202 and distributed by Keith E. Muller.

10.2   POWER FOR UNIVARIATE GLMs

In an experimental design in which the dependent variable is random and the independent variables are fixed, recall that the relationship between the dependent variable y and the matrix of independent variables X is represented by the general linear model

$$y = X\beta + e, \tag{10.1}$$

where y is a $N_t \times 1$ vector, X is the $N_t \times q$ design matrix of rank q, β is a $q \times 1$ unknown vector, and e is a vector of independent random errors $e_i \sim N(0, \sigma^2)$. Given the general linear model in (10.1), unbiased estimators of the parameter vector and the unknown common variance are

$$\hat{\beta} = (X'X)^{-1}X'y \tag{10.2}$$

and

$$s^2 = \frac{(y - X\hat{\beta})'(y - X\hat{\beta})}{N_t - q} = MSE, \tag{10.3}$$

where N_t is a target sample size for the design. The null hypothesis for the general linear model has the general structure

$$H_0 : C\beta = \psi_0, \tag{10.4}$$

where C is a $\nu_h \times q$ hypothesis test matrix with rank $\nu_h \leq q$ and ψ_0 is a vector of unknown constants, often taken to be zero. The alternative hypothesis is $H_a : C\beta \neq \psi_0$. Letting $\Psi = C\beta$, the sum of squares for the hypothesis is

$$SSH(X, \hat{\beta}, N_t) = (\hat{\Psi} - \psi_0)'[C(X'X)^{-1}C']^{-1}(\hat{\Psi} - \psi_0), \tag{10.5}$$

and the observed test statistic for testing the null hypothesis H_0 versus the alternative hypothesis H_a is

$$F_{obs} = \frac{SSH(X, \hat{\beta}, N_t)/\nu_h}{s^2}. \tag{10.6}$$

The test statistic F_{obs} follows a noncentral F distribution with degrees of freedom ν_h and $\nu_e = N_t - q$, $F_\lambda(\nu_h, \nu_e, \lambda)$, with noncentrality parameter λ:

$$\lambda = SSH(X, \beta, N_t) = \frac{(\Psi - \psi_0)'[C(X'X)^{-1}C']^{-1}(\Psi - \psi_0)}{\sigma^2} = \frac{\delta^2}{\sigma^2}. \quad (10.7)$$

Letting $f_{cv} = F_{(\nu_h,\nu_e,0)}^{(1-\alpha)}$ represent the upper $1 - \alpha$ critical value for the central F distribution, the power of the test is

$$\pi = 1 - F_\lambda[f_{cv}] \quad (10.8)$$

where F_λ is the cumulative distribution function of the noncentral $F_\lambda(\nu_h, \nu_e, \lambda)$ distribution. When the null hypothesis is not true, the noncentrality parameter $\lambda > 0$ and hence $\delta^2 > 0$, since $\sigma^2 > 0$.

To evaluate power (π) using SAS, the critical value may be obtained using the FINV function, $f_{cv} = FINV(1 - \alpha, \nu_h, \nu_e, 0)$. The SAS function PROBF is used to calculate power $\pi = 1 - PROBF(f_{cv}, \nu_h, \nu_e, \lambda)$. The function FINV returns the critical value for the central F distribution and the function PROBF returns a p-value for the noncentral distribution.

10.3 ESTIMATING POWER, SAMPLE SIZE, AND EFFECT SIZE FOR THE GLM

10.3.1 Power and Sample Size

Power analysis is used to determine an adequate sample size for a study design specified by the matrix X. To estimate the sample size and associated power for a design, one performs a sensitivity analysis by considering several design factors. From the formula for power, one observes that power depends upon the size of the test α, the unknown population variance σ^2, the unknown parameter β and the choice of C (the hypothesis of interest), the value of ψ_0, and finally the total sample size N_t.

To initiate a power analysis, one first decides upon possible values of the unknown variance. Information is commonly obtained from former studies or from a pilot study for the design under consideration. Next, one specifies various choices of Ψ and ψ_0 and because multiple tests effect the overall size of the test, the test size α is adjusted using, for example, the Bonferroni-Dunn or Dunn-Šidák correction. Finally, one changes the sample size to calculate values for the power of the tests. The process will be illustrated with several examples using the SAS procedure GLMPOWER.

Uncertainty about population parameters is used to define the process associated with a power analysis. In particular, uncertainty about the estimates of σ^2 and β drives the power analysis process. Thus, the noncentrality parameter and power may be considered random variables. This led Taylor and Muller (1995, 1996) to calculate confidence intervals for the noncentrality parameter and power based upon estimates of σ^2 and β. To determine an adequate sample size for a study may require a pilot study based upon ν degrees of freedom to estimate the unknown population variance σ^2. Under normality, the quantity $\frac{\nu s^2}{\sigma^2}$ has a central chi-square distribution. To obtain exact bounds for the noncentrality parameter λ, we follow the procedure proposed by Taylor and Muller (1995, 1996). Letting α_{cL} and α_{cU} represent the lower and upper tail probabilities used to define the confidence coefficient $(1 - \alpha_{cL} - \alpha_{cU})$, and $c_{\text{critical}}(p|\nu)$ the $100 \times (p)$ percentile for a central chi-square distribution with ν degrees of freedom, confidence interval for the noncentrality parameter λ is

$$\hat{\lambda}_L \le \lambda \le \hat{\lambda}_U, \tag{10.9}$$

with the boundaries of the interval defined as

$$
\begin{aligned}
\hat{\lambda}_L &= c_{\text{critical}}(\alpha_{cL}|\nu) \left[\frac{SSH(X, \beta, N_t)}{SSE} \right] \\
\hat{\lambda}_U &= c_{\text{critical}}(1 - \alpha_{cU}|\nu) \left[\frac{SSH(X, \beta, N_t)}{SSE} \right]
\end{aligned}
\tag{10.10}
$$

and $SSH(X, \beta, N_t)$ is fixed under the assumption of known (nonzero) β. Selecting $\alpha_{cL} = \alpha_{cU} = 0.025$, an exact 95% confidence interval for the noncentrality parameter λ is obtained:

$$\hat{\lambda}_{cL} \le \lambda \le \hat{\lambda}_{cU}. \tag{10.11}$$

Using these bounds, an asymptotically correct confidence set for power is realized.

10.3.2 Effect Size

Generally speaking, an effect size (ES) is a scale-free index constructed to evaluate the magnitude of the effect of the independent variable(s) on a dependent variable. In experimental designs with fixed nonrandom parameters, the dependent variable under study is random; however, the independent variables may be random or fixed. If both the dependent and independent variables are jointly sampled from a population (often assumed to be multivariate normal) both variables are random. In this *correlational study*, one may evaluate the effect of the independent variable on the dependent variable by conditioning on random independent variables. However, the converse is not true. The joint relationship between two variables may not be necessarily inferred by investigating only a conditional likelihood, Ericsson (1994).

In correlational designs, study effect sizes may be reported as: (1) scale-free correlation coefficients (e.g. r_{alerting}, r_{contrast}, $r_{\text{effect-size}}$, r_{BESD}, $r_{\text{counternull}}$), Rosenthal, Rosnow, & Rubin, 2000, or $(r_{\text{equivalent}})$, Rosenthal & Rubin, 2003, and (2) correlation ratios (e. g., Pearson-Fisher's η^2, Kelley's ε^2, Hays' ϖ^2), Hedges & Olkin, 1985, p. 100-103. This is not the case when the independent variables are fixed and nonrandom in regression analysis and analysis of variance (ANOVA). The sampling scheme is stratified and depends upon the nonrandomly selected values of the independent variables. Because the dependent variables are obtained using a stratified sampling scheme, the observations may be statistically independent, but they are no longer identically distributed. The sample multiple correlation coefficient, while perhaps a descriptive measure of association, is not an estimate of the population correlation coefficient, whether one adjusts or does not adjust the coefficient, Goldberger (1991, p. 176-179). While some continue to recommend reporting a measure of association (or a correlation ratio) as an effect size measure to evaluate the effect of a fixed independent variable on a dependent variable, the practice has limited value since one variable is fixed by design and not random. Because it is not an estimate of a population correlation coefficient (or its square), it is difficult to interpret. Expanding the range of the independent variable or the sample size does not correct this situation, Goldberger (1991, p. 178). This is not the case for the study effect size measures: Cohen's d, Hedges's g, and Glass's $\hat{\Delta}$; however, these scale-free indices are only defined when comparing two independent population means. For two independent groups, experimental (E) and control (C), let $\hat{\mu}_C$ and $\hat{\mu}_E$ represent estimators of the population means. Then, Glass's effect size measure is $\hat{\Delta} = \frac{\hat{\mu}_C - \hat{\mu}_E}{s_c}$; Hedges's $g = \frac{\hat{\mu}_C - \hat{\mu}_E}{s}$; and Cohen's $d = \frac{\hat{\mu}_C - \hat{\mu}_E}{\sigma_{\text{within}}}$; where N_t is the total sample size, $\nu_e = N_t - 2$ is the error degrees of freedom, s is the square root of the pooled within estimator of σ, s_c is the estimate of the standard deviation for the control group, and $\sigma_{\text{within}} = s\sqrt{\frac{\nu_e}{N_t}}$ (see, e.g., Rosenthal, 1994; Rosenthal et al., 2000).

In an experimental design, a test of significance TS is used to evaluate whether or not the null hypothesis may be supported based upon the sample data. Associated with the test of significance is a target sample size N_t of observations. The unknown effect of the independent variable on the dependent variable is the magnitude of the scale-free effect size (ES). The effect size depends on three unknowns: (1) the sampling plan represented by the design matrix X, (2) the unknown fixed model parameters β used to model the relationship between the dependent and independent variables, and (3) the unknown population variance σ^2. Letting λ represent the unknown nuisance parameter, Rosenthal (1994) shows that the test of significance is related to the study effect size by the relationship:

$$TS(\lambda) = SF(N_t) \times ES(X, \beta, \sigma^2) \qquad (10.12)$$

where $SF(N_t)$ is a size of study factor multiplier that depends on the sample size. By dividing the test statistic by the study multiplier, one is able to isolate the study effect size index. Given the relationship in (1), observe that a confidence interval for an effect size depends on size of the test, α. Even though effect sizes are scale-free their interpretation often depends upon the context of the study, the scale of measurement for the dependent variable, the study design, and the power of the test.

For effect sizes that are distributed symmetrically, Rosenthal (1994) defines a counternull effect size defined by the equation

$$ES_{\text{counternull}} = 2ES_{\text{obtained}} - ES_{\text{null}} \qquad (10.13)$$

which does not depend on the size of the test α. These are likelihood bounds for the effect size and not a confidence interval. Because the effect size under the null hypothesis is very often zero, the counternull effect size bound is two times the observed effect size. The $ES_{\text{counternull}}$ is an alternative non-null effect size that is as equally likely as the null to produce the observed sample effect size. If the P-value of the test is P, the null-counternull bounds: $[ES_{\text{null}}, 2ES_{\text{obtained}} - ES_{\text{null}}]$ define a $(1 - P) \times 100$ percentile interval.

When comparing two or more independent groups, following Cohen (1988, p. 414), the noncentrality parameter λ for a linear regression model may be represented as:

$$\lambda = (ES)^2(\nu_e + \nu_h + 1) = \frac{\delta^2}{\sigma^2},   \tag{10.14}$$

where $(ES)^2$ is the squared effect size for the parameter β being tested. ES is defined on the unbounded interval $[0, \infty)$. It measures the distance between Ψ and ψ_0 in population σ units. Using the relationship in (10.12) and comparing it with (10.7), the quantity $\sqrt{(\nu_e + \nu_h + 1)}$ is Rosenthal's "size of study" multiplier.

To construct a plug-in predictor for ES, one may simply substitute estimators for δ^2 and σ^2 in (10.7), obtaining:

$$\widehat{ES} = \frac{(\hat{\Psi} - \psi_0)'[C(X'X)^{-1}C']^{-1}(\hat{\Psi} - \psi_0)}{s\sqrt{(\nu_e + \nu_h + 1)}}$$

$$= \frac{\hat{\delta}}{s\sqrt{(\nu_e + \nu_h + 1)}}.   \tag{10.15}$$

The metric of $\widehat{ES}$ is expressed in sample standard deviation units s on the unbounded interval $[0, \infty)$. For this plug-in estimator, when comparing the equality of means for two treatment groups with equal sample sizes $(n_1 = n_2 = n = \frac{N_t}{2})$, $\widehat{ES}$ is identical to Hedges's $g = \frac{2t}{\sqrt{N_t}} = \widehat{ES}$, where t is Student's t-statistic for testing the equality of two population means. Or, using the transformation $d = g\sqrt{\frac{N_t}{\nu_e}}$, one obtains Cohen's d. Extensions of these indices to studies with unequal sample sizes are provided by Rosenthal et al. (2000, p. 31). These effect size indices may not be used if one is interested in comparing more than two group means. Following Hedges (1981), one may adjust the estimators to obtain an unbiased estimate for the effect size when comparing two groups. Solving (10.11) for the effect size, a plug-in predictor of the effect in terms of the noncentrality parameter is obtained:

$$\widehat{ES} = \frac{\sqrt{\hat{\lambda}}}{\sqrt{(\nu_e + \nu_h + 1)}}.   \tag{10.16}$$

Dividing (10.9) by $(\nu_e + \nu_h + 1)$ and using the nonlinear square root transformation, an approximate confidence interval for the ubiquitous effect size is:

$$\widehat{ES}_L = \frac{\sqrt{\hat{\lambda}_{cL}}}{\sqrt{(\nu_e + \nu_h + 1)}} \leq ES \leq \frac{\sqrt{\hat{\lambda}_{cU}}}{\sqrt{(\nu_e + \nu_h + 1)}} = \widehat{ES}_U.   \tag{10.17}$$

The one-sided least upper bound and greatest lower bounds for the effect sizes are $[0, \widehat{ES_U}]$ and, $[\widehat{ES_L}, \infty)$, respectively. The accuracy of the bounds increases as the sample size increases. More research is needed to set exact bounds for ES; however, preliminary simulation results for two independent groups suggest that each bound appears to be within $O(N_t^{-1})$ of the exact bound. Because the distribution of ES is not symmetrical, the counternull bounds of Rosenthal (1994) may not be used for the ubiquitous effect size.

Even though the noncentrality parameter is divided by the study multiplier to obtain a scale-free effect size index, the study effect size still depends upon the design matrix. For example, it is well known that if block effects are appreciably larger than zero that a randomized block design is more efficient than a completely randomized design. When comparing effect sizes for the two different design matrices, one must adjust the effect size by the relative efficiency design matrix factor. To illustrate, suppose study A has $ES_A = 0.10$ and that study B has the same effect size, $ES_B = ES_A$, but, suppose study design A is 1.05 times more efficient than study design B (given equal sample sizes and population variances). The effective effect size for study A is not 0.10, but $\frac{0.10}{1.05} = 0.0952$ when compared to study B. When comparing effect sizes for different multifactor designs, adjustments become more difficult (Gillett, 2003).

10.4   POWER AND SAMPLE SIZE BASED ON INTERVAL-ESTIMATION

Sample size determination based upon power and effect size analysis is driven by hypothesis testing. One is interested in a sample size that is sufficient to reject hypotheses that are false in the population with sufficient statistical power. The number of subjects required is driven by the high probability of rejecting the null hypothesis. Traditional interval-based methods led to sample size determination based solely upon the width of a confidence set (Bristol, 1989). A new method proposed by Jiroutek et al. (2003) takes into account power, the width of the interval, and the probability that the estimated interval contains the population parameter. To implement the new procedure, we need some preliminaries. Let ψ represent a true unknown population parameter and ψ_0 its null value. Thus, the null hypothesis may be stated as $H_0 : \psi = \psi_0$. Letting L and U represent the lower and upper (random) bounds for the parameter in the population, the confidence interval is defined as $U - L$. The event width is defined as $U - L \leq \xi$, for fixed $\xi > 0$ chosen a priori. Event validity is defined as $L \leq \psi \leq U$. And finally, event rejection is said to occur if the observed interval excludes ψ_0. The goal of the new procedure is to simultaneously consider width, validity, and rejection. That is in the words of the authors: "Given validity, how many subjects are needed to have a high probability of producing a confidence interval that correctly does not contain the null value when the null hypothesis is false and has a width not greater than ξ?"

To more formally put forth the position of the authors, suppose we desire to test $H_0 : \psi = \psi_0$ versus the two sided alternative $H_a : \psi \neq \psi_0$. Then the probability of rejecting the null given that it should be rejected (the power of the test) may be written as:

$$P\{R\} = P\{(U < \psi_0) \cup (\psi_0 < L)\}. \tag{10.18}$$

For a one sided test $H_0 : \psi = \psi_0$ and $H_a : \psi > \psi_0$ $(\psi < \psi_0)$ event rejection becomes $P\{\psi_0 < L\}$ $(P\{U < \psi_0\})$. With $\xi > 0$ fixed a priori, the event width (W) occurs if $\widehat{U} - \widehat{L} \leq \xi$, so that

$$P\{W\} = P\{U - L \leq \xi\}. \tag{10.19}$$

Finally, event validity (V) occurs if the observed interval contains the true parameter, $\widehat{L} \leq \psi \leq \widehat{U}$, so that

$$P\{V\} = P\{L \leq \psi \leq U\}. \tag{10.20}$$

Setting $P\{V\} = 1 - \alpha$, with α fixed a priori, $[L, U]$ defines an exact $(1 - \alpha)$-size confidence set for ψ. Whether or not the interval does or does not contain the true parameter we say the confidence set has "validity" since as suggested by Lehmann (1991) there is no merit in short intervals that are far from the true value. This suggests that $P\{V\} \geq 1 - \alpha$ for classical non-Bayesian methods. Using this idea, Beal (1989) suggested the determination of sample sizes using the conditional probability

$$P\{W|V\} = P\{(U - L \leq \xi | L \leq \psi \leq U\} = \frac{P\{W \cap V\}}{P\{V\}}. \tag{10.21}$$

While Beal's approach takes into account the width of the confidence set and validity, it ignores power. To take into account rejection (power) for the two sided hypothesis, power, validity, and width, Jiroutek et al. (2003) suggested evaluation of the expression:

$$P\{W \cap R|V\} = P\{[(U - L \leq \xi \cap (U < \psi_0 \cup \psi_0 < L)]|L \leq \psi \leq U\}, \tag{10.22}$$

where $P\{W \cap R|V\}$ is the probability that the length of the interval is less than a fixed constant and the null hypothesis is rejected, given that the interval contains the true value of the population parameter. When considering contrasts of the form

$$H_0 : c'\beta = \psi = \psi_0, \tag{10.23}$$

the F test becomes

$$F_{obs} = \frac{(\hat{\psi} - \psi_0)[c'(X'X)^{-1}c]^{-1}(\hat{\psi} - \psi_0)/1}{s^2} = \frac{(\hat{\psi} - \psi_0)^2}{s^2 m} \tag{10.24}$$

where $m = c'(X'X)^{-1}c$ is a scalar. Given σ^2, the difference $\psi_d = \psi - \psi_0$, and the critical value of the test of size α, and $\xi > 0$, Jiroutek et al. (2003) provide an expression for the evaluation of the probability defined in (10.21) under normality that depends on the cumulative distribution function of the central F distribution. The computer program CISIZE written by Michael R. Jiroutek and Keith E. Muller using SAS/IML with exact numerical integration provided by the QUAD function is available on Muller's Web site: http://www.bios.unc.edu/~muller. We illustrate the software later in this chapter.

10.5   CALCULATING POWER AND SAMPLE SIZE FOR SOME MIXED MODELS

Except for mixed models that include one or two fixed factors and a single random factor, a power analysis for general mixed models is complicated since one must address trade-offs between tests of random components of variance and tests of fixed effects, taking into account the intraclass correlation, and standard errors for the model parameters, subject to budget constraints with and without the inclusion of covariates. In this section we address some of the issues for a few common designs.

10.5.1   Random One-Way ANOVA Design

Overview

The statistical model for the completely randomized one-way ANOVA design is

$$y_{ij} = \mu + a_j + e_{ij}, \qquad (10.25)$$

where y_{ij} is the ith observation for the jth class, μ is a general overall unknown mean, and a_j is the random effect for the jth level for J randomly selected levels of a treatment factor A and $i = 1, 2, \ldots, n; \; j = 1, 2, \ldots, J$. The random effects a_j are selected from a large population of treatment effects where the population is assumed to be normal with mean $\mu_j = \mu + a_j$ and unknown variance component σ_a^2. The overall mean is defined: $\mu = \sum_j \frac{\mu_j}{J}$. If the population is not large, one has to include in all calculations of variance estimates the finite population correction (FPC) factor. A population is small relative to the sample size if the ratio of the number of observations selected relative to the total population size exceeds 10 of the population. In our discussion, we assume that the population is sufficiently large so that FPC factors are not included in the expressions for the calculation of variance estimates for model parameters. The component σ_a^2 is a measure of the variability of the effects a_j in the population. The random errors e_{ij} are assumed to be normally distributed $e_{ij} \sim IN(0, \sigma^2)$ mutually independent of the treatment effects $a_j \sim IN(0, \sigma_a^2)$. Thus, the variance of the y_{ij}, $\sigma_y^2 = \sigma_a^2 + \sigma^2$. Although the elements a_j and e_{ij} are uncorrelated, the observations within the same class j are conditionally independent. The correlation between a pair of observations is the intraclass correlation coefficient

$$\rho_I = \frac{\sigma_a^2}{(\sigma_a^2 + \sigma^2)} = \frac{\sigma_a^2}{\sigma_y^2} = \frac{\theta}{(1+\theta)} = \frac{1}{(1 + 1/\theta)} \qquad (10.26)$$

where the ratio $\theta = \frac{\sigma_a^2}{\sigma^2}$. The intraclass correlation coefficient, introduced by Fisher (1925, p. 190), is the analogue of η^2 for regression models that have random independent variables; the value of the intraclass correlation is between zero and one. It represents the proportion of the variability in the outcome observations y_{ij} that is accounted for by the variability in the treatment effects a_j. Or, for a two level hierarchical model, the intraclass correlation represents the proportion of the variance in the outcome variable (at the first-level) that may be attributed to the second-level factor (Snijders & Bosker, 1999, p. 16). In meta analysis, methods of data analysis that promote the synthesis of research findings across independent studies, the intraclass correlation coefficient is used as an estimate of effect size for a study (Hedges & Olkin, 1985, p. 102).

For the random effects one-way ANOVA model, one is interested in testing for differences in the treatment effects and in estimating the model parameters. The unknown parameters for the model are: μ, σ^2, and σ_a^2. Also of interest are the intraclass correlation coefficient ρ_I, the ratio of variance components $\theta = \frac{\sigma_a^2}{\sigma^2}$, and the population variance $\sigma_y^2 = \sigma_a^2 + \sigma^2$. From (10.26) and the fact that $\sigma_y^2 = \sigma^2(1 + \theta)$, the parameter θ is critical to the study design. Given a total budget T and the associated costs C_o for an observation and C_r for selecting a class, the ratio θ is used to find the optimal sample size for the design so as to minimize the variance estimates of model parameters, for example the variance of the estimate for μ. The test for treatment differences is

$$
\begin{aligned}
H_0 &: \sigma_a^2 = 0 \\
H_A &: \sigma_a^2 > 0.
\end{aligned}
\tag{10.27}
$$

The null hypothesis implies that all μ_j are equal so that $\mu_j \equiv \mu$ while the alternative hypothesis implies that the μ_j differ.

While one may design experiments with an unequal number of observations per class, we only consider balanced designs, designs in which the number of observation per class are equal. An unequal number of observations per class causes difficult analysis problems with both the estimation of variance components and with tests of hypotheses (Searle, Casella, & McCulloch, 1992, p. 69-78). The ANOVA table for the design is given in Searle et al. (1992, p. 60). The test statistic for testing the null hypothesis H_0 is

$$
\frac{\sigma^2 F^*}{(n\sigma_a^2 + \sigma^2)} \sim F[\nu_h = J - 1, \nu_e = J(n-1)]
\tag{10.28}
$$

where $F^* = \frac{MSA}{MSE} = \frac{[SSA/(J-1)]}{[SSE/J(n-1)]}$ and F represents the F distribution with degrees of freedom ν_h and ν_e, the hypothesis and error degrees of freedom, respectively. When the null hypothesis is true, the statistic reduces to F^*. Selecting some level α for the test of H_0, the test statistic F^* is evaluated by comparing it to the upper $1 - \alpha$ critical value of the central F distribution, $F_0^{(1-\alpha)}$. When the null hypothesis is not true, F^* is a multiple of the central F distribution and not the noncentral F distribution as in the fixed effect model. The power of the test is given by

$$
\begin{aligned}
\pi(\theta) &= Prob\left(\frac{\sigma^2 F^*}{(n\sigma_a^2 + \sigma^2)}\right) \geq \frac{F_0}{(1 + n\theta)} \\
&= 1 - Prob\left(\frac{\sigma^2 F^*}{(n\sigma_a^2 + \sigma^2)}\right) < \frac{F_0}{(1 + n\theta)}.
\end{aligned}
\tag{10.29}
$$

Estimation

Assuming the population is sufficiently large, the variance estimators for the overall and class means are

$$V(\hat{\mu}) = \text{var}\left(\sum_i \sum_j \frac{y_{ij}}{Jn}\right) = \frac{(\sigma_a^2 + \sigma^2/n)}{J}$$

$$V(\hat{\mu}_j) = \text{var}\left(\sum_i \frac{y_{ij}}{I}\right) = (\sigma_a^2 + \sigma^2/n).$$

(10.30)

By using the ANOVA procedure, also called the method of moments technique, uniform minimum variance unbiased estimators for the two variance components are

$$\hat{\sigma}^2 = MSE$$

$$\hat{\sigma}_a^2 = \frac{(MSA - MSE)}{n}$$

(10.31)

as illustrated in Milliken and Johnson (1992, p. 235-236). While the estimate for σ_a^2 may be negative, the possibility of this occurring is small provided the number of classes J is not too small (Searle et al., 1992, p. 68). When the estimate is negative, one merely sets its value to zero. For a discussion of obtaining maximum likelihood (ML), restricted maximum likelihood (REML), and minimum variance quadratic estimates (MINQUE) for the variance components, one may consult Searle et al. (1992) or Milliken and Johnson (1992).

Since the ratio $\frac{MSA}{[MSE(1+n\theta)]}$ follows a F distribution with mean $\frac{\nu_e}{(\nu_e-1)}$, an unbiased estimator for the ratio $\theta = \frac{\sigma_a^2}{\sigma^2}$ is

$$\hat{\theta} = \frac{1}{n}\left[\left(\frac{J(n-1)-2}{J(n-1)}\right)\left(\frac{MSA}{MSE}\right) - 1\right].$$

(10.32)

And, an estimator for the intraclass correlation coefficient is

$$\hat{\rho}_I = \frac{MSA - MSE}{MSA + J(n-1)MSE},$$

(10.33)

(Shrout & Fleiss, 1979). While one may construct exact $(1 - \alpha)$ confidence intervals for σ^2, θ, and ρ_I, there is no exact confidence interval for σ_a^2 (Searle et al., 1992, p. 66). To construct an approximate confidence interval, one may use the Satterthwaite procedure as discussed in Section 2.6. Or, one may use the Graybill and Wang (1980) method (2.109). A more precise interval may be calculated by using the modified large sample (MLS) procedure developed by Ting et al. (1990) with details provided by Neter et al. (1996, p. 970-975).

Sample Size Calculations

When designing a random one-way design to investigate the significance of the J random effects, a primary goal is to ensure that one has sufficient power for the test. This depends on the size of the test α and both the number of effect levels J and the sample size n at each level. Secondly, to minimize the magnitude of error in estimating the model parameters μ, σ^2, and σ_a^2, the variance of the estimators should be small. The variance of the estimator for the overall mean is given in (10.30). To obtain the variance of the ANOVA estimators for σ^2 and σ_a^2, one uses the fact that the SSA and SSE are proportional to independent central chi square distributions (Searle et al., 1992, p. 64). Then, one can show that the variance of the ANOVA estimates are

$$V(\sigma^2) = \frac{2\sigma^2}{J(n-J)}$$

$$V(\sigma_a^2) = \frac{2}{n^2}\left[\frac{(n\sigma_a^2 + \sigma^2)^2}{J-1} + \frac{\sigma^2}{J(n-1)}\right].$$

(10.34)

Using the fact that the estimate of θ is related to a central F distribution, the variance of the unbiased estimator $\hat{\theta}$ can be shown to have the structure

$$V(\hat{\theta}) = \left(\theta + \frac{1}{n}\right)^2 \left(\frac{2(Jn-3)}{(J-1)(Jn-J-4)}\right).$$

(10.35)

Each of these estimators is also seen to depend on both J and n. The goals of having sufficient power for the test of the variance component and also for minimizing the estimated variance of model parameters must be accomplished within a total budget T.

To obtain an optimal sample size for the random one-way ANOVA design, we consider the linear cost function defined by the inequality

$$T \geq J(nC_o + C_r),$$

(10.36)

where T is the total variable cost for the study, C_o is the cost of sampling an observation, and C_r is the cost of sampling a random treatment. To obtain an optimal sample size, we minimize the variance of the estimate for the overall mean μ, since its value is seen to be directly related to both J and n, and show that the optimal sample size also minimizes the estimated variances of σ^2, σ_a^2, and θ. Solving (10.36) under equality for J and substituting it into (10.30), we have that the

$$V(\hat{\mu}) = \frac{(\sigma_a^2 + \sigma^2/n)(nC_o + C_r)}{T}.$$

(10.37)

Minimizing (10.37) with respect to the sample size n, the variance estimate for the overall mean is minimized for the optimal sample size

$$n_o = \sqrt{\left(\frac{\sigma^2}{\sigma_a^2}\right)\left(\frac{C_r}{C_o}\right)} = \sqrt{\left(\frac{1}{\theta}\right)\left(\frac{C_r}{C_o}\right)} = \sqrt{\left(\frac{1-\rho_I}{\rho_I}\right)\left(\frac{C_r}{C_o}\right)}.$$

(10.38)

This result is identical to that obtained by Cochran (1977, p. 277, equation 10.8) for minimizing the variance estimate for the overall mean in two-stage cluster sampling, if we set the fpc factors to one in Cochran's formula. The expression for the optimal sample size is seen to be directly proportional to the ratio of the costs $\frac{C_r}{C_o}$ and $\left(\frac{1-\rho_I}{\rho_I}\right)$, and inversely proportional to $\theta = \frac{\sigma_a^2}{\sigma^2}$. In practice, the cost of sampling random treatments is usually larger than the cost of sampling additional observations per treatment. Given values for the cost ratio $\frac{C_r}{C_o}$ and the ratio of the variances $\frac{\sigma_a^2}{\sigma^2}$, one may determine the optimal sample size. And, for a total budget amount T we may determine J since it is related to the sample size and the total cost by the relation: $J \leq \frac{T}{(nC_o+C_r)}$. Then, the total sample size for the experiment is $N = n_o J$. Given a sample size and the number of levels J for the random treatment effect, the power of the test of significance for the variance component may be calculated.

To investigate how the total sample size affects the variance estimates of the variance components, one substitutes N into (10.34). The variance of the variance components is

$$V(\hat{\sigma}^2) = \frac{2\sigma^2}{(N-J)}$$

$$V(\hat{\sigma}_a^2) = \frac{2J^2}{N^2}\left[\frac{(N\sigma_a^2/J+\sigma^2)^2}{J-1} + \frac{\sigma^2}{N-J}\right].$$

(10.39)

If in the population variance $\sigma_a^2 \gg \sigma^2$ or $\sigma^2 \gg \sigma_a^2$, we see that the first term in the estimate for the variance component σ_a^2 dominates the size of the variance estimate. One also observes that large values for J ensure that the variance for $\hat{\sigma}^2$ is small. Thus, we usually want J large to minimize the variance of both variance component estimates. Because the power for testing that $\sigma_a^2 = 0$ is related to n, J, and θ, we study the variance estimate for the ratio estimate for θ. The variance of the estimator in terms of the total sample size is

$$V(\hat{\theta}) = \left(\theta + \frac{1}{n_o}\right)^2 \left(\frac{2(N-3)}{(J-1)(N-J-4)}\right),$$

(10.40)

(Dean & Voss, 1997, p. 608). Provided $\sigma_a^2 \gg \sigma^2$, the variance estimate of this component is dominated by θ^2. If $\sigma^2 \gg \sigma_a^2$, the variance is dominated by $1/n_o^2$. However, provided $J \gg n_o$, the second term in the expression for the variance estimate again remains small. Thus, to minimize the variance estimates for μ, σ_a^2, and the ratio θ, it is better to sample more levels of the treatment effect with an equal number of observations per level than to sample a large number of observations at a small number of treatment levels. This result is consistent with the findings of Lohr (1995). Unfortunately, this may adversely affect the budget for the study since the cost of sampling a random level is usually larger (perhaps 2-5 times) than the cost of sampling an observation.

Calculating Power

The "optimal" sample size for the model is related to the cost ratio $\frac{C_r}{C_o}$ and the inverse of the ratio of variance components $\theta = \frac{\sigma_a^2}{\sigma^2}$ in the population. When testing

Table 10.1: Power Analysis for Random One-Way ANOVA Design.

$\frac{C_r}{C_o}$	θ	ρ_I	n_o	J_1	J_2	J_3	π_1	π_2	π_3
1	.05	.05	4	200	100	40	.48	.31	.18
1	.10	.09	3	250	125	50	.78	.53	.29
1	.20	.17	2	333	167	67	.92	.70	.39
2	.05	.05	6	125	63	25	.61	.40	.23
2	.10	.09	4	167	83	33	.85	.60	.33
2	.20	.17	4	200	100	40	.99	.86	.53
5	.05	.05	10	67	33	13	.73	.49	.28
5	.10	.09	7	83	42	17	.93	.74	.44
5	.20	.17	5	100	50	20	1.00	.92	.63
10	.05	.05	14	42	21	8	.77	.63	.29
10	.10	.09	10	50	25	10	.94	.76	.45
10	.20	.17	7	59	29	12	1.00	.92	.64

for a significant variance component, recall that the intraclass correlation coefficient is an indication of the magnitude of the effect size. Thus, to evaluate the power of the test for treatment differences, it is essential that one understand the relationship between the ratio θ of variance components and the intraclass correlation coefficient. Incrementing the square root of the intraclass correlation $\rho = \sqrt{\rho_I}$ from zero to one by 0.10 (say), relationships between ρ_I, $\theta = \frac{\rho_I}{(1-\rho_I)}$ and θ^{-1} are readily established. To investigate the relationship of J, the number of treatment effects, to the sample size n_o at each level, and power, one needs to specify an effect size. There is no agreed upon value for a large "significant" effect size; however, following guidelines developed by Rosenthal et al. (2000, p. 17) a value of $\rho_I = 0.10$ is a "medium" effect size. This corresponds to a θ ratio of about 0.10 so that $\frac{1}{\theta} = 10$. For $\sigma^2 = 1$ in the population, a medium value for σ_a^2 is 0.10 so the $\sigma_a = 0.32$.

In most studies, cost ratios $\frac{C_r}{C_0}$ usually range between 1 and 5, but no more than 10. To evaluate the power of the test for significant differences in the treatments, low, medium and high values of θ are taken as: $0.05, 0.10$, and 0.20. These correspond to intraclass correlations $\rho_I = \frac{\theta}{(1+\theta)}$: $0.05, 0.09$, and 0.17. The values for n_o, J_j, and power (π_j) for three values of total cost: $T_j = 1000, 500$, and 200 are provided in Table 10.1 for $\alpha = 0.05$.

For $T = 500$ and $\rho_I = .09$, the probability of rejecting $H_0 : \sigma_a^2 = 0$ given that $\sigma_a^2 = 0.10$ in the population is approximately 0.601 provided $J = 83$ and $n_o = 4$. Thus, one would need a total of $N = 332$ observations for the study. We clearly see from the entries in Table 10.1 that the power of the test increases as J increases and that it is bounded by budget constraints. When one has budget constraints, one may consider using a covariate in a design to reduce variability in estimates and to increase power (Milliken & Johnson, 2001, p. 287-304).

10.5.2   Two Factor Mixed Nested ANOVA Design

Overview

The random one-way ANOVA model depends upon the random assignment of observations to the treatment levels. When such assignments are not possible, one often employs a hierarchical (nested) design. For example, when investigating curriculum in schools, one may not randomly assign classrooms to schools, but schools may be randomly assigned to different classroom treatments. In experiments in the health sciences, schools may be replaced by dosage levels or litters of mice. This leads to a nested design in which treatments (A) are fixed and schools (B) within treatments are random and nested within factor A, represented as $B(A)$, a mixed model. A nested design is balanced if there are the same number of observations per treatment and the same number of levels for each random factor. The statistical model for the balanced random two-factor mixed nested design is

$$y_{ijk} = \mu + \alpha_j + b_{k(j)} + e_{ijk} \tag{10.41}$$

where $i = 1, \ldots, n; j = 1, \ldots, J$; and $k = 1, \ldots, K$. The fixed effects $\alpha_j = \mu_{j\cdot} - \mu$ are subject to the side conditions $\sum_j \alpha_j = 0$ where μ is an overall mean and the fixed treatment mean is $\mu_{j\cdot} = \sum_k \frac{\mu_{jk}}{K}$ so that $\mu = \sum_j \sum_k \frac{\mu_{jk}}{JK}$. The random components $b_{k(j)} \sim IN(0, \sigma_b^2)$ are the random errors $e_{ijk} \sim IN(0, \sigma^2)$ and mutually independent. As in the random one-way ANOVA model, pairs of observations y_{ijk} and $y_{i'jk'}$ are correlated with intraclass correlation $\rho_I = \frac{\sigma_b^2}{(\sigma_b^2 + \sigma^2)}$.

The parameters for the random two-factor mixed nested model are μ, α_j, σ_b^2, and σ^2. The two null hypotheses of interest for the design are: (a) the fixed effect hypothesis, H_0 : all $\alpha_j = 0$ and (b) the random component hypothesis, $H_0 : \sigma_b^2 = 0$. The ANOVA table for the design is provided, for example, by Neter et al. (1996, p. 1129, 1133).

Hypothesis Testing

The test statistic for testing the null hypothesis $H_0 : \sigma_b^2 = 0$ is

$$\frac{\sigma^2 F^*}{(n\sigma_b^2 + \sigma^2)} \sim F[\nu_h = J(K-1), \nu_e = JK(n-1)] \tag{10.42}$$

where $F^* = \frac{MSA}{MSB(A)} = \frac{[SSA/(J-1)]}{[SSB(A)/JK(n-1)]}$ and F represents the F distribution with degrees of freedom ν_h and ν_e, the hypothesis and error degrees of freedom. As in the random ANOVA model, when the null hypothesis is not true, the test statistic follows a central F distribution. Thus, the central F distribution is used to evaluate the power of the test.

The test statistic for testing the null hypothesis that all fixed treatments are equal, H_0 : all $\alpha_j = 0$, is tested using the F statistic $F = \frac{MSB(A)}{MSE}$. However, when the null hypothesis is not true, the statistic follows the noncentral F distribution

$$F \sim F[\nu_h = J - 1, \nu_e = J(K - 1), \lambda] \tag{10.43}$$

where λ is the noncentrality parameter defined by

$$\lambda = \frac{Kn\left(\frac{\sum_j \alpha_j^2}{J-1}\right)}{(n\sigma_b^2 + \sigma^2)}. \tag{10.44}$$

To evaluate the power of the fixed treatment effect, the noncentral F distribution is used. When the null hypothesis is true, the noncentrality parameter λ is zero.

Estimation

Using the ANOVA procedure, uniform minimum variance unbiased estimators for the two variance components are

$$\begin{aligned} \hat{\sigma}^2 &= MSE \\ \hat{\sigma}_b^2 &= \frac{MSB(A) - MSE}{n}. \end{aligned} \tag{10.45}$$

If the estimate for σ_b^2 is less than zero, one again sets it to zero. ML and REML estimators for the variance components are given by Searle et al. (1992, p. 148-149). The estimator of the intraclass correlation coefficient is

$$\hat{\rho}_I = \frac{MSB(A) - MSE}{MSB(A) + [JK(n-1)]MSE}. \tag{10.46}$$

Assuming the population is sufficiently large, the variance estimates for the overall mean and treatment means are

$$\begin{aligned} V(\hat{\mu}) &= \operatorname{var}\left(\sum_i \sum_j \sum_k \frac{y_{ijk}}{JKn}\right) = \frac{(\sigma_b^2 + \sigma^2/n)}{KJ} \\ V(\hat{\mu}_j.) &= \operatorname{var}\left(\sum_i \sum_k \frac{y_{ijk}}{Kn}\right) = \frac{(\sigma_b^2 + \sigma^2/n)}{K}. \end{aligned} \tag{10.47}$$

For each value of $j = 1, 2, \ldots, J$, the nested design is seen to be identical to a random one-way ANOVA design with σ_a^2 replaced by σ_b^2 and J replaced by K. To obtain an optimal sample size for the design, we let

$$\frac{T}{K} \geq J(nC_o + C_r) \tag{10.48}$$

where T is the total cost of the study. Substituting (10.45) into the $V(\hat{\mu})$ under equality, we have that

$$V(\hat{\mu}) = \frac{(\sigma_b^2 + \sigma^2/n)(nC_o + C_r)}{T}, \tag{10.49}$$

which is seen to be the same as (10.38). Minimizing the $V(\hat{\mu})$, the optimal sample size for the nested design is

$$n_o = \sqrt{\left(\frac{\sigma^2}{\sigma_b^2}\right)\left(\frac{C_r}{C_o}\right)} = \sqrt{\left(\frac{1}{\theta}\right)\left(\frac{C_r}{C_o}\right)} = \sqrt{\left(\frac{1-\rho_I}{\rho_I}\right)\left(\frac{C_r}{C_o}\right)} \qquad (10.50)$$

where ρ_I is the intraclass correlation coefficient for the design and $\theta = \frac{\sigma^2}{\sigma_b^2}$. However, for this design in addition to considering the power of the test for the random component, one must also evaluate the power of the fixed effect test which depends on the unknown fixed treatment effects α_j. And, both the number of levels J and K enter the total cost for the design. This design is J times more costly than the random one-way ANOVA design.

Calculating Power

To obtain an estimate of power for the fixed effect component in the nested design, as with fixed effect designs, the parameters α_j and σ^2 must be defined. As with random components, there is no agreed upon standard for a "significant" effect size. To illustrate the power calculations, we assume that the fixed effect means are equal, but that at least one differs from the others by σ units. Hence, the noncentrality parameter $\lambda = \frac{Kn}{(J-1)(n\theta+1)}$ does not depend on σ. To evaluate the power of the fixed and random effects for the nested design, we set the cost ratio $\frac{C_r}{C_o} = 2$ and 5 and allow θ to assume the values for the random one-way ANOVA model. The number of fixed levels J are set to 4, 7, and 10 and the values for the total cost are $T = 2000$ and 1000. Given T and J, the values for n_o, K and power for the fixed and random hypotheses (π_f & π_r) are provided in Table 10.2 for the Type I error rate $\alpha = 0.05$ for both the fixed effect and random effect tests. For a study with $J = 7$ fixed levels, a total budget $T = 1000$, a cost ratio of 2, and $\theta = 0.10$, one would randomly select 24 levels for K and have $n_0 = 4$ observations per random effect to ensure that the power of both tests are larger than 0.69. Thus, the total sample size $N = 672$. Again, power is seen to increase with K for fixed levels of J. For small values of J, $K \gg J$. Observe also that the optical allocation of observations for the fixed effect hypothesis may not be best for the random effect test and vis-a-vis.

The random assignment of the K levels of the nested factor to the J levels of the treatment factor A and the random assignment of treatment combinations to the experimental units often requires a large number of subjects for a study. When $C_r \gg C_o$, the cost of the study may be prohibitive. Raudenbush (1997) shows how costs may be reduced by adding a covariate to the design.

In the random one-way design, the primary hypothesis of interest involves the random factor (A). For the two factor nested design, we are primarily interested in the fixed treatment factor (A). The random factor is really a nuisance variable. However, differences in the random factor cannot be ignored since this may result in biased tests for the fixed treatments.

R. L. Anderson (1975, 1981) reviews design strategies for estimating variance components in multi-factor random nested designs, hierarchical nested designs in

Table 10.2: Power Analysis for Two-Factor Mixed Nested Design.

$\frac{C_r}{C_o}$	θ	J	n	K $T = 2000$	K 1000	π_r $T = 2000$	π_r 1000	π_f $T = 2000$	π_f 1000
2	.05	4	6	63	31	.85	.60	1.00	1.00
2	.05	7	6	36	18	.85	.60	.99	.78
2	.05	10	6	25	13	.84	.60	.68	.35
2	.10	4	4	83	42	.98	.85	1.00	1.00
2	.10	7	4	48	24	.98	.84	.96	.69
2	.10	10	4	33	17	.98	.84	.58	.29
2	.20	4	3	100	50	1.00	.98	1.00	1.00
2	.20	7	3	57	29	1.00	.98	.90	.58
2	.20	10	3	40	20	1.00	.98	.46	.23
5	.05	4	10	33	17	.93	.72	1.00	1.00
5	.05	7	10	19	10	.92	.72	.94	.65
5	.05	10	10	13	7	.91	.70	.51	.26
5	.10	4	7	42	21	1.00	.93	1.00	1.00
5	.10	7	7	24	12	1.00	.93	.87	.51
5	.10	10	7	17	8	1.00	.90	.42	.19
5	.20	4	5	50	25	1.00	1.00	1.00	.98
5	.20	7	5	29	14	1.00	1.00	.73	.37
5	.20	10	5	20	10	1.00	.99	.30	.15

which all factors are random. Because he is primarily concerned with the estimation of model parameters, a discussion of power and sampling costs for the designs are not addressed.

Summary

In designing mixed models, one has to ensure that there are a sufficient number of levels for each random factor and that the number of observations per cell are adequate to simultaneously ensure that one has enough power for tests of both fixed and random factors, and that the estimated standard errors of model parameters are small. Furthermore, the two overall goals must be realized within a total budget. These complications make the design of mixed models more difficult than fixed effect models where the primary variable is the number of observations per cell in the design. When designing mixed models, one has to consider trade-offs between the number of random effect levels for each random factor in the design and the number of observations per cell. For mixed models, the estimated standard errors of variance estimates are influenced more by the number of random effect levels than by the number of observations at each level. Because the cost for a random level is usually larger than for an observation, this causes the cost of the study to increase. It is also the case that the total number of observations for a mixed model, using an A-optimal design criterion, ensures that one can find a design such that both the fixed and random tests of hypotheses have adequate power, provided one has a sufficient budget. However, adequate power for a fixed (random) factor test does not simultaneously maximize the power for testing a random (fixed) factor in the design.

10.6 POWER FOR MULTIVARIATE GLMS

Overview

In the population, the $\mathcal{E}(Y_{n \times p}) = X_{n \times k} B_{k \times p}$ for a MGLM with a fixed design matrix. Under multivariate normality, the noncentrality parameter of the Wishart distribution under the null hypothesis $H_0 : C_{\nu_h \times k} B_{k \times p} A_{p \times u} = 0$ is

$$\Gamma = (CBA)'(C(X'X)^{-1}C')^{-1}(CBA)(A'\Sigma A)^{-1} = \Delta \Sigma^{-1}, \qquad (10.51)$$

the population counterpart of HE^{-1}. The nonzero eigenvalues $\lambda_1, \lambda_2, \ldots, \lambda_s$ of $\Delta \Sigma^{-1}$ represent the s-dimensional subspace spanned by rows of the design matrix B where $s = \min(\nu_h, u) = \text{rank}(\Delta)$. An indication of the rank of Δ may be obtained by examining the sample roots of HE^{-1} by solving $|H - \lambda E| = 0$. The population roots $\lambda_1 \geq \lambda_2 \geq \cdots \geq \lambda_s > 0$ lie between two extreme situations: (a) $\lambda_1 > 0$ and $\lambda_i = 0$ for all $i \geq 2$ and (b) $\lambda_1 = \lambda_2 = \cdots = \lambda_s > 0$ where the vectors are equally diffuse in s dimensions. Other configurations correspond to different relations among the roots.

When $\lambda_1 > 0$ and all other roots are zero, Roy's largest root criterion tends to out perform the other criteria since the relationship among the test statistics is as follows: $\theta \geq U_0^{(s)} \geq \Lambda \geq V^{(s)}$. For fixed u and as $\nu_e \to \infty$ the test statistics based upon $U_0^{(s)}$, Λ, and $V^{(s)}$ are equivalent since they all depend on an asymptotic noncentral chi-square distribution. Olson (1974) found that the tests are equivalent, independent of the relationship among the roots in the population if $\nu_e \geq 10\nu_h u$.

When the null hypothesis is false, power for the multivariate test is found by evaluating the noncentral Wishart distribution. For small sample sizes use various approximations to the noncentral Wishart distribution of the roots of HE^{-1}, when the roots $\lambda_1 \geq \lambda_2 \geq \cdots \geq \lambda_s > 0$ have different configurations and when the ordering of power is generally as follows for the four criteria $V^{(s)} \geq \Lambda \geq U_0^{(s)} \geq \theta$, with little differences in power for $V^{(s)}$, Λ and $U_0^{(s)}$ (Pillai & Jaysachandran, 1967; S. Y. Lee, 1971; Olson, 1974; Muller & Peterson, 1984; Muller et al., 1992; Schatzoff, 1966). Even though there is no uniformly most powerful test, as a general rule, one may use Roy's criterion if $\lambda_1 \gg \lambda_2$, use $V^{(s)}$ if it is known that the λ_i are equal, use $U_0^{(s)}$ if it is known that the λ_i are very unequal, and with no knowledge the LRT criterion Λ is often used.

Estimating Power

To estimate power for the MGLM, Muller and Peterson (1984) suggested approximating the noncentral Wishart distribution with a noncental F distribution for the three criteria $V^{(s)}$, Λ and $U_0^{(s)}$. And, they develop a lower bound for power using Roy's criterion. To illustrate, we use Rao's transformation for the F approximation to the distribution of U (Theorem 5.6). For the univariate GLM, recall that the noncentrality parameter λ:

$$\lambda = \frac{(C\beta)'[C(X'X)^{-1}C']^{-1}(C\beta)}{\sigma^2} = \nu_1 F^*(\nu_1, \nu_2, \lambda) \tag{10.52}$$

where F^* is the F statistic with population parameters when testing the hypothesis $H_0 : C\beta = 0$. Using Theorems 5.2 and 5.6 with $\nu_1 = p\nu_h$, and

$$U \equiv \Lambda_* = \frac{|\nu_e\Sigma|}{|\nu_e\Sigma + \nu_h\Sigma|}, \tag{10.53}$$

with population noncentrality parameter

$$\Gamma = (CBA)'(C(X'X)^{-1}C')^{-1}(CBA)(A'\Sigma A)^{-1} = \Delta\Sigma^{-1},$$

the noncentrality parameter for the noncentral F distribution becomes

$$\lambda = \nu_1[(\Delta^*)^{-a} - 1]\left(\frac{\nu_2}{\nu_1}\right) \tag{10.54}$$

where a is defined in (5.49). Hence, with X, B, C, A and Σ known one may calculate λ. Using the SAS function PROBF, one may calculate the power for the MGLM. Muller and Peterson (1984) develop F approximations for the criteria $V^{(s)}$ and $U_0^{(s)}$ that depend on measures of multivariate association. Finally, letting $\nu_1 = \max(\nu_h, p)$ and $\nu_2 = \nu_e - \nu_1 + \nu_h$, $\frac{\nu_2\hat{\lambda}_1}{\nu_1 F^{\max}} \leq F$. Using this result one may approximate the power of Roy's criterion. Muller et al. (1992) develop software to evaluate power and show how it may also be used in repeated measurement design. The software is called Powerlib202 and is available from Muller's Web site: http://www.bios.unc.edu/~muller.

10.7 POWER AND EFFECT SIZE ANALYSIS FOR UNIVARIATE GLMS

10.7.1 One-Way ANOVA

Power

Consider a four group ANOVA design. A researcher is interested in evaluating the effect that four different instructional formats may have on the speed of completing a puzzle. Subjects are provided materials that are: (1) Written, Oral, and Visual, (2) Written only, (3) Oral only, and (4) Visual only. For each group, the time in seconds to complete the task is the dependent variable. Assuming the single treatment formats are about equal, primary interest is on whether the first treatment differs from the other three by $\sigma = 10$ units. Thus, the population values for the treatments are: $\{\mu + \sigma, \mu, \mu, \mu\}$. For this study, assume that the population parameter $\beta = (60, 50, 50, 50)$ and that the power for the study is $\pi = 0.80$. Using the SAS

Output 10.7.1: SAS GLMPOWER for One-Way ANOVA.

			Test DF	Error DF	Actual Power	N Total
Index	**Type**	**Source**	**Test DF**	**Error DF**	**Actual Power**	**N Total**
1	Effect	treat	3	60	0.813	64
2	Contrast	treat1-treat4	1	64	0.819	68
3	Contrast	1 vs 2,3,4	1	40	0.800	44

procedure GLMPOWER, the total sample size for the design is $N = 64$ (Output 10.7.1). However, obtaining subjects for the study is costly. The number of subjects required to detect the specific contrast $\psi = 3\mu_1 - \mu_2 - \mu_3 - \mu_4$ is only 44. Thus the researcher sets the target sample at $N_t = 40$ subjects for the study. The code for the example is provided in Program 10_1.sas.

Effect Size

Assigning $N_t = 40$ subjects to the four treatment groups at random, the estimates for β and σ^2 for the study are: $\hat{\beta}' = (68.8, 52.1, 52.4, 57.0)$ and $s^2 = 70.36$. Letting C be a hypothesis test matrix of rank 3, $C\beta = \Psi$, to test the overall null hypothesis $H_0 : \Psi = 0$ versus the alternative $H_a : \Psi \neq 0$, the degrees of freedom for the hypothesis and error are $\nu_h = 3$ and $\nu_e = 36$, respectively. For $\alpha = 0.05$, the critical value for the test is: $f_{cv} = 2.87$. Since $F_{obs} = 8.67$, the P-value for the test is $P = 0.0002$. The estimate for the contrast is $\hat{\psi} = 44.9$ and using Scheffé's (1953) S-method, a 95% confidence interval for the contrast ψ is: $[17.96, 71.84]$. The estimate of the ubiquitous effect size for the test is $\widehat{ES} = 0.81$. An approximate 95% confidence interval for ES for the omnibus test is: $0.62 \leq ES \leq 0.99$. The corresponding greatest lower bound interval for ES is: $[0.648, \infty]$. This suggests that on average the group means differ by at least 0.65 standard deviation units over all possible contrasts. The observed ordered pairwise, $\{(1,2), (1,3), (1,4), (2,3), (2,4), (3,4)\}$, mean differences in standard deviation units are: $\{1.99, 1.95, 1.40, -0.04, -0.58, -0.55\}$.

Because this example compares four group means, Cohen's d and Hedges's g effect size indices are only defined for pairwise comparisons. Thus, some authors suggest reporting the goodness of fit indices: $\tilde{\eta}^2 = \frac{SS_B}{SS_B + SS_W}$ and $\tilde{\epsilon}^2 = 1 - (N_t - 1)\left(\frac{1 - \tilde{\eta}^2}{\nu_e}\right)$, and $\tilde{\omega}^2 = (\nu_h - 1)\left(\frac{MS_B - MS_W}{(\nu_h - 1)MS_B + (\nu_e + 1)MS_W}\right)$ due to Pearson-Fisher, Kelley, and Hays, respectively. Because these effect sizes are nondirectional and can have the same value with different orderings of the population means, they are not useful indicators of an effect size (Hedges & Olkin, 1985, p. 100). The square roots of these indices of fit are: $\tilde{\eta} = 0.65$, $\tilde{\epsilon} = 0.61$, and $\tilde{\omega} = 0.51$. Cohen (1988, p. 283) claims that $\tilde{\eta}$ is a ratio of standard deviations. As shown by Goldberger (1991, p. 179), this is not the case when the independent variables are fixed.

Replacing the hypothesis test matrix C with the contrast vector

$$c' = \begin{pmatrix} 3 & -1 & -1 & -1 \end{pmatrix},$$

the effect size estimate for the contrast is: $\widehat{ES}_\psi = 0.773$. The approximate 95% confidence interval and the greatest lower bound interval for the contrast are: $0.61 \leq ES_\psi \leq 0.99$, and $[0.67, \infty]$, respectively. This suggests that the first group mean differs by at least 0.67 standard deviation units from the other three.

10.7.2 Three-Way ANOVA

For this example, we calculate effect size and power for diffuse or omnibus tests and for a linear contrast in a completely randomized three-factor fixed effect factorial design. So that the effect size discussed in this chapter may be compared to indices put forth by others, we consider the hypothetical $2 \times 2 \times 5$ factorial ANOVA design discussed by Rosenthal et al. (2000, p. 85) which has three factors (levels): age (2), sex (2), treatment (5).

Power

First, using the SAS procedure GLMPOWER, we evaluate power for the main effect tests and the linear treatment contrast. Using $\alpha = 0.05$ and the standard deviation in the population equal to 0.025, a sample of size $N = 40$ yields power of 0.830, 0.971, and 0.818 for the main effect tests (Output 10.7.2), and the power for the linear contrast for treatments is equal to 0.936. Thus, we conclude that the design should have at most two observations per cell. The code for the example is provided in Program 10_2.sas.

Effect Size

For the Rosenthal example data, the estimated standard deviation for the design is $s = 2.68$. Using $\alpha = 0.05$ for all tests of significance, the overall test of interaction is not significant (P-value $= 0.17$). This is as expected since from the power analysis for the design the power for the three-way test was only $\pi = 0.057$. The P-values for the main effect tests are: 0.0013, 0.0021, and 0.0086, respectively. Fitting a linear trend contrast $\psi_{\text{linear}} = -2\mu_1 - \mu_2 + 0\mu_3 + \mu_4 + 2\mu_5$ to the five treatment means, the P-value for linear trend is 0.0023.

The estimate of the linear contrast is $\hat{\psi}_{\text{linear}} = 10.5$ and the associated 95% simultaneous S-interval for the contrast is: $[0.342, 20.66]$. Estimates for ES for each omnibus test of significance and the contrast and approximate 95% confidence sets are provided in Output 10.7.3.

Because the estimated ES effect sizes are defined in standard deviation units they are very easy to interpret. The effect size for the contrast indicates that the means differ by 0.75 s units. Alternatively, Rosenthal et al. (2000, p. 83) calculate three association measures ($r_{\text{effect-size}}$, r_{contrast}, $r_{\text{effect-size}|\text{NS}}$) for the contrast. Their values

Output 10.7.2: SAS GLMPOWER for Three-Way ANOVA.

Computed Power						
Index	Type	Source	N Total	Test DF	Error DF	Power
1	Effect	s	40	1	20	0.830
2	Effect	s	60	1	40	0.965
3	Effect	s	80	1	60	0.994
4	Effect	a	40	1	20	0.971
5	Effect	a	60	1	40	0.999
6	Effect	a	80	1	60	>.999
7	Effect	s*a	40	1	20	0.026
8	Effect	s*a	60	1	40	0.027
9	Effect	s*a	80	1	60	0.027
10	Effect	t	40	4	20	0.818
11	Effect	t	60	4	40	0.979
12	Effect	t	80	4	60	0.998
13	Effect	s*t	40	4	20	0.455
14	Effect	s*t	60	4	40	0.746

Output 10.7.3: SAS IML — Estimate of Effect Size for Three-Way ANOVA.

RES				
	ES	ES_L	ES_U	GLB
sex	0.754	0.522	0.985	0.555
age	1.005	0.696	1.314	0.740
treatment	0.856	0.593	1.119	0.631
contrast	0.746	0.517	0.975	0.550
treatment x sex	0.537	0.372	0.703	0.396

ES_L = 95% CI ES_L; ES_U = 95% CI ES_U; GLB = greatest lower bound

are: 0.46, 0.62, and 0.52, respectively. The r effect sizes are difficult to interpret. They are not estimates of population correlation coefficients since the independent variable is fixed and not random. The summary ES information in Output 10.7.3 may be used to evaluate the size of the effects within the study as well as across other completely randomized three-factor fixed effect factorial designs.

10.7.3 One-Way ANCOVA Design with Two Covariates

In Chapter 3 Section 3.6.2, we analyzed a one factor ANCOVA design with two covariates. We now show how to perform a power analysis to (a) evaluate the effect of sample size and the correlation between the covariate and the dependent variable changes on power and to (b) determine sample size given a desired value for power and different correlations between the covariate and the dependent variable. SAS only allows one to evaluate only one correlation, even though we have two covariates. The covariates are expected to reduce the common residual variance σ^2 by a common factor $1 - \rho^2$. The code for the example is included in Program 10_3.sas. For (a), we also include power curve plots.

For the ANCOVA design, we compare three treatments and allow the standard deviation to assume two population values: $\{0.70, 1.2\}$, allow the correlation between the covariates and the dependent variable to vary: $\{0.2, 0.5, 0.6, 0.8\}$, and finally the total sample size assumes the values: $\{11, 15, 18\}$. To see the results of these choices for the design parameters, power curves are generated (Graph 10.7.1). Inspection of the output or the power curves indicate that a sample size of about $N = 18$ subjects is more than adequate for the design.

Alternatively, we may use SAS to determine the sample size given a power value of about 0.85 across the same domain of design parameters (Output 10.7.4). The sample size ranges from a minimum of 8 to a maximum of 17 subjects for the study. The sample weights for the design were chosen proportional to population values. When the weight parameter is not used, SAS assumes equal sample sizes per treatment group.

10.8 POWER AND SAMPLE SIZE BASED ON INTERVAL-ESTIMATION

To illustrate the calculation of power and sample size based upon interval estimation, the program CISIZE, available from Muller's Web site: http://www.bios.unc.edu/~muller, is used. Also available from the Web site is Manual1.pdf which describes how to use the program. Before working an example, we briefly describe the necessary input for the program. The program assumes the multivariate linear model $Y = XB + E$, where the hypothesis test matrix is of the form $CBU = \Theta_0$. Defining the six parameter matrices: Σ (SIGMA), X, B (BETA), C, U, and Θ_0 (THETA-0) and the constants α (ALPHA) and confidence set width δ (DELTA), the program calculates the probabilities $P\{W \cap R|V\}$, $P\{W|V\}$, $P\{W\}$, and $P\{R\}$: EVENTS=\{"WARGV" "WGV" "WIDTH" "REJECT"\} simultaneously for two-sided confidence interval and two-sided test combinations. The matrix X is input as an essence design matrix (ESSENCEX) which can be defined as a full

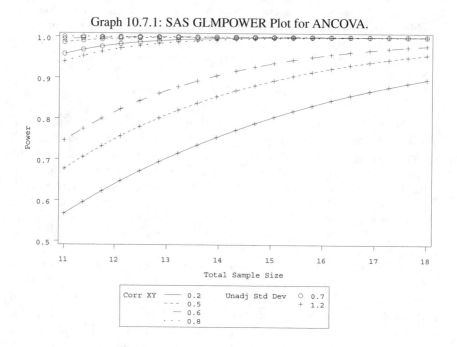

Graph 10.7.1: SAS GLMPOWER Plot for ANCOVA.

Output 10.7.4: SAS GLMPOWER Sample Size Estimate for ANCOVA with Power = .85.

					Computed Ceiling N Total		
Index	**Corr XY**	**Unadj Std Dev**	**Adj Std Dev**	**Error DF**	**Fractional N Total**	**Actual Power**	**Ceiling N Total**
1	0.2	0.7	0.686	5	9.297895	0.910	10
2	0.2	1.2	1.176	12	16.325340	0.870	17
3	0.5	0.7	0.606	4	8.614737	0.896	9
4	0.5	1.2	1.039	9	13.854766	0.856	14
5	0.6	0.7	0.560	4	8.267112	0.934	9
6	0.6	1.2	0.960	8	12.602900	0.869	13
7	0.8	0.7	0.420	3	7.397216	0.953	8
8	0.8	1.2	0.720	5	9.626181	0.882	10

rank cell means matrix (cell mean coding), a full rank regression matrix (reference coding), a reparameterized full rank matrix (effect coding), or a less than full rank matrix (LFR coding). The essence matrix is created from a design matrix by deleting duplicate rows from the matrix. Details are provided, for example, in Muller and Fetterman (2002, p. 300-310). For a univariate linear model with only three groups, the ESSENCE design matrices can be defined as:

$$Es(X_{\text{cell}}) = \begin{pmatrix} 1 & 0 & 0 \\ 0 & 1 & 0 \\ 0 & 0 & 1 \end{pmatrix}$$

$$Es(X_{\text{ref}}) = \begin{pmatrix} 1 & 0 & 0 \\ 1 & 1 & 0 \\ 1 & 0 & 1 \end{pmatrix}$$

$$Es(X_{\text{effect}}) = \begin{pmatrix} 1 & -1 & -1 \\ 1 & 1 & 0 \\ 1 & 0 & 1 \end{pmatrix}$$

$$Es(X_{\text{LFR}}) = \begin{pmatrix} 1 & 1 & 0 & 0 \\ 1 & 0 & 1 & 0 \\ 1 & 0 & 0 & 1 \end{pmatrix}$$

where the vectors are of dimension n_i, with $N = \sum n_i$. The corresponding univariate BETA parameter vectors for the designs take the form:

$$\beta_{\text{cell}} = \begin{pmatrix} \mu_1 \\ \mu_2 \\ \mu_3 \end{pmatrix} \qquad \beta_{\text{ref}} = \begin{pmatrix} \mu_1 \\ \delta_2 \\ \delta_3 \end{pmatrix}$$

$$\beta_{\text{effect}} = \begin{pmatrix} \mu \\ \zeta_2 \\ \zeta_3 \end{pmatrix} \qquad \beta_{\text{LFR}} = \begin{pmatrix} \mu \\ \alpha_1 \\ \alpha_2 \\ \alpha_3 \end{pmatrix}$$

Hence, the corresponding hypotheses are $H_{\text{cell}} : \mu_1 = \mu_2 = \mu_3$; $H_{\text{ref}} :_1 \delta_2 = \delta_3 = 0$; $H_{\text{effect}} : \zeta_2 = \zeta_3 = 0$; $H_{\text{LFR}} : \alpha_1 = \alpha_2 = \alpha_3$. Finally, the hypothesis test matrices (C) for the corresponding parameters are:

$$C_{\text{cell}} = \begin{pmatrix} 1 & 0 & -1 \\ 0 & 1 & -1 \end{pmatrix} \qquad C_{\text{ref}} = \begin{pmatrix} 0 & 1 & 0 \\ 0 & 0 & 1 \end{pmatrix}$$

$$C_{\text{effect}} = \begin{pmatrix} 0 & 1 & 0 \\ 0 & 0 & 1 \end{pmatrix} \qquad C_{\text{LFR}} = \begin{pmatrix} 1 & 0 & -1 \\ 0 & 1 & -1 \end{pmatrix}$$

all with rank 2, the degrees of freedom for the hypothesis. For a contrast $\psi = c'Bu = \theta$.

10.8.1 One-Way ANOVA

We again consider a four group ANOVA design. A researcher is interested in evaluating the effect that four different instructional formats may have on the speed

Output 10.8.1: SAS IML — CISIZE for One-Way ANOVA with n=10.

	ALPHA	BETASCAL	DELTSCAL	SIGSCAL	TOTAL_N	ICISIDE	IALTHYP	WARGV	WGV	WIDTH	REJECT
					HOLDPR						
ROW1	0.05	1	30	60	40	2	2	0.143	0.143	0.15	0.931
ROW2	0.05	1	35	60	40	2	2	0.565	0.577	0.588	0.931
ROW3	0.05	1	45	60	40	2	2	0.946	0.995	0.995	0.931
ROW4	0.05	1	55	60	40	2	2	0.95	1	1	0.931
ROW5	0.05	1	30	70	40	2	2	0.049	0.049	0.053	0.89
ROW6	0.05	1	35	70	40	2	2	0.315	0.325	0.336	0.89
ROW7	0.05	1	45	70	40	2	2	0.881	0.963	0.965	0.89
ROW8	0.05	1	55	70	40	2	2	0.909	1	1	0.89
ROW9	0.05	1	30	80	40	2	2	0.016	0.016	0.018	0.846
ROW10	0.05	1	35	80	40	2	2	0.155	0.162	0.17	0.846
ROW11	0.05	1	45	80	40	2	2	0.772	0.876	0.88	0.846
ROW12	0.05	1	55	80	40	2	2	0.863	0.999	0.999	0.846

Deltscal = δ; Sigscal = σ^2; Reject = Power

Output 10.8.2: SAS IML — CISIZE for One-Way ANOVA with n=5.

	ALPHA	BETASCAL	DELTSCAL	SIGSCAL	TOTAL_N	ICISIDE	IALTHYP	WARGV	WGV	WIDTH	REJECT
					HOLDPR						
ROW1	0.05	1	30	60	20	2	2	0.006	0.006	0.008	0.651
ROW2	0.05	1	35	60	20	2	2	0.032	0.034	0.039	0.651
ROW3	0.05	1	45	60	20	2	2	0.229	0.276	0.292	0.651
ROW4	0.05	1	55	60	20	2	2	0.515	0.704	0.715	0.651
ROW5	0.05	1	30	70	20	2	2	0.002	0.002	0.003	0.585
ROW6	0.05	1	35	70	20	2	2	0.014	0.015	0.018	0.585
ROW7	0.05	1	45	70	20	2	2	0.13	0.16	0.174	0.585
ROW8	0.05	1	55	70	20	2	2	0.374	0.534	0.549	0.585
ROW9	0.05	1	30	80	20	2	2	0.001	0.001	0.001	0.53
ROW10	0.05	1	35	80	20	2	2	0.007	0.007	0.009	0.53
ROW11	0.05	1	45	80	20	2	2	0.074	0.093	0.103	0.53
ROW12	0.05	1	55	80	20	2	2	0.263	0.386	0.403	0.53

Deltscal = δ; Sigscal = σ^2; Reject = Power

of completing a puzzle. Subjects are provided materials that are: (1) Written, Oral, and Visual, (2) Written only, (3) Oral only, and (4) Visual only. For each group, the time in seconds to complete the task is the dependent variable. We are concerned with the contrast $\psi = 3\mu_1 - \mu_2 - \mu_3 - \mu_4$. For the population parameter vector $\beta = (60, 50, 50, 50)$, we investigate the power of the study for contrast sets of width $\delta = \{30, 35, 45, 55\}$ for $\alpha = 0.05$ and $\sigma^2 = \{60, 70, 80\}$. Given validity $\alpha = 0.05$, are 40 subjects adequate to ensure that the confidence interval that correctly does not contain the null value when the null hypothesis is false has a width not greater than δ? The code for the example is provided in Program 10_4.sas. Over all parameters, the power is seen to vary between 0.846 and 0.931 (Output 10.8.1). Changing the sample size to only 5 subjects per group, the power only ranges between 0.53 and 0.651 (Output 10.8.2). Thus, using confidence-based inference, a sample of ten subjects per group appears adequate to attain the study goals.

Output 10.9.1: SAS Powerlib202 - Two-groups MANOVA.

ALPHA	SIGSCAL	BETASCAL	TOTAL_N	POWER_MULT
0.01	1	1	10	0.977

10.9   POWER ANALYSIS FOR MULTIVARIATE GLMs

To illustrate the calculation of power and sample size for the MGLM, the program POWERLIB202 available from Muller's Web site, http://www.bios.unc.edu/~muller, is used (please check the website for the most recent version). The program Manual202.pdf written by Jacqueline L. Johnson, Keith E. Muller, James C. Slaughter, and Matthew J. Gurka describes how to use the program and is also available from the Web site. The program assumes the multivariate linear model $Y = XB + E$, where the hypothesis test matrix is of the form $CBU = \Theta_0$. Defining the six parameter matrices: Σ (SIGMA), X, B (BETA), C, U, and Θ_0 (THETA-0) and the constant α, the program calculates the power for the MGLM. The program works in a manner similar to the CISIZE program in that one defines the design matrix using an ESSENCEX matrix. The new program has some new features and replaces the program power.pro used in the first edition of this book.

10.9.1   Two Groups

To illustrate the setup for the program, we use data from Timm (1975, p. 264) for a simple two-group MANOVA design. The associated theory for the procedure is included in Muller et al. (1992). To perform a power analysis for a multivariate design requires substantial calculations since one wants to vary the covariance matrix Σ and the the test matrices. The example considered here is straightforward; we are testing the hypothesis

$$H_0 : \mu_1 - \mu_2 = \begin{pmatrix} 20.38 \\ -38.33 \end{pmatrix} \quad \text{with} \quad \Sigma = \begin{pmatrix} 307.08 & 280.83 \\ 280.83 & 421.67 \end{pmatrix}$$

and $\alpha = 0.01$ with a total sample size of $N = 10$, five observations per group. The power for the design is $\pi = 0.977$ (Output 10.9.1). For this example the power is exact.

10.9.2   Repeated Measures Design

In Chapter 5, Section 5.10 we analyzed a repeated measures design (Program 5_10.sas). We calculate the power for the design for each of the main tests. The parameter matrix B is

$$B = \begin{pmatrix} 4 & 7 & 10 \\ 7 & 8 & 11 \\ 5 & 7 & 9 \end{pmatrix},$$

and based upon a pilot study the unknown covariance matrix Σ is

$$\Sigma = \begin{pmatrix} 74 & 65 & 35 \\ 65 & 68 & 38 \\ 35 & 38 & 26 \end{pmatrix}.$$

For $n_i = \{5, 10, 15\}$ subjects in each of the groups or a total sample size of $N = \{15, 30, 45\}$, the researcher wants to evaluate the associated power of several multivariate tests using $\alpha = 0.05$. The program calculates the approximate power for the Geisser-Greenhouse (POWER_GG) adjusted test, and the three multivariate criteria: Hotelling-Lawley Trace (POWER_HLT), Pillai-Bartlett Trace (POWER_PBT), and Wilks (POWER_WLK). When all the criteria are equal, the output is represented by POWER. The power results for the example are provided in Output 10.9.2.

For a repeated measurement design, the primary hypothesis of interest is the test of interaction, and given no interaction, the test of differences in condition. While 10 subjects per group appears adequate for the test of conditions given parallelism, the test of interaction has low power. For adequate power, one would need over 60 subjects per treatment group.

When performing a power analysis for a MGLM, the unknown covariance matrix is critical to the analysis. Changing Σ to a value that is closer to the covariance matrix for the sample:

$$\Sigma = \begin{pmatrix} 7.4 & 6.5 & 3.5 \\ 6.5 & 6.8 & 3.8 \\ 3.5 & 3.8 & 2.6 \end{pmatrix},$$

a study with only 10 subjects per group is adequate.

Output 10.9.2: SAS Powerlib202 for Repeated Measures Design.
ANOVA means

ALPHA	SIGSCAL	BETASCAL	TOTAL_N	POWER
0.05	1	1	15	0.061
0.05	1	1	30	0.076
0.05	1	1	45	0.092

Conditions

ALPHA	SIGSCAL	BETASCAL	TOTAL_N	POWER_MULT	EPSILON	EXEPS_GG	POWER_GG
0.05	1	1	15	0.731	0.781	0.712	0.847
0.05	1	1	30	0.982	0.781	0.747	0.994
0.05	1	1	45	0.999	0.781	0.759	1

Interaction

ALPHA	SIGSCAL	BETASCAL	TOTAL_N	POWER_HLT	POWER_PBT	POWER_WLK
0.05	1	1	15	0.089	0.099	0.095
0.05	1	1	30	0.154	0.165	0.161
0.05	1	1	45	0.227	0.239	0.234
0.05	1	1	180	0.801	0.802	0.802

EPSILON	EXEPS_GG	POWER_GG
0.781	0.712	0.074
0.781	0.747	0.125
0.781	0.759	0.179
0.781	0.776	0.669

Group mean vectors

ALPHA	SIGSCAL	BETASCAL	TOTAL_N	POWER_HLT	POWER_PBT	POWER_WLK
0.05	1	1	15	0.513	0.497	0.532
0.05	1	1	30	0.938	0.885	0.921
0.05	1	1	45	0.996	0.983	0.992

Condition mean vectors

ALPHA	SIGSCAL	BETASCAL	TOTAL_N	POWER_HLT	POWER_PBT	POWER_WLK
0.05	1	1	15	0.115	0.13	0.129
0.05	1	1	30	0.252	0.267	0.267
0.05	1	1	45	0.403	0.415	0.416

CHAPTER 11

Two-Level Hierarchical Linear Models

11.1 INTRODUCTION

In our formulation of a growth curve model for a single group, the same model is fit to each subject and the variation within subjects has a structured covariance matrix. The growth curve model does not model random variation among subjects in terms of model parameters. Because subjects are really not the same within a group, we may allow the model for each subject to have two components: fixed effects that are common to subjects and random effects that are unique to subjects. In this way, we may formulate a general linear mixed model (GLMM) for each subject.

More generally suppose we have n groups indexed by j where subjects are nested within groups. Then for an observation y_{ij}, the subjects are indexed by i and groups are indexed by j where subjects are nested within groups. Then for any group level, even though the β_j are not observed, one may model the β_j hierarchically, using observed data to structure models conditionally on fixed parameters. Such models are called hierarchical linear models or multilevel models. In this chapter, we discuss estimation and hypothesis testing for two-level hierarchical linear models (2LHLM). Analysis of these models is illustrated using PROC MIXED.

11.2 TWO-LEVEL HIERARCHICAL LINEAR MODELS

The classical approach to the analysis of repeated measures data is to use growth curve models or repeated measures models for groups of subjects over a fixed number of trials. A more general approach is to formulate growth as a 2LHLM where population parameters, individual effects and within-subject variation are defined at the level-1 of model development, and between-subject variation is modeled at the level-2 (Laird & Ware, 1982; Bock, 1989; Bryk & Raudenbush, 1992; Raudenbush & Bryk, 2002; Longford, 1993; Diggle, Liang, & Zeger, 1994; Ware, 1985; Vonesh & Chinchilli, 1997).

For a simple curvilinear regression model, one may postulate a quadratic model of the form

$$y_i = \beta_0 + \beta_1 x + \beta_2 x^2 + e_i \qquad i = 1, 2, \ldots, n \qquad (11.1)$$

where $e_i \sim N(0, \sigma^2)$ and $\beta_i' = \begin{pmatrix} \beta_{0i} & \beta_{1i} & \beta_{2i} \end{pmatrix}$ are the regression parameters for the ith subject. The the ordinary least squares estimator (OLSE) of β_i is

$$\hat{\beta}_i = (X_i'X_i)^{-1}X_i'y_i \tag{11.2}$$

where $y_{i(n_i \times 1)}$ is a data vector of measurements, $X_{i(n_i \times 3)}$ is a matrix with rows $(1, x, x^2)$ for $x = 1, 2, \ldots, n_i$. The parameters β_i contain information about the variation between individuals that is not specified in (11.2). Each β_i contains information regarding an overall population mean $\beta' = \begin{pmatrix} \beta_0 & \beta_1 & \beta_2 \end{pmatrix}$ and b_i the random departures of the individual variation from the population mean

$$\beta_i = \beta + b_i \tag{11.3}$$

where $b_i' = \begin{pmatrix} b_{0i} & b_{1i} & b_{2i} \end{pmatrix}$. The model for y_i given b_i is for $i = 1, 2, \ldots, n$:

$$y_i | b_i = (\beta_0 + b_{0i}) + (\beta_1 + b_{1i})x + (\beta_2 + b_{2i})x^2 + e_i. \tag{11.4}$$

Further, we may assume a distribution for b_i, for example

$$b_i \sim N_3(0, D). \tag{11.5}$$

This regression model may be formulated as a 2LHLM. The population parameters, individual effects and within-subject variation are specified in level-1 (11.4), and the between-person variation in level-2, (11.5).

Level-1. Conditionally on β and $b_1, b_2, \ldots, b_n$ the vectors y_i are independent with structure

$$y_i | (\beta, b_i) = X_i(\beta + b_i) + e_i \tag{11.6}$$

where $e_i \sim N_{n_i}(0, \sigma^2 I_{n_i})$.
Level-2. The vectors $b_i \sim N_k(0, D)$ and are independent of e_i.

For this illustration, the design matrix X_i was common to β and b_i. More generally we may let $\beta_{p \times 1}$ be a vector of unknown population parameters and $X_{i(n_i \times p)}$ the design matrix linking β to y_i. Let $b_{i(k \times 1)}$ be a vector of unknown individual effects and $Z_{i(n_i \times k)}$ a design matrix linking b_i and y_i. For this general case we have

Level-1. For each subject i,

$$y_i = X_i\beta + Z_i b_i + e_i \tag{11.7}$$

where $e_i \sim N_{n_i}(0, \Psi_i)$. At this level β and b_i are considered fixed and Ψ_i depends on i through n_i. When $\Psi_i = \sigma^2 I_{n_i}$, the 2LHLM is said to be *conditional-independent* since the responses for each individual are independent, conditional on β and b_i and
Level-2. The $b_i \sim N_k(0, D)$, are independent of each other and of e_i. The population parameters β are treated as fixed.

Thus it follows from Level-1 and Level-2 that for $i = 1, 2, \ldots, n$

$$\mathcal{E}(y_i) = X_i\beta$$
$$\mathrm{var}(y_i) = Z_iDZ_i' + \Psi_i = \Omega_i$$
$$\mathrm{cov}(y_i, y_j) = 0 \qquad\qquad i \neq j. \qquad (11.8)$$

Assuming all covariances in (11.8) are known and β is fixed, the best linear unbiased estimator (BLUE) of β by (4.6) is

$$\hat{\beta} = \left(\sum_{i=1}^{n} X_i'\Omega_i^{-1}X_i \right)^{-1} \sum_{i=1}^{n} X_i'\Omega_i^{-1}y_i. \qquad (11.9)$$

To estimate b_i is more complicated. By an extension of the Gauss-Markov theorem for general linear mixed models, Harville (1976) showed that the BLUE of b_i is

$$\hat{b}_i = DZ_i'\Omega_i^{-1}(y_i - X_i\hat{\beta}) \qquad (11.10)$$

and that the

$$\mathrm{var}(\hat{\beta}) = \left(\sum_{i=1}^{n} X_i'\Omega_i^{-1}X_i \right)^{-1}$$
$$\mathrm{var}(\hat{b}_i) = D - DZ_i'P_iZ_iD$$
$$P_i = \Omega_i^{-1} - \Omega_i^{-1}X_i \left(\sum_{i=1}^{n} X_i'\Omega_i^{-1}X_i \right)^{-1} X_i'\Omega_i^{-1}. \qquad (11.11)$$

Motivation for (11.10) follows from the fact that if

$$y = X\beta + Zb + e$$
$$\mathrm{var}(y) = ZDZ' + \Psi = \Omega$$
$$\mathrm{cov}(y, b) = ZD \qquad (11.12)$$

and

$$\begin{pmatrix} b \\ y \end{pmatrix} \sim N\left[\begin{pmatrix} 0 \\ X\beta \end{pmatrix}, \begin{pmatrix} V & VZ' \\ ZV & \Omega \end{pmatrix} \right]$$

then the conditional mean of the distribution of $b|y$ is

$$\mathcal{E}(b|y) = V'Z\Omega^{-1}(y - X\beta), \qquad (11.13)$$

the best linear unbiased predictor (BLUP) of b. Furthermore, the variance of $b|y$ is

$$\mathrm{var}(b|y) = V - V'Z(Z'VZ + \Psi)^{-1}ZV. \qquad (11.14)$$

Using the matrix identity

$$(Z'VZ + \Psi)^{-1} = \Psi^{-1} - \Psi^{-1}Z(Z'\Psi^{-1}Z + V^{-1})^{-1}Z'\Psi^{-1} \qquad (11.15)$$

in (11.14), and letting $\Phi = Z'\Psi^{-1}Z$, the

$$
\begin{aligned}
\mathrm{var}(b|y) &= V - V\Phi[I - (\Phi + V^{-1})^{-1}\Phi]V \\
&= V - V\Phi[(\Phi + V^{-1})(\Phi + V^{-1})^{-1} - (\Phi + V^{-1})^{-1}\Phi]V \\
&= V - V\Phi[(\Phi + V^{-1})^{-1}(\Phi + V^{-1} - \Phi)V] \\
&= V - V\Phi(\Phi + V^{-1})^{-1} \\
&= V[I - \Phi(\Phi + V^{-1})^{-1}] \\
&= V[(\Phi + V^{-1})(\Phi + V^{-1})^{-1} - \Phi(\Phi + V^{-1})^{-1}] \\
&= [V(\Phi + V^{-1} - \Phi)](\Phi + V^{-1})^{-1} \\
&= (\Phi + V^{-1})^{-1} \\
&= (Z'\Psi^{-1}Z + V^{-1})^{-1}
\end{aligned}
\tag{11.16}
$$

has a simple form using Ψ^{-1} and V^{-1}. If $\Psi = \sigma^2 I$, then the $\mathrm{var}(b|y) = \sigma^2(Z'Z + \sigma^2 V^{-1})^{-1}$.

Because the population covariance matrices Ψ_i and D are unknown, we may not directly use (11.9) and (11.10) to estimate β and b_i. There are two strategies for the estimation of model parameters for the 2LHLM, either classical inference using least squares and maximum likelihood or the empirical Bayes method using as prior information the distribution of the parameters defined in level-2.

To obtain parameter estimates it is convenient to let $\theta_{q\times 1}$ be a vector of the non-redundant variance and covariances in $\Psi_i, i = 1, \ldots, n$ and D. With a consistent estimator of θ and hence Ψ_i and D, we may let

$$
\widehat{\Omega}_i^{-1} = (\widehat{\Psi}_i + Z_i \widehat{D} Z_i')^{-1}
\tag{11.17}
$$

and estimate β and b_i using the least squares equations (11.9) and (11.10) by replacing Ω_i^{-1} with $\widehat{\Omega}_i^{-1}$ (Harville, 1977). Let these estimates be represented by $\hat{\beta}(\hat{\theta})$ and $\hat{b}(\hat{\theta})$.

To estimate θ and β using maximum likelihood (ML), the marginal normal distribution of $y' = \begin{pmatrix} y_1' & y_2' & \cdots & y_n' \end{pmatrix}$ is used where

$$
y_{N\times 1} = N_N(\mu, \Omega)
\tag{11.18}
$$

where $\Omega = \mathrm{diag}\begin{pmatrix} \Omega_1 & \Omega_2 & \cdots & \Omega_n \end{pmatrix}$, $\mu' = \begin{pmatrix} \mu_1 & \mu_2 & \cdots & \mu_n \end{pmatrix}$, $\mu_i = X_i\beta$ and $N = \sum_i n_i$. By maximizing the likelihood function given in (11.14) or equivalently

$$
L = L(\beta, \Omega|y) = (2\pi)^{N/2}|\Omega|^{-1/2} \exp\left[-\frac{1}{2}(y - \mu)'\Omega^{-1}(y - \mu)\right],
\tag{11.19}
$$

the ML estimates of β and θ, $\hat{\beta}_{ML}$ and $\hat{\theta}_{ML}$, satisfy the relationship $\hat{\beta}_{ML} = \hat{\beta}(\hat{\theta}_{ML})$ for the parameter β (see, Searle et al., 1992, Chapter 6).

The ML approach provides an estimate of β and θ; however, in balanced ANOVA designs, ML estimates of variance components fail to take into account the degrees of freedom lost in estimating β, and are hence biased downward, Searle et al. (1992, p. 250). This is overcome using classical methods by employing a restricted (residual) maximum likelihood estimator (REML) for θ, $\hat{\theta}_R$. Using (11.9) and (11.10) estimates $\hat{\beta}_{ML}$ and $\hat{b}_{ML} = \hat{b}_i(\theta_{ML})$ are realized. The REML estimate is obtained by maximizing the likelihood (11.17) of θ based on independent residual contrasts of y, $\psi = Cy$ where the rows of C are in the orthocomplement of the design space (Searle et al., 1992, p. 252 and 323). These estimators $\hat{\beta}(\hat{\theta}_R)$ and $\hat{b}_i(\hat{\theta}_R)$, provide alternative estimates of β and θ. Letting $b' = \begin{pmatrix} b'_1 & b'_2 & \dots & b'_n \end{pmatrix}$ and equating $\hat{b}_{ML} = \mathcal{E}(b|y, \hat{\beta}_{ML}, \hat{\theta}_{ML})$ gives $\hat{b}_{ML} = \hat{b}(\hat{\theta}_{ML})$, yields the empirical Bayes estimate of b when $\hat{\theta}_{ML}$ is ML. Hence, for the 2LHLM one may estimate β using the ML method and b by the empirical Bayes procedure which is equivalent to the REML estimate, $\hat{b}(\hat{\theta}_R)$ (Harville, 1976).

An alternative approach is to use the full Bayes method of estimation. Then we have to introduce a prior distribution on the location parameter in level-2. Recall that if θ and Y are two random variables, then the density of θ given Y is

$$f(\theta|Y) = \frac{f(Y, \theta)}{f(Y)} = \frac{f(Y|\theta)f(\theta)}{\int f(Y, \theta)d\theta}$$
$$= \frac{f(Y|\theta)f(\theta)}{\int f(Y|\theta)f(\theta)d\theta}. \tag{11.20}$$

However, $f(\theta|Y)$ may be regarded as a function of θ called the likelihood where if $Y_i \sim f(Y_i|\theta)$, $L(\theta|Y) = \prod_i f(Y_i|\theta)$. Hence, $f(\theta|Y)$ in (11.20) is proportional to a likelihood times a prior:

$$f(\theta|Y) \propto L(\theta|Y)f(\theta). \tag{11.21}$$

Knowing the posterior density $f(\theta|Y)$, the Bayes estimator for θ is defined as

$$\mathcal{E}(\theta|Y) = \int \theta f(\theta|Y)d\theta. \tag{11.22}$$

Suppose $\theta' = (\theta_1, \theta_2)$, then to find the joint density $f(\theta_1, \theta_2|Y)$ we have that

$$f(\theta_1, \theta_2|Y) = \frac{f(Y, \theta_1, \theta_2)}{f(Y)} = \frac{f(Y|\theta_1, \theta_2)f(\theta_1|\theta_2)f(\theta_2)}{\int f(Y|\theta_1, \theta_2)f(\theta_1|\theta_2)f(\theta_2)d\theta_1 d\theta_2}. \tag{11.23}$$

Hence to estimate θ using the exact Bayes procedure requires specification of a prior for θ_2 and the density of $f(\theta_1|\theta_2)$ where θ_2 represents parameters of the distribution of θ_1 called hyperparameters to calculate the joint posterior of θ_1 and θ_2 given Y, which may be problematic. However, suppose we can specify the distribution $f(\theta_1|\theta_2)$, then the posterior density for θ_1 is

$$f(\theta_1|Y, \theta_2) = \frac{f(Y|\theta_1, \theta_2)f(\theta_1|\theta_2)}{\int f(Y|\theta_1, \theta_2)f(\theta_1|\theta_2)d\theta_1} \tag{11.24}$$

and the Bayes estimator of θ_1 is

$$\theta_1 = \mathcal{E}(\theta_1|Y_1, \theta_1). \tag{11.25}$$

Because θ_2 is unknown, we may estimate θ_2 from the marginal distribution $f(Y|\theta_2)$ where

$$f(Y|\theta_2) = \int f(Y|\theta_1, \theta_2) f(\theta_1|\theta_2) d\theta_1 \qquad (11.26)$$

and the empirical Bayes estimate of θ_1 is

$$\hat{\theta}_1 = \mathcal{E}(\theta_1|Y, \hat{\theta}_2). \qquad (11.27)$$

For a Bayesian formulation of the 2LHLM, level-1 remains unchanged. For level-2, we let β and b_i be independent and normally distributed with means 0 and the $\text{var}(\beta) = \Gamma$, $\text{var}(b_i) = D$ and the $\text{cov}(\beta, b_i) = 0$. Then the marginal distribution of y_i is

$$y_i \sim N_{n_i}(0, X_i \Gamma X_i' + Z_i D Z_i' + \Psi_i) \qquad (11.28)$$

Again, θ is equal to the unknown nonredundant parameters in D and Ψ_i, $i = 1, 2, \ldots, n$.

If θ and Γ were known, Bayes estimates of β and b could be obtained as the posterior expectations given y, θ and Γ^{-1}.

$$\mathcal{E}(\beta|y, \Gamma^{-1}, \theta) = \left(\sum_{i=1}^{n} X_i' \Omega_i^{-1} X_i + \Gamma^{-1} \right)^{-1} \sum_{i=1}^{n} X_i' \Omega_i^{-1} y_i = \hat{\beta}$$

$$\mathcal{E}(b_i|y_i, \Gamma^{-1}, \theta) = D Z_i' \Omega_i^{-1}(y_i - X_i \hat{\beta}) = \hat{b}_i \qquad (11.29)$$

where

$$\Omega_i^{-1} = (Z_i D Z_i' + \Psi_i + \Gamma^{-1})^{-1}$$

$$\text{var}(\beta|y, \theta) = \left(\sum_{i=1}^{m} X_i' \Omega_i^{-1} X_i + \Gamma^{-1} \right)^{-1}$$

$$\text{var}(b_i|y_i, \theta) = D - D Z_i' P_i Z_i D$$

$$P_i = \Omega_i^{-1} - \Omega_i^{-1} X_i \left(\Gamma^{-1} + \sum_{i=1}^{n} X_i' \Omega_i^{-1} X_i \right)^{-1} X_i' \Omega_i^{-1}. \qquad (11.30)$$

Because Γ and θ are unknown, the empirical Bayes approach is to replace Γ and θ with maximum likelihood estimates obtained by maximizing their marginal normal likelihoods based on y, integrating over β and b.

For the Bayes formulation of the 2LHLM we have no information about the variation in Γ. We only have information regarding the between and within variation. Hence, Harville (1976); Laird and Ware (1982); Dempster, Rubin, and Tsutakawa (1981) propose setting $\Gamma^{-1} = 0$, indicating vague prior information about β. Then, from the limiting marginal distribution of θ given y, the estimate of θ is REML, $\hat{\theta}_R$

(Harville, 1976). Thus, we can replace θ for $\hat{\theta}_R$ in (11.29) and (11.30) to obtain $\hat{\beta}(\hat{\theta}_R)$ and $\hat{b}(\hat{\theta}_R)$, the empirical Bayes estimates:

$$\mathcal{E}(\beta|y, \Gamma^{-1} = 0, \hat{\theta}_R) = \hat{\beta}(\hat{\theta}_R)$$
$$\mathcal{E}(b_i|y_i, \Gamma^{-1} = 0, \hat{\theta}_R) = \mathcal{E}[b_i|y_i, \hat{\beta}(\hat{\theta}_R), \hat{\theta}_R] = \hat{b}_i(\hat{\theta}_R). \qquad (11.31)$$

Substituting $\hat{\theta}_R$ in (11.30) causes the variances to be underestimated. This problem was studied by Kackar and Harville (1984); Kass and Steffey (1989). The revised estimate involves the estimated variance as the first term plus a quadratic form $\delta' F_\theta^{-1} \delta$ where δ are first-order partial derivatives of a real valued function $G(y, \hat{\theta})$ with regard to θ evaluated at $\theta = \hat{\theta}$ and $F^{-1}(\theta)$ is the inverse of the second-order partial derivatives of the log-likelihood of the REML estimate (Searle et al., 1992, p. 341-343),

$$\text{var}[G(y, \hat{\theta})] \doteq \text{var}[G(y, \theta)|\hat{\theta}] + \delta' F^{-1}(\theta)\delta. \qquad (11.32)$$

To compute the ML or REML estimates of θ, the expectation-maximization (EM) algorithm was developed by Dempster, Laird, and Rubin (1977) to obtain ML estimates of μ and Σ when data are missing at random. The EM algorithm consists of two steps, the expectation or E-step and the maximization or M-step of sufficient statistics t of the parameters:

$$E\text{-step}: \hat{t} \equiv \mathcal{E}(t|y, \hat{\theta})$$
$$M\text{-step}: \hat{\theta} \equiv M(t). \qquad (11.33)$$

The algorithm treats the actual data as the *incomplete* set and the "actual + missing" data as the *complete* (augmented) set. The algorithm alternates between calculating conditional expected values (predictions) and maximizing likelihoods. If we use the procedure for mixed model estimation, the observed y are the *incomplete* data, and the complete data are y and the unobserved parameters β, b_i, and θ (Dempster et al., 1981; Laird, Lange, & Stram, 1987).

To utilize the EM algorithm for the 2LHLM, we let

$$\Psi_i = \sigma^2 I_{n_i} \qquad\qquad \Omega_i = \sigma^2 I_{n_i} + Z_i D Z_i' \qquad (11.34)$$

Then by (11.15) and (11.16),

$$\Omega_i^{-1} = \frac{[I - Z_i(\sigma^2 D^{-1} + Z_i'Z_i)^{-1}Z_i']}{\sigma^2}$$
$$b_i|y_i \sim N[DZ_i'\Omega_i^{-1}(y_i - X_i\hat{\beta}_i), \sigma^2 Z_i(\sigma^2 D^{-1} + Z_i'Z_i)^{-1}Z_i'] \qquad (11.35)$$

and

$$\mathcal{E}(b_i'b_i|y_i) = [\mathcal{E}(b_i|y_i)]'[\mathcal{E}(b_i|y_i)] + \sigma^2 \text{tr}[Z_i(Z_i'Z_i + \sigma^2 D^{-1})^{-1}Z_i'] \qquad (11.36)$$

using the fact that if $y \sim N(\mu, \Sigma)$, then $\mathcal{E}(y'Ay) = \mu'A\mu + \text{tr}(A\Sigma)$, Timm (1975, p. 112). Furthermore, observe that by (11.35)

$$\sigma^2 \text{tr}[Z_i(Z_i'Z_i + \sigma^2 D^{-1})^{-1} Z_i'] = \sigma^2 \text{tr}(I - \sigma^2 \Omega_i^{-1}) \qquad (11.37)$$

and that sufficient statistics for θ are t_1 and the $k(k+1)/2$ nonredundant elements of T_2:

$$\hat{\sigma}^2 = \frac{\sum_{i=1}^n e_i'e_i}{\sum_{i=1}^n n_i} = \frac{t_1}{N}$$

$$\hat{D} = n^{-1} \sum_{i=1}^n b_i b_i' = \frac{T_2}{n}. \qquad (11.38)$$

If an estimate of θ is available, we may calculate estimates of the missing sufficient statistics be equating them to their expectations conditioned on the incomplete, observed data y. Letting $\hat{\theta}$ be an estimate of θ, $\hat{\beta}(\theta)$, and $\hat{b}_i(\hat{\theta})$ the estimates of β and b_i, the estimated sufficient statistics are

$$\hat{t}_1 = \mathcal{E}\left[\sum_{i=1}^n e_i'e_i | y_i, \hat{\beta}(\hat{\theta}), \hat{\theta}\right]$$

$$= \sum_{i=1}^n \left[\hat{e}_i'(\hat{\theta})\hat{e}_i(\hat{\theta}) + \text{tr}\left(\text{var}(e_i|y_i, \hat{\beta}(\hat{\theta}), \hat{\theta})\right)\right]$$

$$\hat{T}_2 = \mathcal{E}\left[\sum_{i=1}^n b_i b_i' | y_i, \hat{\beta}(\hat{\theta}), \hat{\theta}\right]$$

$$= \sum_{i=1}^n \left[\hat{b}_i(\hat{\theta})\hat{b}_i'(\hat{\theta}) + \text{var}\left(b_i|y_i, \hat{\beta}(\hat{\theta}), \hat{\theta}\right)\right] \qquad (11.39)$$

where $\hat{e}_i = \mathcal{E}(e_i|y_i, \hat{\beta}(\hat{\theta}), \hat{\theta}) = y_i - X_i\hat{\beta}(\hat{\theta}) - Z_i\hat{b}_i(\hat{\theta})$.

To convert these expressions into useful computational form following Laird et al. (1987), let ω ($\omega = 0, 1, 2, \ldots, \infty$) index the iterations where $\omega = 0$ is a starting value. For

$$\beta^{(\omega)} = \left(\sum_{i=1}^n X_i'W_i^{(\omega)}X_i\right)^{-1} \sum_{i=1}^n X_i'W_i^{(\omega)}y_i$$

$$W_i^{(\omega)} = \left[\Omega_i^{(\omega)}\right]^{-1}$$

$$\Omega_i^{(\omega)} = \sigma^{(\omega)2}I_{n_i} + Z_iD^{(\omega)}Z_i'$$

$$b_i^{(\omega)} = D^{(\omega)}Z_i'W_i^{(\omega)}r_i^{(\omega)}$$

$$r_i^{(\omega)} = y_i - X_i\beta^{(\omega)} \qquad (11.40)$$

by (11.37) and (11.36), we have, substituting into (11.39), that the ML estimates of θ are

$$\sigma^{2(\omega+1)} = \frac{\sum_{i=1}^{n}\left[\left(r_i^{(\omega)} - z_i b_i^{(\omega)}\right)'\left(r_i^{(\omega)} - z_i b_i^{(\omega)}\right) + \sigma^{(\omega)2}\operatorname{tr}\left(I - \sigma^{(\omega)2}W_i^{(\omega)}\right)\right]}{N}$$

$$D^{(\omega+1)} = \frac{\sum_{i=1}^{n}\left[b_i^{(\omega)}b_i^{'(\omega)} + D^{(\omega)}\left(I - Z_i'W_i^{(\omega)}Z_i D^{(\omega)}\right)\right]}{n}. \tag{11.41}$$

To obtain the REML estimates of θ, $\hat{\theta}_R$, the matrix $W_i^{(\omega)}$ in (11.41) with $P_i^{(\omega)}$ where

$$P_i^{(\omega)} = W_i^{(\omega)}\left[I - X_i\left(\sum_{i=1}^{n}X_i'W_i^{(\omega)}X_i\right)^{-1}X_i'W_i^{(\omega)}\right]. \tag{11.42}$$

For starting values, one uses the OLSE of β, b_i, σ^2, and D:

$$\hat{\beta}_0 = \left(\sum_{i=1}^{n}X_i'X_i\right)^{-1}\sum_{i=1}^{n}X_i'y_i$$

$$\hat{b}_i = (Z_i'Z_i)^{-1}Z_i'(y_i - X_i\hat{\beta}_0)$$

$$\hat{\sigma}_0^2 = \frac{\sum_{i=1}^{n}y_i'y_i - \hat{\beta}_0'\sum_{i=1}^{n}X_i'y_i - \sum_{i=1}^{n}b_i'Z_i'(y_i - X_i\hat{\beta}_0)}{N - (n-1)k - p}$$

$$\widehat{D}_0 = \frac{\sum_{i=1}^{n}b_ib_i'}{n} - \frac{\hat{\sigma}_0^2\sum_{i=1}^{n}(Z_i'Z_i)^{-1}}{n}. \tag{11.43}$$

The EM algorithm allows one to obtain parametric empirical Bayes (PEB) estimates of θ which results in a REML estimate of θ, $\hat{\theta}_R$, or a ML estimate of θ, $\hat{\theta}_{ML}$, for the variance and covariance components of the mixed model. With $\hat{\theta}_{ML}$ substituted into (11.9), $\hat{\beta}_{ML} = \hat{\beta}(\hat{\theta}_{ML})$ is the maximum likelihood estimate of the fixed effect β; $\hat{b}_{ML}^{(i)} = \hat{b}_i(\hat{\theta}_{ML})$ is the empirical Bayes estimate of b_i by (11.10). While the ML procedure yields estimates of β and b_i, they are not unbiased for the components of variance; however, the REML estimate $\hat{\theta}_R$ is unbiased for the components, but does not directly include a procedure for estimating β and b_i. Harville (1976) showed then an estimate of θ obtained by maximizing the limiting marginal likelihood of θ given y is equivalent to the REML likelihood in the classical case. Thus, the effects $\hat{\beta} = \hat{\beta}(\hat{\theta}_R)$ and $\hat{b}_i = \hat{b}_i(\hat{\theta}_R)$ are called REML estimates or PEB estimates. More importantly Harville (1976) showed that if we use the $\operatorname{var}(\beta|y, \hat{\theta}_R)$ and the $\operatorname{var}(b|y, \hat{\theta}_R)$ to estimate the variances of $\hat{\beta}(\hat{\theta}_R)$ and $\hat{b}(\hat{\theta}_R)$ that these are the same as the sample-theory variances of $\operatorname{var}(\hat{\beta})$ and the $\operatorname{var}(\hat{b} - b)$ where from (11.11) the

$$\operatorname{var}(\hat{\beta}) = \left(\sum_{i=1}^{n}X_i'\Omega_i^{-1}X_i\right)^{-1}$$

$$\operatorname{var}(\hat{b}_i - b_i) = D - DZ_i'\Omega_i^{-1}Z_iD + DZ_i'\Omega_i^{-1}X_i\left(\sum_{i=1}^{n}X_i'\Omega_i^{-1}X_i\right)^{-1}X_i'\Omega_iZ_iD. \tag{11.44}$$

If the $\mathrm{var}(\hat{b}_i)$ is used, the variation in $\hat{b}_i - b_i$ would be underestimated since it ignores the variation in b_i. However, as with the weighted linear model with fixed effects and unknown covariance matrix, substituting $\hat{\Omega}_i$ based on $\hat{\theta}_R$ for Ω_i^{-1} in (11.44) underestimates the variance of $\hat{\beta}$ and the variance of $\hat{b}_i - b_i$ since for example, if $\Gamma^{-1} = 0$, the

$$\mathrm{var}(\beta|y) = \mathcal{E}[\mathrm{var}(\beta|y,\theta)] + \mathrm{var}[\mathcal{E}(\beta|y,\theta)] \qquad (11.45)$$

and only the first term in the expression is being estimated by the "substitution principle". Because the second term is ignored, variances are underestimated (Searle et al., 1992, p. 339-343).

Estimation of model parameters using the EM algorithm has been improved upon by Longford (1987) which incorporates Fisher scoring. An iterative generalized least squares procedure has been developed by Goldstein (1986). Park (1993) shows that the EM algorithm and the iterative two-stage Zellner's iterative Aitken estimator (IZEF) algorithm are equivalent. Kreft, De Leeuw, and Leeden (1994) compared several computer packages using various methods of estimation. Lindstrom and Bates (1988) develop a Newton-Raphson (NR) algorithm to estimate the model parameters. Their new algorithm is efficient, effective and consistently converges, and it shows that NR is better than EM. For comprehensive discussion of the EM method see Liu and Rubin (1994); McLachlan and Krishnaiah (1997). Wolfinger, Tobias, and Sall (1994) developed an NR sweep algorithm that is employed in PROC MIXED.

For the 2LHLM we have three groups of parameters that are estimated: the fixed effects β, the random effects, b_i, $i = 1, 2, \ldots, n$, and the variance components θ. To test hypotheses, we write the model as a general linear mixed model:

$$y_{N \times 1} = X_{N \times p}\beta_{p \times 1} + Z_{N \times k}b_{k \times 1} + e_{N \times 1} \qquad (11.46)$$

where

$$y'_{1 \times N} = \begin{pmatrix} y'_1 & y'_2 & \cdots & y'_n \end{pmatrix}$$
$$b'_{1 \times N} = \begin{pmatrix} b'_1 & b'_2 & \cdots & b'_n \end{pmatrix}$$
$$X'_{p \times N} = \begin{pmatrix} X'_1 & X'_2 & \cdots & X'_n \end{pmatrix}$$

and $X_{i(p \times n_i)}$, $\sum_{i=1}^{n} n_i = N$, $\sum_{i=1}^{n} k_i = k$. For random coefficient models, $Z_{n \times k}$ takes the general form

$$Z = \begin{pmatrix} Z_1 & 0 & \cdots & 0 \\ 0 & Z_2 & \cdots & 0 \\ \vdots & \vdots & \ddots & \vdots \\ 0 & 0 & \cdots & Z_n \end{pmatrix} \qquad (11.47)$$

where $Z_{i(n_i \times k_i)}$. Alternatively, for variance component models, Z is partitioned such that

$$Z = \begin{pmatrix} Z_1 & Z_2 & \cdots & Z_n \end{pmatrix} \qquad (11.48)$$

then

$$y = X\beta + \sum_{i=1}^{n} Z_i B_i + e$$

$$\text{var}(y) = ZVZ' + \sigma^2 I = \sum_{i=1}^{n} \sigma_i^2 Z_i Z_i' + \sigma_e^2 I_N.$$

More generally,

$$y = X\beta + Zb + e$$
$$\text{cov}(y) = ZVZ' + \sigma^2 I_N \qquad\qquad V = I_N \otimes D$$
$$\text{cov}(b_i, b_j) = 0$$
$$\text{cov}(b, e) = 0$$
$$\text{cov}(bi) = D \qquad\qquad (11.49)$$

For developing tests of hypotheses for the mixed model, it is convenient to use the equivalent hierarchical normal form of the model

$$(y|\beta, b) \sim N(X\beta + Zb, \Omega = ZVZ' + \sigma^2 I_N)$$
$$\beta \sim N(\beta_0, \Gamma)$$
$$b \sim N(0, V = I_n \otimes D) \qquad\qquad (11.50)$$

and assume $\Gamma^{-1}=0$, then we have that

$$\mathcal{E}(\beta|y) = (X'\Omega^{-1}X + \Gamma^{-1})^{-1}(X'\Omega^{-1}y + \Gamma^{-1}\beta_0)$$
$$= (X'\Omega^{-1}X)^{-1}X'\Omega^{-1}y$$
$$\text{var}(\beta|y) = (X'\Omega^{-1}X + \Gamma^{-1})^{-1}$$
$$= (X'\Omega^{-1}X)^{-1}$$
$$\mathcal{E}(b|y) = VZ'(\Omega + X\Gamma X')^{-1}(y - X\beta_0)$$
$$= VZ'\Omega^{-1}(y - X\hat{\beta})$$
$$= \hat{b} \qquad\qquad (BLUP)$$
$$\text{var}(b|y) = V - VZ'(X\Gamma XV + ZVZ' + \sigma^2 I_N)^{-1}ZV$$
$$= [Z'(X\Gamma X' + \sigma^2 I_N)^{-1}Z + V^{-1}]^{-1}$$
$$= \sigma^2(Z'Z + \sigma^2 V^{-1})^{-1}. \qquad\qquad (11.51)$$

Testing hypotheses using the mixed model or the parametric empirical Bayes method is complicated by the estimation of the variance components θ. Tests depend on the large sample covariance matrix of the maximum likelihood or empirical Bayes estimates.

If one tests the hypothesis regarding the fixed effects β, $H : C\beta = 0$, the statistic

$$X^2 = (C\hat{\beta})'(C(X'\hat{\Omega}^{-1}X)^{-1}C)^{-1}C\hat{\beta} \qquad\qquad (11.52)$$

has a chi-square distribution with $\nu = \text{rank}(C)$ when H is true. Alternatively, one may also use an F-approximation, X^2/ν.

A very approximate test of $H : Cb = 0$ is to use the empirical Bayes estimate for b. Then to test H, one again forms a chi-square statistic

$$X^2 = (C\hat{b})'(C\hat{\Omega}^*C')^{-1}C\hat{b}$$
$$\hat{\Omega}^* = \hat{V} - \hat{V}Z'[\hat{\Omega}^{-1} - \hat{\Omega}^{-1}X(X'\hat{\Omega}^{-1}X)^{-1}X'\hat{\Omega}^{-1}]Z\hat{V}, \qquad (11.53)$$

where X^2 has a chi-square distribution under the null hypothesis with degrees of freedom $\nu = \text{rank}(C)$.

The last hypothesis of interest is whether some variances or covariance components of θ are perhaps zero. Clearly whether the variances are zero is of interest since it indicates whether there is random variation in b_i of b. However, there is no adequate test of $H_\theta : C\theta = 0$, even for large samples since the normal approximation is very poor. Bryk and Raudenbush (1992, p. 55) propose a goodness-of-fit test.

11.3   RANDOM COEFFICIENT MODEL: ONE POPULATION

Given a group of n_j subjects, the classical form of the general linear model

$$y = X\beta + e \qquad \mathcal{E}(e) = 0 \quad \text{var}(e) = \Omega = \sigma^2 I_{n_j} \qquad (11.54)$$

assumes a known design matrix $X_{n_j \times p}$ that is under the control of the experimenter, where the elements are measured without error, homogeneity of variance, and a fixed parameter vector β. Until now, most of our discussion regarding GLM has allowed the structure Ω to change, but β has remained fixed, the same for all subjects.

Because subjects are really different within a group, we may allow β to be random and different among subjects. Then,

$$y_i = x_i'\beta_i + e_i$$
$$\beta_i = \beta + b_i$$

where the vector b_i are assumed to be independent and identically distributed with zero mean and covariance structure D. Then, we have a fixed mean

$$\mathcal{E}(y) = X\beta$$

but heteroscedasticity among the y_i:

$$\text{var}(y_i) = x_i'Dx_i + \sigma^2 = \sigma_i^2.$$

More generally, suppose we have n groups indexed by j. Then (11.54) becomes

$$y_j = X_j\beta_j + e_j \qquad (11.55)$$

where y_j and e_j are vectors of length n_j, the number of subjects in the jth group. Furthermore, assume

$$\mathcal{E}(e_j) = 0$$
$$\text{var}(e_j) = \sigma_j^2 I$$

where the e_j are again independent, but not identically distributed since the variances are not homogeneous across groups. Letting index i represent classes and index j subjects within classes, (11.55) and (11.56) ignore the fact that classes are part of the same school, a level of commonality for the model equations.

To model group differences, one may assume an ANCOVA model where the β_j in (11.55) have equal slopes with different intercepts and homogeneous variances. That is

$$\beta_j = Z_j\gamma = \begin{pmatrix} 1'_j & 0 \\ 0 & I \end{pmatrix} \begin{pmatrix} \alpha \\ \beta \end{pmatrix} \tag{11.56}$$

where 1_j is the jth unit vector, and α is a vector of n fixed intercepts. Combining (11.56) with (11.55), we have

$$y_j = X_j Z_j \gamma + e_j \tag{11.57}$$

where the $\text{var}(y_j) = \sigma_j^2 I$, the classical linear model, results if $\sigma_1^2 = \sigma_2^2 = \cdots = \sigma_n^2 = \sigma^2$.

Associating a random coefficient model to (11.55), we may postulate a different model for each classroom where the structure of the coefficients is not fixed but random, varying over replications,

$$\begin{aligned} y_j &= X_j\beta_j + e_j \\ \beta_j &= \beta + b_j \end{aligned} \tag{11.58}$$

so that

$$\begin{aligned} y_j &= X_j\beta + X_j b_j + e_j \\ \text{cov}(y_j) &= X_j' D X_j + \sigma_j^2 I. \end{aligned} \tag{11.59}$$

In (11.59), we assume the b_j are independent of one another, independent of the errors e_j, have zero mean, and have common covariance matrix D. Model (11.59) has the form of the mixed model. If the random errors at the second level are all zero, (11.59) reduces to (11.55).

Model (11.59) is a hierarchical linear model, sometimes called a multilevel model or a nested mixed linear model (Goldstein, 1986; Longford, 1987). In particular, we have that

$$\begin{aligned} y_j &= X_j\beta_j + e_j \\ \beta_j &= Z_j\gamma + b_j \end{aligned}$$

so that

$$\begin{aligned} y_j &= (X_j Z_j)\gamma + X_j b_j + e_j \\ \mathcal{E}(y_j) &= (X_j Z_j)\gamma \\ \text{cov}(y_j) &= X_j D X_j' + \sigma_j^2 I, \end{aligned} \tag{11.60}$$

a mixed linear model.

Given (11.60), an estimate of β_j is

$$\hat{\beta}_j = (X'_j X_j)^{-1} X'_j y_j \tag{11.61}$$

so that

$$\hat{\gamma} = \left(\sum_{j=1}^{n} Z'_j Z_j \right)^{-1} \sum_{j=1}^{n} Z'_j \beta_j.$$

Alternatively, using (11.57),

$$\hat{\gamma} = \left(\sum_{j=1}^{n} Z'_j X'_j X_j Z_j \right)^{-1} \sum_{j=1}^{n} A_j X_j y_j$$

$$= \left(\sum_{j=1}^{n} Z'_j X'_j X_j Z_j \right)^{-1} \sum_{j=1}^{n} Z_j (X'_j X_j) \hat{\beta}_j \tag{11.62}$$

by (11.61). However, this estimator of γ is neither a BLUE nor a best linear unbiased prediction (BLUP) estimator (De Leeuw & Kreft, 1986; Vonesh & Carter, 1987). The BLUE of γ, assuming σ_j^2 and D are known, is

$$\hat{\gamma} = \left(\sum_{j=1}^{n} Z'_j W_j^{-1} Z_j \right)^{-1} \sum_{j=1}^{n} Z'_j W_j \hat{\beta}_j \tag{11.63}$$

where $W_j = D + \sigma_j^2 (X'_j X_j)^{-1}$, the covariance matrix for the OLSE $\hat{\beta}_j$. Because D and σ_j^2 are unknown, estimation of model parameters for hierarchical linear models is complicated with the analysis dependent upon the algorithm and the computer program (Kreft et al., 1994; Gelfand, Sahu, & Carlin, 1995).

In the Section 11.2, the two-level hierarchical linear model was introduced as a repeated measurement experiment in which n_i repeated observations are obtained on $i = 1, 2, \ldots, n$ subjects which permitted one to estimate individual models by modeling within-subject and between-subject variation as a mixed model. From (11.60), we see that the presentation of the HLM of Bryk and Raudenbush (1992) is also a mixed linear model.

There are many issues associated with the analysis of hierarchical linear models in the social sciences. Articles in volume 20, Number 2 of the *Journal of Educational and Behavioral Statistics* review several problems and concerns. For example, the article by D. Draper (1995) reviews inference issues. A discussion of nonlinear models is examined by Davidian and Giltinan (1995). Longford (1993) discusses numerous algorithms. Everson and Morris (2000) develop a procedure for making inferences about the parameters of the two-level normal HLM which outperforms REML and

Gibbs sampling procedures. Their procedure is based on independent draws from an exact posterior distribution, a full Bayesian model. Thum (1997) develops an HLM for multivariate observations.

We now turn to the analysis of several data sets using the MIXED procedure, documented in SAS (1992) and PROC GENMOD, discussed in SAS (1993) to analyze the random coefficient model, with repeated measures and hierarchical data models.

In Chapter 8, we used the Elston and Grizzle (1962) Ramus heights data to illustrate the SUR model in the analysis of cross-sectional time series data. Recall that the data are the Ramus heights (in mm) of 20 boys measured at four time points (8, 8.5, 9, and 9.5 years of age). The SUR model fit a model to the entire group with an unknown covariance structure $\Sigma_{p \times p}$. With a consistent estimator for Σ, we used the generalized least squares procedure to estimate the model parameters for the group.

Plotting the data for each boy (see, Graph 11.3.1), observe that two boys show rapid growth during the period 8.5 to 9 years, while the growth of the others is more gradual. Hence, the variation among the boys suggests fitting a model that considers stochastic variation of the n slopes, a random coefficient or mixed model. For this analysis, we have $i = 1, 2, \ldots, 20$ boys and $j = 1, 2, 3, 4$ time points. For each boy, we fit the linear model,

$$y_i = \beta_0 + \beta_i x_i + e_i \qquad i = 1, 2, \ldots, n \tag{11.64}$$

where $e_i \sim IN(0, \sigma^2)$, β_0 is the group mean common to all boys and β_i is specific to a boy. That is, β_i is random with linear structure $\beta_i = \beta_1 + b_i$ where $\mathcal{E}(\beta_i) = \beta_1$, $\mathcal{E}(b_i) = 0$ and the variance of b_i is σ_b^2. Using the random coefficient model with orthogonal polynomials for a linear model, the model in (9.64) becomes

$$y_{i(4 \times 1)} = X_{i(4 \times 2)} \beta_{(2 \times 1)} + Z_{i(4 \times 1)} b_{i(1 \times 1)} + e_{i(4 \times 1)} \tag{11.65}$$

$$= \begin{pmatrix} 1 & -3 \\ 1 & -1 \\ 1 & 1 \\ 1 & 3 \end{pmatrix} \begin{pmatrix} \beta_0 \\ \beta_1 \end{pmatrix} + \begin{pmatrix} 1 \\ 1 \\ 1 \\ 1 \end{pmatrix} b_i + e_i$$

$$\mathrm{cov}(y_i) = Z_i D Z_i' + \Psi_i = \sigma_b^2 J_4 + \sigma^2 I_4.$$

Stacking the vectors y_i into a vector y, the GLM for the problem is

$$y = X\beta + Zb + e \tag{11.66}$$

$$\begin{pmatrix} y_1 \\ y_2 \\ \vdots \\ y_n \end{pmatrix} = \begin{pmatrix} X_1 \\ X_2 \\ \vdots \\ X_n \end{pmatrix} \beta + \begin{pmatrix} Z_1 & 0 & \cdots & 0 \\ 0 & Z_2 & \cdots & 0 \\ \vdots & \vdots & \ddots & \vdots \\ 0 & 0 & \cdots & Z_n \end{pmatrix} \begin{pmatrix} b_1 \\ b_2 \\ \vdots \\ b_n \end{pmatrix} + \begin{pmatrix} e_1 \\ e_2 \\ \vdots \\ e_n \end{pmatrix}$$

$$\text{cov}(b) = \begin{pmatrix} D & 0 & \cdots & 0 \\ 0 & D & \cdots & 0 \\ \vdots & \vdots & \ddots & \vdots \\ 0 & 0 & \cdots & D \end{pmatrix} = I_n \otimes \sigma_b^2 = \sigma_b^2 I_{20} = V$$

$$\text{cov}(e) = I_n \otimes \Psi_i = \Psi = \sigma^2 I_{80} \tag{11.67}$$

so that the $\text{cov}(y) = ZVZ' + \Psi = \Omega_{80 \times 80} = I_n \otimes \Omega_i$ where $\Omega_i = Z_i DZ_i' + \Psi_i = \sigma_b^2 J_4 + \sigma^2 I_4$ and σ_b^2 and σ^2 are the variance components.

To analyze the data using PROC MIXED, we reorganize the data for the SUR model so that each boy has four repeated measurements at four ages, and the design matrix of orthogonal polynomials for a linear model is input. PROC MIXED has three essential statements: MODEL, RANDOM, and REPEATED. PROC MIXED provides both REML and ML estimates of model parameters, as well as minimum variance quadratic unbiased estimates (MINQUE). The latter procedure is useful for large data sets when REML and ML methods fail to converge. The SAS code to perform the analysis is provided in Program 11_3.sas.

The MODEL statement of PROC MIXED is used to specify the design matrix X of the mixed model, where the intercept in the model is always included by default. The SOLUTION option requests the fixed effect estimate $\hat{\beta}$ to be printed and the OUTP= option creates predicted and residual scores as a new data file. The RANDOM statement defines the random effects in the vector b and is used to define the Z matrix in the mixed model. The structure of V is defined by using the TYPE= option. In our example, $V = \sigma_b^2 I_{20}$ so that TYPE=VC, for variance components. The REPEATED statement is used to establish the structure of Ψ and the R option requests the first block of Ψ, Ψ_1 to be printed. In PROC MIXED, the $\text{cov}(e) = I \otimes \Psi_i \equiv R$ and the $\text{cov}(b) = I \otimes D \equiv G$.

The REML estimate of β is

$$\begin{aligned} \hat{\beta}_{REML} &= \left(\sum_{i=1}^{n} X_i' \hat{\Omega}_i^{-1} X_i \right)^{-1} \left(\sum_{i=1}^{n} X_i' \hat{\Omega}_i^{-1} y_i \right) \\ &= \begin{pmatrix} 50.0750 \\ .4665 \end{pmatrix} \end{aligned}$$

where $\hat{\Omega}_i = \hat{\sigma}_b^2 J + \sigma^2 I$, and $\hat{\sigma}_b^2 = 6.0953$ and $\hat{\sigma}^2 = .6779$ are REML estimates of variance components (Output 11.3.1). The corresponding ML estimates are 5.7849

Graph 11.3.1: Random coefficient model: one population - individual profiles.

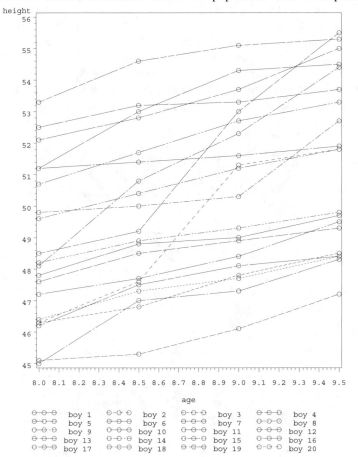

and .6666, respectively, which are biased downward (Output 11.3.2). The ML estimates can be requested using **PROC MIXED** by changing `METHOD=REML` option to `METHOD=ML`. Comparing $\hat{\beta}_{REML}$ with the estimates of $\hat{\beta}_{FGLS}$ using the SUR model, we have close agreement between the two sets of parameter estimates (Output 11.3.3). To obtain the SUR model result using PROC MIXED, we remove the RANDOM statement which models D and use the `TYPE=UN` option in the `REPEATED` statement. Because $S = \sum_{i=1}^{n} \hat{\Psi}_i/n$, $\hat{\beta}'_{REML} = \hat{\beta}'_{FGLS} = (50.0496, .4654)$. To invoke compound symmetry structure for the diagonal block Ψ_i of Ψ, we use the `TYPE=CS` option in the `RANDOM` statement. The common covariance matrix is outputted with the R option. In the Output 11.3.4, `CS` contains the common covariance REML estimate; `Residual` contains the estimate of common variance. Even though the random coefficient and compound symmetry models are not the same, in general, since $\Sigma = ZDZ' + \Psi \neq Psi$ we have that $Z_iDZ'_i + \sigma^2 I = \Psi_i$ and $\text{diag}(\Psi_i) = \Psi$ so that the random coefficient model and the model fit assuming compound symmetry of Σ give identical results in this example.

Output 11.3.1: REML Estimates of the Fixed Effects and Variance Components.

Solution for Fixed Effects					
Effect	**Estimate**	**Standard Error**	**DF**	**t Value**	**Pr > |t|**
Intercept	50.0750	0.5597	19	89.47	<.0001
x	0.4665	0.04117	59	11.33	<.0001

Estimated R Matrix for boys 1				
Row	**Col1**	**Col2**	**Col3**	**Col4**
1	0.6779			
2		0.6779		
3			0.6779	
4				0.6779

Covariance Parameter Estimates						
Cov Parm	**Subject**	**Ratio**	**Estimate**	**Standard Error**	**Z Value**	**Pr Z**
boys	boys	8.9915	6.0953	2.0328	3.00	0.0014
Residual	boys	1.0000	0.6779	0.1248	5.43	<.0001

To evaluate the model fit to the data using PROC MIXED, numerous model fitting statistics are provided. The likelihood ratio test (LRT) compares the null model $\Omega = \sigma^2 I$ with the structure model input. For the Ramus example, the SUR model appears to be the best model (Output 11.3.5). This follows from the observation that the model-fit criteria provided by PROC MIXED, Akaike's Information Criterion (AIC), Corrected Akaike's Information Criterion (AICC), and Bayesian Information

Output 11.3.2: ML Estimates of the Variance Components.

Estimated R Matrix for boys 1				
Row	Col1	Col2	Col3	Col4
1	0.6666			
2		0.6666		
3			0.6666	
4				0.6666

Covariance Parameter Estimates						
Cov Parm	Subject	Ratio	Estimate	Standard Error	Z Value	Pr Z
boys	boys	8.6782	5.7849	1.8823	3.07	0.0011
Residual	boys	1.0000	0.6666	0.1217	5.48	<.0001

Criterion (BIC), are smallest. There was no difference in model fit between the variance component and compound symmetry models. For discussion regarding "fit" statistics, see Littell et al. (1996); SAS (1992).

Alternatively, one may fit a model to the Ramus data using the generalized linear model and the PROC GENMOD. The generalized linear model extends GLM to a wider class of models. The components of the model are

1. $\mathcal{E}(y_i) = \mu_i$

2. $g(\mu_i) = x_i'\beta = \eta_i$

3. $\mathrm{var}(y_i) = \frac{\phi \mathrm{var}(\mu_i)}{w_i}$

where ϕ is a known or must be estimated, w_i are known weights for each observation, and g is a monotonic differentiable link function (McCullagh & Nelder, 1989). For additional detail, see SAS (1993).

11.4 RANDOM COEFFICIENT MODEL: SEVERAL POPULATIONS

In Chapter 7, we analyzed the Grizzle and Allen (1969) dog data by fitting a third-degree polynomial (i.e., cubic trend) to each group using a full multivariate growth curve model of the form $\mathcal{E}(Y) = XBP$ with large unknown covariance matrix Σ. For such a model, the matrix Σ may be poorly estimated given its size. In addition, the growth curve model does not permit the specification and estimation of random individual characteristics or permit one to specify a known structure for Σ. We now analyze the same data using PROC MIXED which permits the specification of fixed and random effects and the characterization of Σ using a hierarchical linear model. The SAS code to perform the analysis is given in Program 11_4.sas.

Output 11.3.3: SUR Estimates of the Fixed Effects and Variance Components.

Solution for Fixed Effects					
Effect	Estimate	Standard Error	DF	t Value	Pr > |t|
Intercept	50.0496	0.5563	19	89.97	<.0001
x	0.4654	0.05223	19	8.91	<.0001

Estimated R Matrix for boys 1				
Row	Col1	Col2	Col3	Col4
1	6.3269	6.1836	5.7827	5.5557
2	6.1836	6.4339	6.1547	5.9356
3	5.7827	6.1547	6.8942	6.9306
4	5.5557	5.9356	6.9306	7.4461

Covariance Parameter Estimates					
Cov Parm	Subject	Estimate	Standard Error	Z Value	Pr Z
UN(1,1)	boys	6.3269	2.0518	3.08	0.0010
UN(2,1)	boys	6.1836	2.0361	3.04	0.0024
UN(2,2)	boys	6.4339	2.0827	3.09	0.0010
UN(3,1)	boys	5.7827	2.0126	2.87	0.0041
UN(3,2)	boys	6.1547	2.0776	2.96	0.0031
UN(3,3)	boys	6.8942	2.2296	3.09	0.0010
UN(4,1)	boys	5.5557	2.0252	2.74	0.0061
UN(4,2)	boys	5.9356	2.0908	2.84	0.0045
UN(4,3)	boys	6.9306	2.2813	3.04	0.0024
UN(4,4)	boys	7.4461	2.4102	3.09	0.0010

Output 11.3.4: REML Estimates of the Fixed Effects and Variance Components Assuming Compound Symmetry.

Solution for Fixed Effects					
Effect	Estimate	Standard Error	DF	t Value	Pr > |t|
Intercept	50.0750	0.5597	19	89.47	<.0001
x	0.4665	0.04117	59	11.33	<.0001

Estimated R Matrix for boys 1				
Row	Col1	Col2	Col3	Col4
1	6.7732	6.0953	6.0953	6.0953
2	6.0953	6.7732	6.0953	6.0953
3	6.0953	6.0953	6.7732	6.0953
4	6.0953	6.0953	6.0953	6.7732

Covariance Parameter Estimates						
Cov Parm	Subject	Ratio	Estimate	Standard Error	Z Value	Pr Z
CS	boys	8.9915	6.0953	2.0328	3.00	0.0027
Residual		1.0000	0.6779	0.1248	5.43	<.0001

Output 11.3.5: Model Fit of Ramus Data Using Various Methods.
Variance component (TYPE=VC)

Fit Statistics	
-2 Res Log Likelihood	270.0
AIC (smaller is better)	274.0
AICC (smaller is better)	274.2
BIC (smaller is better)	276.0

SUR Model (TYPE=UN)

Fit Statistics	
-2 Res Log Likelihood	227.9
AIC (smaller is better)	247.9
AICC (smaller is better)	251.2
BIC (smaller is better)	257.9

Compound Symmetry (TYPE=CS)

Fit Statistics	
-2 Res Log Likelihood	270.0
AIC (smaller is better)	274.0
AICC (smaller is better)	274.2
BIC (smaller is better)	276.0

For Level 1 of the TLHLM, we assume the 7×1 vector of repeated measures for each dog has the general form

$$y_{i(7\times1)} = X_{i(7\times7)}\beta_{(7\times1)} + Z_{i(7\times4)}b_{i(4\times1)} + e_{i(7\times1)} \qquad (11.68)$$

where X_i is the design matrix of the fixed effects. β contains the intercept, independent treatment effects and common slopes represented by linear (L), quadratic (Q), and cubic (C) polynomials; b_i is a random vector of trends, intercept (I), (L), (Q), (C); e_i is a random vector of errors with structure $\Psi_i = \sigma^2 I$ and Z_i are covariates. For Level 2 of the model, we assume that b_i has mean zero and covariance matrix $D_{4\times4}$ so that the $\mathrm{cov}(b) = I_n \otimes D = V \equiv G$, $b' = (b'_1, \ldots, b'_n)$ and $\Psi = R$. Model (11.68) is a random coefficient growth curve model. To analyze this model using PROC MIXED, we create the matrices

$$X_{N\times p} = \begin{pmatrix} X_1 \\ X_2 \\ \vdots \\ X_n \end{pmatrix} \qquad Z_{N\times k} = \begin{pmatrix} Z_1 & 0 & \cdots & 0 \\ 0 & Z_2 & \cdots & 0 \\ \vdots & \vdots & \ddots & \vdots \\ 0 & 0 & \cdots & Z_n \end{pmatrix} \quad \text{and } ZGZ' + R. \quad (11.69)$$

The MODEL statement of PROC MIXED defines X. To remove the intercept, we use the NOINT option. The RANDOM statement defines the random effects in b by creating the matrix Z and the structure G. The SOLUTION option on the RANDOM statement prints the EBLUPs of the individual dogs expressed as deviations about the population estimate. The structure of G is defined using the TYPE= option where G requests that the estimate of G be printed. PROC MIXED does not include an intercept in the RANDOM statement by default. The structure of $G = I_n \otimes D$ and the $\mathrm{cov}(e) = I_n \otimes \sigma^2 I$, so that $\Omega_i = Z_i D Z'_i + \sigma^2 I$ and $R = \sigma^2 I$.

From Output 11.4.1, we see that the 4×4 block-diagonal matrix of D is printed, the first block of G. The estimate of the residual variance $\hat{\sigma}^2$ is found in the Residual row of the "Covariance Parameter Estimates" table; the value is $\hat{\sigma}^2 = .05887$. The asymptotic z-tests show that the variance components in D are significantly different from zero. The random coefficient growth curve model is better than the null model, $\Psi \equiv R = \sigma^2 I$. This is supported by the significant LRT, $\chi^2 = 216.58$, $p < .0001$.

The fixed effects for the model are the intercept (μ), treatment effects $\hat{\alpha}_i = \hat{\mu}_i - \hat{\mu}$ for $i = 1, 2, 3$ and slopes (linear, quadratic, and cubic). PROC MIXED follows McLean, Sanders, and Stroup (1991) to test hypotheses regarding fixed effects using approximate F-tests. In particular, under the null hypotheses $H_0 : L\mu = 0$ is tested using

$$F = \frac{(L\hat{\mu})'(L\widehat{C}L')^-(L\hat{\mu})}{\nu_h} \qquad (11.70)$$

where

$$\widehat{C} = \begin{pmatrix} X'\widehat{R}^{-1}X & X'\widehat{R}^{-1}Z \\ Z'\widehat{R}^{-1}X & Z'\widehat{R}^{-1}Z + \widehat{G}^{-1} \end{pmatrix}^{-1}$$

Output 11.4.1: Random Coefficient Model — Several Population: Variance Components.

Estimated G Matrix						
Row	Effect	dog	Col1	Col2	Col3	Col4
1	Intercept	1	0.2886	0.03629	-0.00753	-0.02382
2	x1	1	0.03629	0.01439	0.000133	-0.01008
3	x2	1	-0.00753	0.000133	0.002010	0.000554
4	x3	1	-0.02382	-0.01008	0.000554	0.01545

Covariance Parameter Estimates						
Cov Parm	Subject	Ratio	Estimate	Standard Error	Z Value	Pr Z
UN(1,1)	dog	4.9027	0.2886	0.08299	3.48	0.0003
UN(2,1)	dog	0.6165	0.03629	0.01505	2.41	0.0159
UN(2,2)	dog	0.2445	0.01439	0.003954	3.64	0.0001
UN(3,1)	dog	-0.1279	-0.00753	0.005251	-1.43	0.1517
UN(3,2)	dog	0.002260	0.000133	0.001131	0.12	0.9063
UN(3,3)	dog	0.03414	0.002010	0.000655	3.07	0.0011
UN(4,1)	dog	-0.4046	-0.02382	0.01686	-1.41	0.1577
UN(4,2)	dog	-0.1713	-0.01008	0.003849	-2.62	0.0088
UN(4,3)	dog	0.009418	0.000554	0.001402	0.40	0.6925
UN(4,4)	dog	0.2625	0.01545	0.006186	2.50	0.0062
Residual		1.0000	0.05887	0.008012	7.35	<.0001

Fit Statistics	
-2 Res Log Likelihood	281.2
AIC (smaller is better)	303.2
AICC (smaller is better)	304.4
BIC (smaller is better)	320.6

Null Model Likelihood Ratio Test		
DF	Chi-Square	Pr > ChiSq
10	216.58	<.0001

with $\nu_h = \text{rank}(L)$, and $\hat{\mu}' = (\hat{\beta}', \hat{b}')$. The approximate t-test and p-values show that the control differs from the treatments (treat) and that the slopes (x1-x3) are significantly different from zero (Output 11.4.2).

Output 11.4.2: Random Coefficient Model — Several Population: Fixed Effects.

Type 3 Tests of Fixed Effects				
Effect	Num DF	Den DF	F Value	Pr > F
treat	3	108	3.21	0.0260
x1	1	35	13.10	0.0009
x2	1	35	12.53	0.0012
x3	1	35	2.10	0.1558

The matrix $\hat{C}$ is an approximate estimate of the covariance matrix of $\hat{\mu}$ which tends to underestimate the true sampling variability (Kass & Steffey, 1989). The test statistic in (11.70) is a Wald-like statistic. In large samples, results using PROC MIXED and the SUR model are equivalent.

For the random coefficient growth curve model, we modeled variation in the dogs by varying the design Z_i across dogs. An alternative specification is to assume $Z_1 = Z_2 = \cdots = Z_n = 0$ and to model R. In particular, assume that we have a repeated measures design where R is unknown or satisfied the structure of compound symmetry. As part of the model, we must estimate R. Thus we have a restricted growth curve model with restriction on Σ. To analyze this design in PROC MIXED, we remove the RANDOM statement and use the REPEATED statement with the TYPE=UN or TYPE=CS options. Comparing AIC for the three models, the constrained model with unstructured Σ is best where $AIC = 259.5$ (Output 11.4.1 and 11.4.3).

Finally by using the GROUP option in REPEATED, we may fit a model that permits heterogeneity of compound symmetry across groups. The AIC statistic is larger for the heterogeneous model (Output 11.4.4). This is also supported by approximate LRT using -2 REML likelihood (-2 Res Log Likelihood). If we subtract the two values $(371.5 - 321.7 = 49.8)$ and compare that result with a chi-square distribution with degree of freedom $8 - 2 = 6$, the number of parameters estimated in Ω minus the number estimated under the hypothesis ω. This result favors the heterogeneous model over the model with compound symmetry. The model with unstructured Σ appears to be optimal. The test of group diff is also significant for the heterogeneous model. This was tested using a contrast since the NOINT option was used in the MODEL statement. In the "Tests of Fixed Effects" table, Treat test the null hypothesis that means are equal.

Output 11.4.3: Random Coefficient Model — Several Population: Model Fit.
Unstructured (TYPE=UN)

Fit Statistics	
-2 Res Log Likelihood	203.5
AIC (smaller is better)	259.5
AICC (smaller is better)	267.0
BIC (smaller is better)	303.8

Null Model Likelihood Ratio Test		
DF	Chi-Square	Pr > ChiSq
27	294.33	<.0001

Compound Symmetry (TYPE=CS)

Fit Statistics	
-2 Res Log Likelihood	371.5
AIC (smaller is better)	375.5
AICC (smaller is better)	375.6
BIC (smaller is better)	378.7

Null Model Likelihood Ratio Test		
DF	Chi-Square	Pr > ChiSq
1	126.28	<.0001

Output 11.4.4: Random Coefficient Model — Several Population: Heterogeneous Model.

Fit Statistics	
-2 Res Log Likelihood	321.7
AIC (smaller is better)	337.7
AICC (smaller is better)	338.4
BIC (smaller is better)	350.4

Null Model Likelihood Ratio Test		
DF	Chi-Square	Pr > ChiSq
7	176.05	<.0001

Type 3 Tests of Fixed Effects				
Effect	Num DF	Den DF	F Value	Pr > F
treat	4	32	611.31	<.0001
x1	1	213	0.61	0.4358
x2	1	213	9.30	0.0026
x3	1	213	0.11	0.7438

Contrasts				
Label	Num DF	Den DF	F Value	Pr > F
group diff	3	32	8.92	0.0002

11.5   MIXED MODEL REPEATED MEASURES

In Chapter 3, we analyzed a split-plot repeated measures design with PROC REG and PROC GLM using an RGLM and the GLM. Both the full-rank model and the overparameterized less-than-full-rank model required the within-subjects covariance matrix Σ to be homogeneous and have circularity structure for exact F-tests. While some authors recommend using adjusted F-tests when the circularity conditions does not hold, we recommend that a multivariate analysis be used as illustrated in Chapter 5.

The split-plot repeated measures design may also be analyzed using PROC MIXED. This approach allows one to jointly test hypotheses regarding fixed effects and to evaluate the structure of the within-subjects covariance matrix by estimating its components. To illustrate, we reanalyze 3_5_2.dat from using PROC MIXED. Recall that these data were used to evaluate memory capacity based on probe-word position. In Program 11_5.sas, we analyze the problem using two structures for the within-subjects covariance matrix: compound symmetry and unstructure. In Section 3.5.3, we found the structure to satisfy the circularity condition.

To analyze the data using PROC MIXED, we organize the data as a three-factor design with three class variables: GROUP, SUBJ, and PROBE. The MODEL statement specifies the linear model and the REPEATED statement models the within-subjects covariance matrix. For this design, subjects are nested within groups. The REPEATED statement defines the block diagonal matrix $\Psi \equiv R$. The two covariance structures are specified using the TYPE= option where CS and UN defines compound symmetry and general unstructure respectively.

Comparing the results for the two models, observe that the F-tests are identical for the between-subjects factor group (Output 11.5.1), and that the results agree with PROC GLM from Section 3.5.3. Between tests do not depend on the structure of Σ; only the within-subjects tests depend on Σ. For this example, the within-subjects test under compound symmetry for PROC GLM and PROC MIXED is identical. This is the case by Theorem 4.1 since we have an equal number of subjects for each group and compound symmetry (McLean & Sanders, 1988). When Σ is unstructured, one may use Satterthwaite's formula to adjust p-values by using the DDFM=SATTERTH option on the MODEL statement as illustrated in Program 11_5.sas. The adjusted p-values for the interaction test is .5047. Without Satterthwaite's adjustment, the p-value is .8479. For large data sets, Satterthwaite's formula significantly increases the execution time.

Finally, comparing the AIC values for each model, we see that the model with compound symmetry appears to fit the data better (670.3 vs. 681.1). These results are consistent with those obtained in Chapter 3. The multivariate analysis in Chapter 5 is very conservative for these data. When analyzing repeated measures data one should consider several models.

Hess and Iyer (2002) macro was used to estimate the confidence intervals of the variance components. The SAS code to perform the analysis is given in Program 11_5ci.sas. The variance component estimates from PROC MIXED is displayed along with the confidence interval computed using the macro (Output 11.5.2). The

Output 11.5.1: Mixed Model Repeated Measures.
Compound Symmetry (TYPE=CS)

Fit Statistics	
-2 Res Log Likelihood	666.3
AIC (smaller is better)	670.3
AICC (smaller is better)	670.5
BIC (smaller is better)	672.3

Null Model Likelihood Ratio Test		
DF	Chi-Square	Pr > ChiSq
1	13.39	0.0003

Type 3 Tests of Fixed Effects				
Effect	Num DF	Den DF	F Value	Pr > F
group	1	18	8.90	0.0080
probe	4	72	14.48	<.0001
group*probe	4	72	0.34	0.8479

Unstructured (TYPE=UN)

Fit Statistics	
-2 Res Log Likelihood	651.1
AIC (smaller is better)	681.1
AICC (smaller is better)	687.6
BIC (smaller is better)	696.0

Null Model Likelihood Ratio Test		
DF	Chi-Square	Pr > ChiSq
14	28.64	0.0117

Type 3 Tests of Fixed Effects				
Effect	Num DF	Den DF	F Value	Pr > F
group	1	18	8.90	0.0080
probe	4	18	15.97	<.0001
group*probe	4	18	0.86	0.5047

compound symmetry estimate in PROC MIXED is equal to subj(group) from the macros. The confidence interval around this variance component is (10.286, 75.354). The residual variance component estimate in PROC MIXED is equal to probe*subj(group). The confidence interval around the residual variance component was (43.059, 83.128). The covariance estimates tests produced by PROC MIXED agree with the macro results. Both variance estimates were significant at $\alpha = .05$. The Residual from the macro output do not apply to this model and therefore has been set to N/A. There are only two error variances in the model.

Output 11.5.2: Mixed Model Repeated Measures: Confidence Intervals.
Compound Symmetry (TYPE=CS)

Covariance Parameter Estimates						
Cov Parm	Subject	Ratio	Estimate	Standard Error	Z Value	Pr Z
CS	subj(group)	0.4838	28.1683	13.4120	2.10	0.0357
Residual		1.0000	58.2217	9.7036	6.00	<.0001

Hess and Iyer's macro

PARAMETER	METHOD	LB	ESTIMATE	UB
subj(group)	TBGJL	10.28613	28.168333	75.354202
probe*subj(group)	Exact	43.059358	58.221667	83.127768
Residual	N/A	-99	1.11E-12	-99

11.6 MIXED MODEL REPEATED MEASURES WITH CHANGING COVARIATES

Mixed model analysis of repeated measures data allows the researcher to build alternatives models for the within-subjects covariance matrix. To evaluate model fit, we use the AIC, $AICC$, and BIC criteria. The model with a smallest value is usually the best model. To illustrate, we now analyze with PROC MIXED the Winer data with time-varying covariates, previously analyzed using the SUR model in Chapter 8. The SAS code to perform the analysis is given in Program 11_6.sas.

The analysis is performed for three models: (1) variance components, (2) unstructured covariance, and (3) compound symmetry. For the variance components model, the RANDOM and REPEATED statements generate spherical structure for each subject. The components of variance are $\sigma^2_{\text{subjects}}$ and σ^2. To generate compound symmetry, there are no random effects in the model; the REPEATED statement generates a covariance matrix, $\text{cov}(y) = \Psi \equiv R$, with compound symmetry. This is not the same as the variance components model where the $\text{cov}(y) = ZDZ' + \Psi$ and $\Psi = \sigma^2 I$. Finally, we allow the $\text{cov}(y) = \Omega = ZDZ' + \Psi$ for arbitrary Ψ, and unstructured for Ω.

Comparing the AIC statistics, observe that the fit for the variance components model and compound symmetry model are equal, both better than the model with unstructured (Output 11.6.1-11.6.3). Also, the null hypothesis that $\Psi = \sigma^2 I$ is rejected. It is more likely that Ω has compound symmetry structure, since $ZDZ' + \sigma^2 I$ may not have a simple structure.

Selecting the ML method, $\hat{\sigma}^2_{\text{subjects}} = 2.2985$ and $\hat{\sigma} = .3331$ (the corresponding REML estimates are 3.7406 and .5477 respectively). The ANCOVA table for the fixed effect approximate F-tests using both REML and ML method is shown in Output 11.6.1. The results are not identical since $\hat{C}$ in (11.70) is not equal for the two estimation procedures.

Using OLS and the ANCOVA model, we can obtain the results given in Winer (1971, p. 801) using PROC GLM as illustrated in Output 11.6.4. The results are similar to Output 11.6.1. They would be identical for an orthogonal design by Theorem 4.1.

In Section 8.11, we used the Winer data to evaluate the difference in intercepts using the SUR model to obtain tests. Because F in (11.70) is a Wald-like statistic, SUR and PROC MIXED give equivalent results in large samples. To compare the two procedures, we use PROC MIXED to analyze the Patel data given in Section 8.10, with unstructured covariance matrix. The SAS statements are included in Program 11_6.sas.

The ML estimates of B (treat*period estimates) and Ω (covariance parameter estimates) using PROC MIXED are displayed in Output 11.6.5, which agree with Patel's result using a different algorithm. The treatment effect $\hat{\delta} = -.232$ is in close agreement with the value $-.257$ obtained by Patel and the SUR model. The p-value for the approximate F-tests for treatment and interaction are .3927 and .0228, respectively, using the SUR model. These p-values are .3861 and .0170, respectively, using the PROC MIXED (Output 11.6.6). For very small samples, both procedures should be avoided.

The analysis of repeated measures data is complicated. If one does not impose any structure on the covariance matrix, a multivariate analysis is used. If one wants to model the covariance matrix and the mixed effects, a mixed model analysis is used. For an excellent discussion of the problems associated with the analysis of repeated measurement data using PROC MIXED and PROC GLM, see Littell et al. (1996).

11.7   APPLICATION: TWO-LEVEL HIERARCHICAL LINEAR MODELS

The hierarchical linear model (HLM) examples we have considered so far have been in the context of repeated measurement designs with subjects random and nested within fixed factors. We now turn out attention to the analysis of two-level nested experimental designs (e.g., classrooms nested within schools, schools nested within districts, departments nested within firms) where the nested factor is always random. To illustrate these models, we use PROC MIXED to obtain results discussed by Raudenbush (1993). The SAS code to perform the analysis of several data sets is provided in Program 11_7.sas.

To begin, we consider the HLM representation for the one-way random effects ANOVA model. For such a model, the observation of subject i in group j is repre-

Output 11.6.1: Mixed Model Repeated Measures with Changing Covariates: Variance Components.

ML — Variance Components (TYPE=VC)

Covariance Parameter Estimates						
Cov Parm	Subject	Ratio	Estimate	Standard Error	Z Value	Pr Z
person	person	6.9000	2.2985	1.1647	1.97	0.0242
Residual	person	1.0000	0.3331	0.1570	2.12	0.0169

Fit Statistics	
-2 Log Likelihood	55.5
AIC (smaller is better)	73.5
AICC (smaller is better)	96.0
BIC (smaller is better)	75.3

Type 3 Tests of Fixed Effects				
Effect	Num DF	Den DF	F Value	Pr > F
g	2	5	5.63	0.0525
b	1	5	106.89	0.0001
g*b	2	5	4.10	0.0883
x	1	5	56.93	0.0006

REML — Variance Components (TYPE=VC)

Covariance Parameter Estimates						
Cov Parm	Subject	Ratio	Estimate	Standard Error	Z Value	Pr Z
person	person	6.8294	3.7406	2.3631	1.58	0.0567
Residual	person	1.0000	0.5477	0.3241	1.69	0.0455

Fit Statistics	
-2 Res Log Likelihood	50.6
AIC (smaller is better)	54.6
AICC (smaller is better)	56.1
BIC (smaller is better)	55.0

Type 3 Tests of Fixed Effects				
Effect	Num DF	Den DF	F Value	Pr > F
g	2	5	3.46	0.1142
b	1	5	65.05	0.0005
g*b	2	5	2.50	0.1771
x	1	5	34.78	0.0020

Output 11.6.2: Mixed Model Repeated Measures with Changing Covariates: Compound Symmetry.

Covariance Parameter Estimates						
Cov Parm	Subject	Ratio	Estimate	Standard Error	Z Value	Pr Z
CS	person	6.9000	2.2985	1.1647	1.97	0.0484
Residual		1.0000	0.3331	0.1570	2.12	0.0169

Fit Statistics	
-2 Log Likelihood	55.5
AIC (smaller is better)	73.5
AICC (smaller is better)	96.0
BIC (smaller is better)	75.3

Null Model Likelihood Ratio Test		
DF	Chi-Square	Pr > ChiSq
1	12.95	0.0003

Type 3 Tests of Fixed Effects				
Effect	Num DF	Den DF	F Value	Pr > F
g*b	6	2	21.29	0.0455
x	1	2	56.93	0.0171

Contrasts				
Label	Num DF	Den DF	F Value	Pr > F
G	2	2	5.63	0.1509
B	1	2	106.89	0.0092
BG	2	2	4.10	0.1961

Output 11.6.3: Mixed Model Repeated Measures with Changing Covariates: Unstructured.

Covariance Parameter Estimates					
Cov Parm	Subject	Estimate	Standard Error	Z Value	Pr Z
UN(1,1)	person	2.2477	1.0609	2.12	0.0171
UN(2,1)	person	2.3078	1.1642	1.98	0.0475
UN(2,2)	person	3.0366	1.4483	2.10	0.0180

Fit Statistics	
-2 Log Likelihood	54.7
AIC (smaller is better)	74.7
AICC (smaller is better)	106.2
BIC (smaller is better)	76.7

Null Model Likelihood Ratio Test		
DF	Chi-Square	Pr > ChiSq
2	13.77	0.0010

Type 3 Tests of Fixed Effects				
Effect	Num DF	Den DF	F Value	Pr > F
g	2	6	5.72	0.0407
b	1	6	107.19	<.0001
g*b	2	6	4.35	0.0680
x	1	6	48.42	0.0004

Output 11.6.4: Mixed Model Repeated Measures with Changing Covariates: PROC GLM.

Source	DF	Sum of Squares	Mean Square	F Value	Pr > F
Model	12	371.5019841	30.9584987	51.63	0.0002
Error	5	2.9980159	0.5996032		
Corrected Total	17	374.5000000			

R-Square	Coeff Var	Root MSE	y Mean
0.991995	5.881067	0.774340	13.16667

Source	DF	Type I SS	Mean Square	F Value	Pr > F
g	2	100.0000000	50.0000000	83.39	0.0001
person(g)	6	177.0000000	29.5000000	49.20	0.0003
b	1	68.0555556	68.0555556	113.50	0.0001
g*b	2	16.4444444	8.2222222	13.71	0.0093
x	1	10.0019841	10.0019841	16.68	0.0095

Source	DF	Type III SS	Mean Square	F Value	Pr > F
g	2	38.08303436	19.04151718	31.76	0.0014
person(g)	6	44.37055236	7.39509206	12.33	0.0072
b	1	31.54702917	31.54702917	52.61	0.0008
g*b	2	2.33928571	1.16964286	1.95	0.2365
x	1	10.00198413	10.00198413	16.68	0.0095

Tests of Hypotheses Using the Type III MS for person(g) as an Error Term					
Source	DF	Type III SS	Mean Square	F Value	Pr > F
g	2	38.08303436	19.04151718	2.57	0.1558

Output 11.6.5: Mixed Model Repeated Measures with Changing Covariates: Patel Data — Model Fit.

Covariance Parameter Estimates					
Cov Parm	Subject	Estimate	Standard Error	Z Value	Pr Z
UN(1,1)	person	0.1467	0.05035	2.91	0.0018
UN(2,1)	person	0.02934	0.02229	1.32	0.1880
UN(2,2)	person	0.05046	0.01736	2.91	0.0018

Fit Statistics	
-2 Log Likelihood	11.0
AIC (smaller is better)	29.0
AICC (smaller is better)	36.5
BIC (smaller is better)	36.5

Null Model Likelihood Ratio Test		
DF	Chi-Square	Pr > ChiSq
2	6.68	0.0354

Output 11.6.6: Mixed Model Repeated Measures with Changing Covariates: Patel Data — Fixed Effects.

Solution for Fixed Effects							
Effect	treat	period	Estimate	Standard Error	DF	t Value	Pr > |t|
treat*period	1	1	0.4419	0.2474	17	1.79	0.0919
treat*period	1	2	0.09413	0.1564	17	0.60	0.5553
treat*period	2	1	0.8148	0.3072	17	2.65	0.0168
treat*period	2	2	-0.04660	0.1841	17	-0.25	0.8032
z1			0.8453	0.1548	17	5.46	<.0001
z2			1.1150	0.09416	17	11.84	<.0001

Type 3 Tests of Fixed Effects				
Effect	Num DF	Den DF	F Value	Pr > F
treat*period	4	17	2.69	0.0666
z1	1	17	29.83	<.0001
z2	1	17	140.22	<.0001

Estimates					
Label	Estimate	Standard Error	DF	t Value	Pr > |t|
treat(delta)	-0.2322	0.2610	17	-0.89	0.3861

Contrasts				
Label	Num DF	Den DF	F Value	Pr > F
treat	1	17	0.79	0.3861
inter	1	17	7.00	0.0170

sented

$$\begin{pmatrix} y_{1j} \\ y_{2j} \\ \vdots \\ y_{n_j j} \end{pmatrix} = \begin{pmatrix} 1 \\ 1 \\ \vdots \\ 1 \end{pmatrix} \beta_j + \begin{pmatrix} e_{1j} \\ e_{2j} \\ \vdots \\ e_{n_j j} \end{pmatrix} \tag{11.71}$$

and

$$\beta_j = \mu + b_j$$

where there are n_j subjects per group, $j = 1, \ldots, J$ groups and $b_j \sim N(0, \sigma_b^2)$. The fixed j^{th} group mean β_j at level 2 is represented as an overall mean μ plus the random effect for the jth group. Substituting β_j into the vector equation (11.71), we obtain

$$y_j = 1_{n_j}\mu + 1_{n_j} b_j + e_j \qquad j = 1, 2, \ldots, J$$

a mixed linear model with $X_j = Z_j = 1_{n_j}$. Each element y_{ij} of y_j may be represented as a random effects ANOVA model

$$y_{ij} = \mu + b_j + e_{ij}$$

where $e_{ij} \sim N(0, \sigma^2)$, $b_j \sim N(0, \sigma_b^2)$, and σ^2 and σ_b^2 are the variance components to be estimated.

To illustrate the analysis of this design using PROC MIXED, we use data for 50 children randomly assigned to 5 training methods developed to discriminate among blocks (Raudenbush, 1993, p. 466). The responses are the number of blocks correctly identified. Program 11_7.sas contains the SAS code to analyze this data set.

The G matrix provides the REML estimate of σ_b^2, $\hat{\sigma}_b^2 = .4452$ (Output 11.7.1). The residual variance estimate is $\hat{\sigma}^2 = 1.0178$. The ANOVA table for the fixed effects parameter for the null hypothesis $H_0 : \mu = 0$ is also provided ($p = .0033$). The intraclass correlation ρ which represents the proportion of the variance in the dependent variable Y (discrimination) accounted for by the second level (method) variable is $\hat{\rho} = \frac{.4452}{.4452+1.0178} = .3043$. Hence, approximately 30% of block discrimination is due to method.

For our second example, the two-factor nested design analyzed by (Raudenbush, 1993, p. 475) is reanalyzed. Four classrooms are nested within two methods where the classroom factor is random and methods are fixed effects. As with the ANOVA model, the first level of the model is identical to (11.71) where y_{ij} is the observation of the th subject for the jth classroom. The fixed parameter β_j is the mean of the nested factor (class j). For level 2, we let

$$\beta_j = \mu + \gamma_j + b_j \qquad b_j \sim N(0, \sigma_b^2)$$

so that the jth classroom mean equals an overall mean plus the effect due to the jth classroom and a random error term. Using (11.56), SAS forms a matrix Z_j

Output 11.7.1: Two-Level Hierarchical Linear Model: Example 1.

Covariance Parameter Estimates						
Cov Parm	Subject	Ratio	Estimate	Standard Error	Z Value	Pr Z
Intercept	treat	0.4374	0.4452	0.3874	1.15	0.1252
Residual		1.0000	1.0178	0.2146	4.74	<.0001

Fit Statistics	
-2 Res Log Likelihood	150.6
AIC (smaller is better)	154.6
AICC (smaller is better)	154.8
BIC (smaller is better)	153.8

Solution for Fixed Effects					
Effect	Estimate	Standard Error	DF	t Value	Pr > |t|
Intercept	2.0800	0.3308	4	6.29	0.0033

orthogonal to $X_j = I_{n_j}$, a matrix of contrasts. The SAS code to analyze these data is given in Program 11_7.sas.

The variance component REML estimates are $\hat{\sigma}_b^2 = 4.1615$ and $\hat{\sigma}_e^2 = .7708$ (Output 11.7.2). The approximate z-tests for variance components are significant. Furthermore, the overall mean estimate $\hat{\mu} = 5.3750$ and the contrast $\hat{\psi} = \hat{\gamma}_1 - \hat{\gamma}_2 = -3.7500$ are also significant.

For our third example, we consider a two-factor crossed design (with replications within cells) (Raudenbush, 1993, p. 482). Forty subjects were assigned to 10 tutors with the subjects then randomized to two instructional tasks (no practice, practice). Writing this model as a HLM, we let

$$y_{ij} = \beta_j + X_j \gamma_j + e_{ij} \qquad i = 1, \ldots, n_j; j = 1, \ldots, J$$

where X_j is a contrast vector comparing subjects in tutor j at $K - 1$ fixed effect levels. For example, with $K = 2$ levels the linear model is

$$y_j = X_j \beta_j + e_j$$

$$\begin{pmatrix} y_{1j} \\ y_{2j} \\ y_{3j} \\ y_{4j} \end{pmatrix} = \begin{pmatrix} 1 & 1 \\ 1 & 1 \\ 1 & -1 \\ 1 & -1 \end{pmatrix} \begin{pmatrix} \beta_{0j} \\ \beta_{1j} \end{pmatrix} + \begin{pmatrix} e_{1j} \\ e_{2j} \\ e_{3j} \\ e_{4j} \end{pmatrix}$$

where β_{0j} is the mean for the jth tutor and β_{1j} is the practice contrast. For level two

Output 11.7.2: Two-Level Hierarchical Linear Model: Example 2.

Covariance Parameter Estimates						
Cov Parm	Subject	Ratio	Estimate	Standard Error	Z Value	Pr Z
classes	method	5.3986	4.1615	2.5145	1.65	0.0490
Residual		1.0000	0.7708	0.2225	3.46	0.0003

Estimates					
Label	Estimate	Standard Error	DF	t Value	Pr > |t|
overall mean	5.3750	0.7377	6	7.29	0.0003
contrast diff method	-3.7500	1.4755	6	-2.54	0.0440

of the design, we assume a random coefficient model

$$\beta_{0j} = \mu + b_{0j}$$
$$\beta_{1j} = \mu_c + b_{1j}$$

where μ is the overall mean, μ_c is a treatment contrast (slope), and b_{0j} and b_{1j} are randomly correlated errors. Thus, the mixed model for the design is

$$y_j = X_j\gamma + Z_jb_j + e_j \tag{11.72}$$

$$= \begin{pmatrix} 1 & 1 \\ 1 & 1 \\ 1 & -1 \\ 1 & -1 \end{pmatrix} \begin{pmatrix} \mu \\ \mu_c \end{pmatrix} + \begin{pmatrix} 1 & 1 \\ 1 & 1 \\ 1 & -1 \\ 1 & -1 \end{pmatrix} b_j + e_j$$

where the $\text{cov}(y_j) = Z_j D Z_j' + \sigma^2 I$. The SAS code to perform this analysis is given in Program 11_7.sas. To analyze (11.72) using PROC MIXED, the MODEL statement again specifies the mean model effects. The intercept is included by default and the variable PRACTICE models μ_c. Because we exclude PRACTICE from the CLASS statement, it is considered as a continuous variable. The RANDOM statement models the covariance matrix and the matrix Z. The INTERCEPT option is required since RANDOM does not include the intercept by default. The TYPE=UN option specifies that covariance matrix.

The variance of b_{0j} is equal to $\hat{\sigma}_{b_0}^2 = 111.65$ (Output 11.7.3). The variance of b_{1j} is equal to $\hat{\sigma}_{b_1}^2 = 44.94$. The covariance between b_{0j} and b_{1j} was not significant, $\hat{\sigma}_{b_0b_1} = -12.39$, $p = .6466$. The overall mean on the DV was $\hat{\mu} = 106.25$. There was a significant difference on DV between the two instructional tasks, $\hat{\mu}_c = -39.95$. The practice factor was coded 1 and -1. This is the same as group-centering in HLM.

The last HLM example considered is a standard split-plot repeated measures design with constant, linear, quadratic, and cubic polynomials fit to the within-subjects

Output 11.7.3: Two-Level Hierarchical Linear Model: Example 3.

Covariance Parameter Estimates						
Cov Parm	Subject	Ratio	Estimate	Standard Error	Z Value	Pr Z
UN(1,1)	tutor	3.2887	111.65	56.6976	1.97	0.0245
UN(2,1)	tutor	-0.3649	-12.3889	27.0231	-0.46	0.6466
UN(2,2)	tutor	1.3237	44.9403	25.3287	1.77	0.0380
Residual		1.0000	33.9500	10.7359	3.16	0.0008

Solution for Fixed Effects					
Effect	Estimate	Standard Error	DF	t Value	Pr > |t|
Intercept	106.25	3.4661	9	30.65	<.0001
practice	-39.9500	2.3114	9	-17.28	<.0001

Type 3 Tests of Fixed Effects				
Effect	Num DF	Den DF	F Value	Pr > F
practice	1	9	298.72	<.0001

effect. The data are taken from Kirk (1982, p. 244) where eight subjects (blocks) are observed over four treatment conditions. The SAS code to analyze these data are found in Program 11_7.sas.

The results indicate the estimate of variance components for error and block are 1.3571 and .1071, respectively ($\hat{\sigma}^2 = 1.3571$ and $\hat{\sigma}_b^2 = .1071$; Output 11.7.4). From the test of fixed effects, there was a significant linear trend, $t(21) = 11.67$, $p < .0001$. The quadratic trend was also significant, but not cubic trend.

Output 11.7.4: Two-Level Hierarchical Linear Model: Example 4.

Covariance Parameter Estimates						
Cov Parm	Subject	Ratio	Estimate	Standard Error	Z Value	Pr Z
CS	person	0.07895	0.1071	0.2606	0.41	0.6810
Residual		1.0000	1.3571	0.4188	3.24	0.0006

Type 3 Tests of Fixed Effects				
Effect	Num DF	Den DF	F Value	Pr > F
x1	1	21	136.24	<.0001
x2	1	21	5.89	0.0243
x3	1	21	1.18	0.2899

CHAPTER 12

Incomplete Repeated Measurement Data

12.1 INTRODUCTION

In our discussion of the general linear model (GLM), we have assumed that the data matrix $Y_{n \times p}$ is complete, contains no missing values. However, in many studies involving the analysis of repeated measurement designs and longitudinal growth data, the data matrix is not complete. When data are incomplete, estimating model parameters and testing linear hypotheses becomes more difficult since bias may be introduced into the estimation procedure, and test statistics with known exact small sample distributions are not easily obtained. A review of procedures for dealing with regression models where data are missing at random or by design is provided by Little (1992).

If the number of missing data points is very small, the simplest method of analysis is to discard the incomplete response vectors and employ only the complete cases in the analysis; this is called the listwise deletion method. The primary concern with this procedure is whether bias is introduced into the sample estimates of population parameters (Timm, 1970; Little, 1992). Furthermore, data are discarded, a practice many would find undesirable. There are four general approaches to the missing data problem for growth or repeated-measures studies:

1. imputation techniques which involve estimating the missing values and then adjusting the estimates for bias and the degrees of freedom of the sampling distribution for the test statistic for the imputed values.

2. maximum likelihood methods which estimate model parameters using maximum likelihood (ML) or restricted maximum likelihood (REML) estimates from the incomplete data followed by LR tests, large sample chi-square tests, or goodness-of-fit tests.

3. feasible generalized least squares (FGLS) approaches which involve finding a consistent estimate of the covariance matrix Σ from the incomplete data and estimating model parameters using the GLSE with hypothesis testing performed by an asymptotic Wald-type statistic.

455

4. Bayesian, parametric empirical Bayes (PEB) and complex simulation techniques such as the Gibbs sampling and multiple imputation methods.

In general, maximum likelihood procedures under multivariate normality use three general algorithms: Newton-Raphson, Fisher scoring, and the EM algorithm. FGLS procedures and Bayesian methods are superior to single-value imputation procedures (Little, 1992).

When missing data were encountered in the early years of data analysis, the approach was to estimate the missing data and, after inserting these estimates into the data matrix Y, to use this modified matrix for the analysis (Timm, 1970). Although this approach is not incorrect, iterative procedures that estimate the model parameters from the complete data, then use the estimated parameters to estimate the missing data, estimate the model parameter again, and repeat the process until some convergence criterion is met work much better. This iterative approach was first proposed by Orchard and Woodbury (1972); the process is called the missing information principle. Dempster, Laird, and Rubin (1977) expanded upon the principle to include a larger class of problems and called their approach the EM algorithm. Under general regularity conditions, the approach leads to parameter estimates that converge to the ML estimates.

Kleinbaum (1973) proposed a large sample iterative method employing a generalized least squares procedure to solve missing data problems in growth curve studies. He showed that the resulting estimates were best asymptotically normal (BAN) and developed a Wald statistics to test hypotheses. The difference between this approach and the EM algorithm approach is that, under multivariate normality, the EM algorithm procedure is an exact method whereas Kleinbaum (1973) is asymptotic. Liski (1985) developed an EM procedure for the generalized multivariate analysis of variance (GMANOVA) model. In large samples, both are essentially equivalent. In this chapter, we show how to analyze repeated measures design with missing data using PROC MIXED which uses a Newton-Raphson algorithm (Wolfinger et al., 1994). We also illustrate Kleinbaum's procedure using PROC IML code.

12.2 MISSING MECHANISMS

If the missingness is independent of both missing values and the observed values of other variables, the data are said to be missing completely at random (MCAR). However, if the missingness is only independent of the missing values and not on the observed values of the variable then the data is said to be missing at random (MAR; Little & Rubin, 1987). When data are MCAR or MAR, likelihood based methods can yield acceptable statistical results (e.g., Enders & Bandalos, 2001; Gold & Bentler, 2000; Yuan & Bentler, 2000). However, when data are neither MCAR nor MAR, missing not at random (MNAR), they cause serious problems no matter how few of them there are because they affect the generalizability of results. There are three major problems created by incomplete data: (a) if the missingness is neither MCAR or MAR, then any analysis that ignores the nonrandomness may be biased, (b) the existence of incomplete data usually implies a loss of information, so that estimates will be less efficient than planned, and (c) standard statistical methods are designed for complete data; thus any incomplete data analysis is usually more complicated (e.g., Bernaards & Sijtsma, 2000; Little & Schenker, 1995). There are currently two tests of MCAR based on ML and GLS (Little, 1988; K. H. Kim & Bentler, 2002).

12.3   FGLS PROCEDURE

One of the first solutions to the analysis of the growth curve model with missing data was proposed by Kleinbaum (1973). Kleinbaum's approach involved the use of an FGLS estimate for the model parameters. The FGLS estimate depends on obtaining a consistent estimate of Σ which is positive definite. With this approach one partitions the GMANOVA model (7.23) data matrix $Y_{n \times p}$ into m mutually exclusive subsets according to the pattern of missing data where the jth subset is the data matrix $Y_{j(n_j \times p_j)}$ with $p_j \leq p$. For each data pattern Y_j, one formulates the model

$$\mathcal{E}(Y_j) = X_j B P A_j$$
$$\text{cov}(Y_j) = I_{n_j} \otimes A'_{j'} \Sigma A_j \qquad\qquad j = 1, 2, \ldots, m \qquad (12.1)$$

where $X_{j(n_j \times k)}$, $B_{k \times q}$, $P \equiv X_{2(q \times p)}$ and Σ are as defined in the GMANOVA model (7.23) and $A_{j(p \times p_j)}$ is an incidence matrix of 0's and 1's such that $A_j = (a_{rs}^{(j)})$ where $a_{rs}^{(j)} = 1$ if the rth ordered observation over time in the jth data set corresponds to the sth ordered observation over time for the complete data set, and $a_{rs}^{(j)} = 1$ otherwise. Alternatively, using the vec($\cdot$) operator, we may let $y_{N \times 1} = \begin{pmatrix} y'_1 & y'_2 & \cdots & y'_m \end{pmatrix}' = \begin{bmatrix} \text{vec}'(Y_1) & \text{vec}'(Y_2) & \cdots & \text{vec}'(Y_m) \end{bmatrix}'$ then

$$\mathcal{E}(y) = D\beta$$
$$\text{cov}(y) = \Omega$$

$$D_{N \times kq} = \begin{pmatrix} A'_1 P' \otimes X_1 \\ A'_2 P' \otimes X_2 \\ \vdots \\ A'_m P' \otimes X_m \end{pmatrix}$$

$$\Omega_{N \times N} = \text{diag} \begin{pmatrix} A'_1 \Sigma A_1 \otimes I_{n_1} & A'_2 \Sigma A_2 \otimes I_{n_2} & \cdots & A'_m \Sigma A_m \otimes I_{n_m} \end{pmatrix} \quad (12.2)$$

where $N = \sum_{i=1}^{m} n_i p_i$ and $\beta = \text{vec}(B)$.

To obtain the incidence matrix A_j, we take the identity matrix of order p_j, and insert a row vector of zeros between rows of the identity matrix whenever a missing value occurs. This results in a $p \times p_j$ matrix such that whenever the kth ordered observation is missing, the kth row of A_j will be all zeros.

Some examples are:

1. No missing data, $A_j = I_p$

2. Only the last observation is missing, $A_j = \begin{pmatrix} I_{p-1} \\ 0' \end{pmatrix}$

3. Only the ith and j^{th} observation is missing $(i < j)$

$$A_j = \begin{pmatrix} I_{(i-1)\times(i-1)} & 0_{(i-1)\times(p-i-1)} & \\ 0_{1\times(p-2)} & & \\ 0_{(j-i-1)\times(i-1)} & I_{(j-i-1)\times(j-i-1)} & 0_{(j-i-1)\times(p-j)} \\ 0_{1\times(p-2)} & & \\ & 0_{(p-j)\times(j-2)} & I_{(p-j)\times(p-j)} \end{pmatrix}.$$

If $\widehat{\Sigma}$ is a consistent estimate of Σ, then the FGLS of β is

$$\check{\beta} = (D'\widehat{\Omega}^{-1}D)^{-1}D'\widehat{\Omega}^{-1}y$$

$$= \left[\sum_{j=1}^{m} PA_j(A'_j\widehat{\Sigma}A_j)^{-1}A'_jP' \otimes X'_jX_j \right]^{-1} \sum_{j=1}^{m} \mathrm{vec}\left[X'_jY_j(A'_j\widehat{\Sigma}A_j)^{-1}A_jP' \right]$$

$$= \left[\sum_{j=1}^{m} PA_j(A'_j\widehat{\Sigma}A_j)^{-1}A'_jP' \otimes X'_jX_j \right]^{-1} \sum_{j=1}^{m} \left[PA_j(A'_j\widehat{\Sigma}A_j)^{-1} \otimes X'_j \right] y_j.$$

$$(12.3)$$

Letting $C_* \equiv A' \otimes C$ for parametric functions of the form $C_*\beta$ where $\beta = \mathrm{vec}(B)$, Kleinbaum showed that the asymptotic covariance matrix of $C_*\beta$ is computed to be

$$C_*(D'\widehat{\Omega}^{-1}D)^{-1}C'_* = C_* \left[\sum_{j=1}^{m} PA_j(A'_j\widehat{\Sigma}A_j)^{-1}A'_jP' \otimes X'_jX_j \right]^{-1} C'_* \quad (12.4)$$

and that the Wald statistic for testing $H : C_*\beta = \gamma$ is given by

$$X^2 = (C_*\check{\beta} - \gamma)'(C_*(D'\widehat{\Omega}^{-1}D)^{-1}C'_*)^{-1}(C_*\check{\beta} - \gamma) \quad (12.5)$$

and converges to a chi-square distribution with $\nu = \mathrm{rank}(C_*)$ degrees of freedom under the null hypothesis.

Kleinbaum (1973) suggested estimating Σ with the unbiased and consistent estimate $\widehat{\Sigma} = (\widehat{\Sigma}_{rs})$ where $\widehat{\Sigma}_{rs}$ is estimated from the data matrix Y using variables r and s where both pairs are observed. Then

$$\widehat{\Sigma}_{rs} = \left(\frac{1}{N_{rs} - k} \right) y'_{rs}(I_{N_{rs}} - D_{rs}(D'_{rs}D_{rs})^{-1}D'_{rs})y_{sr} \quad r,s = 1,2,\ldots,p$$

$$(12.6)$$

where N_{rs} is the number of observations for which both variables r and s are observed; $y_{rs(N_{rs}\times1)}$ is the observation vector for variable r corresponding to those in which both variables r and s are observed; and $D_{rs(N_{rs}\times k)}$ is the across variables design matrix consisting of a row of X_j matrices corresponding to y_{rs}.

Unfortunately, the sample multivariate analysis of variance (MANOVA) estimator for Σ is not necessarily nonsingular (positive definite) in small samples. While one may replace ordinary inverses with generalized inverses $(AA^- A = A)$ to obtain estimates of model parameters, the Wald statistic may lead to spurious results or even be negative. To address this problem, Woolson and Clarke (1987) recommend writing the incomplete MANOVA model as a seemingly unrelated regression (SUR) model and using an ordinary least squares (OLS) to estimate β. With β estimated, an estimate of Σ is obtained from the vector of residuals, $r_j = y_j - X_j\hat{\beta}_j$, which is adjusted so as to ensure a positive definite estimate of Σ. More recently, Mensah, Elswick, and Chinchilli (1993) conducted a Monte Carlo simulation experiment to evaluate estimators of Σ using the GMANOVA model instead of the MANOVA model.

Using only complete data vectors Y_j $(j = 1, 2, \ldots, p)$ and corresponding design matrices $Z_{j(n_j \times k)}$, Mensah et al. (1993) constructed an unweighted and fully-weighted GMANOVA estimator of B:

$$\hat{B}_{UW} = \left[(Z_1'Z_1)^{-1}Z_1Y_1 \quad \ldots \quad (Z_p'Z_p)^{-1}Z_p'Y_p\right] P'(PP')^{-1}$$

$$\hat{B}_{FW} = \left[(Z_1'Z_1)^{-1}Z_1Y_1 \quad \ldots \quad (Z_p'Z_p)^{-1}Z_p'Y_p\right] \hat{\Sigma}_{KL}P'(P\hat{\Sigma}_{KL}^{-1}P)^{-1} \quad (12.7)$$

where $\hat{\Sigma}_{KL} = (\hat{\Sigma}_{rs})$ is Kleinbaum's positive definite estimator of Σ. Following Woolson and Clarke (1987), unweighted and fully-weighted estimators of Σ were constructed utilizing residual matrices for observation vectors of complete pairs

$$\hat{\sigma}_{UW,rs} = \left(\frac{1}{N_{rs} - k}\right)(y_{rs} - D_{rs}\breve{\beta}_{UW})'(y_{sr} - D_{rs}\breve{\beta}_{UW})$$

$$\hat{\sigma}_{FW,rs} = \left(\frac{1}{N_{rs} - k}\right)(y_{rs} - D_{rs}\breve{\beta}_{FW})'(y_{sr} - D_{rs}\breve{\beta}_{FW}) \quad (12.8)$$

where $\hat{\Sigma}_{UW} = (\hat{\Sigma}_{UW,rs})$, $\hat{\Sigma}_{FW} = (\hat{\Sigma}_{FW,rs})$ and $r, s = 1, 2, \ldots, p$. Recognizing that the ML estimate of Σ for the GMANOVA model has the form

$$\hat{\Sigma} = \hat{\Sigma}_{KL} + W'\left(\frac{Y'Y}{n} - \hat{\Sigma}_{KL}\right)W \quad (12.9)$$

where $W = I_p - \hat{\Sigma}_{KW}^{-1}P'(P\hat{\Sigma}_{KL}^{-1}P')^{-1}P$ as shown in (8.6) for the complete data matrix $Y_{n \times p}$, they suggested two other consistent missing data estimators of Σ:

$$\hat{\Sigma}_{FW}^* = \hat{\Sigma}_{KL} + W_{FW}'(T_{KL} - \hat{\Sigma}_{KL})W_{FW}$$

$$\hat{\Sigma}_{UW}^* = \hat{\Sigma}_{KL} + W_{UW}'(T_{KL} - \hat{\Sigma}_{KL})W_{UW} \quad (12.10)$$

where

$$W_{FW} = I_p - \hat{\Sigma}_{KL}P'(P\hat{\Sigma}_{KL}^{-1}P')^{-1}P$$

$$W_{UW} = I_p - P'(P'P)P$$

$$T_{KL} = \frac{y_{rs}'y_{rs}}{N_{rs}} \qquad r, s = 1, 2, \ldots, p. \quad (12.11)$$

While Mensah et al. (1993) only investigated one covariance matrix and two missing data patterns (10% and 40%), they found that the estimators $\widehat{\Sigma}_{KL}$ and $\widehat{\Sigma}^*_{UW}$ were better than $\widehat{\Sigma}_{FW}$, $\widehat{\Sigma}^*_{FW}$, and $\widehat{\Sigma}_{UW}$, although both tended to underestimate Σ. The poorest estimator was Σ_{UW}, with $\widehat{\Sigma}_{FW}$ and $\widehat{\Sigma}^*_{FW}$ overestimating Σ. Tests of hypotheses were also compared, with $\widehat{\Sigma}_{KL}$ and $\widehat{\Sigma}^*_{UW}$ yielding consistent results with some gains for the unweighted estimator over Kleinbaum's estimate. With 40% or more data missing, asymptotic tests using Wald statistics were found to be very unreliable and should not be used.

12.4   ML PROCEDURE

The FGLS methods fit a model using the incomplete data and between subject variation for j patterns of missing data for the data matrix $Y_{n \times p}$. An alternative approach is to develop a model that represents within-subject variation and to use maximum likelihood methods to estimate Σ_i, a submatrix of Σ; this is the approach taken by Jennrich and Schluchter (1986); Wolfinger et al. (1994). They considered the n_i responses of each subject $y_{n_i \times 1}$ where the model for the vector of responses is

$$y_i = X_i\beta + e_i \qquad\qquad i = 1, 2, \ldots, n \qquad (12.12)$$

where $X_{i(n_i \times p)}$ is a known matrix, $\beta_{p \times 1}$ is a vector of unknown parameters, and $e_i \sim IN(0, \Sigma_i)$. In addition, the structure of Σ_i is assumed to be known functions of q unknown covariance parameters of the general form

$$\Sigma_i(\theta) = Z_i D Z'_i + \Psi_i \qquad\qquad (12.13)$$

contained in a vector θ.

To estimate the components θ_i of θ and β using the method of maximum likelihood, recall that the normal density for y_i is

$$f(y_i) = (2\pi)^{-n_i/2} |\Sigma_i|^{-1/2} \exp\left[-\frac{1}{2}(y_i - X_i\beta)'\Sigma_i^{-1}(y_i - X_i\beta) \right] \qquad (12.14)$$

so that the log-likelihood function is

$$L^*_{ML} = -\frac{1}{2}\sum_{i=1}^n n_i \log(2\pi) - \frac{1}{2}\sum_{i=1}^n \log|\Sigma_i| + \sum_{i=1}^n (y_i - X_i\beta)'\Sigma_i^{-1}(y_i - X_i\beta)$$

$$(12.15)$$

for obtaining ML estimates of $\Sigma_i(\theta)$ and β. To obtain restricted maximum likelihood (REML) estimates, the OLSE of β is given by

$$\hat{\beta} = \left(\sum_{i=1}^n X'_i X_i \right)^{-1} \left(\sum_{i=1}^n X_i y_i \right) \qquad (12.16)$$

so that the least squares residuals are $r_i = y_i - X_i \hat{\beta}$, $i = 1, 2, \ldots, n$. The log-likelihood of the residuals is

$$L^*_{REML} = -\frac{N-p}{2} \log(2\pi) - \frac{1}{2} \sum_{i=1}^{n} \log |\Sigma_i| - \frac{1}{2} \log \left| \sum_{i=1}^{n} X_i' \Sigma_i^{-1} X_i \right| - \frac{1}{2} \sum_{i=1}^{n} r_i' \Sigma_i^{-1} r_i$$

(12.17)

where $N = \sum_{i=1}^{n} n_i$. If $\hat{\theta}$ is the ML or REML estimate of θ and $\widehat{\Sigma}_i = \Sigma_i(\hat{\theta})$, then the FGLS estimate of β is

$$\check{\beta} = \left(\sum_{i=1}^{n} X_i' \widehat{\Sigma}_i^{-1} X_i \right)^{-1} \left(\sum_{i=1}^{n} X_i' \widehat{\Sigma}_i^{-1} y_i \right).$$

(12.18)

The algorithms for computing ML estimates (Newton-Raphson, Fisher scoring, and hybrid scoring/generalized EM) are described in Jennrich and Schluchter (1986). The ML procedure used in PROC MIXED is described in Wolfinger et al. (1994). In calculating the estimator of β given in (12.18), standard errors of the elements of β are obtained using the inverse of Fisher's expected information matrix

$$\mathrm{var}(\check{\beta}) = \left(\sum_{i=1}^{n} X_i' \widehat{\Sigma}_i^{-1} X_i \right)^{-1}.$$

(12.19)

To test hypotheses regarding the elements of β, we again may use the Wald statistic. If we want to test $H : C\beta = \beta_0$:

$$X^2 = (C\check{\beta} - \xi_0)' \left[C \left(\sum_{i=1}^{n} X_i' \widehat{\Sigma}_i^{-1} X_i \right)^{-1} C' \right]^{-1} (C\check{\beta} - \xi_0)$$

(12.20)

has a chi-square distribution under the null hypothesis for large n. Alternatively, an F-distribution approximation may be used.

12.5 IMPUTATIONS

Little and Rubin (1989) distinguished three strategies for dealing with the incomplete data: imputation, weighting, and direct analysis. The first two methods, imputation and weighting, seek to restore the rectangular form of the data matrix for subsequent analysis using complete data routine. Imputation accomplishes this by substituting a reasonable estimate for each missing datum. Imputed values are commonly chosen by mean, regression, or by sampling from other cases in the same data set (hot-deck imputation).

The simplest form of weighting approaches is listwise deletion (complete-case analysis), which assumes that cases are a random subsample of the original sample (i.e., missingness is MCAR) and therefore weighs all complete cases equally (Gold & Bentler, 2000; Gold, Bentler, & Kim, 2003). Parameter estimation in the listwise deletion, although unbiased for MCAR, is inefficient due to the waste of measured

data. Moreover, even with a fixed number of cases in the data set and a fixed proportion of data missing, a greater proportion of cases will need to be deleted to conduct
listwise deletion as the number of variables in the data set increases (Little & Rubin,
1987, 2002).

An example of direct analysis method with missing data is pairwise deletion
(available-case analysis), in which each quantity required for a statistical analysis is
simply computed from all data available to calculate it. Pairwise deletion assumes
MCAR. In addition, pairwise deletion will often produce a covariance matrix of
indeterminate sample size (Brown, 1994; Marsh, 1998), because the number of observations involved in calculations can vary from variable to variable. According to
Little and Rubin (1987, 1989), pairwise deletion were sometimes even less efficient
than listwise deletion and are otherwise unreliable.

12.5.1   EM Algorithm

Expectation-Maximization (EM) algorithm formalized by Dempster, Laird, and
Rubin (1977) is a single imputation method for missing data. The EM algorithm is
comprised of two steps: an expectation step (E-step) and a maximization step (M-
step). It is a very general iterative algorithm for obtaining MLEs with missing data
(iteratively reweighted regression). At a point θ, the E-step consists of computing

$$F(\theta', \theta) = \left(\frac{N}{2}\right) \left(p \log(2\pi) + \log |\Sigma(\theta')| \right.$$
$$\left. + \operatorname{tr} \left\{ \Sigma^{-1}(\theta') \left[S^* - \mu(\theta')\bar{y}^{*'} - \bar{y}^*\mu'(\theta') + \mu(\theta')\mu'(\theta') \right] \right\} \right) \quad (12.21)$$

where

$$S^* = \frac{1}{N} \sum_{i=1}^{N} \mathcal{E}(y_i y_i') \qquad (12.22)$$

and

$$\bar{y}^* = \frac{1}{N} \sum_{i=1}^{N} \mathcal{E}(y_i). \qquad (12.23)$$

The M-step consists of maximizing (12.21) with respect to θ' to obtain a new point,
$\tilde{\theta}$. The iteration process continually replaces θ by $\tilde{\theta}$ and repeats the E and M steps
until the sequence of values of θ converges to $\hat{\theta}$.

The EM algorithm assumes multivariate normality and MAR (Dempster, Laird,
& Rubin, 1977; Gold & Bentler, 2000). Even though, it assumes multivariate normality, it has also been shown to work well with nonnormal data (Gold & Bentler,
2000; Yuan & Bentler, 2000).

12.5.2   Multiple Imputation

Multiple imputation is a Monte Carlo approach to the analysis of missing data proposed by Rubin (1987). In multiple imputation, each missing value is replaced by a list of $d > 1$ simulated values from posterior probability distributions. A posterior probability distribution can be estimated in multiple imputation by Markov Chain Monte Carlo (MCMC) methods. The two most popular MCMC methods are Gibbs sampling and Metropolis-Hastings algorithm. Like single imputation, multiple imputation allows the analyst to proceed with complete data technique and software. Unlike other Monte Carlo methods, multiple imputation does not need a large number of repetitions for precise estimates. Rubin (1987) showed that the efficiency of an estimated based on d imputations, relative to one based on an infinite number, is $(1 + \lambda/d)^{-1}$, where λ is the rate of missing. For example, with 25% missing and $d = 10$ imputations, the efficiency is $(1 + \cdot^{25}/_{10})^{-1} = .976$ or 97.6%. Only small numbers of multiple imputation are necessary (about 5 to 10).

The simplest method for combining the results of d analyses is Rubin (1987) method for a scalar parameter. Suppose that Q represents a population quantity to be estimated. Let $\widehat{Q}$ and $\sqrt{SW_{\widehat{Q}}}$ denote the estimate of Q and the standard error respectively. The method assumes that the sample is large enough so that $\sqrt{SW_{\widehat{Q}}}\left(\widehat{Q} - Q\right)$ has approximately a standard normal distribution. We can not compute $\widehat{Q}$ and $SW_{\widehat{Q}}$; rather, there are d different versions, $(\widehat{Q}^{(i)}, SW_{\widehat{Q}}^{(i)})$, $i = 1, \ldots, d$. Rubin (1987) overall estimate is simply the average of the d estimates,

$$\bar{Q} = \frac{1}{d}\sum_i^d \widehat{Q}^{(i)}. \tag{12.24}$$

The uncertainty in $\bar{Q}$ has two parts: the average within-imputation variance,

$$\overline{SW}_{\widehat{Q}} = \frac{1}{d}\sum_{i=1}^d SW_{\widehat{Q}}^{(i)}, \tag{12.25}$$

and the between-imputation variance,

$$SB_{\widehat{Q}} = \frac{1}{d-1}\sum_{i=1}^d \left[\widehat{Q}^{(i)} - \bar{Q}\right]^2. \tag{12.26}$$

The total variance is a weighted sum of the two components,

$$T = \overline{SW}_{\widehat{Q}} + (1+d)^{-1}SB_{\widehat{Q}}, \tag{12.27}$$

and the square root of T is the overall standard error. For confidence intervals and tests, Rubin (1987) recommended the use of a Student's t approximation $T^{-1/2}(\bar{Q} - Q) \sim t_\nu$, where the degrees of freedom are given by

$$\nu = (d-1)\left[1 + \frac{\overline{SW}_{\widehat{Q}}}{(1+d)^{-1}SB_{\widehat{Q}}}\right]^2. \tag{12.28}$$

12.6   REPEATED MEASURES ANALYSIS

In the analysis of repeated measures data, one frequently has data missing in varying degrees at the end of a series of repeated measurements. In clinical experiments, missing data may be due to death or exhaustion. To illustrate the analysis of such data, we utilize the data by Cole and Grizzle (1966). The study involves the analysis of the drugs morphine and trimethophon, each at two levels of histamine release, and their effect on hypertension in dogs. To eliminate or minimize the positive association between means and variances, the logarithm of blood histamine levels ($\mu g/ml$) at four time points was used in the analysis. The data are incomplete since the 6th dog has a missing observation.

To analyze these data, Cole and Grizzle (1966) performed a multivariate analysis with unstructured Σ since by the Box (1950) test, they rejected the hypothesis that Σ satisfied compound symmetry structure. In Program 12_6.sas, we fit two models to the data: compound symmetry and unstructured matrices.

To test the hypothesis that Σ has compound symmetry structure versus unstructured, we form a chi-square difference test. The chi-square difference test is equal to $\Delta\chi^2 = 129.88 - 78.08 = 51.08$ with $\Delta df = 9 - 1 = 8$ (Output 12.6.1 and 12.6.2). Since $\Delta\chi^2$ test was significant, Σ does not have a compound symmetry. Thus, an analysis with unstructured Σ is appropriate. We may use either a mixed model analysis employing PROC MIXED to estimate Σ or a MANOVA analysis using PROC GLM.

For the mixed model analysis, we estimated the unknown covariance matrix by obtaining an ML estimate for Σ. Using all the observations, we performed approximate tests of drug effects, time difference, and the test of interaction. There was a significant difference on blood histamine levels among drugs averaged across time, $F(3, 12) = 5.84, p = .0107$. The other tests were also significant.

For the analysis using PROC GLM, SAS performed a listwise deletion. The missing value can be imputed using various methods (e.g., EM, MI). There was also significant difference on blood histamine levels among drugs averaged across time using listwise deletion, $F(3, 11) = 4.05, p = .0364$ (Output 12.6.3). There is only a slight difference between PROC MIXED and PROC GLM because there is only one case with a missing value.

12.7   REPEATED MEASURES WITH CHANGING COVARIATES

To illustrate how to analyze repeated measures data with changing covariates and MCAR, we reanalyze the Winer data previously analyzed in Section 11.6. Missing values were created randomly to demonstrate purpose. The SAS code to perform the analysis is given in Program 12_7.sas. Two data sets are present (Winer1 — data with missing and Winer2 — complete data).

PROC MI was used to examine the pattern of missing (Output 12.7.1). The number of imputed data sets was set to 0 using a keyword NIMPUTE. Group in the "Missing Data Patterns" table indicates the pattern of missing groups. It shows that 14 cases (77.78%) had no missing values (i.e., complete cases); 3 cases (16.67%)

Output 12.6.1: Repeated Measures Analysis: PROC MIXED — Compound Symmetry.

Covariance Parameter Estimates					
Cov Parm	Subject	Estimate	Standard Error	Z Value	Pr Z
CS	dog	0.3729	0.1366	2.73	0.0063
Residual		0.05319	0.01097	4.85	<.0001

Fit Statistics	
-2 Log Likelihood	47.6
AIC (smaller is better)	83.6
AICC (smaller is better)	99.1
BIC (smaller is better)	97.5

Null Model Likelihood Ratio Test		
DF	Chi-Square	Pr > ChiSq
1	78.08	<.0001

Type 3 Tests of Fixed Effects				
Effect	Num DF	Den DF	F Value	Pr > F
drug	3	12	5.74	0.0113
time	3	35	80.39	<.0001
drug*time	9	35	38.57	<.0001

Output 12.6.2: Repeated Measures Analysis: PROC MIXED — Unstructured.

Covariance Parameter Estimates					
Cov Parm	Subject	Estimate	Standard Error	Z Value	Pr Z
UN(1,1)	dog	0.2632	0.09305	2.83	0.0023
UN(2,1)	dog	0.2623	0.1095	2.40	0.0166
UN(2,2)	dog	0.4671	0.1651	2.83	0.0023
UN(3,1)	dog	0.3305	0.1264	2.61	0.0089
UN(3,2)	dog	0.4825	0.1755	2.75	0.0060
UN(3,3)	dog	0.5563	0.1967	2.83	0.0023
UN(4,1)	dog	0.2824	0.1082	2.61	0.0091
UN(4,2)	dog	0.3946	0.1472	2.68	0.0073
UN(4,3)	dog	0.4710	0.1675	2.81	0.0049
UN(4,4)	dog	0.4084	0.1444	2.83	0.0023

Fit Statistics	
-2 Log Likelihood	-4.2
AIC (smaller is better)	47.8
AICC (smaller is better)	86.8
BIC (smaller is better)	67.9

Null Model Likelihood Ratio Test		
DF	Chi-Square	Pr > ChiSq
9	129.88	<.0001

Type 3 Tests of Fixed Effects				
Effect	Num DF	Den DF	F Value	Pr > F
drug	3	12	5.84	0.0107
time	3	12	40.38	<.0001
drug*time	9	12	22.92	<.0001

Output 12.6.3: Repeated Measures Analysis: PROC GLM.

Source	DF	Type III SS	Mean Square	F Value	Pr > F
drug	3	26.84996622	8.94998874	4.05	0.0364
Error	11	24.30614532	2.20964957		

						Adj Pr > F	
Source	DF	Type III SS	Mean Square	F Value	Pr > F	G - G	H - F
time	3	12.13740554	4.04580185	54.28	<.0001	<.0001	<.0001
time*drug	9	17.46077605	1.94008623	26.03	<.0001	<.0001	<.0001
Error(time)	33	2.45946015	0.07452910				

Greenhouse-Geisser Epsilon	0.5675
Huynh-Feldt Epsilon	0.8438

had missing values on y; and 1 case (5.56%) had missing values on both x and y. The means of x and y are reported for each pattern of missing values.

Comparing the results from these two data sets shows slight differences in the estimate of compound symmetry variance (2.5672 vs. 2.2985) (Output 12.7.2 and 12.7.3). There is more noticeable difference in the estimate of the residual variance, $\hat{\sigma}^2$ (.0730 vs. .3331). The fixed effect results show similar p-values.

Output 12.7.1: Repeated Measures with Changing Covariates: Pattern of Missing.

					Group Means	
Group	x	y	Freq	Percent	x	y
1	X	X	14	77.78	7.357143	12.928571
2	X	.	3	16.67	7.000000	.
3	O	O	1	5.56	.	.

12.8   RANDOM COEFFICIENT MODEL

For our next example, data provided by (Lindsey, 1993, p. 223) are utilized to show how to fit a linear regression model with both a random intercept and slope with missing data. The data are the cumulative number of operating hours for successive failures of air-conditioning equipment in 13 Boeing 720 aircraft. The SAS code to perform the analyses is provided in 12_8.sas.

PROC MI was used to examine the pattern of missing in the data set. It shows a monotonic missing pattern (Output 12.8.1). The variable y30 was not included in the analysis because there was only 1 case. The percent of cases in each pattern

Output 12.7.2: Repeated Measures with Changing Covariates: PROC MIXED —
with Missing.

Covariance Parameter Estimates					
Cov Parm	Subject	Estimate	Standard Error	Z Value	Pr Z
CS	person	2.5672	1.3054	1.97	0.0492
Residual		0.07300	0.04210	1.73	0.0415

Fit Statistics	
-2 Log Likelihood	35.9
AIC (smaller is better)	53.9
AICC (smaller is better)	98.9
BIC (smaller is better)	55.6

Null Model Likelihood Ratio Test		
DF	Chi-Square	Pr > ChiSq
1	17.62	<.0001

Type 3 Tests of Fixed Effects				
Effect	Num DF	Den DF	F Value	Pr > F
a	2	5	6.26	0.0435
b	1	2	275.11	0.0036
a*b	2	2	14.43	0.0648
x	1	2	291.47	0.0034

Output 12.7.3: Repeated Measures with Changing Covariates: PROC MIXED —
Complete Data.

Covariance Parameter Estimates					
Cov Parm	Subject	Estimate	Standard Error	Z Value	Pr Z
CS	person	2.2985	1.1647	1.97	0.0484
Residual		0.3331	0.1570	2.12	0.0169

Fit Statistics	
-2 Log Likelihood	55.5
AIC (smaller is better)	73.5
AICC (smaller is better)	96.0
BIC (smaller is better)	75.3

Null Model Likelihood Ratio Test		
DF	Chi-Square	Pr > ChiSq
1	12.95	0.0003

Type 3 Tests of Fixed Effects				
Effect	Num DF	Den DF	F Value	Pr > F
a	2	6	5.63	0.0420
b	1	5	106.89	0.0001
a*b	2	5	4.10	0.0883
x	1	5	56.93	0.0006

of missing is also reported along with the means of the variables (only first table is printed). The EM nor MI can be performed with the data set because there are more variables than the number of subjects (i.e., 30 variables and 10 subjects).

Output 12.8.1: Random Coefficient Model: Pattern of Missing.

Missing Data Patterns

Group	y1	y2	y3	y4	y5	y6	y7	y8	y9	y10	y11	y12	y13	y14	y15	y16	y17	y18	y19	y20	y21	y22	y23	y24	y25	y26	y27	y28	y29
1	X	X	X	X	X	X	X	X	X	X	X	X	X	X	X	X	X	X	X	X	X	X	X	X	X	X	X	X	X
2	X	X	X	X	X	X	X	X	X	X	X	X	X	X	X	X	X	X	X	X	X	X	X	X	X	X	X	.	.
3	X	X	X	X	X	X	X	X	X	X	X	X	X	X	X	X	X	X	X	X	X	X	X	X	X	.	.	.	.
4	X	X	X	X	X	X	X	X	X	X	X	X	X	X	X	X	X	X	X	X	X	X	.	.	.	.	.	.	.
5	X	X	X	X	X	X	X	X	X	X	X	X	X	X	.	.	.	.	.	.	.	.	.	.	.	.	.	.	.
6	X	X	X	X	X	X	X	X	X	X	X	X	X	.	.	.	.	.	.	.	.	.	.	.	.	.	.	.	.
7	X	X	X	X	X	X	X	X	X	X	X	X	.	.	.	.	.	.	.	.	.	.	.	.	.	.	.	.	.
8	X	X	X	X	X	X	X	X	X	.	.	.	.	.	.	.	.	.	.	.	.	.	.	.	.	.	.	.	.
9	X	X	X	X	X	X	.	.	.	.	.	.	.	.	.	.	.	.	.	.	.	.	.	.	.	.	.	.	.
10	X	X	.	.	.	.	.	.	.	.	.	.	.	.	.	.	.	.	.	.	.	.	.	.	.	.	.	.	.

Missing Data Patterns

Group Means

Group	Freq	Percent	y1	y2	y3	y4	y5	y6	y7	y8	y9	y10
1	2	15.38	56.500000	192.000000	265.500000	362.000000	452.500000	484.000000	522.000000	557.500000	698.000000	743.500000
2	1	7.69	97.000000	148.000000	159.000000	163.000000	304.000000	322.000000	464.000000	532.000000	609.000000	689.000000
3	1	7.69	50.000000	94.000000	196.000000	268.000000	290.000000	329.000000	332.000000	347.000000	544.000000	732.000000
4	1	7.69	413.000000	427.000000	485.000000	522.000000	622.000000	687.000000	696.000000	865.000000	1312.000000	1496.000000
5	2	15.38	88.000000	221.000000	252.000000	295.000000	573.000000	595.000000	663.500000	704.000000	789.500000	1058.500000
6	1	7.69	55.000000	375.000000	440.000000	544.000000	764.000000	1003.000000	1050.000000	1296.000000	1472.000000	1654.000000
7	1	7.69	487.000000	505.000000	605.000000	612.000000	710.000000	715.000000	800.000000	891.000000	934.000000	1164.000000
8	1	7.69	359.000000	368.000000	380.000000	650.000000	1253.000000	1256.000000	1360.000000	1362.000000	1800.000000	
9	2	15.38	122.000000	256.500000	279.500000	435.500000	469.500000	566.000000				
10	1	7.69	130.000000	623.000000								

The data were converted into one row per observation (i.e., subjects scores appear in multiple rows; data=lindsey2). PROC MIXED was used to fit this data. Since each observation is in its own row, PROC MIXED will delete an observation if it is missing, but will retain all other observed scores from the same subject. This is basically a pairwise deletion. PROC GLM would have deleted the entire case (i.e., listwise deletion) by default.

We have fit a linear model in which both the slope and intercept are random. Furthermore, we have specified the covariance matrix for the random slope and intercept effects to be unstructured, in order to estimate the correlation between the random coefficients. The covariance between the random coefficients are negative, -681.76. There was a significant difference in the slope (rate of failure of air conditioners) among the aircrafts, $\sigma^2 = 1295.20$, $z = 2.00$, $p = .0299$. The fixed effect parameters in the model represent the estimated mean for the random intercept and slope parameters across the aircrafts. Thus, the average number of hours to failure is given by the linear model:

$$y = 65.02 + 99.92t \qquad t = 1, 2, \ldots, 30;$$

t is represented in the SAS syntax by fail.

Output 12.8.2: Random Coefficient Model: PROC MIXED.

Covariance Parameter Estimates					
Cov Parm	Subject	Estimate	Standard Error	Z Value	Pr Z
UN(1,1)	aircraft	24920	11726	2.13	0.0168
UN(2,1)	aircraft	-681.76	1938.33	-0.35	0.7250
UN(2,2)	aircraft	1295.20	648.79	2.00	0.0229
Residual		15437	1597.57	9.66	<.0001

Solution for Fixed Effects					
Effect	Estimate	Standard Error	DF	t Value	Pr > |t|
Intercept	65.0201	48.2739	12	1.35	0.2029
fail	99.9200	10.8973	12	9.17	<.0001

Table 12.1: Example of Missing Data Pattern.

	Time				
Subject	1	2	3	4	Pattern
1	✓	✓	✓	✓	1
2	✓	✓	✓	✓	1
3	✓	✓	−	✓	2
4	✓	−	✓	✓	3
5	✓	−	−	✓	4

12.9 GROWTH CURVE ANALYSIS

We now illustrate Kleinbaum's approach to the missing data problem. From (12.1), recall that Kleinbaum associates a linear model with each missing data pattern. To illustrate, suppose we have five observations over four time points where ($\checkmark$) denotes an observed data point and ($-$) represents a missing element. Suppose we have data presented in Table 12.1. Hence, we have four missing data patterns. Now for each pattern of missing data, we formulate a model of the form

$$\mathcal{E}(Y_j) = X_j B P A_j \qquad \text{cov}(Y_j) = I_{n_j} \otimes A_j' \Sigma A_j \qquad j = 1, 2, \ldots, m \quad (12.29)$$

where m is the number of patterns of missing data, n_j is the number of subjects with the jth patterns, p_j is the number of nonmissing elements in the jth pattern, $N = \sum_{i=1}^{m} n_j p_j$ is the number of nonmissing observations. $Y_{j(n_j \times p_j)}$ is the data matrix for the jth patterns, $X_{j(n_j \times k)}$ is the design matrix for the jth pattern, and $A_{j(p \times p_j)}$ is a matrix of 0's and 1's that represent the jth pattern.

For our example, the A_j's for $j = 1, 2, 3, 4$ are

$$A_1 = \begin{pmatrix} 1 & 0 & 0 & 0 \\ 0 & 1 & 0 & 0 \\ 0 & 0 & 1 & 0 \\ 0 & 0 & 0 & 1 \end{pmatrix} A_2 = \begin{pmatrix} 1 & 0 & 0 \\ 0 & 1 & 0 \\ 0 & 0 & 0 \\ 0 & 0 & 1 \end{pmatrix} A_3 = \begin{pmatrix} 1 & 0 & 0 \\ 0 & 0 & 0 \\ 0 & 1 & 0 \\ 0 & 0 & 1 \end{pmatrix} A_4 = \begin{pmatrix} 1 & 0 \\ 0 & 0 \\ 0 & 0 \\ 0 & 1 \end{pmatrix}.$$

$$(12.30)$$

Using (12.3), we see that the FGLS estimate of $\beta = \text{vec}(B)$ is

$$\check{\beta} = (D'\widehat{\Omega}^{-1}D)^{-1}D'\widehat{\Omega}^{-1}y \qquad (12.31)$$

where D and $\widehat{\Omega}$ are defined in (12.2), with any consistent estimator $\widehat{\Sigma}$ replacing Σ in Ω.

Although this approach is not iterative, an iterative process may be developed to estimate $\check{\beta}$. To define the process, we construct $\check{\beta}$ as defined in (12.4) and call this estimate $\check{\beta}_1$. Then we define

$$\widehat{\Omega}_1 = (y - D\check{\beta})(y - D\check{\beta})' \qquad (12.32)$$

and substitute this value into (12.31) to construct $\check{\beta}_2$. This process continues until some convergence criterion is met. This process is equivalent to Zellner's iterative process proposed for the SUR model. Under normality the estimator converges to the ML estimate.

To test hypotheses, we use the Wald statistic defined in (12.5). Critical to the Kleinbaum procedure is the process employed to estimate Σ when one does not utilize the iterative process. The initial estimate of Σ is less critical when one employs iterations. While considerable research has been done on how to estimate Σ, most procedures result in a $\widehat{\Sigma}$ that is not positive definite. To make the matrix positive semi-definite, we use the smoothing procedure suggested by Bock and Peterson (1975). They recommend obtaining a spectral decomposition of $\widehat{\Sigma}$ so that

$$\widehat{\Sigma} = P\Lambda P' \qquad (12.33)$$

where Λ is the diagonal matrix of eigenvalues and P is an orthogonal matrix of eigenvectors of $\widehat{\Sigma}$. The smoothed estimator of the matrix Σ is

$$\widehat{\Sigma}^* = P\Lambda^* P' \qquad (12.34)$$

where Λ^* is the matrix Λ with all negative eigenvalues set to zero. Under multivariate normality, $\widehat{\Sigma}^*$ is the maximum likelihood estimate under the constraint that Σ is positive semi-definite (Bock & Peterson, 1975).

To illustrate Kleinbaum's procedure to analyze growth curve data, we use the Grizzle and Allen (1969) dog data with missing data patterns used by Kleinbaum (1970). Kleinbaum used two patterns of incomplete data and one complete data pattern (Output 12.9.1). The missing data pattern 1, complete data, had 3 dogs from group 1; 5 dogs from group 2; 2 dogs from group 3; and 4 dogs from group 4, a total of 14 dogs. The X_1 and A_1 matrices are

$$\underset{14\times4}{X_1} = \begin{pmatrix} 1_3 & 0 & 0 & 0 \\ 0 & 1_5 & 0 & 0 \\ 0 & 0 & 1_2 & 0 \\ 0 & 0 & 0 & 1_4 \end{pmatrix} \quad \text{and} \quad A_1 = I_7. \tag{12.35}$$

For the missing data pattern 2, the first observation was deleted. There were 3 dogs from group 1; 3 dogs from group 2; 3 dogs from group 3; and 3 dogs from group 4, a total of 12 dogs. The X_2 and A_2 matrice are

$$\underset{12\times4}{X_2} = \begin{pmatrix} 1_3 & 0 & 0 & 0 \\ 0 & 1_3 & 0 & 0 \\ 0 & 0 & 1_3 & 0 \\ 0 & 0 & 0 & 1_3 \end{pmatrix} \quad \text{and} \quad \underset{7\times6}{A_2} = \begin{pmatrix} 0_{(1\times6)} \\ I_6 \end{pmatrix}. \tag{12.36}$$

For the missing data pattern 3, Kleinbaum removed the first three observations for the remaining dogs. There were 3 dogs from group 1; 2 dogs from group 2; 3 dogs from group 3; and 2 dogs from group 4, a total of 10 dogs. The X_3 and A_3 matrices are

$$\underset{10\times4}{X_3} = \begin{pmatrix} 1_3 & 0 & 0 & 0 \\ 0 & 1_2 & 0 & 0 \\ 0 & 0 & 1_3 & 0 \\ 0 & 0 & 0 & 1_2 \end{pmatrix} \quad \text{and} \quad \underset{7\times4}{A_3} = \begin{pmatrix} 0_{(3\times4)} \\ I_6 \end{pmatrix}. \tag{12.37}$$

Output 12.9.1: Growth Curve Analysis: Missing Data Pattern.

										Group Means						
Group	y1	y2	y3	y4	y5	y6	y7	Freq	Percent	y1	y2	y3	y4	y5	y6	y7
1	X	X	X	X	X	X	X	14	38.89	3.714286	3.850000	4.014286	4.042857	4.142857	4.278571	4.057143
2	.	X	X	X	X	X	X	12	33.33	.	3.783333	4.041667	4.166667	4.316667	4.375000	4.308333
3	.	.	X	X	X	X	X	10	27.78	.	.	4.180000	4.400000	4.320000	4.070000	3.910000

In Section 11.4, we analyzed these dog data using a mixed model, fitting a random coefficient growth curve model, a growth curve model with unstructured covariance matrix, a growth curve model with compound symmetry, and a heterogeneous model under compound symmetry. Using the AIC criterion, we saw the model that best fit the data was the growth curve model with an unstructured covariance matrix. It yielded smallest AIC statistics; however, the random coefficient model also provided a reasonable fit. In Program 12_9.sas, we reanalyzed the example with missing data. The growth curve model with unstructured Σ again performed better than the random coefficient model. Comparing the AIC values for the complete data and the missing data pattern, we see that the unstructured covariance matrix model seems to be more robust for the missing data since AIC decreased in value from 259.5 to

Output 12.9.2: Growth Curve Analysis: PROC MIXED.
Unstructured Covariance Matrix:

Fit Statistics	
-2 Res Log Likelihood	198.6
AIC (smaller is better)	254.6
AICC (smaller is better)	263.4
BIC (smaller is better)	299.0

Random coefficient model:

Fit Statistics	
-2 Res Log Likelihood	198.6
AIC (smaller is better)	254.6
AICC (smaller is better)	263.4
BIC (smaller is better)	299.0

254.6. The AIC for the random coefficient model also decreased from 303.2 to 273.4
(Output 12.9.2).

In the second part of Program 12_9.sas, we analyze the data using Kleinbaum's
model, using PROC IML. We first estimate Σ; the obvious choice is to use the estimate obtained from using only the complete data. While such an estimate is always
positive semidefinite, it may not be a reasonable estimate unless few observations
are missing or an iterative procedure is employed to estimate the fixed effects. Alternatively, we may use (12.6) to compute $\widehat{\Sigma}$ which employs all data for (r, s) pairs of
variables.

For our example,

$$(N_{rs}) = \begin{pmatrix} 14 & 14 & 14 & 14 & 14 & 14 & 14 \\ 14 & 26 & 26 & 26 & 26 & 26 & 26 \\ 14 & 26 & 26 & 26 & 26 & 26 & 26 \\ 14 & 26 & 26 & 36 & 36 & 36 & 36 \\ 14 & 26 & 26 & 36 & 36 & 36 & 36 \\ 14 & 26 & 26 & 36 & 36 & 36 & 36 \\ 14 & 26 & 26 & 36 & 36 & 36 & 36 \end{pmatrix}$$

$$D_{1s} = X_1 \qquad \text{for } s = 1, \ldots, 7$$

$$D_{rs} = \begin{pmatrix} X_1 \\ X_2 \end{pmatrix} \qquad \text{for } r = 2, 3 \text{ and } s = r, \ldots, 7$$

$$D_{rs} = \begin{pmatrix} X_1 \\ X_2 \\ X_3 \end{pmatrix} \qquad \text{for } r = 4, \ldots, 7 \text{ and } s = r, \ldots, 7$$

where $D_{rs} = D_{sr}$ for all r and s. Thus,

$$\hat{\sigma}_{rs} = \left(\frac{1}{N_{rs} - 4} \right) y'_{rs}[I_{N_{rs}} - D_{rs}(D'_{rs}D_{rs})^{-1}D'_{rs}]y_{rs} \tag{12.38}$$

and $r, s = 1, \ldots, 7$. We estimate Σ using several approaches. Because we have all 36 observations for this example, we calculate the unbiased estimate of Σ using all observations. Following Kleinbaum (1970), we estimate Σ using only the 14 complete observations. This leads to a matrix that is positive definite. An alternative is to use (12.6). This matrix is indefinite since it contains a negative value. Setting the negative value to zero, the smoothing procedure, we may reestimate Σ with a matrix that is positive semidefinite. However, its inverse does not exist so that one must use a generalized inverse. Thus, we have three estimates for Σ: $\hat{\Sigma}_{GA}$, $\hat{\Sigma}_C$, and $\hat{\Sigma}_{SM}$ where GA denotes the Grizzle-Allen estimate, C represents the complete or listwise deletion estimate, and SM the estimate obtained by setting negative eigenvalues to zero. The three estimates are found in Output 12.9.3.

To find the maximum likelihood estimate for B using the complete model, we employ (7.43) since weighting by $\hat{\Sigma}_{GA}$ is equivalent to using the higher order terms as covariates in the Rao-Khatri conditional model. To estimate B with missing data, we use the vector version (12.3). The estimate $\check{\beta}$ and B in the format of Grizzle and Allen (1969) is provided in Output 12.9.4. The $\hat{B}$ for the missing data cases using the complete and smoothed estimates is also displayed. An estimate for B using (12.6) was not calculated since this "usual" estimate leads to negative test statistics whenever a matrix is indefinite. For this example, both complete data and the smoothed estimate provide reasonable estimates for B. This may not be the case with more missing data (Kleinbaum, 1970).

Having found parameter estimates, we next illustrate hypothesis testing. To test hypotheses in the form $H : CBA = 0$, we let $C_* = A' \otimes C$ to transform the hypothesis to the form $H : C_*\beta = 0$, where $\beta = \text{vec}(B)$ and use the Wald statistic given in (12.5) for hypothesis testing. That is

$$X^2 = (C_*\check{\beta})'[C_*(D'\hat{\Omega}^{-1}D)^-C'_*]^{-1}(C_*\check{\beta}) \tag{12.39}$$

has any asymptotic χ^2 distribution with degrees of freedom $\nu = \text{rank}(C_*)$. Kleinbaum (1970) computed Wald statistics for six hypothesis in the form $H : CB = 0$. The matrices C_i are

$$C_1 = \begin{pmatrix} 1 & 0 & 0 & -1 \end{pmatrix}$$
$$C_2 = \begin{pmatrix} 1 & 0 & -1 & 0 \end{pmatrix}$$
$$C_3 = \begin{pmatrix} 1 & -1 & 0 & 0 \end{pmatrix}$$
$$C_4 = \begin{pmatrix} 0 & 1 & -1 & 0 \end{pmatrix}$$
$$C_5 = \begin{pmatrix} 0 & 1 & 0 & -1 \end{pmatrix}$$
$$C_6 = \begin{pmatrix} 0 & 0 & 1 & -1 \end{pmatrix}$$

which are all pairwise comparisons of regression profiles, a test of homogeneity of profiles. The χ^2 statistics for each estimate of Σ are given in Output 12.9.5.

Output 12.9.3: Growth Curve Analysis: PROC IML — $\widehat{\Sigma}$.

Complete Data 36 Obs Estimate of Sigma

SIGMA_GA

0.226	0.172	0.172	0.205	0.171	0.196	0.182
0.172	0.170	0.184	0.192	0.163	0.170	0.164
0.172	0.184	0.392	0.347	0.237	0.188	0.222
0.205	0.192	0.347	0.441	0.369	0.287	0.258
0.171	0.163	0.237	0.369	0.434	0.373	0.318
0.196	0.170	0.188	0.287	0.373	0.524	0.461
0.182	0.164	0.222	0.258	0.318	0.461	0.513

Listwise Deletion Estimate of Sigma

SIGMA_C

0.158	0.154	0.213	0.199	0.174	0.170	0.189
0.155	0.190	0.275	0.274	0.257	0.265	0.244
0.213	0.275	0.592	0.542	0.472	0.510	0.507
0.199	0.274	0.542	0.640	0.624	0.651	0.628
0.174	0.257	0.472	0.624	0.674	0.690	0.632
0.170	0.265	0.510	0.651	0.690	0.788	0.704
0.189	0.244	0.507	0.628	0.632	0.704	0.726

smoothed Estimate of Sigma

SIGMA_SM

0.158	0.154	0.213	0.199	0.173	0.170	0.189
0.154	0.179	0.208	0.211	0.197	0.219	0.230
0.213	0.208	0.416	0.367	0.278	0.298	0.341
0.199	0.211	0.367	0.441	0.369	0.287	0.258
0.173	0.197	0.278	0.369	0.434	0.373	0.318
0.170	0.219	0.298	0.287	0.373	0.524	0.461
0.189	0.230	0.341	0.258	0.318	0.461	0.513

Output 12.9.4: Growth Curve Analysis: PROC IML — $\widehat{B}$.

Grizzle and Allen Beta = beta*(pp`)

BETAT_F			
32.702	4.136	-4.680	-1.142
25.071	-0.687	-0.995	0.110
28.585	4.101	-2.402	-0.480
27.116	1.908	-1.549	0.004

Beta based on 14 complete obs to estimate sigma

BETAT_SIGMA_C			
33.250	4.196	-2.922	-1.042
25.295	-0.452	-0.379	0.022
28.473	4.375	-5.548	-0.093
26.819	0.832	-1.373	0.107

Beta based on sigma_smoothed estimate of sigma

BETAT_SIGMA_SM			
32.690	3.772	-3.070	-1.184
24.908	-0.872	-0.380	-0.001
28.134	5.681	-4.493	-0.227
27.300	2.553	-0.422	-0.097

Output 12.9.5: Growth Curve Analysis: PROC IML — χ^2.

Comparison of H with various covariance matrices

CH			
	GA	C	SM
H1	23.16	31.30	24.48
H2	13.03	34.31	35.98
H3	34.58	30.66	33.20
H4	12.95	15.44	47.88
H5	4.66	2.49	15.98
H6	3.54	8.80	13.08

For $\alpha = .05$, the $\chi^{@}$ critical value with 4 degrees of freedom is 9.49. For this example, the conclusions reached using $\widehat{\Sigma}_C$ are in agreement with the $\widehat{\Sigma}_{GA}$ complete data; however, the use of $\widehat{\Sigma}_{SM}$ finds all comparisons significant. If the estimate of Σ is not positive definite, even though $\widehat{B}$ may appear reasonable, the Wald tests may lead to spurious results. To use the large sample procedure recommended by Kleinbaum requires very large data sets and an estimate of Σ that is positive definite to be reliable. However, the "effective size" with missing data is unknown.

Tests of hypotheses using either ML or FGLS procedures discussed in this chapter are valid if data are MCAR. Missing data is a complex issue in the analysis of repeated measurements; articles by Little and Schenker (1995) and Robins, Rotnitzky, and Zhao (1995) discuss classes of models for analyzing data that are not MCAR. An excellent overview of missing data problems in the analysis of repeated measurement data is provided by Vonesh and Chinchilli (1997, p. 264-274).

CHAPTER 13

Structural Equation Modeling

13.1 INTRODUCTION

Structural equation modeling (SEM) is a method of multivariate analysis concerned with representing population moments on the basis of relatively few parameters that are hypothesized on the basis of a substantive theory. Historically, the main emphasis in SEM has been to model centered second-order moments, namely covariances, and the associated field is often called covariance structure analysis. Because a variety of different methods can be cast as covariance structure methods, SEM amounts to a family of techniques that encompasses many other statistical methods such as confirmatory factor analysis (CFA), path analysis, and simultaneous equations modeling. When population means are additionally modeled on the basis of fewer parameters, such as in growth curve modeling, the field is sometimes called mean and covariance structure analysis. SEM is used in many disciplines (behavioral and social sciences, education, information science, health, marketing, etc.) and its popularity is due to its ability to model complex behaviors of variables according to a theory, especially in contexts where it is hard or impossible to use randomized experiments to evaluate causal theories. In this chapter, we provide an overview of basic ideas that are elaborated fully in the chapters of this volume. We concentrate especially on covariance structure models, and discuss mean structures in the final paragraphs.

In covariance structure modeling, the null hypothesis is that the population covariance matrix is structured in terms of fewer parameters, which we may write abstractly as

$$H_0 : \Sigma = \Sigma(\theta) \tag{13.1}$$

where Σ is the population covariance matrix and $\Sigma(\theta)$ is a model-based covariance matrix under the structural hypothesis. As is explained below, the null hypothesis is evaluated by comparing an optimal estimate of $\Sigma(\theta)$ to the sample covariance matrix S; the model is a good one if the discrepancy between these matrices is small. Unlike other statistical methods, SEM models a sample covariance matrix and not the raw sample data.

There are several major parts to any SEM analysis. Usually the model is represented as a path diagram, because this is a convenient nontechnical way for substantive researchers to describe their models. Path diagrams, and the associated path tracing rules that are equivalent to matrix algebra, were introduced by Wright (1918, 1921, 1934, 1960). Path diagrams and the tracing of influences of variables on each other immediately imply that variables can have direct and/or indirect effects on other variables. A direct effect is akin to a β coefficient in an equation, while indirect effects operate through at least one intervening variable and represent various products of such coefficients. Models also typically involve latent variables, first conceived by Spearman (1904), which represent hypothesized underlying sources of variance generating the observed variables. Since latent variables are figments of the researcher's imagination, an important part of model specification involves precise hypotheses about how latent variables relate to certain but not other observed variable indicators. And, as a general multivariate statistical method, any implementation of SEM involves general estimation and testing procedures (Bollen, 1989). These aspects — path analysis, covariance structure algebra possibly involving latent variables, and statistical methodology — still make up essential parts of SEM today.

The broad attraction of SEM is that it incorporates methodology from psychometrics, econometrics, and sociometrics. This unity grew out of the 1960's attempts in sociology (e.g., Blalock, 1964) to understand the causal modeling of nonexperimental variables. The method of path analysis for the representation of systems of simultaneous linear equations was rediscovered (e.g., Duncan, 1966; Wright, 1960), and interest grew in a variety of confirmatory models from factor analysis (Jöreskog, 1969) to econometric simultaneous equation models (e.g., Blalock, 1971; Werts & Linn, 1970). Arthur Goldberger, an econometrician, was one of the first to sense that an integration of some basic ideas in psychometrics, econometrics, and biometric path analysis might be achievable (e.g., Goldberger, 1971; Goldberger & Duncan, 1973). Bentler (1986) provides a more detailed historical review. In this chapter, we will emphasize the integration that was achieved by the Jöreskog-Keesling-Wiley model, described further below.

An important consequence of the early developments summarized above was an emphasis on specifying the correlational and covariance implications of specific models. Today's statistical theory for evaluating a covariance structure hypothesis is largely based on the distributions of sample covariances. The maximum likelihood (ML) estimation procedure, most popular in SEM, was introduced by Jöreskog (1973) building upon earlier works by other researchers (T. W. Anderson & Rubin, 1956; Bock & Borgman, 1966; Goldberger, 1964; Jöreskog, 1969; Lawley, 1940). Another commonly used estimation method is generalized least squares (GLS) introduced by Jöreskog and Goldberger (1972) and M. W. Browne (1974, 1982, 1984). Both ML and GLS are based on the assumption that a sample arises from independent and identically distributed variables that are multivariate normally distributed. An important further method, the asymptotically distribution free approach (ADF; M. W. Browne, 1982, 1984), does not assume any specific distribution for the variables whose covariances are modeled. There is also a scaling method that accepts normal theory estimators, but adjusts any test statistics and standard error for nonnormality (Satorra & Bentler, 1988, 1994). While these are some of the key ideas we need to expand on in this introductory chapter, there are, of course, many variants and extensions of models and methods, as is clarified in subsequent chapters.

13.2 MODEL NOTATION

Although the covariance structure model (13.1) is written in a very abstract form, implying that basically any function of model parameters is of interest, in practice most SEM models are much more restricted and represent models for specifying linear structural relations among variables. Although model variants exist, the two main model structures in the field are those of Jöreskog-Keesing-Wiley (Jöreskog, 1973; Keesing, 1972; Wiley, 1973) and Bentler-Weeks (Bentler & Weeks, 1980). The former model, often known by the name of the computer program LISREL, is a model involving 3 matrix equations (1 for the structural model and 2 for the measurement model). The structural model represents relationships among latent variables:

$$\eta = B\eta + \Gamma\xi + \zeta \tag{13.2}$$

where

$\underset{m \times 1}{\eta}$ (eta) is a vector of latent endogenous variables

$\underset{n \times 1}{\xi}$ (xi) is a vector of latent exogenous variables

$\underset{m \times m}{B}$ (beta) is a matrix of regression coefficients among latent endogenous variables. The diagonal elements of the matrix are 0.

$\underset{m \times n}{\Gamma}$ (gamma) is a matrix of regression coefficients predicting η by ξ

$\underset{m \times 1}{\zeta}$ (zeta) is a vector of latent errors.

A latent variable is an unobserved variable, often referred to as a factor or an unobserved construct (see Bollen, 2002, for a review of alternative ways to define latent variables). In the model (13.2), there are two types of latent variables, endogenous and exogenous. An endogenous latent variable is a function of exogenous latent variable(s). In other words, an endogenous variable is a dependent variable. In contrast, exogenous variables are independent variables. There are two covariance matrices associated with the structural model, Φ (phi) and Ψ (psi). Φ is a $n \times n$ covariance matrix of the exogenous factors ξ, $\mathcal{E}(\xi\xi')$, while Ψ is a $m \times m$ covariance matrix of the latent residuals ζ, $\mathcal{E}(\zeta\zeta')$. These matrices are covariance matrices because it is assumed that all variables in the model have zero means, that is, (1) $\mathcal{E}(\eta) = 0$, (2) $\mathcal{E}(\xi) = 0$, (3) $\mathcal{E}(\zeta) = 0$. In addition, it is assumed that residuals and predictors are uncorrelated, that is, (4) $\text{cov}(\zeta, \xi) = 0$, and (5) $(I - B)$ is nonsingular where I is a $m \times m$ identity matrix.

There are two factor analytic matrix equations in the measurement model. The measurement model represents relationships between observed and latent variables. There is one matrix equation for endogenous variables and another for exogenous variables.

$$x = \Lambda_x \xi + \delta \tag{13.3}$$

$$y = \Lambda_y \eta + \epsilon \tag{13.4}$$

where

$\underset{q \times 1}{x}$ is a vector of observed exogenous variables

$\underset{p \times 1}{y}$ is a vector of observed endogenous variables

$\underset{q \times n}{\Lambda_x}$ (lambda x) is a matrix of regression coefficients predicting x by ξ

$\underset{p \times m}{\Lambda_y}$ (lambda y) is a matrix of regression coefficients predicting y by η

$\underset{q \times 1}{\delta}$ (delta) is a vector of measurement errors of x

$\underset{p \times 1}{\epsilon}$ (epsilon) is a vector of measurement errors of y.

x and y are indicators of latent variables ξ and η respectively. Standardized Λ_x and Λ_y are better known as factor loading matrices in the factor analysis literature. There are also two covariance matrices associated with the residual variables in the measurement model, Θ_δ and Θ_ϵ. Θ_δ is a $q \times q$ covariance matrix of δ, $\mathcal{E}(\delta\delta')$, while Θ_ϵ is a $p \times p$ covariance matrix of ϵ, $\mathcal{E}(\epsilon\epsilon')$. In standard factor analysis, these matrices are diagonal matrices containing unique variances associated with the variables, but in SEM, selected covariances are allowed. The matrices are covariance matrices because it is similarly assumed that: (1) $\mathcal{E}(\epsilon) = 0$ and (2) $E(\delta) = 0$. It is further assumed that factors and errors, and errors with other errors, are uncorrelated, that is, (3) $\text{cov}(\epsilon, \eta) = 0$, (4) $\text{cov}(\epsilon, \xi) = 0$, (5) $\text{cov}(\epsilon, \delta) = 0$, (6) $\text{cov}(\delta, \xi) = 0$, and (7) $\text{cov}(\delta, \eta) = 0$. This model structure implies a particular covariance structure that is a special case of (13.1). It is given below in (13.29)-(13.32).

The second major way of specifying linear structural relations models is the Bentler-Weeks model. In this model, although latent variables are allowed, the distinction between observed and latent variables is not highlighted in the model structure. Rather, the model emphasizes that any variable can be predicted by other variables, that some of the predictors may themselves be predicted by other variables, and that some variables exist as independent variables with no further explanation within the system. In a notation that is somewhat distinct from the LISREL emphasis on endogenous and exogenous variables, in the Bentler-Weeks system any variable (whether observed or latent) that is predicted by at least one other variable is called a dependent variable, and all other variables (whether observed or latent) are called independent variables. This simple notation allows all 3 matrix equations of the LISREL notation to be specified with one matrix equation. For didactic purposes, we map the LISREL notation into the Bentler-Weeks notation.

$$\eta^* = B^*\eta^* + \Gamma^*\xi^* \tag{13.5}$$

where

$\eta^*_{(m+p+q)\times 1}$ is a vector of dependent variables. $\eta^{*'} = \begin{pmatrix} \eta & y & x \end{pmatrix}$.

$\xi^*_{(n+p+q)\times 1}$ is a vector of independent variables. $\xi^{*'} = \begin{pmatrix} \xi & \epsilon & \delta & \zeta \end{pmatrix}$.

$B^*_{(m+p+q)\times(m+p+q)}$ is a matrix of regression coefficients among η^*. The diagonal elements of the matrix are 0.

$\Gamma^*_{(m+p+q)\times(n+p+q)}$ is a matrix of regression coefficients predicting η^* by ξ^*.

The covariance matrix of the independent variables ξ^* is Φ^*:

$$\Phi^* = \begin{pmatrix} \Phi & 0 & 0 \\ 0 & \Theta_\epsilon & 0 \\ 0 & 0 & \Theta_\delta \end{pmatrix}. \tag{13.6}$$

It can be shown that the LISREL model structure is a special case of the Bentler-Weeks model structure, e.g., notice that Γ^* permits any of the four LISREL independent variable types to predict any of the three LISREL dependent variable types, or that Φ^* allows covariances that are forced to zero in LISREL. In other words, the Bentler-Weeks model allows influences and covariances for which there is no matrix counterpart in the LISREL model, and various distinctions within LISREL, such as differentiating observed variables y and x, are not important. In the practical implementation of this model in EQS (Bentler, in press), the variables in (13.5) are specialized to *Variables* (y and x), *Factors* (η and ξ), *Errors* (ϵ and δ), and *Disturbances* (ζ).

Irregardless of model structure chosen to be used, in order for any latent variable to be identified in SEM, the model structure (e.g., (13.2)-(13.4)) must imply a unique parameterization. In the LISREL structure, it is important that the factors defined by (13.3) and (13.4) are unique. However, it will be seen that the factor loading structure remains identical if one inserts a full rank matrix D and its inverse into the equation, e.g., $x = \Lambda_x DD^{-1}\xi + \delta$. Thus it would be possible to write $x = \Lambda_{x*}\xi^* + \delta$, where $\Lambda_{x*} = \Lambda_x D$ and $\xi^* = D^{-1}\xi$. The elements in the factor loading matrix and the particular latent variables chosen for analysis are jointly arbitrary. In most applications of SEM, as will be illustrated further below, there are enough zero restrictions in the factor loading matrices so that D can no longer be an arbitrary transformation matrix, but rather must be a diagonal matrix. In this case, identification can be achieved by either fixing one of the factor loadings on a factor to 1.0, or fixing the variance of the corresponding factor to 1.0. If a factor loading in each column is fixed to 1.0, the corresponding factor cannot be rescaled. The alternative of fixing the variance of a factor to 1.0 is limited to exogenous factors, since endogenous factors do not have variances as parameters. Specifically, for η_i, there is no choice but to fix one of its factor loadings to one. With ξ_i, one has a choice, and their factor variances can be fixed to 1.0. In fact, this is the traditional choice in exploratory factor analysis, where the factors are scaled to have unit variances. The interpretation of these identification conditions is as follows: If a factor loading is fixed to 1, the scaling of the latent variable is that of the observed variable, i.e., if an observed variable was on a 5 point scale then so will be the latent variable, and large scores on the latent variable can be interpreted in the same direction as large scores on the corresponding observed variable. If the factor variance is fixed to 1, the factor is standardized, i.e., the latent factor scores are on a standardized scale with a standard deviation of 1. The question of which variable to fix to 1, in order to identify a latent variable, is often asked by novice researchers. In general, it makes no difference in overall fit or model estimation as to which variable is fixed to 1, as long as that variable is truly related to the factor.

As noted earlier, a particular SEM model is often represented using a path diagram, which presents an easily grasped picture of how various variables influence each other. For many researchers, a diagram is the preferred method of displaying their model. An example of a path diagram is displayed in Figure 13.1. This diagram uses the usual convention that observed variables are shown in squares or rectangles, latent variables or factors are in circles or ovals, one-way arrows represent coefficients in equations, and two-way arrows represent covariances. The data consists of 9 observed variables, and everything else represents our hypotheses on how these variables are generated by hypothetical latent variables, and how the latent variables relate to each other. Evidently, each of the 3 factors has three indicators, meaning that those 3 indicators are hypothesized to be highly correlated among themselves due to their being generated by a given factor, and furthermore, if those factors were to be controlled, the observed variables would no longer be correlated. The measurement model (13.3)-(13.4) will be used to relate the observed variables to the factors. The hypothesized causal influences among the factors, the one-way arrows relating factors, will be described with the linear relations model (13.2). The variables and matrices involved, and their specific interpretation from the LISREL viewpoint, are summarized in Table 13.1.

There are 3 latent variables (2 η's and 1 ξ). Each latent variable has 3 indicators (either x's or y's). The structural model for the example in Figure 13.1 is:

$$\eta = B\eta + \Gamma\xi + \zeta$$

$$\begin{pmatrix} \eta_1 \\ \eta_2 \end{pmatrix} = \begin{pmatrix} 0 & 0 \\ b_{21} & 0 \end{pmatrix} \begin{pmatrix} \eta_1 \\ \eta_2 \end{pmatrix} + \begin{pmatrix} \gamma_{11} \\ \gamma_{21} \end{pmatrix} (\xi_1) + \begin{pmatrix} \zeta_1 \\ \zeta_2 \end{pmatrix} \tag{13.7}$$

where b_{21} is the regression coefficient predicting η_2 by η_1. The first subscript denotes a dependent variable and second subscript denotes its predictor variable. γ_{11} and γ_{21} are the regression coefficients predicting η_1 and η_2 by ξ_1 respectively. If B is a lower-triangle matrix, as in this example, then a model is said to be recursive. If B is a full matrix (including parameters in the lower and upper triangle), then a model is said to be nonrecursive. A nonrecursive model is harder to estimate than a recursive model (Bollen, 1989). The two covariance matrices associated with the structural model are:

$$\Phi = \left(\phi_{11} \right) \tag{13.8}$$

$$\Psi = \begin{pmatrix} \psi_{11} & 0 \\ 0 & \psi_{22} \end{pmatrix} \tag{13.9}$$

where ϕ_{11} represents a variance of ξ_1. ψ_{11} is the variance of ζ_1 and ψ_{22} is the variance of ζ_2. The element ψ_{12} is set to 0; i.e., ζ_1 and ζ_2 are uncorrelated.

The measurement model for Figure 13.1 is

$$x = \Lambda_x \xi + \delta$$

$$\begin{pmatrix} x_1 \\ x_2 \\ x_3 \end{pmatrix} = \begin{pmatrix} 1 \\ \lambda_{x_{21}} \\ \lambda_{x_{31}} \end{pmatrix} \left(\xi_1 \right) + \begin{pmatrix} \delta_1 \\ \delta_2 \\ \delta_3 \end{pmatrix} \tag{13.10}$$

$$y = \Lambda_y \eta + \epsilon$$

$$\begin{pmatrix} y_1 \\ y_2 \\ y_3 \\ y_4 \\ y_5 \\ y_6 \end{pmatrix} = \begin{pmatrix} 1 & 0 \\ \lambda_{y_{21}} & 0 \\ \lambda_{y_{31}} & 0 \\ 0 & 1 \\ 0 & \lambda_{y_{52}} \\ 0 & \lambda_{y_{62}} \end{pmatrix} \begin{pmatrix} \eta_1 \\ \eta_2 \end{pmatrix} + \begin{pmatrix} \epsilon_1 \\ \epsilon_2 \\ \epsilon_3 \\ \epsilon_4 \\ \epsilon_5 \\ \epsilon_6 \end{pmatrix}. \tag{13.11}$$

The factor loadings of x_1 on ξ_1, y_1 on η_1, and y_4 on η_2 have been set to 1 for factor identification. The two covariance matrices associated with the measurement model are

$$\Theta_\delta = \begin{pmatrix} \theta_{\delta_{11}} & 0 & 0 \\ 0 & \theta_{\delta_{22}} & 0 \\ 0 & 0 & \theta_{\delta_{33}} \end{pmatrix} \tag{13.12}$$

$$\Theta_\epsilon = \begin{pmatrix} \theta_{\epsilon_{11}} & 0 & 0 & 0 & 0 & 0 \\ 0 & \theta_{\epsilon_{22}} & 0 & 0 & 0 & 0 \\ 0 & 0 & \theta_{\epsilon_{33}} & 0 & 0 & 0 \\ 0 & 0 & 0 & \theta_{\epsilon_{44}} & 0 & 0 \\ 0 & 0 & 0 & 0 & \theta_{\epsilon_{55}} & 0 \\ 0 & 0 & 0 & 0 & 0 & \theta_{\epsilon_{66}} \end{pmatrix}. \tag{13.13}$$

The elements of $\{B, \Gamma, \Phi, \Psi, \Lambda_x, \Lambda_y, \Theta_\epsilon, \Theta_\delta\}$ can be: (1) freely estimated, (2) fixed to a certain value, e.g., 1 or 0, or (3) constrained to be equal to another parameter or to be equal to a function of other parameters. In the example, Greek letters represent free parameters, and 1's and 0's are fixed. An illustrative equality restriction would be that all error variances in Θ_δ are equal to each other; but in this model, we assume no functional relations among parameters. Restricted models require specialized computational machinery (e.g., S.-Y. Lee, 1985; S.-Y. Lee & Poon, 1985).

Table 13.1: Symbols Used in Path Diagrams.

Items		Symbol
Observed variables	$\{x, y\}$	rectangle or square
Latent variables	$\{\eta, \xi\}$	circle or oval
Errors	$\{\zeta, \epsilon, \delta\}$	no symbol (unenclosed)
Regression coefficients	$\{B, \Gamma, \Lambda_x, \Lambda_y\}$	one-head arrow
Covariances	$\{\Phi, \Psi, \Theta_\epsilon, \Theta_\delta\}$	two-head arrow

Figure 13.1: An example of SEM path diagram.

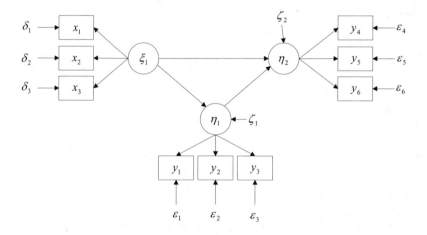

The example in Figure 13.1 uses all of the matrices in the Jöreskog-Keesling-Wiley model. Various specializations of this general model structure provide for other model structures that are useful in their own right. The well-known confirmatory factor analysis (CFA) model can be expressed by just using one matrix equation from the measurement model,

$$x = \Lambda_x \xi + \delta. \tag{13.14}$$

The other measurement equation and structural model are not needed. Λ_x is a matrix of factor loading of x_i on ξ_j. An example of a CFA model is displayed in Figure 13.2, which describes 6 observed variables that are hypothesized to be generated by two latent factors plus residuals. The matrix notation of the example is

$$x = \Lambda_x \xi + \delta$$

$$
\begin{pmatrix} x_1 \\ x_2 \\ x_3 \\ x_4 \\ x_5 \\ x_6 \end{pmatrix} = \begin{pmatrix} \lambda_{x_{11}} & 0 \\ \lambda_{x_{21}} & 0 \\ \lambda_{x_{31}} & 0 \\ 0 & \lambda_{x_{42}} \\ 0 & \lambda_{x_{52}} \\ 0 & \lambda_{x_{62}} \end{pmatrix} + \begin{pmatrix} \delta_1 \\ \delta_2 \\ \delta_3 \\ \delta_4 \\ \delta_5 \\ \delta_6 \end{pmatrix} \tag{13.15}
$$

and the covariance matrices are

$$\Phi = \begin{pmatrix} 1 & \phi_{12} \\ \phi_{21} & 1 \end{pmatrix} \tag{13.16}$$

$$
\Theta_\delta = \begin{pmatrix}
\theta_{\delta_{11}} & 0 & 0 & 0 & 0 & 0 \\
0 & \theta_{\delta_{22}} & 0 & 0 & 0 & 0 \\
0 & 0 & \theta_{\delta_{33}} & 0 & 0 & 0 \\
0 & 0 & 0 & \theta_{\delta_{44}} & 0 & 0 \\
0 & 0 & 0 & 0 & \theta_{\delta_{55}} & 0 \\
0 & 0 & 0 & 0 & 0 & \theta_{\delta_{66}}
\end{pmatrix}. \tag{13.17}
$$

In this example, the variances of ξ_1 and ξ_2 are fixed at 1 for factor identification, $\phi_{11} = \phi_{22} = 1$. All the nonzero factor loadings are estimated. In a CFA model, it is typical that identification is dealt with by having all factor loadings estimated and factor variances fixed to 1. $\phi_{12} = \phi_{21}$ represents correlation between ξ_1 and ξ_2 since they are standardized.

The econometric simultaneous equation model, a model without any latent variables, is a special case of SEM. Sometimes such a model is simply called a path analysis model (without latent variables). A matrix equation for a path analysis model can be derived from the general Jöreskog-Keesling-Wiley model by setting $y = \eta$, $x = \xi$, and $\epsilon = \zeta$:

$$y = By + \Gamma x + \epsilon. \tag{13.18}$$

An example of a path analysis model is displayed in Figure 13.3, which hypothesizes that the influence of the x variables on y_2 is mediated by y_1. That is, as there is no

Figure 13.2: An example of CFA model.

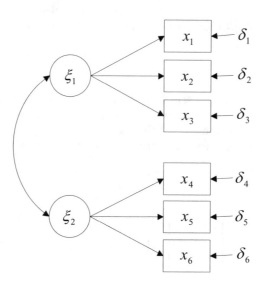

one-way arrow shown, there is no direct influence of any x variable on y_2, and the entire influence of x on y_2 is indirect. In terms of the model matrices, there are two y's and three x's in the model. The structural equation is

$$y = By + \Gamma x + \epsilon$$

$$\begin{pmatrix} y_1 \\ y_2 \end{pmatrix} = \begin{pmatrix} 0 & 0 \\ b_{21} & 0 \end{pmatrix} \begin{pmatrix} y_1 \\ y_2 \end{pmatrix} + \begin{pmatrix} \gamma_{11} & \gamma_{12} & \gamma_{13} \\ 0 & 0 & 0 \end{pmatrix} \begin{pmatrix} x_1 \\ x_2 \\ x_3 \end{pmatrix} + \begin{pmatrix} \epsilon_1 \\ \epsilon_2 \end{pmatrix} \qquad (13.19)$$

and the two covariance matrices are

$$\Phi = \begin{pmatrix} \phi_{11} & \phi_{12} & \phi_{13} \\ \phi_{21} & \phi_{22} & \phi_{23} \\ \phi_{31} & \phi_{32} & \phi_{33} \end{pmatrix} \qquad (13.20)$$

$$\Theta_\epsilon = \begin{pmatrix} \theta_{\epsilon_{11}} & 0 \\ 0 & \theta_{\epsilon_{22}} \end{pmatrix}. \qquad (13.21)$$

Φ is now a covariance matrix of x. For example, ϕ_{11} is the variance of x_1 and $\phi_{12} = \phi_{21}$ is the covariance between x_1 and x_2.

Another interesting special case is the multiple regression model. It presents a special case of path analysis where there is only one y and $B = 0$:

$$y = \Gamma x + \epsilon. \qquad (13.22)$$

Figure 13.3: An example of a path analysis model.

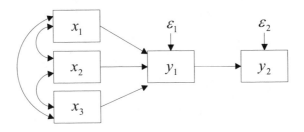

Figure 13.4 is an example of a multiple regression model specified as a path diagram. There are 3 predictors in the model and 1 dependent variable. In multiple regression, the predictors are allowed to covary, but they do not covary with the dependent variable or its residual. The matrix representation of the model is

$$y = \Gamma x + \epsilon$$

$$\begin{pmatrix} y_1 \end{pmatrix} = \begin{pmatrix} \gamma_{11} & \gamma_{12} & \gamma_{13} \end{pmatrix} \begin{pmatrix} x_1 \\ x_2 \\ x_3 \end{pmatrix} + \begin{pmatrix} \epsilon_1 \end{pmatrix} \tag{13.23}$$

where the covariance matrices are

$$\Phi = \begin{pmatrix} \phi_{11} & \phi_{12} & \phi_{13} \\ \phi_{21} & \phi_{22} & \phi_{23} \\ \phi_{31} & \phi_{32} & \phi_{33} \end{pmatrix} \tag{13.24}$$

$$\Theta_\epsilon = \begin{pmatrix} \theta_{\epsilon_{11}} \end{pmatrix}. \tag{13.25}$$

Note that since we have been discussing covariance structure models, i.e., models in which variable means are assumed to be zero, this model does not compute a y-intercept. We would need a mean structure model to include the intercept. However, a multiple regression model is not an interesting model from the SEM point of view. It is a saturated model (see below) having as many parameters as data points to be fitted, so that it can not be rejected by any data. In multiple regression, R^2, the proportion of variance of a dependent variable explained by the predictors is more important than whether or not the sample covariance matrix is correctly reproduced. R^2 can be computed in an SEM multiple regression model by $1 - \frac{\theta_{\epsilon_{11}}}{\mathrm{var}(y_1)}$.

13.3 ESTIMATION

The information modeled in the covariance structure version of SEM is the observed covariance matrix, $\underset{p+q \times p+q}{S}$. Let $\underset{p^* \times 1}{s} = \mathrm{vech}(S)$ represent a vector of nonredundant elements of S (Magnus & Neudecker, 1988) where $p^* = \frac{(p+q)(p+q+1)}{2}$. The

Figure 13.4: An example of a multiple regression model.

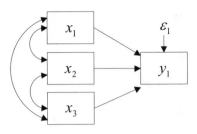

parameters estimated in SEM are the nonredundant elements of the regression coefficients and covariance matrices,

$$\theta_{r \times 1} = \{B, \Gamma, \Lambda_y, \Lambda_x, \Phi, \Psi, \Theta_\epsilon, \Theta_\delta\}. \tag{13.26}$$

In order for the model to be identified, there are technical conditions such as that the partial derivative matrix of Σ with respect to θ is of full rank r (see Δ below). This requires at a minimum that $r \leq p^*$. We do not discuss such technical conditions in this chapter, but assume they are met. A model with $r < p^*$ is called an over-identified model. The degrees of freedom for such a model is

$$df = p^* - r. \tag{13.27}$$

A model with $r = p^*$ is called a just identified or a saturated model, and it has the feature that such a model perfectly reproduces S. Since a saturated model will always perfectly reproduce a sample covariance matrix, no simplification is gained by explaining the relationships among the variables using a saturated SEM model. In order to have a testable model — a potentially rejectable model — there have to be fewer parameters estimated than data points, i.e., the df needs to be positive. Sometimes degrees of freedom can be gained by imposing restrictions on parameters and adjusting the testing machinery appropriately (e.g., S.-Y. Lee & Bentler, 1980). A model with $r > p^*$ is called an under-identified model. This model can not be tested, as there are negative df. The difficulty is similar to solving a system of simultaneous equations with more unknowns than the number of unique equations.

We now turn to the specific covariance structure implied by the Jöreskog-Keesling-Wiley model. The model-implied covariance matrix, $\Sigma(\theta)$, can be partitioned into four submatrices:

$$\Sigma(\theta) = \begin{pmatrix} \Sigma_{yy}(\theta) & \Sigma_{yx}(\theta) \\ \Sigma_{xy}(\theta) & \Sigma_{xx}(\theta) \end{pmatrix}. \tag{13.28}$$

Each submatrix can be defined in terms of matrices in the structural and measurement models in SEM by performing covariance algebra.

$$\Sigma_{yy}(\theta) = E(yy')$$
$$= E\left[(\Lambda_y\eta + \epsilon)(\Lambda_y\eta + \epsilon)'\right]$$

substituting $\eta = (I - B)^{-1}(\Gamma\xi + \zeta)$ and solving the expectation,

$$= \Lambda_y(I - B)^{-1}(\Gamma\Phi\Gamma' + \Psi)\left[(I - B)^{-1}\right]'\Lambda_y' + \Theta_\epsilon. \qquad (13.29)$$

Likewise, solving for the other submatrices,

$$\Sigma_{yx}(\theta) = E(yx')$$
$$= E\left[(\Lambda_y\eta + \epsilon)(\Lambda_x\xi + \delta)'\right]$$
$$= \Lambda_y(I - B)^{-1}\Gamma\Phi\Lambda_x', \qquad (13.30)$$
$$\Sigma_{xy}(\theta) = \Sigma_{yx}'(\theta)$$
$$= \Lambda_x\Phi\Gamma'\left[(I - B)^{-1}\right]'\Lambda_y' \qquad (13.31)$$

and

$$\Sigma_{xx}(\theta) = E(xx')$$
$$= E\left[(\Lambda_x\xi + \delta)(\Lambda_x\xi + \delta)'\right]$$
$$= \Lambda_x\Phi\Lambda_x' + \Theta_\delta. \qquad (13.32)$$

Notice that having estimates of the parameters in each of the matrices yields, by these formulas, the model-implied covariances among the variables. It is for this reason that the methodology is sometimes called covariance structure analysis.

Let, $\sigma(\theta) = \text{vech}[\Sigma(\theta)]$, then given $\hat{\theta}$ one can test model fit by examining the discrepancy between s and $\sigma(\hat{\theta})$. If a model perfectly reproduces the relationships among the observed variables, then the residuals $s - \sigma(\hat{\theta}) = 0$. In general the residuals will not be precisely zero, and the statistical issue associated with model fitting involves determining whether the residuals are small enough from a statistical point of view. Large sample theory has been invoked to determine these and other statistical issues in SEM. In turn, this is based on the discrepancy functions chosen to optimize when estimating the parameters of a model. There are several typically used discrepancy functions. These functions can be written in generalized least squares (GLS) form

$$F = [s - \sigma(\theta)]' W [s - \sigma(\theta)] \qquad (13.33)$$

where W is an appropriate weight matrix. The weight matrix W can be specialized to yield a number of different estimators depending on: (1) distribution assumption (e.g., normal, arbitrary) and (2) computation invoked (e.g., identity, optimal reweighting) (see, e.g., Bentler & Dijkstra, 1985). For example, as shown by S.-Y.

Lee and Jennrich (1979), if one assumes the observed variables come from a multivariate normal distribution and utilize computations in which W is updated each iteration, this produces ML estimates. The weight matrix for any normal theory estimator is

$$W_{ML} = \frac{1}{2}D'(A^{-1} \otimes A^{-1})D \qquad (13.34)$$

where now D is the 0-1 duplication matrix (e.g., Magnus & Neudecker, 1988) and A is any matrix that converges to the population covariance matrix, Σ, with probability equal to 1, e.g., S for normal theory GLS and $\Sigma(\hat{\theta})$ iteratively updated for ML. A least squares estimator can be obtained by setting $A = I$. Of course, the ML discrepancy function can also be written directly as a function of S and $\Sigma(\theta)$,

$$F_{ML} = \log|\Sigma(\theta)| + tr[S\Sigma(\theta)^{-1}] - \log|S| - (p+q). \qquad (13.35)$$

Minimizing F at an optimal estimate $\hat{\theta}$ produces $\widehat{F}$. The null hypothesis in SEM, $H_0 : \sigma = \sigma(\theta)$, is tested with a test statistic obtained by multiplying $\widehat{F}$ with number of subjects minus 1,

$$T = (N-1)\widehat{F} \qquad (13.36)$$

where $\sigma = \text{vech}(\Sigma)$. If the null hypothesis is true, then $T \sim \chi^2_{df}$, and we accept the null hypothesis if the statistic does not exceed a standard cutoff in the chi-square distribution. In such a case, we may say that the model is consistent with the data, i.e., it reproduces the data well. But, since we can never prove the null hypothesis, we can not say that our model is the true model for this population. All that we can say is that the model is not rejected by the data. Of course, if T is large compared to degrees of freedom, the probability associated with the model is very small, and we reject the null hypothesis.

An estimation method that is ideal in principle is the asymptotically distribution free or ADF method (M. W. Browne, 1982, 1984). As a minimum-χ^2 method (Ferguson, 1958), unlike ML, it does not require any assumption that the variables are multivariate normally distributed. Let $z = \begin{pmatrix} y & x \end{pmatrix}$, then the weight matrix for ADF is equal to $W_{ADF} = V_{ADF}^{-1}$ where

$$V_{ADF} = \frac{1}{N-1}\sum_{i=1}^{N}(b_i - \bar{b})(b_i - \bar{b})', \qquad (13.37)$$

$b_i = D^+\text{vec}(z_i - \bar{z})(z_i - \bar{z})'$, $\bar{b}$ and $\bar{z}$ are the mean vectors of the b_i's and z_i's respectively, D^+ is the unique Moore-Penrose inverse of D (see, Timm, 2002, p. 47), and $\text{vec}(\cdot)$ vectorizes a matrix. Unfortunately, because the weight matrix becomes huge, the ADF method can not be used with models containing a large number of variables, and due to its implicit use of 4th order moments of the data in the weight matrix, it has been found to lack robustness in small to moderate sample sizes. Yuan and Bentler (1997) have introduced an adjustment to the ADF test statistic to improve its behavior with small to moderate sample sizes.

A method that is intermediate between ML and ADF is to accept the ML estimator, but to correct the test statistic and standard errors for nonnormality. This approach was introduced by Satorra and Bentler (1988, 1994) and has become very popular. The test statistic is called the scaled χ^2. The scaled statistic can be applied to various estimation methods, but it is typically used when the initial estimator is ML. The scaled statistic is obtained by correcting T,

$$T_{SB} = c^{-1}T \tag{13.38}$$

where

$$c = \frac{1}{df}\left[tr\left(UV_{ADF}\right)\right], \tag{13.39}$$

and U depends on the estimation method. For ML,

$$U_{ML} = W_{ML} - W_{ML}\Delta(\Delta'W_{ML}\Delta)^{-1}\Delta'W_{ML} \tag{13.40}$$

where $\Delta = \frac{\partial\sigma(\theta)}{\partial\theta'}$, a Jacobian matrix. If data are normally distributed, then $tr(UV_{ADF}) = df$ and therefore $c = 1$; i.e., $T_{SB} = T$. Under the null hypothesis the expected value of T_{SB} is that of the χ^2 variate, so the method amounts to correcting the mean of the distribution of T. Nonetheless, this correction yields very good behavior at the tail of the distribution where accept/reject decisions are made. Other methods for this situation are described by Bentler and Yuan (1999) and Yuan and Bentler (1998).

In any model, the statistical significance of any specific parameter $\hat{\theta}_i$ is evaluated using the usual machinery of ML or optimal GLS estimation, that is, sampling variances can be obtained from the inverse of the observed or expected information matrices. Specifically, if the discrepancy function is correctly specified, the standard error of θ can be obtained from the information matrix,

$$se(\theta) = \sqrt{\text{diag}\left[\frac{(\Delta'V\Delta)^{-1}}{N}\right]} \tag{13.41}$$

where $V = \frac{\partial^2 F}{\partial\sigma\partial\sigma'}$ evaluated at Σ, and $\text{diag}(\cdot)$ is a diagonal function of a matrix. However, this formula is not appropriate under distributional misspecification. When the discrepancy function is incorrect for the actual distribution of the data, i.e., W is not equal to V^{-1}, the sandwich-type covariance matrix

$$(\Delta'W\Delta)^{-1}\Delta'WVW\Delta(\Delta'W\Delta)^{-1}$$

replaces the inverse information matrix $(\Delta'V\Delta)^{-1}$ (e.g., Bentler & Dijkstra, 1985; M. W. Browne, 1984). This is part of the Satorra-Bentler correction when normal theory ML is used with nonnormal data. A test of individual θ_i can be performed using a z-test,

$$z = \frac{\theta_i - \theta_{i_0}}{se(\theta_i)} \tag{13.42}$$

where θ_{i_0} is a value of θ_i under the null hypothesis, $H_0 : \theta_i = \theta_{i_0}$, usually 0.

The literature also contains discussions that question the appropriateness of classical tests of exact fit $\Sigma = \Sigma(\theta)$ such as are summarized above. For example, Serlin and Lapsley (1985, 1993) suggested a "good-enough" principle and recommended that tests of close fit replace tests of exact fit. In SEM, MacCallum, Browne, and Sugawara (1996) developed such a test of close fit $\Sigma \approx \Sigma(\theta)$ as a substitute for the exact fit null hypothesis. We do not provide details here, but such a conceptualization is related to our next topic.

13.4 MODEL FIT IN PRACTICE

Test statistics T in SEM have been shown to be biased against a sample size (Jackson, 2003; MacCallum, Widaman, Preacher, & Hong, 2001). Such observations have led to the viewpoint that test statistics often need to be augmented by other means of evaluating model fit. Problems with T arise from many sources. For example, the distributional assumption underlying T may be false, and the distribution of the statistic may not be robust to violation of this assumption; no specific model $\Sigma(\theta)$ may be assumed to exist in the population, and T is intended to provide a summary regarding closeness of $\Sigma(\hat{\theta})$ to S; the sample size may not be large enough for T to be χ^2 distributed and the probability values used to evaluate the null hypothesis may not be correct; and in large enough samples, any a priori hypothesis $\Sigma = \Sigma(\theta)$, although only trivially false, may be rejected (Bentler, 1990). The latter two points also involve issues of power. Excessive power implies that even a small discrepancy $(s - \sigma(\hat{\theta}))$ may lead to a significant T in a large enough sample. Insufficient power arises where a large discrepancy may not be significant because of a sample size that is too small. Therefore, a number of different fit indices designed to measure the extent to which the variation and covariation in the data are accounted for by a model have been developed. There are probably about two dozen such fit indices, of which a dozen are sometimes used (Hu & Bentler, 1998). We discuss only a selected few.

Practical fit indices can be grouped into either incremental or absolute. The absolute fit indices directly assess how well a model reproduces the sample covariance matrix. Although no reference model is specified, a saturated model that exactly reproduces a sample covariance matrix can be used as a reference model. Incremental fit indices measure the improvement in fit over a nested baseline model, often an independence model. An independence model or a model of uncorrelated variables states that $\Sigma(\theta)$ is a diagonal matrix, i.e., involves variances of the variables but no covariances. Some equations for computation of fit indices depend on the choice of estimator. Only ML-based fit indexes will be shown (see, Hu & Bentler, 1998, 1999 for complete discussion).

One of the first fit indices introduced into SEM is the normed-fit-index (NFI; Bentler & Bonnett, 1980). NFI is an incremental fit index that ranges from 0-1, with

1 indicating perfect fit

$$NFI = \frac{T_B - T}{T_B} \tag{13.43}$$

where T_B is the test statistic for a baseline model. NFI can be interpreted as the proportion of relationships among the variables explained by a model. NFI has a small downward bias in small samples which has been eliminated by another incremental fit index, the comparative fit index (CFI; Bentler, 1990), defined as

$$CFI = 1 - \frac{\max[(T - df), 0]}{\max[(T_B - df_B), (T - df), 0]} \tag{13.44}$$

where df_B is the degrees of freedom of the baseline model. One way of thinking about CFI vs. NFI is that CFI is a better estimator of the population counterpart to (13.43), i.e., population fit based on noncentrality parameters of the noncentral χ^2 distribution. Thus CFI can be interpreted as the relative reduction in noncentrality of the hypothesized model as compared to the model of uncorrelated variables. Values of .90-.95 or greater in incremental fit indices represent a good fit.

Two of the most popular absolute fit indices are the goodness-of-fit-index (GFI) and adjusted GFI (AGFI) (Jöreskog & Sörbom, 1984)

$$GFI = 1 - \frac{tr(\Sigma^{-1}(\hat{\theta})S - I)^2}{tr(\Sigma^{-1}(\hat{\theta})S)^2} \tag{13.45}$$

$$AGFI = 1 - \frac{p^*/df}{1 - GFI}. \tag{13.46}$$

AGFI, as its name sounds, adjusts GFI for model complexity. Both GFI and AGFI have a maximum value of 1, but they can be less than 0. A higher value also indicates a better fit. A value of .90 or greater on GFI or AGFI represents a good fit. Two problems with these indices is that they are sensitive to sample size (Hu & Bentler, 1998, 1999) and may yield surprisingly large values even for the uncorrelated variables model (Bentler, in press).

Two residual-based absolute fit indices are the root mean square error of approximation (RMSEA; Steiger & Lind, 1980) and the standardized-root-mean-square-residual (SRMR). The former measures the quality of fit per degree of freedom, i.e., it has a parsimony rationale. The latter is an intuitively attractive measure of how well, on average, sample covariances are being reproduced by the model. SRMR was first introduced as the root-mean-square residual (RMR; Jöreskog & Sörbom, 1981) and was later modified to be based on residuals standardized to be measured in the correlation metric (Bentler, 1995). Unlike other fit indices, the size of these indices is directly proportional to the amount of error between observed and model implied covariance matrices,

$$RMSEA = \sqrt{\frac{\max\left[\frac{T-df}{N-1}, 0\right]}{df}} \tag{13.47}$$

$$SRMR = \sqrt{\frac{1}{p^*}\sum_{i=1}^{p+q}\sum_{j=1}^{i}\left(\frac{s_{ij} - \hat{\sigma}_{ij}}{s_{ii}s_{jj}}\right)^2} \tag{13.48}$$

where $S = (s_{ij})$ and $\Sigma(\hat{\theta}) = (\hat{\sigma}_{ij})$ for $i, j = 1, \ldots, (p + q)$. These fit indices range from 0 to 1, with a low value indicating a better fit. An RMSEA value of .06 or less indicates a good fit, while an SRMR value of .08 or less indicates a good fit.

Many authors have studied the behavior of fit indices (e.g., Lei & Lomax, 2005; Marsh, Hau, & Wen, 2004; McDonald, 1989; Tanaka, 1993). The only theoretical study is by Ogasawara (2001), who showed in an example that asymptotically various fit indices are very highly correlated. In their large empirical studies, Hu and Bentler (1998, 1999) found that the indices are not all very highly correlated. It seems that CFI, RMSEA, and SRMR provide somewhat different information about model fit, and hence some combined use may be beneficial. This conclusion is consistent with results of Beauducel and Wittman (2005); Fan and Sivo (2005); Kenny and McCoach (2003). See Bentler (in press, Chapter 14) for a fuller review.

Yuan (2005) provided a serious challenge to the value of fit indices. It has been known for a long time, and superbly verified by Yuan, that fit indices are not just a measure of model fit, but also of choice of T as well as uncontrolled factors such as sample size and the distribution of the data. It follows from Yuan that one should not expect conclusions based on one fit index to precisely mirror the conclusions based on another fit index (see also Weng & Cheng, 1997; Schmukle & Hardt, 2005).

> There can never be a best coefficient for assessing fit ... any more than there is a single best automobile (Steiger, 1990, p. 179).

13.5   MODEL MODIFICATION

An initial model specified according to a theory often is inadequate at reproducing a sample covariance matrix, that is, the residuals $s - \sigma(\theta) \neq 0$ are large and the χ^2 test clearly suggests model rejection. Fortunately, like any model building enterprise, a model can be respecified to yield potentially improved fit. Model modification is a somewhat controversial enterprise, because if it is not theoretically motivated independently of the data, there is liable to be some capitalization on chance associations in the data, and hence the modified model may not replicate (see, e.g., MacCallum, Roznowski, & Necowitz, 1992). Currently available procedures for avoiding capitalization on chance in this situation (e.g., Green, Thompson, & Babyak, 1998; Hancock, 1999) tend to be overly conservative. Nonetheless, in our opinion, when a model does not fit an expensive data set, it is important to try to locate the sources of misspecification even if resulting conclusions have to be somewhat provisional.

Model respecification can involve freeing fixed parameters and/or fixing free parameters. Freeing a fixed parameter, i.e., allowing a formerly fixed parameter to be freely estimated, will increase the number of parameters estimated in the model, r, and hence will decrease df. The consequence is that the model with the new parameter will fit better — hopefully, substantially better — than the original model. On the other hand, fixing a free parameter, that is, setting a parameter to a fixed value such as zero, or adding a new constraint, will decrease r and increase df. A model with larger df (hence, smaller r) is seen as more parsimonious, which is desirable primarily if the decrease in fit that results from the new restriction is small relative to df. Usually, models are modified by first finding enough free parameters to add until the overall model fit is acceptable, and only then considering which of the free parameters are unnecessary and could be fixed to zero.

There are three asymptotically equivalent procedures for accomplishing model modification, namely, the χ^2 difference, LaGrange Multiplier (LM), and Wald tests (e.g., Buse, 1982). These can be applied in univariate or multivariate contexts, that is, to test single or multiple simultaneous restrictions. The general theory is simply applied to SEM in an appropriate way (e.g., Chou & Bentler, 1990, 2002; Sörbom, 1989). All three will generally lead to a same final model; however, they have a sufficiently different ease of use so that these tests are not used interchangeably. Typically, the χ^2 difference test is used when a set of a priori model restrictions are being tested; the LM test, also known as the score test in ML estimation, is used especially when one has few ideas on what restrictions to release in a model; and the Wald test is used to simplify a final model that is otherwise acceptable. The χ^2 difference test, $\Delta\chi^2$, is used to compare two models, M_1 and M_2, where M_1 is nested within M_2. More specifically, let r_1 and r_2 denote the number of parameters estimated in each model respectively, then, $r_1 < r_2$. A comparison of two models is performed by

$$\Delta\chi^2 = T(M_1) - T(M_2) \qquad (13.49)$$

where $T(M_1)$ and $T(M_2)$ are test statistics for M_1 and M_2 respectively. $\Delta\chi^2 \sim \chi^2$ with $df = r_2 - r_1 = df(M_1) - df(M_2)$ where $df(M_1)$ and $df(M_2)$ are degrees of freedom of M_1 and M_2 respectively. This is testing a null hypothesis,

$$H_0 : \sigma(\theta_1) = \sigma(\theta_2) \qquad (13.50)$$

where $\sigma(\theta_1)$ and $\sigma(\theta_2)$ are model implied covariance matrices for M_1 and M_2 respectively.

The LM and Wald tests do not require estimation of another model, and they have the advantage that they can be implemented in a blind search computerized methodology. For example, an SEM program can be asked to search all fixed elements of any matrices involved in a given model specification, and to compute an LM test for each and every current model restriction and perhaps a multivariate test for the set of statistically unacceptable restrictions. Similarly, an SEM program can be asked to search each of the free parameters to see if their associated z-statistics imply that the parameter is not significant, and sets of such nonsignificant parameters can be tested for simultaneously being unnecessary (e.g., set to zero). As a result, these model modifications are local in nature, that is, they accept the current model structure and work within it to yield an improved model. What they can not accomplish is to inform the researcher that the model structure is itself fundamentally incorrect. For example, if the given model is a path model with no latent variables, these methods can not inform the user to add, say, two latent factors that might be needed to achieve an adequate model fit.

Like the χ^2 difference test, LM and Wald tests are distributed as χ^2 with df equal to a number of parameters freed or fixed. The LM test for ML estimation is defined as

$$LM = \left(\frac{N-1}{2}\right)\left(\frac{\partial F_{ML}}{\partial \theta}\right)'\left[E\left(\frac{\partial^2 F_{ML}}{\partial\theta\partial\theta'}\right)\right]^{-1}\left(\frac{\partial F_{ML}}{\partial \theta}\right) \qquad (13.51)$$

where F_{ML} is the ML minimum discrepancy function. Let $c(\theta)$ be a $u \times 1$ vector of any set of fixed or constrained parameters to be freed, where $u \leq r$, then

$$Wald = [c(\theta)]'\{acov[c(\theta)]\}^{-1}[c(\theta)] \qquad (13.52)$$

where

$$acov(c(\theta)) = \left[\frac{\partial c(\theta)}{\partial \theta}\right]'[acov(\theta)]\left[\frac{\partial c(\theta)}{\partial \theta}\right] \qquad (13.53)$$

and $acov(\cdot)$ is an asymptotic covariance matrix. Note that c could refer to a given free parameter and that, more generally, $c(\hat{\theta}) \neq 0$, that is, the Wald test evaluates functions of free parameters.

13.6 SUMMARY

SEM is a very flexible technique for testing a model based on a theory. A brief introduction to SEM has been provided in this chapter, but there are many advanced topics and statistical intricacies that we did not have space to cover. These advanced topics build upon the foundations discussed here.

In closing, we illustrate three extensions to the basic ideas given above, including mean and covariance structure models, growth curve modeling, and the treatment of missing data. In the multiple regression example discussed earlier, we noted that the y-intercept was not part of the model. Including the y-intercept in an SEM model requires a mean and covariance structure model. When the means of observed or latent variables are also of interest in an SEM model, the model structure has to be extended so that intercept and mean parameters are part of the model structure. In case of the Jöreskog-Keesling-Wiley model, these can appear as follows,

$$\eta = \alpha + B\eta + \Gamma\xi + \zeta \qquad (13.54)$$

$$x = \tau_x + \Lambda_x\xi + \delta \qquad (13.55)$$

$$y = \tau_y + \Lambda_y\eta + \epsilon \qquad (13.56)$$

where α, τ_x, τ_y and $E(\xi)$ are the structural and measurement intercepts of appropriate dimensions. As usual, it is necessary to assure that the model structure is identified. This requires the new concept that there are fewer mean or intercept parameters in the model than there are means of observed variables to have an overidentified mean structure. When there are as many mean and intercept parameters as observed means, the mean structure is just identified and is not testable.

Growth curve modeling (aka latent growth curve modeling) is a special application of mean and covariance structure modeling. This method is concerned with individual trajectories (rates of change) on a repeatedly measured variable over time, where each individual is allowed to have a different rate of growth. Both linear and nonlinear rates of change can be modeled using a number of different coding techniques (e.g., polynomial, spline). A simple example may be given with the use of (13.55) when $\tau_x = 0$ and Λ_x codes time. In that case the factors ξ are interpreted relative to the coding of time, for example, one factor may represent an initial value and a second factor may represent a linear slope of growth across time. In that case the factors are interpreted as representing individual differences in initial value and individual differences in slope. In such a model setup, the factor means $E(\xi)$ represent the mean initial value and mean slope. This is a mean structure model because, taking expectations of both sides of (13.55) under the assumption that $E(\delta) = 0$, $E(x) = \Lambda_x E(\xi)$. In other words, the means of the observed variables are a parametric function of the latent variable means. Growth curve modeling can be performed in either a multilevel modeling (i.e., hierarchical linear models) or SEM framework (Chou, Bentler, & Pentz, 1998; Curran, 2003).

A final example we will discuss involves the ever-present problem of missing data. We have discussed covariance structure analysis as concerned with modeling the sample covariance matrix S, yet with missing data, such a matrix is not even defined. An obvious solution is to use listwise, pairwise deletion, or imputation of missing data to create a full data matrix that can be analyzed in the usual way. However, these are not ideal procedures. Better procedures involve the use of all available scores to estimate parameters using the direct ML method based upon a casewise definition of the likelihood. This approach prevents loss of power due to loss of subjects and uncertainty introduced by imputing missing data. Two different approaches to ML estimation and testing with missing data are given by Arbuckle (1996); Jamshidian and Bentler (1999). These basic approaches have been extended in several ways to permit correct inference in the presence of nonnormal missing data (Yuan & Bentler, 2000).

13.7 PATH ANALYSIS

Roth, Wiebe, Fillingim, and Shay (1989) performed a study measuring exercise, hardiness, fitness, stress, and illness of 373 university students. A path model is displayed in Figure 13.5. They hypothesized (1) exercise predicts fitness, (2) hardiness predicts stress, and (3) fitness and stress predicts illness.

A SEM analysis can be performed without raw data. A minimum requirement is a covariance (or correlation) matrix. The correlation, standard deviation, and mean of

Figure 13.5: Path Analysis: Roth Data.

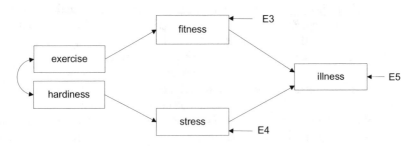

the variables from Roth et al. (1989) were entered into SAS using DATA command. PROC CALIS was used to performed the analysis. PROC CALIS can use a raw data or covariance matrix. The SAS code to perform the analysis is given in Program 13_7.sas.

By default, PROC CALIS analyzes a correlation matrix. In SEM, in order to estimate the standard errors of parameters correctly, a covariance matrix should be analyzed. The keyword COVARIANCE can be added to a syntax to analyze a covariance matrix. There are several estimation methods that can be used in SEM. The most popular is the maximum likelihood (ML). The estimation method is defined by keyword METHOD=ML. Some of the other estimation methods available in SAS are LS, GLS, WLS (ADF). PROC CALIS uses several different optimization algorithms. TECH=NR requests that Newton-Raphson algorithm be used. MODIFICATION requests model modification be performed. It will produce LaGrange Multiplier and Wald tests. The number of observations is defined by NOBS=373.

There are three different ways to define a model in PROC CALIS: (1) COSAN, (2) RAM, and (3) LINEQS. All three will produce the same results. The LINEQS method is easiest to use for an applied researcher. It is based on EQS software model specification. There is an equation for each "dependent variable" in the model. The variance and covariance parameters are specified in the STD and COV sections. Each parameter to be estimated is labeled. For example, the prediction of fitness by exercise (v3 ← v1) was labeled b31.

There was a significant difference between the observed and model covariance matrices, $\chi^2(5) = 14.805$, $p = .011$ (Output 13.7.1). However, since the model χ^2 is sensitive against sample size, a model fit is tested by examining the fit indices. According to the GFI and AGFI, the model is "good". The CFI for the model is slightly below the cutoff of .95. The RMSEA for the model is above the cutoff of .06.

PROC CALIS produces model modification for each of the matrices in SEM. It produces model modifications in both matrix and rank order formats. According to the model modifications for the Roth data, fitness → stress and stress → fitness show highest significance (Output 13.7.2-13.7.4). The model modification Φ table exam-

Output 13.7.1: Path Analysis: Roth Data.

Fit Function	0.0398
Goodness of Fit Index (GFI)	0.9843
GFI Adjusted for Degrees of Freedom (AGFI)	0.9528
Root Mean Square Residual (RMR)	10887.4099
Parsimonious GFI (Mulaik, 1989)	0.4921
Chi-Square	14.8049
Chi-Square DF	5
Pr > Chi-Square	0.0112
Independence Model Chi-Square	165.50
Independence Model Chi-Square DF	10
RMSEA Estimate	0.0726
RMSEA 90% Lower Confidence Limit	0.0315
RMSEA 90% Upper Confidence Limit	0.1168
ECVI Estimate	0.0944
ECVI 90% Lower Confidence Limit	0.0730
ECVI 90% Upper Confidence Limit	0.1366
Probability of Close Fit	0.1591
Bentler's Comparative Fit Index	0.9369
Normal Theory Reweighted LS Chi-Square	14.0604
Akaike's Information Criterion	4.8049
Bozdogan's (1987) CAIC	-19.8030
Schwarz's Bayesian Criterion	-14.8030
McDonald's (1989) Centrality	0.9869
Bentler & Bonett's (1980) Non-normed Index	0.8739
Bentler & Bonett's (1980) NFI	0.9105
James, Mulaik, & Brett (1982) Parsimonious NFI	0.4553
Z-Test of Wilson & Hilferty (1931)	2.2788
Bollen (1986) Normed Index Rho1	0.8211
Bollen (1988) Non-normed Index Delta2	0.9389
Hoelter's (1983) Critical N	280

ines relations among the "independent" variables including errors. The model modification Γ examines relations between the "independent" and "dependent" variables. Finally, model modification B examines relations among "dependent" variables. Both paths should not be added. A path that can only be interpreted should be added. SEM software will produces model modifications regardless of conceptual understanding of a model. It is up to a researcher to decide which parameters should be added regardless of which parameter is most significant. Roth et al. (1989) hypothesized that stress would be predicted by fitness. This parameter is added and the model was rerun. For the revised model, there was still a significant difference between the observed and model covariance matrices, $\chi(4) = 9.650$, $p = .0474$. This model was significantly better than the earlier model, $\Delta \chi^2(1) = 14.805 - 9.650 = 5.155$, $p = .023$. The CFI for the revised model was above the cutoff of .95. The RMSEA was closer to the cutoff of .06.

Output 13.7.2: Path Analysis: Model Modification — Φ.

Rank Order of the 9 Largest Lagrange Multipliers in _PHI_			
Row	Column	Chi-Square	Pr > ChiSq
e4	e3	3.89607	0.0484
e3	v1	2.93112	0.0869
e3	v2	2.93112	0.0869
e5	e4	2.14100	0.1434
e5	v2	2.07004	0.1502
e4	v1	1.27281	0.2592
e4	v2	1.27281	0.2592
e5	e3	0.50247	0.4784
e5	v1	0.43525	0.5094

Output 13.7.3: Path Analysis: Model Modification — Γ.

Rank Order of the 4 Largest Lagrange Multipliers in _GAMMA_			
Row	Column	Chi-Square	Pr > ChiSq
v3	v2	2.93112	0.0869
v5	v2	2.12657	0.1448
v4	v1	1.27281	0.2592
v5	v1	0.49807	0.4804

The unstandardized and standardized parameter estimates are displayed in Output 13.7.5-13.7.6. According to the output, all regression coefficients were significant. The significance test of a parameter is listed in the unstandardized solu-

Output 13.7.4: Path Analysis: Model Modification — B.

Rank Order of the 4 Largest Lagrange Multipliers in _BETA_			
Row	Column	Chi-Square	Pr > ChiSq
v3	v4	5.35727	0.0206
v4	v3	5.09629	0.0240
v3	v5	4.02591	0.0448
v4	v5	0.49388	0.4822

tions. For example, there was a significant positive prediction of fitness by exercise, $B = .108, t = 8.169$. Since t-critical value is approximately equal to the z-critical value when $df \geq 30$, a t-test can be treated as a z-test. In fact, since most SEM software compute asymptotic standard errors, t-tests are really z-tests. The standardized parameter for the regression coefficient is .390 and about 15% of variability of the fitness is explained by exercise. There was no significant correlation between exercise and hardiness, $r = -.03$.

13.8 CONFIRMATORY FACTOR ANALYSIS

The CFA example is taken from Williams, Eaves, and Cox (2002). There were 10 variables from the Visual Similes Test II in the data set. They hypothesized a two-factor CFA model. The first 5 variables loaded on the first factor, affective form, while the remaining 5 variables loaded on the second factor, cognitive form. The SAS code to perform the analysis is given in Program 13_8.sas.

All 10 factor loadings were freely estimated. The factor variances were fixed at 1 for factor identification. There was a significant difference between the observed and model covariance matrices, $\chi^2(34) = 88.159, p < .0001$ (Output 13.8.1). Once again, the fit indices should be examined to evaluate model fit instead of model χ^2. According to the CFI and GFI, the model fit is "good". For AGFI and RMSEA, the model is "adequate".

All the factor loadings were significant (Output 13.8.2). According to the standardized solution, all the factor loadings were above .85. The correlation between the factors was .811.

13.9 GENERAL SEM

A 6 variables general SEM example is from Wheaton, Muthén, Alwin, and Summers (1977) study. They hypothesized a 3 factors model. The factor 1, alienation 1967, has two items loading on it: anomia and powerlessness 1967. The factor 2, alienation 1977, has also two items loading on it: anomia and powerlessness 1977. The factor 3, SES, has two items loading on it: education and occupation status index. The SES factor predicts both alienation 1967 and 1977 factors. The alienation

Output 13.7.5: Path Analysis: Unstandardized Solution.
Manifest Variable Equations with Estimates

```
v3        =  0.1079 * v1   +   1.0000   e3
Std Err      0.0132   b31
t Value      8.1689


v4        = -0.0417 * v3   +  -0.3914 * v2   + 1.0000 e4
Std Err      0.0182   b43      0.0884   b42
t Value     -2.2841           -4.4298


v5        = -8.4897 * v3   + 28.6760 * v4   + 1.0000 e5
Std Err      1.7281   b53      4.7557   b54
t Value     -4.9127            6.0298
```

Variances of Exogenous Variables				
Variable	Parameter	Estimate	Standard Error	t Value
v1	v1	4422	324.25510	13.64
v2	v2	14.44000	1.05879	13.64
e3	e3	287.06502	21.04863	13.64
e4	e4	41.93008	3.07446	13.64
e5	e5	371389	0	Infty

Covariances Among Exogenous Variables					
Var1	Var2	Parameter	Estimate	Standard Error	t Value
v1	v2	v1v2	-7.58100	13.10778	-0.58

Output 13.7.6: Path Analysis: Standardized Solution
Manifest Variable Equations with Standardized Estimates

$$v3 = 0.3900 * v1 + 0.9208 \quad e3$$
$$ b31$$

$$v4 = -0.1147 * v3 + -0.2225 * v2 + 0.9685 \ e4$$
$$ b43 b42$$

$$v5 = -0.2357 * v3 + 0.2893 * v4 + 0.9195 \ e5$$
$$ b53 b54$$

Squared Multiple Correlations				
	Variable	Error Variance	Total Variance	R-Square
1	v3	287.06502	338.56000	0.1521
2	v4	41.93008	44.70363	0.0620
3	v5	371389	439266	0.1545

Correlations Among Exogenous Variables			
Var1	Var2	Parameter	Estimate
v1	v2	v1v2	-0.03000

Output 13.8.1: Confirmatory Factor Analysis: Model Fit.

Fit Function	0.4100
Goodness of Fit Index (GFI)	0.9269
GFI Adjusted for Degrees of Freedom (AGFI)	0.8817
Root Mean Square Residual (RMR)	29.4151
Parsimonious GFI (Mulaik, 1989)	0.7003
Chi-Square	88.1592
Chi-Square DF	34
Pr > Chi-Square	<.0001
Independence Model Chi-Square	2455.1
Independence Model Chi-Square DF	45
RMSEA Estimate	0.0861
RMSEA 90% Lower Confidence Limit	0.0642
RMSEA 90% Upper Confidence Limit	0.1084
ECVI Estimate	0.6159
ECVI 90% Lower Confidence Limit	0.5027
ECVI 90% Upper Confidence Limit	0.7669
Probability of Close Fit	0.0045
Bentler's Comparative Fit Index	0.9775
Normal Theory Reweighted LS Chi-Square	84.7903
Akaike's Information Criterion	20.1592
Bozdogan's (1987) CAIC	-128.6003
Schwarz's Bayesian Criterion	-94.6003
McDonald's (1989) Centrality	0.8822
Bentler & Bonett's (1980) Non-normed Index	0.9703
Bentler & Bonett's (1980) NFI	0.9641
James, Mulaik, & Brett (1982) Parsimonious NFI	0.7284
Z-Test of Wilson & Hilferty (1931)	4.7047
Bollen (1986) Normed Index Rho1	0.9525
Bollen (1988) Non-normed Index Delta2	0.9776
Hoelter's (1983) Critical N	120

Output 13.8.2: Confirmatory Factor Analysis: Unstandardized Solution.

Manifest Variable Equations with Estimates

```
v1      = 40.6640 * f1    + 1.0000 e1
Std Err    2.3904   l11
t Value   17.0112

v2      = 39.6677 * f1    + 1.0000 e2
Std Err    2.3236   l21
t Value   17.0720

v3      = 34.0281 * f1    + 1.0000 e3
Std Err    1.8593   l31
t Value   18.3016

v4      = 39.1895 * f1    + 1.0000 e4
Std Err    2.4456   l41
t Value   16.0245

v5      = 43.2246 * f1    + 1.0000 e5
Std Err    2.3026   l51
t Value   18.7720

v6      = 21.4022 * f2    + 1.0000 e6
Std Err    1.3886   l62
t Value   15.4126

v7      = 23.8532 * f2    + 1.0000 e7
Std Err    1.4807   l72
t Value   16.1092

v8      = 28.4652 * f2    + 1.0000 e8
Std Err    1.8095   l82
t Value   15.7309

v9      = 25.1009 * f2    + 1.0000 e9
Std Err    1.6105   l92
t Value   15.5861

v10     = 24.1057 * f2    + 1.0000 e10
Std Err    1.4486   l102
t Value   16.6404
```

Output 13.8.3: Confirmatory Factor Analysis: Standardized Solution.

Manifest Variable Equations with Standardized Estimates

$$v1 = 0.9006 * f1 + 0.4347 \, e1$$
$$l11$$

$$v2 = 0.9024 * f1 + 0.4308 \, e2$$
$$l21$$

$$v3 = 0.9388 * f1 + 0.3445 \, e3$$
$$l31$$

$$v4 = 0.8689 * f1 + 0.4950 \, e4$$
$$l41$$

$$v5 = 0.9519 * f1 + 0.3064 \, e5$$
$$l51$$

$$v6 = 0.8523 * f2 + 0.5230 \, e6$$
$$l62$$

$$v7 = 0.8758 * f2 + 0.4826 \, e7$$
$$l72$$

$$v8 = 0.8632 * f2 + 0.5048 \, e8$$
$$l82$$

$$v9 = 0.8583 * f2 + 0.5132 \, e9$$
$$l92$$

$$v10 = 0.8930 * f2 + 0.4501 \, e10$$
$$l102$$

Output 13.8.4: Confirmatory Factor Analysis: Factor Correlation.

Correlations Among Exogenous Variables			
Var1	Var2	Parameter	Estimate
f1	f2	phi12	0.81077

Figure 13.6: General SEM: Conceptual Model.

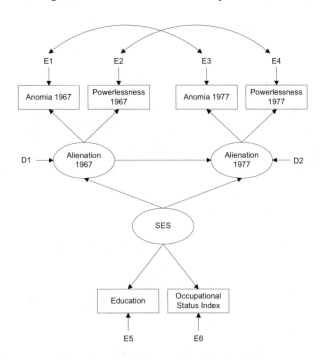

1967 factor also predicts the alienation 1977 factor. The conceptual model is displayed in Figure 13.6. The SAS code to perform the analysis is given in Program 13.9.sas.

Since alienation factors are the same just measured at two different time points (1967 and 1977), the factor loadings were constrained to be equal. To constrain parameters to be equal in PROC CALIS, just give the parameters the same label (in the example, `lp`). The factor loadings for anomia do not have to be constrained because they were both fixed at 1. By definition, they are the same.

There was a significant negative prediction of alienation 1967 by SES, $B = .5844$, $t = -10.407$, $R^2 = .318$ (Output 13.9.1). There was a significant positive prediction of alienation 1977 by alienation 1967 after adjusting for SES, $B = .5913$, $t = 12.474$. There was a significant negative prediction of alienation 1977 by SES after adjusting for alienation 1967, $B = -.222$, $t = 4.306$.

Output 13.9.1: General SEM: Unstandardized and Standardized Solutions.
Latent Variable Equations with Estimates

```
f1       = -0.5844 * f3   +   1.0000   d1
Std Err     0.0562   b13
t Value   -10.4070


f2       =  0.5913 * f1   + -0.2219 * f3   + 1.0000 d2
Std Err     0.0474   b21     0.0515   b23
t Value    12.4743          -4.3058
```

Latent Variable Equations with Standardized Estimates

```
f1 = -0.5641 * f3   +   0.8257   d1
                b13


f2 =  0.5685 * f1   + -0.2059 * f3   + 0.7088 d2
             b21               b23
```

Squared Multiple Correlations				
	Variable	Error Variance	Total Variance	R-Square
1	V1	4.61917	11.91331	0.6123
2	V2	2.68859	9.29211	0.7107
3	V3	4.56635	12.45786	0.6335
4	V4	2.92021	10.06455	0.7099
5	V5	2.81361	9.61000	0.7072
6	V6	2.64629	4.50288	0.4123
7	f1	4.97285	7.29414	0.3182
8	f2	3.96447	7.89151	0.4976

References

Agresti, A. (1990). *Categorical data analysis*. New York: Wiley. (Alan Agresti. ill. ; 24 cm. "A Wiley-Interscience publication.")

Akaike, H. (1969). Statistical predictor identification. *Annals of the Institute of Statistical Mathematics*, *22*, 203-217.

Akaike, H. (1973). Information theory and an extension of the maximum likelihood principle. In B. N. Petrov & F. Csaki (Eds.), *Second international symposium on information theory* (p. 267-281). Budapest: Akademia Kiado.

Akaike, H. (1974). A new look at the statistical model identification. *IEEE Transactions on Automatic Control*, *19*, 716-723.

Akaike, H. (1978). A bayesian analysis of the minimum aic procedure. *Annals of the Institute of Statistical Mathematics*, *30*, 9-14.

Al-Subaihi, A. A. (2000). *Utilizing the multivariate reduced risk rank regression model for selecting predictors in linear multivariate regression*. Phd thesis, University of Pittsburgh.

Anderson, R. L. (1975). Designs and estimators for variance components. In J. N. Strivastava (Ed.), *A survey of statistical designs and linear models* (p. 1-29). Amsterdam: North-Holland.

Anderson, R. L. (1981). Recent developments in designs and estimators for variance components. In M. Csörgo (Ed.), *Statistics and related topics* (p. 3-22). Amsterdam: North-Holland.

Anderson, T. W. (1984). *An introduction to multivariate statistical analysis* (2nd ed.). New York: Wiley.

Anderson, T. W., & Rubin, H. (1956). Statistical inference in factor analysis. *Proceedings of the Third Berkeley Symposium for Mathematical Statistics Problems*, *5*, 115-150.

Andrews, D. F., Gnanadesikan, R., & Warner, J. L. (1971). Transformation of multivariate data. *Biometrics*, *27*, 825-840.

Arbuckle, J. L. (1996). Full information estimation in the presence of incomplete data. In G. A. Marcoulides & R. E. Schumacker (Eds.), *Advanced structural equation modeling: issues and techniques* (p. 243-277). Mahwah, NJ: Lawrence Erlbaum Associates.

Arnold, S. F. (1981). *The theory of linear models and multivariate analysis*. New York: Wiley.

Barrett, B. E., & Ling, R. F. (1992). General classes of influence measures for multivariate regression. *Journal of the American Statistical Association, 87*(184-191).

Bartlett, M. S. (1939). A note on tests of significance in multivariate analysis. *In Proceedings of the Cambridge Philosphical Society, 35*(180-185).

Beal, S. L. (1989). Sample size determination for confidence intervals on the population mean and on the difference between two population means. *Biometrics, 45*, 969-977.

Beauducel, A., & Wittman, W. W. (2005). Simulation study on fit indexes in cfa based on data with slightly distorted simple structure. *Structural Equation Modeling, 12*, 41-75.

Bentler, P. M. (1986). Structural modeling and psychometrika: an historical perspective on growth and achievements. *Psychometrika, 51*, 35-51.

Bentler, P. M. (1990). Comparative fit indexes in structural models. *Psychological Bulletin, 107*, 238-246.

Bentler, P. M. (1995). *Eqs structural equations program manual*. Encino, CA: Multivariate Software.

Bentler, P. M. (in press). *Eqs 6 structural equations program manual*. Encino, CA: Multivariate Software.

Bentler, P. M., & Bonnett, D. G. (1980). Significance tests and goodness-of-fit in the analysis of covariance structures. *Psychological Bulletin, 88*, 558-600.

Bentler, P. M., & Dijkstra, T. (1985). Efficient estimation via linearization in structural models. In P. R. Krishnaiah (Ed.), *Multivariate analysis vi* (p. 9-42). Amsterdam: North-Holland.

Bentler, P. M., & Lee, S. Y. (1978). Statistical aspects of a three-mode factor analysis model. *Psychometrika, 43*, 343-352.

Bentler, P. M., & Weeks, D. G. (1980). Multivariate analysis with latent variables. In P. R. Krishnaiah & L. Kanal (Eds.), *Handbook of statistics* (Vol. 2, p. 747-771). Amsterdam: North-Holland.

Bentler, P. M., & Yuan, K. H. (1999). Structural equation modeling with small samples: test statistics. *Multivariate Behavioral Research, 34*, 181-197.

Bernaards, C. A., & Sijtsma, K. (2000). Influence of imputation and em methods on factor analysis when item nonresponse in questionnaire data is nonignorable. *Multivariate Behavioral Research, 35*, 321-364.

Besley, D. A., Kuh, E., & Welch, R. C. (1980). *Regression diagnostics*. New York: John Wiley and Sons.

Bhapkar, V. P. (1966). A note on the equivalence of two test criteria for hypotheses in categorical data. *Journal of the American Statistical Association, 61*, 228-235.

Bilodeau, M. (2002). Multivariate flattening for better predictions. *Canadian Journal of Statistics, 28*, 159-170.

Bilodeau, M., & Brenner, D. (1999). *Theory of multivariate statistics*. New York: Springer.

Bilodeau, M., & Kariya, T. (1989). Minimax estimators in the normal manova model. *Journal of multivariate analysis, 28*, 260-270.

Bishop, Y. M. M., Fienberg, S. E., & Holland, P. W. (1975). *Discrete multivariate analysis: theory and practice.* Cambridge, MA: MIT Press.

Blalock, H. M. (1964). *Causal inferences in nonexperimental research.* Chapel Hill, NC: University of North Carolina.

Blalock, H. M. (Ed.). (1971). *Causal models in the social sciences.* Chicago, IL: Aldine-Atherton.

Bock, R. D. (1989). *Multilevel analysis of educational data.* New York: Academic Press.

Bock, R. D., & Borgman, R. E. (1966). Analysis of covariance structures. *Psychometrika, 31*, 507-534.

Bock, R. D., & Peterson, A. C. (1975). A multivariate correction for attenuation. *Biometrika, 62*, 673-678.

Boik, R. J. (1988). The mixed model for multivariate repeated measures: validity conditions and an approximate test. *Psychometrika, 53*(4), 469-486.

Boik, R. J. (1991). Scheffé's mixed model for multivariate repeated measures: a relative efficiency evaluation. *Communications in Statistics, Part A - Theory and Methods, 20*, 1233-1255.

Boik, R. J. (1993). The analysis of two-factor interactions in fixed effects linear-models. *Journal of Educational Statistics, 18*(1), 1-40. (Times Cited: 16 Article English Cited References Count: 31 Nj434)

Bollen, K. A. (1989). *Structural equations with latent variables.* New York: Wiley.

Bollen, K. A. (2002). Latent variables in psychology and the social sciences. *Annual Review of Psychology, 53*, 605-634.

Box, G. E. P. (1949). A general distribution theory for a class of likelihood criteria. *Biometrika, 36*, 317-346.

Box, G. E. P. (1950). Problems in the analysis of growth and wear curves. *Biometrics, 6*, 362-389.

Box, G. E. P. (1954). Some theorems on quadratic forms applied in the study of analysis of variance problems, ii. effects of inequality of variance and correlation between errors in the two-way classification. *Annals of Mathematical Statistics, 25*, 484-498.

Box, G. E. P., & Cox, D. R. (1964). An analysis of transformation. *Journal of the Royal Statistical Society, Series B, Methodological, 26*, 211-243.

Box, G. E. P., Hunter, W. G., & Hunter, J. S. (1978). *Statistics for experimenters : an introduction to design, data analysis, and model building.* New York: Wiley.

Bradu, D., & Gabriel, K. R. (1974). Simultaneous statistical inference on interactions in two-way analysis of variance. *Journal of the American Statistical Association, 69*, 428-436.

Breiman, L., & Friedman, J. H. (1997). Predicting multivariate responses in multiple linear regression. *Journal of the Royal Statistical Society, Series B, Methodological, 59*, 3-37.

Breusch, T. S. (1979). Conflict among criteria for testing hypotheses: extensions and comments. *Econometrica, 47*, 203-207.

Bristol, D. R. (1989). Sample sizes for constructing confidence intervals and testing hypotheses. *Statistics in Medicine, 8*, 803-811.

Brown, R. L. (1994). Efficacy of the indirect approach for estimating structural equation models with missing data: a comparison of five methods. *Structural Equation Modeling, 1*, 287-316.

Browne, M. B., & Forsythe, A. B. (1974). The small sample behavior of some statistics which test the equality of several means. *Technometrics, 16*, 129-132.

Browne, M. W. (1974). Generalized least-squares estimators in the analysis of covariance structures. *South African Statistical Journal, 8*, 1-24.

Browne, M. W. (1975a). A comparison of single sample and cross-validation methods for estimating the mean squared error of prediction in multiple linear regression. *British Journal of Mathematical and Statistical Psychology, 28*(1), 112-120.

Browne, M. W. (1975b). Predictive validity of a linear regression equation. *British Journal of Mathematical and Statistical Psychology, 28*(1), 79-87.

Browne, M. W. (1982). Covariance structures. In D. M. Hawkins (Ed.), *Topics in multivariate analysis* (p. 72-141). Cambridge: Cambridge University Press.

Browne, M. W. (1984). Asymptotically distribution-free methods for the analysis of covariance structures. *British Journal of Mathematical & Statistical Psychology, 37*(1), 62-83.

Bryk, A. S., & Raudenbush, S. W. (1992). *Hierarchical linear models: applications and data analaysis methods.* Newbury Park, CA: Sage Publications.

Burdick, R. K., & Graybill, F. A. (1992). *Confidence intervals on variance components.* New York: Marcel Dekker.

Burket, G. R. (1963). A study of reduced rank models for multiple prediction. *Dissertation Abstracts, 23*(12, pt. 1), 4740.

Burnham, K. P., & Anderson, D. R. (1998). *Model selection and inference: a practical information-theoretic approach.* New York: Springer.

Buse, A. (1982). The likelihod ratio, wald, and lagrange multiplier tests: an expository note. *American Statistician, 36*, 153-157.

Casella, G., & Berger, R. L. (1990). *Statistical inference.* North Scituate, MA: Duxbury Press.

Casella, G., & Berger, R. L. (1994). Estimation with selected binomial information or do you really believe that dave winfield is batting .471? *Journal of the American Statistical Association, 89*, 1080-1090.

Chambers, J. M. (1977). *Computational methods for data analysis.* New York: John Wiley.

Chatterjee, S., & Hadi, A. S. (1988). *Sensitivity analysis in linear regression.* New York: Wiley.

Chattogpadhyay, A. K., & Pillai, K. C. S. (1971). Asymptotic formulae for the distributions of some criteria for tests of equality of covariance matrices. *Journal of multivariate analysis, 1*, 215-231.

Chinchilli, V. M., & Elswick, R. K. (1985). A mixture of the manova and gmanova models. *Communications in Statistics, Part A - Theory and Methods, 14*, 3075-3089.

Chou, C. P., & Bentler, P. M. (1990). Model modification in covariance structure modeling: a comparison among likelihood ratio, lagrange multiplier, and wald tests. *Multivariate Behavioral Research, 25*(1), 115-136.

Chou, C. P., & Bentler, P. M. (2002). Model modification in structural equation modeling by imposing constraints. *Computational Statistics & Data Analysis, 41*, 271-287.

Chou, C. P., Bentler, P. M., & Pentz, M. A. (1998). Comparisons of two statistical approaches to study growth curves: The multilevel model and the latent curve analysis. *Structural Equation Modeling, 5*(3), 247-266.

Cochran, W. C. (1977). *Sampling techniques* (3rd ed.). New York: John Wiley.

Cohen, J. (1988). *Statistical power analysis for the behavioral sciences.* Hillsdale, NJ: Lawrence Erlbaum Associates.

Cole, J. W. L., & Grizzle, J. E. (1966). Applications of multivariate analysis of variance to repeated measurements experiments. *Biometrics, 22*, 810-828.

Cook, R. D., & Weisberg, S. (1994). *An introduction to regression graphics.* New York: Wiley.

Cornell, J. E., Young, D. M., Seaman, S. L., & Kirk, R. E. (1992). Power comparisons of eight tests for sphericity in repeated measure design. *Journal of Educational Statistics, 17*, 233-249.

Cox, C. M., Fang, C., Boudreau, R. M., & Timm, N. H. (1994). *Computer program for krishnaiah's finite intersection tests for multiple comparisons for mean vectors* (Interim Report No. 94-04). University of Pittsburgh.

Cox, C. M., Krishnaiah, P. R., Lee, J. C., Reising, J., & Schuurmann, F. J. (1980). A study of finite intersection tests for multiple comparisons of means. In P. R. Krishnaiah (Ed.), *Multivariate analysis* (Vol. V, p. 435-466). New York: North-Holland.

Coxhead, P. (1974). Measuring the relationship between two sets of variables. *British Journal of Mathematical and Statistical Psychology, 34*, 205-212.

Cramer, E. M., & Nicewander, W. (1979). Some symmetric, invariant measures of multivariate association. *Psychometrika, 44*(1), 43-54.

Curran, P. J. (2003). Have multilevel models been structural equation models all along? *Multivariate Behavioral Research, 38*(4), 529-569.

Danford, M. B., Hughes, H. M., & McNee, R. C. (1960). On the analysis of repeated measurements experiments. *Biometrics, 16*, 547-565.

Davidian, M., & Giltinan, D. M. (1995). *Nonlinear models for repeated measurement data.* New York: Chapman and Hall.

Davidson, M. L. (1988). *The multivariate approach to repeated measures, bmdp technical report #75* (Tech. Rep.). BMDP Statistical Software, Inc.

Davisson, L. D. (1965). The prediction error of stationary gaussian time series of unknown covariance. *IEEE Transactions Information Theory, 11*, 527-532.

Dean, A., & Voss, D. T. (1997). *Design and analysis of experiments.* New York: Springer-Verlag.

De Leeuw, J., & Kreft, I. G. G. (1986). Random coefficient models for multilevel analysis. *Journal of Educational Statistics, 11*, 57-85.

Dempster, A. P., Laird, N. M., & Rubin, D. B. (1977). Maximum likelihood from incomplete data via the em algorithm. *Journal of the Royal Statistical Society, Series B, Methodological, 39*, 1-22.

Dempster, A. P., Rubin, D. B., & Tsutakawa, R. K. (1981). Estimation in covariance components models. *Journal of the American Statistical Association, 76*, 341-353.

Dempster, A. P., Schatzoff, M., & Wermuth, M. (1977). A simulation study of alternatives to ordinary least-squares. *Journal of the American Statistical Association, 72*, 77-106.

Dickey, J. M. (1967). Matricvariate generalizations of the multivariate t distribution and the inverted multivariate t distribution. *The Annals of Mathematical Statistics, 38*, 511-518.

Diggle, P. J., Liang, K.-Y., & Zeger, S. L. (1994). *Analysis of longitudinal data.* Oxford: Clarendon Press.

Dobson, A. J. (1990). *An introduction to generalized linear models* (1st ed.). London; New York: Chapman and Hall.

Draper, D. (1995). Inference and hierarchical modeling in the social sciences. *Journal of Educational and Behavioral Statistics, 20*, 115-147.

Draper, N. R., & Smith, H. (1998). *Applied regression analysis* (3rd ed.). New York: Wiley.

Duncan, O. D. (1966). Path analysis: sociological examples. *American Journal of Sociology, 72*, 1-16.

Dunlop, D. D. (1994). Regression for longitudinal data: A bridge from least square regression. *The American Statistician, 48*, 299-303.

Dunn, O. J. (1958). Estimation of means of dependent variables. *Annals of Mathematical Statistics, 29*, 1095-1111.

Dunnett, C. W. (1955). A multiple comparison procedure for comparing several treatments with a control. *Journal of the American Statistical Association, 50*, 1096-1121.

Dunnett, C. W., & Tamhane, A. C. (1992). A step-up multiple test procedure. *Journal of the American Statistical Association, 87*(417), 162-170.

Elston, R. C., & Grizzle, J. E. (1962). Estimation of time response curves and their confidence bands. *Biometrics, 18*, 148-159.

Enders, C. K., & Bandalos, D. L. (2001). The relative performance of full information maximum likelihood estimation for missing data in structural equation models. *Structural Equation Modeling, 8*, 430-457.

Ericsson, N. R. (1994). Testing exogeneity: an introduction. In N. R. Ericsson & J. S. Irons (Eds.), *Testing exogeneity* (p. 3-38). Oxford: Oxford University Press.

Everson, P. J., & Morris, C. N. (2000). Inference for multivariate normal hierarchical models. *Journal of the Royal Statistical Society, Series B, Methodological, 62*, 399-412.

Fai, A. H.-T., & Cornelius, P. L. (1996). Approximate f-tests of multiple degrees of freedom hypotheses in generalized least squares analyses of unbalanced split-plot experiments. *Journal of Statistical Computation and Simulation, 54,* 363-378.

Fan, X., & Sivo, S. A. (2005). Sensitivity of fit indexes to misspecified structural or measurement model components: rationale of two-index strategy revisited. *Structural Equation Modeling, 12,* 343-367.

Feldstein, M. S. (1971). The error of forecast in econometric models when the forecast-period exogenous variables are stochastic. *Econometrica, 39,* 55-60.

Feller, W. (1967). *An introduction to probability theory and its applications* (3rd ed.). New York: Wiley.

Ferguson, T. S. (1958). A method of generating best asymptotically normal estimates with application to the estimation of bacteria densities. *Annals of mathematical statistics, 29,* 1046-1062.

Fisher, R. A. (1924). The influence of rainfall on the yield of wheat at rothamstead. *Philosophical Transactions of the Royal Society of London, Series B, 213,* 89-142.

Fisher, R. A. (1925). *Statistical methods for research workers.* Edinburgh and London: Oliver and Boyd.

Fisher, R. A. (1935). *The design of experiments.* Edinburgh and London: Oliver and Boyd.

Fisher, R. A. (1936). The use of multiple measurements in taxonomic problems. *Annals of Eugenics, 7,* 179-188.

Freund, R. J., & Littell, R. C. (1991). *Sas system for regression* (2nd ed.). Cary, NC: SAS Institute. ([Rudolf J. Freund, Ramon C. Littell]. ill. ; 28 cm. "56141"– Cover.)

Fujikoshi, Y. (1974). On the asymptotic non-null distributions of the lr criterion in a general manova. *The Canadian Journal of Statistics, 2,* 1-12.

Fujikoshi, Y. (1982). A test for additional information in canonical correlation analysis. *Annals of the Institute of Statistical Mathematics, 34,* 523-530.

Fujikoshi, Y. (1985). Selection of variables in discriminant analysis and canonical correlation analysis. In P. R. Krishnaiah (Ed.), *Multivariate analysis - iv.* New York: Elsevier Science Publishers.

Fujikoshi, Y., & Satoh, K. (1997). Modified aic and c_p in multivariate linear regression. *Biometrika, 84,* 707-716.

Gabriel, K. R. (1969). Simultaneous test procedures-sum theory of multiple comparison. *Annals of Mathematical Statistics, 40,* 224-250.

Galecki, A. T. (1994). General-class of covariance-structures for 2 or more repeated factors in longitudinal data-analysis. *Communications in Statistics-Theory and Methods, 23*(11), 3105-3119.

Gatsonis, C., & Sampson, A. R. (1989). Multiple correlation: exact power and sample size calculations. *Psychological Bulletin, 106*(3), 516-524.

Gelfand, A. E., Sahu, S. K., & Carlin, B. P. (1995). Efficient parameterisations for normal linear mixed models. *Biometrika, 82,* 479-488.

Ghosh, M., & Sinha, B. K. (2002). A simple derivation of the wishart distribution. *The American Statistician, 56*(2), 100-101.

Gillett, R. (2003). The metric comparability of meta-analytic effect size estimators from factorial designs. *Psychological Methods*, *8*, 419-433.

Gleser, L. J., & Olkin, I. (1970). Linear models in multivariate analysis. In *Essays in probability and statistics* (p. 267-292). New York: Wiley.

Glueck, D. H., & Muller, K. E. (2003). Adjusting power for a baseline covariate in linear models. *Statistics in Medicine*, *22*(16), 2535-2551.

Gnanadesikan, R., & Kettenring, J. R. (1972). Robust estimates, residuals, and outlier detection with multiresponse data. *Biometrics*, *28*, 81-124.

Gold, M. S., & Bentler, P. M. (2000). Treatments of missing data: A monte carlo comparison of rbhdi, iterative stochastic regression imputation, and expectation-maximization. *Structural Equation Modeling*, *7*(3), 319-355.

Gold, M. S., Bentler, P. M., & Kim, K. H. (2003). A comparison of maximum-likelihood and asymptotically distribution-free methods of treating incomplete nonnormal data. *Structural Equation Modeling*, *10*(1), 47-79.

Goldberger, A. S. (1964). *Econometric theory*. New York: Wiley.

Goldberger, A. S. (1971). Econometrics and psychometrics: a survey of communalities. *Psychometrika*, *36*, 83-107.

Goldberger, A. S. (1991). *A course in econometrics*. Cambridge, MA: Harvard University Press.

Goldberger, A. S., & Duncan, O. D. (Eds.). (1973). *Structural equation models in the social sciences*. New York: Seminar.

Goldstein, H. (1986). Multilevel mixed linear model analysis using iterative generalized least squares. *Biometrika*, *73*, 43-56.

Graybill, F. A. (1976). *Theory and application of the linear model*. North Scituate, MA: Duxbury Press.

Graybill, F. A., & Wang, C. M. (1980). Confidence intervals for nonnegative linear combinations of variances. *Journal of the American Statistical Association*, *75*, 869-873.

Green, S. B., Thompson, M. S., & Babyak, M. A. (1998). A monte carlo investigation of methods for controlling type i errors with specification searches in structural equation modeling. *Multivariate Behavioral Research*, *33*(3), 365-383.

Greene, W. H. (1992). *Econometric analysis* (2nd ed.). New York: MacMillan Publishing, Inc.

Grizzle, J. E., & Allen, D. M. (1969). Analysis of growth and dose response curves. *Biometrics*, *25*, 357-381.

Grizzle, J. E., Starmer, C. F., & Koch, G. G. (1969). Analysis of categorical data by linear models. *Biometrics*, *25*, 489-504.

Gupta, A. K., & Nagar, D. K. (2000). *Matrix variate distributions*. Boca Raton, FL: Chapman and Hall.

Hancock, G. R. (1999). A sequential scheffé-type respecification procedure for controlling type i error in exploratory structural equation model modification. *Structural Equation Modeling*, *6*, 158-168.

Hannan, E. J., & Quinn, B. G. (1979). The determination of the order of an autoregression. *Journal of the Royal Statistical Society, Series B, Methodological, 41*, 190-195.

Harville, D. A. (1976). Extension of the gauss-markov theorem to include the estimation of random effects. *The Annals of Statistics, 4*, 384-395.

Harville, D. A. (1977). Maximum likelihood approaches to variance component estimation and to related problems. *Journal of the American Statistical Association, 72*, 320-340.

Harville, D. A. (1997). *Matrix algebra from a statistician's perspective*. New York: Springer.

Hayter, A. J., & Hsu, J. C. (1994). On the relationship between stepwise decision procedures and confidence sets. *Journal of the American Statistical Association, 89*(425), 128-136.

Hecker, H. (1987). A generalization of the gmanova-model. *Biometrical Journal, 29*, 763-770.

Hedges, L. V. (1981). Distribution theory for glass's estimator of effect size and related estimators. *Journal of Educational Statistics, 6*, 107-128.

Hedges, L. V., & Olkin, I. (1985). *Statistical methods for meta-analysis*. New York: Academic Press.

Hedges, L. V., & Vevea, J. L. (1998). Fixed- and random-effects models in meta-analysis. *psychological Methods, 3*(4), 486-504.

Helland, I. S., & Almoy, T. (1994). Comparison of prediction methods when only a few components are relevant. *Journal of the American Statistical Association, 89*, 583-591.

Hess, A. M., & Iyer, H. (2002). *A sas/iml macro for computation of confidence intervals for variance components for mixed models*. SAS Users Group International.

Hochberg, Y. (1988). A sharper bonferroni procedure for multiple tests of significance. *Biometrika, 75*, 800-802.

Hochberg, Y., & Tamhane, A. C. (1987). *Multiple comparison procedures*. New York: Wiley.

Holm, S. (1979). A simple sequentially rejective multiple test procedure. *Scandinavian Journal of Statistics, 6*, 65-70.

Hooper, J. W. (1959). Simultaneous equations and canonical correlation theory. *Econometrica, 27*, 245-256.

Horst, P. (1966). *Psychological measurement and prediction*. Belmont, CA: Wadsworth Publishing Co.

Hossain, A., & Naik, D. N. (1989). Detection of influential observations in multivariate regression. *Journal of Applied Statistics, 16*, 25-37.

Hotelling, H. (1931). The generalization of student's ratio. *Annals of Mathematical Statistics, 2*, 360-378.

Hotelling, H. (1936). Relations between two sets of variates. *Biometrika, 28*, 3221-377.

Hotelling, H. (1975). The generalization of student's ratio. In E. H. Bryant & W. R. Atchley (Eds.), *Multivariate statistical methods: among-groups covariation; within-group covariation* (p. 19-37). New York: Halstead Press.

Howe, W. G. (1974). Approximate confidence limits on the mean of x+y where x and y are two tabled independent random variables. *Journal of the American Statistical Association*, *69*, 789-794.

Hu, L.-t., & Bentler, P. M. (1998). Fit indices in covariance structure modeling: Sensitivity to underparameterized model misspecification. *Psychological Methods*, *3*(4), 424-453.

Hu, L.-t., & Bentler, P. M. (1999). Cutoff criteria for fit indexes in covariance structure analysis: Conventional criteria versus new alternatives. *Structural Equation Modeling*, *6*(1), 1-55.

Huberty, C. J. (1994). *Applied discriminant analysis*. New York: Wiley.

Huynh, H., & Feldt, L. S. (1970). Conditions under which mean square ratios in repeated measurements design have exact f distributions. *Journal of the American Statistical Association*, *65*, 1582-1589.

Jackson, D. L. (2003). Revisiting sample size and number of parameter estimates: some support for the n:q hypothesis. *Structural Equation Modeling*, *10*, 128-141.

Jamshidian, M., & Bentler, P. M. (1999). Ml estimation of mean and covariance structures with missing data using complete data routines. *Journal of Educational and Behavioral Statistics*, *24*(1), 21-41.

Jennrich, R. I., & Schluchter, M. D. (1986). Unbalanced repeated-measures models with structured covariance matrices. *Biometrics*, *42*, 805-820.

Jiroutek, M. R., Muller, K. E., Kupper, L. L., & Stewart, P. W. (2003). A new method for choosing sample size for confidence interval-based inferences. *Biometrics*, *59*(3), 580-590.

Jobson, J. D. (1991). *Applied multivariate data analysis*. New York: Springer-Verlag.

John, S. (1971). Some optimal multivariate tests. *Biometrika*, *58*, 123-127.

Johnson, N. L., & Kotz, S. (1972). *Continuous multivariate distributions*. New York: Wiley.

Jöreskog, K. G. (1969). A general approach to confirmatory maximum likelihood factor analysis. *Psychometrika*, *34*, 183-202.

Jöreskog, K. G. (1973). A general method for estimating a linear structural equation system. In A. S. Goldberger & O. D. Duncan (Eds.), *Structural equation models in the social sciences* (p. 85 - 112). New York: Academic Press.

Jöreskog, K. G., & Goldberger, A. S. (1972). Factor analysis by generalized least squares. *Psychometrika*, *37*, 243-260.

Jöreskog, K. G., & Sörbom, D. (1981). *Lisrel v: analysis of linear structural relationships by the method of maximum likelihood*. Chicago: National Educational Resources.

Jöreskog, K. G., & Sörbom, D. (1984). *Lisrel vi user's guide* (3rd ed.). Mooresville, IN: Scientific Software.

Kabe, D. G. (1975). Generalized manova double linear hypothesis with double linear restrictions. *The Canadian Journal of Statistics*, *3*, 35-44.

Kabe, D. G. (1981). Manova double linear hypothesis with double linear restrictions. *Communications in Statistics, Part A - Theory and Methods*, *10*, 2545-2550.

Kackar, R. N., & Harville, D. A. (1984). Approximations for standard errors of estimators of fixed and random effects in mixed linear models. *Journal of the American Statistical Association, 79*, 853-862.

Kandane, J. B., & Lazar, N. A. (2004). Methods and criteria for model selection. *Journal of the American Statistical Association, 99*, 279-290.

Kariya, T. (1985). *Testing in the multivariate general linear model.* Tokyo, Japan: Kinokuniya Co.

Kass, R. E., & Steffey, D. (1989). Approximation bayesian inference in conditionally independent hierarchical models (parametric empirical bayes models). *Journal of the American Statistical Association, 84*, 717-726.

Keesing, J. W. (1972). *Maximum likelihood approaches to causal analysis.* Phd dissertation, University of Chicago.

Kenny, D. A., & McCoach, D. B. (2003). Effect of the number of variables on measures of fit in structural equation modeling. *Structural Equation Modeling, 10*, 333-351.

Kenward, M. G., & Roger, J. H. (1997). Small sample inference for fixed effects from restriction maximum likelihood. *Biometrics, 53*, 983-997.

Keppel, G., & Wickens, T. D. (2004). *Design and analysis : a researcher's handbook.* Upper Saddle River, N.J.: Pearson Prentice Hall.

Kerridge, D. (1967). Errors of prediction in multiple linear regression with stochastic regressor variables. *Technometrics, 9*, 309-311.

Khatri, C. G. (1966). A note on a manova model applied to problems in growth curve. *Annal Inst. Statistical Mathematics, 18*, 75-86.

Khattree, R., & Naik, D. N. (1995). *Applied multivariate statistics with sas software.* Carey, NC: SAS Institute.

Kim, K. H., & Bentler, P. M. (2002). Tests of homogeneity of means and covariance matrices for multivariate incomplete data. *Psychometrika, 67*(4), 609-624.

Kim, S. H., & Cohen, A. S. (1998). On the behrens-fisher problem: a review. *Journal of Educational and Behavioral Statistics, 23*, 356-377.

Kimball, A. W. (1951). On dependent tests of significance in the analysis of variance. *Annals of Mathematical Statistics, 22*, 600-602.

Kirk, R. E. (1982). *Experimental design: procedure for the behavioral sciences* (2nd ed.). Belmont, CA: Brooks/Cole Publishing Company.

Kirk, R. E. (1995). *Experimental design: procedure for the behavioral sciences* (3rd ed.). Belmont, CA: Brooks/Cole Publishing Company.

Kleinbaum, D. G. (1970). *Estimation and hypothesis listing for generalized multivariate linear models.* Ph.d., University of North Carolina, Chapel Hill.

Kleinbaum, D. G. (1973). A generalization of the growth curve model which allows missing data. *Journal of Multivariate Analysis, 3*, 117-124.

Kleinbaum, D. G., Kupper, L. L., & Muller, K. E. (1998). *Applied regression analysis and other mutlivariate methods* (3rd ed.). Belmont, CA: Duxbury Press.

Klockars, A. J., Hancock, G. R., & Krishnaiah, P. R. (2000). Scheffé's more powerful f-protected post hoc procedure. *Journal of Educational and Behavioral Statistics, 25*, 13-19.

Kmenta, J., & Gilbert, R. F. (1968). Small sample properties of alternative estimators of seemingly unrelated regressions. *Journal of the American Statistical Association, 63,* 1180-1200.

Kreft, I. G. G., De Leeuw, J., & Leeden, R. van der. (1994). Review of five multilevel analysis program: Bmdp-5v, genmod, hlm, ml3, varcl. *The American Statistician, 48,* 324-335.

Krishnaiah, P. R. (1964). *Multiple comparisons test in the multivariate case* (ARL No. 64-124). Aerospace Research Laboratories, Air Force System Command, U.S. Air Force.

Krishnaiah, P. R. (1965a). Multiple comparison tests in multiresponse experiments. *Sankhya, Series A, 21,* 65-72.

Krishnaiah, P. R. (1965b). On multivariate generalizations of the simultaneous analysis of variance test. *Annals of the Institute of Statistical Mathematics, 17,* 167-173.

Krishnaiah, P. R. (1965c). On the simultaneous anova and mancova tests. *Annals of the Institute of Statistical Mathematics, 17,* 35-53.

Krishnaiah, P. R. (1969). Simultaneous test procedures under general manova models. In P. R. Krishnaiah (Ed.), *Multivariate analysis ii* (p. 121-143). New York: Academic Press.

Krishnaiah, P. R. (1979). Some developments on simultaneous test procedure: Anova and mancova tests. In P. R. Krishnaiah (Ed.), *Developments in statistics* (Vol. 2, p. 157-201). New York: North-Holland.

Krishnaiah, P. R. (1982). *Handbook of statistics, vol 1: analysis of variance.* New York: North-Holland.

Krishnaiah, P. R., & Armitage, J. V. (1970). On a multivariate f distribution. In I. M. Chakravarti, P. C. Mahalanobis, C. R. Rao, & K. J. C. Smith (Eds.), *Essays in probability and statistics* (p. 439-468). Chapel Hill, NC: University of North Carolina Press.

Krishnaiah, P. R., Mudholkar, G. S., & Subbaiah, P. (1980). Simultaneous test procedures for mean vectors and covariance matrices. In P. R. Krishnaiah (Ed.), *Handbook of statistics vol 1: analysis of variance* (p. 631-671). New York: North-Holland.

Kshirsagar, A. M., & Smith, W. B. (1995). *Growth curves.* New York: M. Dekker.

Laird, N. M., Lange, N., & Stram, D. (1987). Maximum likelihood computations with repeated measures: application of the em algorithm. *Journal of the American Statistical Association, 82,* 97-105.

Laird, N. M., & Ware, J. H. (1982). Random-effects models for longitudinal data. *Biometrics, 38,* 963-974.

Lawley, D. N. (1938). A generalization of fisher's z-test. *Biometrika, 30,* 180-187.

Lawley, D. N. (1940). The estimation of factor loadings by the method of maximum likelihood. *Proceedings of the Royal Society of Edinburgh, 60,* 64-82.

Lee, S. Y. (1971). Some results on the sampling distribution of the multiple correlation coefficient. *Journal of the Royal Statistical Society, Series B, Methodological, 33,* 117-130.

Lee, S. Y. (1972). Tables of upper percentage points of the multiple correlation coefficient. *Biometrika, 59,* 805-820.

Lee, S.-Y. (1985). On testing functional constraints in structural equation models. *Biometrika, 72*, 125-131.

Lee, S.-Y., & Bentler, P. M. (1980). Some asymptotic properties of constrained generalized least squares estimation in covariance structure models. *South African Statistical Journal, 14*, 121-136.

Lee, S.-Y., & Jennrich, R. I. (1979). A study of algorithms for covariance structure analysis with specific comparisons using factor analysis. *Psychometrika, 44*, 99-113.

Lee, S.-Y., & Poon, W.-Y. (1985). Further developments on constrained estimation in the analysis of covariance structures. *The Statistician, 34*, 305-316.

Lehmann, E. L. (1991). *Testing statistical hypotheses* (2nd ed.). Pacific Grove, CA: Wadsworth and Brooks/Cole Advanced Books and Software.

Lei, M., & Lomax, R. G. (2005). The effect of varying degrees of nonnormality in structural equation modeling. *Structural Equation Modeling, 12*, 1-27.

Lewis, T., & Taylor, L. R. (1967). *Introduction to experimental ecology*. New York: Academic Press.

Liang, K.-Y., & Zeger, S. L. (1986). Longitudinal data analysis using generalized linear models. *Biometrika, 73*, 13-22.

Lindsey, J. K. (1993). *Models for repeated measurements*. Oxford: Clarendon Press.

Lindstrom, M. J., & Bates, D. M. (1988). Newton-raphson and em algorithm for linear mixed effects models for repeated-measures data. *Journal of the American Statistical Association, 83*, 1014-1022.

Liski, E. P. (1985). Estimation from incomplete data in growth curves models. *Communications in Statistics, Part B - Simulation and Computation, 14*, 13-27.

Littell, R. C., Milliken, G. A., Stroup, W. W., & Wolfinger, R. D. (1996). *Sas system for mixed models*. Cary, N.C.: SAS Institute Inc. (Ramon C. Littell ... [et al.]. ill. ; 28 cm. Includes index.)

Littell, R. C., & Portier, K. M. (1992). Robust parametric confidence bands for distribution functions. *Journal of Statistical Computation and Simulation, 41*, 187-200.

Little, R. J. A. (1988). A test of missing completely at random for multivariate data with missing values. *Journal of the American Statistical Association, 83*, 1198-1202.

Little, R. J. A. (1992). Regression with missing x's: a review. *Journal of the American Statistical Association, 87*, 1227-1237.

Little, R. J. A., & Rubin, D. B. (1987). *Statistical analysis with missing data*. New York: Wiley.

Little, R. J. A., & Rubin, D. B. (1989). The analysis of social science data with missing values. *Sociological Methods and Research, 18*, 292-326.

Little, R. J. A., & Rubin, D. B. (2002). *Statistical analysis with missing data* (2nd ed.). Hoboken, N.J.: Wiley.

Little, R. J. A., & Schenker, N. (1995). Missing data. In G. Arminger, C. C. Clogg, & M. E. Sobel (Eds.), *Handbook of statistical modeling for the social and behavioral sciences* (p. 39-75). New York: Plenum Press.

Liu, C., & Rubin, D. B. (1994). The ecme algorithm: a simple extension of em and ecm with faster monotone convergence. *Biometrika, 81*, 633-648.

Lohr, S. L. (1995). Optimal bayesian design of experiments for the one-way random effect model. *Biometrika, 26*, 175-186.

Longford, N. T. (1987). A fast scoring algorithm for maximum likelihood estimation in unbalanced mixed models with nested random effects. *Biometrika, 74*, 817-827.

Longford, N. T. (1993). *Random coefficient models*. Oxford: Clarendon Press.

Lord, F. M. (1950). Efficiency of prediction when a regression equation from a sample is used in a new sample. In *Research bulletin* (p. 50-54). Princeton, NJ: Educational Testing Services.

MacCallum, R. C., Browne, M. W., & Sugawara, H. M. (1996). Power analysis and determination of sample size for covariance structural modeling. *Psychological Methods, 1*, 130-149.

MacCallum, R. C., Roznowski, M., & Necowitz, L. B. (1992). Model modification in covariance structural analysis: The problem of capitalization on chance. *Psychological Bulletin, 111*, 490-504.

MacCallum, R. C., Widaman, K. F., Preacher, K. J., & Hong, S. (2001). Sample size in factor analysis: the role of model error. *Multivariate Behavioral Research, 36*, 611-637.

Magnus, J. R., & Neudecker, H. (1988). *Matrix differential calculus with applications in statistics and econometrics*. New York: Wiley.

Mallows, C. L. (1973). Some comments on c_p. *Technometrics, 15*, 661-675.

Mallows, C. L. (1995). More comments on c_p. *Technometrics, 37*, 362-372.

Mantel, N. (1970). Why stepdown procedures in variable selection. *Technometrics, 12*, 621-625.

Marascuilo, L. A., & McSweeney, M. (1977). *Nonparametric and distribution-free methods for the social sciences*. Monterey, CA: Brooks/Cole Pub. Co.

Marchand, E. (1997). On moments of beta mixtures, the noncentral beta distribution, and the coefficient of determination. *Journal of Statistical Computation and Simulation, 59*, 161-178.

Marcus, R., Peritz, E., & Gabriel, K. R. (1976). On closed testing procedures with special reference to ordered analysis of variance. *Biometrika, 63*, 655-660.

Mardia, K. V. (1970). Measures of multivariate skewness and kurtosis with applications. *Biometrika, 57*, 519-530.

Mardia, K. V. (1974). Applications of some measures of multivariate skewness and kurtosis in testing normality and robustness studies. *Sankhyā Ser. B, 36*(2), 115–128.

Marsh, H. W. (1998). Pairwise deletion for missing data in structural equation models: nonpositive definite matrices, parameter estimates, goodness of fit, and adjusted sample size. *Structural Equation Modeling, 5*, 22-36.

Marsh, H. W., Hau, K. T., & Wen, Z. (2004). In search of golden rules: comment on hypothesis-testing approaches to setting cutoff values for fit indexes and danger in overgeneralizing hu and bentler's (1999) findings. *Structural Equation Modeling, 11*, 391-410.

Mauchly, J. W. (1940). Significance test for sphericity of a normal n-variate distribution. *Annals of Mathematical Statistics*, *11*, 204-209.

Maxwell, S. E. (2000). Sample size and multiple regression analysis. *Psychological Methods*, *5*(4), 434-458.

McAleer, M. (1995). The significance of testing empirical non-nested models. *Journal of Econometrics*, *67*, 149-171.

McBride, J. B. (2000). *Adequacy of aproximations to distributions of test statistics in complex mixed linear models*. Phd thesis, Brigham Young University.

McCullagh, P., & Nelder, J. A. (1989). *Generalized linear models* (2nd ed.). New York: Chapman & Hall.

McDonald, R. P. (1989). An index of goodness-of-fit based on noncentrality. *Journal of Classification*, *6*, 97-103.

McElroy, F. W. (1967). A necessary and sufficient condition that ordinary least-squares estimators be best linear unbiased. *Journal of the American Statistical Association*, *62*, 1302-1304.

McKay, R. J. (1977). Variable selection in multivariate regression: an application of simultaneous test procedures. *Journal of the Royal Statistical Society, Series B, Methodological*, *39*, 371-380.

McLachlan, G. J., & Krishnaiah, P. R. (1997). *The em algorithm and extensions.* New York: Wiley.

McLean, R. M., & Sanders, W. L. (1988). Approximating degrees of freedom for standard errors in mixed linear models. In *Proceedings of the statistical computing section* (p. 50-59). New Orleans: American Statistical Association.

McLean, R. M., Sanders, W. L., & Stroup, W. W. (1991). A unified approach to mixed linear models. *The American Statistician*, *45*, 54-64.

McQuarrie, A. D. R., & Tsai, C.-L. (1998). *Regression and time series model selection*. Singapore River Edge, NJ: World Scientific.

Mendoza, J. L. (1980). A significance test for multisample sphericity. *Psychometrika*, *45*, 495-498.

Mensah, R. D., Elswick, R. K., & Chinchilli, V. M. (1993). Consistent estimators of the variance-covariance matrix of the gmanova model with missing data. *Communications in Statistics, Part A - Theory and Methods*, *22*, 1495-1514.

Merwe, A. van der, & Zidek, J. V. (1980). Multivariate regression analysis and canonical variates. *The Canadian Journal of Statistics*, *8*, 27-39.

Miller, A. J. (2002). *Subset selection in regression* (2nd ed.). Boca Raton, FL: Chapman and Hall.

Miller, R. G. J. (1981). *Simultaneous statistical inference* (2nd ed.). New York: Springer-Verlag.

Milliken, G. A. (2003). Multilevel designs and their analysis. In *Sas user's group international*.

Milliken, G. A., & Johnson, D. E. (1984). *Analysis of messy data*. Belmont, CA: Lifetime Learning Publications.

Milliken, G. A., & Johnson, D. E. (1992). *Analysis of messy data: Designed experiments* (Vol. 1). New York: Chapman and Hall.

Milliken, G. A., & Johnson, D. E. (2001). *Analysis of messy data: analysis of covariance* (Vol. 3). New York: Chapman & Hall.

Mudholkar, G. S., Davidson, M. L., & Subbaiah, P. (1974). Extended linear hypotheses and simultaneous tests in multivariate analysis of variance. *Biometrika, 61*, 467-477.

Mudholkar, G. S., & Subbaiah, P. (1980a). Manova multiple comparisons associated with finite intersection tests. In P. R. Krishnaiah (Ed.), *Multivariate analysis* (Vol. V, p. 467-482). New York: North-Holland.

Mudholkar, G. S., & Subbaiah, P. (1980b). A review of step-down procedures for multivariate analysis of variance. In R. P. Gupa (Ed.), *Multivariate statistical analysis* (p. 161-179). New York: North-Holland.

Muirhead, R. J. (1982). *Aspects of multivariate statistical theory*. New York: Wiley.

Muller, K. E., & Fetterman, B. A. (2002). *Regression and anova: an integrated approach to using sas software*. Cary, NC: SAS Institute.

Muller, K. E., LaVange, L. M., Ramey, S. L., & Ramey, C. T. (1992). Power calculations for general linear multivariate models including repeated measures applications. *Journal of the American Statistical Association, 87*, 1209-1226.

Muller, K. E., & Peterson, B. L. (1984). Practical methods for computing power in testing the multivariate general linear hypothesis. *Computational Statistics and Data Analysis, 2*, 143-158.

Naik, D. N., & Khattree, R. (1996). Revisiting olympic track records: Some practical considerations in the principal component analysis. *The American Statistician, 50*, 140-144.

Nelson, P. R. (1993). Additional uses for the analysis of means and extended tables of critical values. *Technometrics, 35*, 61-71.

Neter, J., Kutner, M. H., Nachtsheim, C. J., & Wasserman, W. (1996). *Applied linear statistical models* (4th ed.). Boston, MA: McGraw-Hill.

Neter, J., Wasserman, W., & Kutner, M. H. (1990). *Applied linear statistical models: regression, analysis of variance, and experimental designs* (3rd ed.). Homewood, IL: Irwin. (John Neter, William Wasserman, Michael H. Kutner. ill. ; 25 cm. "Free instructor's copy ... not for sale"–Cover.)

Neyman, J. (1949). Contribution to the theory of the chi-square test. In *Proceedings of the berkeley symposium on mathematical statistics and probability* (p. 239-273).

Nicholson, G. E. (1960). Prediction in future samples. In I. Olkin, S. G. Ghurye, W. Hoeffding, W. G. Madow, & H. B. Mann (Eds.), *Contributions to probability and statistics* (p. 322-330). Stanford, CA: Stanford University Press.

Ogasawara, H. (1999). Standard errors for matrix correlations. *Multivariate Behavioral Research, 34*(1), 103-122.

Ogasawara, H. (2001). Approximations to the distributions of fit indices for misspecified structural equation models. *Structural Equation Modeling, 8*, 556-574.

Olson, C. L. (1974). Comparative robustness of six tests in multivariate analysis of variance. *Journal of the American Statistical Association, 69*, 894-908.

Orchard, T., & Woodbury, M. A. (1972). A missing information principle: theory and application. In *Sixth berkeley symposium on mathematical statistics and probability* (p. 697-715). Berkeley: University of California Press.

Park, T. (1993). Equivalence of maximum likelihood estimation and iterative two-stage estimation for seemingly unrelated regression. *Communications in Statistics, Part A - Theory and Methods, 22*, 2285-2296.

Park, T., & Woolson, R. F. (1992). Generalized multivariate models for longitudinal data. *Communications in Statistics, Part B - Simulation and Computation, 21*, 925-946.

Patel, H. I. (1986). Analysis of repeated measures designs with changing covariates in clinical trials. *Biometrika, 73*, 707-715.

Picard, R. R., & Berk, K. N. (1990). Data splitting. *The American Statistician, 44*, 140-147.

Picard, R. R., & Cook, R. D. (1984). Cross-validation of regression models. *Journal of the American Statistical Association, 79*, 575-583.

Pillai, K. C. S., & Jaysachandran, K. (1967). Power comparisons of tests of two multivariate hypotheses based on four criteria. *Biometrika, 54*, 195-210.

Potthoff, R. F., & Roy, S. N. (1964). A generalized multivariate analysis of variance model useful especially for growth curve problems. *Biometrika, 51*, 313–326.

Raju, N. S., Bilgic, R., Edwards, J. E., & Fleer, P. F. (1997). Methodology review: estimation of population validity and cross-validity, and the use of equal weights in prediction. *Applied Psychological Measurement, 21*(4), 291-305.

Rao, C. R. (1965). *Linear statistical inference and its applications.* New York: John Wiley & Sons Inc.

Rao, C. R. (1966). Covariance adjustment and related problems in multivariate analysis. *Multivariate Analysis (Proc. Internat. Sympos., Dayton, Ohio, 1965)*, 87–103.

Rao, C. R. (1967). Least squares theory using an estimated dispersion matrix and its application to measurement of signals. In *Fifth berkeley symposium on mathematical statistics and probability* (Vol. 1, p. 335-371). Berkeley: University of California Press.

Rao, C. R. (1973). *Linear statistical inference and its application* (2nd ed.). New York: Wiley.

Rao, C. R. (1987). Prediction of future observations in growth curve models. *Statistical Sciences, 2*, 434-471.

Raudenbush, S. W. (1993). Hierarchical linear modes and experimental design. In L. K. Edwards (Ed.), *Applied analysis of variance in behavioral science* (p. 459-496). New York: Marcel Dekker, Inc.

Raudenbush, S. W. (1997). Statistical analysis and optimal design for cluster randomized trials. *Psychological Methods, 2*, 173-185.

Raudenbush, S. W., & Bryk, A. S. (2002). *Hierarchical linear models: applications and data analysis methods* (2nd ed.). Thousand Oaks, CA: Sage Publications.

Reinsel, G. C. (1982). Multivariate repeated-measurement or growth curve models with multivariate random-effects covariance structure. *Journal of the American Statistical Association, 77*, 190-195.

Reinsel, G. C. (1984). Estimation and prediction in a multivariate random effects generalized linear model. *Journal of the American Statistical Association, 79,* 406-414.

Reinsel, G. C., & Velu, R. P. (1998). *Multivariate reduced-rank regression: theory and application.* New York: Springer.

Rencher, A. C. (1998). *Multivariate statistical inference and applications.* New York: Wiley.

Rencher, A. C., & Scott, D. T. (1990). Assessing the contributions of individual variables following rejection of a multivariate hypothesis. *Communications in Statistics, Part B - Simulation and Computation, 19,* 535-553.

Robert, P., & Escoufier, Y. (1976). A unifying tool for linear multivariate statistical methods: the rv-coefficient. *Applied Statistics, 25,* 257-265.

Robins, J. M., Rotnitzky, A., & Zhao, L. P. (1995). Analysis of semiparametric regression models for repeated outcomes in the presence of missing data. *Journal of the American Statistical Association, 90,* 106-121.

Robinson, P. M. (1974). Identification, estimation and large-sample theory for regression containing unobservable variables. *International Economic Review, 15,* 680-692.

Roecker, E. B. (1991). Prediction error and its estimation for subset-selected models. *Technometrics, 33,* 459-468.

Romeu, J. L., & Ozturk, A. (1993). A comparative study of goodness-of-fit tests for multivariate normality. *Journal of Multivariate Analysis, 46,* 309-334.

Rosen, D. von. (1989). Maximum likelihood estimators in multivariate linear normal models. *Journal of Multivariate Analysis, 31,* 187-200.

Rosen, D. von. (1991). The growth curve model: a overview. *Communications in Statistics, Part A - Theory and Methods, 20,* 2791-2822.

Rosenthal, R. (1994). The counternull value of an effect size - a new statistic. *Psychological Science, 5(6),* 329-334.

Rosenthal, R., Rosnow, R. L., & Rubin, D. B. (2000). *Contrasts and effect sizes in behaviroal research: a correlational approach.* New York: Cambridge University Press.

Rosenthal, R., & Rubin, D. B. (2003). r(equivalent): A simple effect size indicator. *Psychological Methods, 8(4),* 492-496.

Roth, D. L., Wiebe, D. J., Fillingim, R. B., & Shay, K. A. (1989). Life events, fitness, hardiness, and health: a simultaneous analysis of proposed stress-resistance effects. *Journal of Personality and Social Psychology, 57,* 136-142.

Roy, S. N. (1953). On a heuristic method of test construction and its use in multivariate analysis. *The Annals of Mathematical Statistics, 24,* 220-238.

Roy, S. N. (1958). Step-down procedure in multivariate analysis. *Annals of Mathematical Statistics, 29,* 1177-1187.

Roy, S. N., & Bose, R. C. (1953). Simultaneous confidence interval estimation. *Annals of Mathematical Statistics, 24,* 513-536.

Royston, J. P. (1982). An extension of shapiro and wilk's w test for normality to large samples. *Applied Statistics, 31,* 115-124.

Royston, J. P. (1992). Which measures of skewness and kurtosis are best? *Statistics in Medicine, 11,* 333-343.

Rozeboom, W. W. (1965). Linear correlations between sets of variables. *Psychometrika, 30*(1), 57-71.

Rubin, D. B. (1987). *Multiple imputation for nonresponse in surveys*. New York: John Wiley & Sons, Inc.

Sampson, A. R. (1974). A tale of two regressions. *Journal of the American Statistical Association, 69*, 682-689.

SAS. (1992). *Sas/stat software: changes and enhacements release 6.07, the mixed procedure* (Tech. Rep. No. P-229). SAS Institute Inc.

SAS. (1993). *Sas/stat software: the genmod procedure release 6.09* (Tech. Rep. No. P-243). SAS Institute Inc.

Satorra, A., & Bentler, P. M. (1988). Scaling corrections for chi-square statistics in covariance structural analysis. In *Proceedings of the business and economic statistics section of the american statistical association* (p. 308-313).

Satorra, A., & Bentler, P. M. (1994). Corrections to test statistics and standard errors in covariance structure analysis. In A. von Eye & C. C. Clogg (Eds.), *Latent variables analysis: Applications for developmental research* (p. 399-419). Thousand Oaks, CA: Sage Publications, Inc.

Satterthwaite, F. E. (1941). Synthesis of variance. *Psychometrika, 6*, 309-316.

Schatzoff, M. (1966). Sensitivity comparisons among tests of the general linear hypothesis. *Journal of the American Statistical Association, 61*, 415-435.

Scheffé, H. (1947). The relationship of control charts to analysis of variance and chi-square tests. *Journal of the American Statistical Association, 42*, 425-431.

Scheffé, H. (1953). A method for judging all contrasts in the analysis of variance. *Biometrika, 40*, 87–104.

Scheffé, H. (1959). *The analysis of variance*. New York: Wiley.

Scheffé, H. (1970). Multiple testing versus multiple estimation, improper confidence sets, estimation of direction and ratios. *The Annals of Mathematical Statistics, 41*, 1-29.

Schmidhammer, J. L. (1982). On the selection of variables under regression models using krishnaiah's finite intersection tests. In P. R. Krishnaiah & L. N. Kanal (Eds.), *Handbook of statistics* (Vol. 2, p. 821-833). Amsterdam: North-Holland.

Schmukle, S. C., & Hardt, J. (2005). A cautionary note on incremental fit indices reported by lisrel. *Methodology, 1*, 81-85.

Schott, J. R. (1997). *Matrix analysis of statistics*. New York: Wiley.

Schuurmann, F. J., Krishnaiah, P. R., & Chattogpadhyay, A. K. (1975). Tables for a multivariate f distribution. *Sankhya, Series B, 37*, 308-331.

Schwarz, G. (1978). Estimating the dimension of a model. *The Annals of Statistics, 6*, 461-464.

Searle, S. R. (1971). *Linear models*. New York: Wiley.

Searle, S. R. (1987). *Linear models for unbalanced data*. New York: Wiley.

Searle, S. R. (1994). Analysis of variance computing package output for unbalanced data from fixed effects models with nested factors. *The American Statistician, 48*, 148-153.

Searle, S. R., Casella, G., & McCulloch, C. E. (1992). *Variance components*. New York: Wiley.

Seber, G. A. F. (1984). *Multivariate observations*. New York: Wiley.

Serlin, R. C., & Lapsley, D. K. (1985). Rationality in psychological theory: the good-enough principle. *American Psychologist, 40*, 73-82.

Serlin, R. C., & Lapsley, D. K. (1993). Rational appraisal of psychological research and the good-enough principle. In G. Keren & C. Lewis (Eds.), *A handbook for data analysis in the behavioral sciences: methodological issues* (p. 129-228). Hillsdale, NJ: Erlbaum.

Shaffer, J. P. (1977). Multiple comparisons emphasizing selected contrasts: an extension and generalization of dunnett's procedure. *Biometrics, 33*, 293-304.

Shaffer, J. P. (1986). Modified sequentially rejective multiple test procedures. *Journal of the American Statistical Association, 81*, 826-831.

Shaffer, J. P. (2002). Multiplicity, directional (type iii) errors, and the null hypothesis. *Psychological Methods, 7*(3), 356-369.

Shaffer, J. P., & Gillo, M. W. (1974). A multivariate extension of the correlation ratio. *Educational Psychological Measurement, 34*, 521-524.

Shapiro, S. S., & Wilk, M. B. (1965). An analysis of variance test for normality. *Biometrika, 52*(591-611).

Shapiro, S. S., Wilk, M. B., & Chen, H. (1968). A comparative study of various test for normality. *Journal of the American Statistical Association, 63*, 1343-1372.

Shrout, P. E., & Fleiss, J. L. (1979). Intraclass correlation: uses in assessing interrater reliability. *Psychological Bulletin, 86*, 420-428.

Singh, A. (1993). Omnibus robust procedures for assessment of multivariate ormlaity and detection of multivariate outliers. In G. P. Patil & C. R. Rao (Eds.), *Multivariate environmental statistics* (p. 445-488).

Small, C. (1978). Sums of powers in arithmetic progressions. *Canad. Math. Bull., 21*(4), 505–506.

Smith, H., Gnanadesikan, R., & Hughes, J. B. (1962). Multivariate analysis of variance (manova). *Biometrics, 18*, 22-41.

Snijders, T. A. B., & Bosker, R. L. (1999). *Multilevel analysis: an introduction to basic and advanced multilevel modeling*. Thousand Oaks, CA: Sage Publication.

Sörbom, D. (1989). Model modification. *Psychometrika, 54*, 371-384.

Sparks, R. S., Coutsourides, D., Troskie, L., & Sugiura, N. (1983). The multivariate c_p. *Communications in Statistics, Part A - Theory and Methods, 12*, 1775-1793.

Spearman, C. (1904). General intelligence, objectively determined and measured. *American Journal of Psychology, 15*, 201-293.

Srikantan, K. S. (1970). Canonical association between nominal measurements. *Journal of the American Statistical Association, 65*, 284-292.

Srivastava, M. S. (2001). Nested growth curve models. *Sankhya, Series A, 64*, 379-408.

Srivastava, M. S., & Bilodeau, M. (1989). Stein estimation under elliptical distributions. *Journal of multivariate analysis, 28*, 247-259.

Srivastava, M. S., & Carter, E. M. (1983). *An introduction to applied multivariate statistics*. New York: North-Holland.

Srivastava, M. S., & Khatri, C. G. (1979). *An introduction to multivariate statistics*. New York: North-Holland/New York.

Srivastava, M. S., & Rosen, D. von. (1999). Growth curve models. *Multivariate analysis, design of experiments, and survey sampling, 159*, 547–578.

Stanek, E. J. I., & Koch, G. G. (1985). The equivalence of parameter estimates from growth curve models and seemingly unrelated regression models. *The American Statistician, 39*, 149-152.

Stapleton, J. H. (1995). *Linear statistical models*. New York: Wiley.

Stefánsson, G., Kim, W. C., & Hsu, J. C. (1988). On confidence sets in multiple comparisons. In S. S. Gupta & J. O. Berger (Eds.), *Statistical decision theory and related topics iv* (p. 89-104). New York: Academic Press.

Steiger, J. H. (1990). Structural model evaluation and modification: an internal estimation approach. *Multivariate Behavioral Research, 25*, 173-180.

Steiger, J. H., & Lind, J. C. (1980). Statistically based tests for the number of common factors. In *The annual meeting of the psychometric society.* Iowa City, IA.

Stein, C. (1960). Multiple regression. In I. Olkin, S. G. Ghurye, W. Hoeffding, W. G. Madow, & H. B. Mann (Eds.), *Contributions to probability and statistics*. Stanford, CA: Stanford University Press.

Stevens, J. (1996). *Applied multivariate statistics for the social sciences* (3rd ed.). Mahwah, N.J.: Lawrence Erlbaum Associates.

Stewart, D., & Love, W. (1968). A general canonical correlation index. *Psychological Bulletin, 70*(3), 160-163.

Stokes, M. E., Davis, C. S., & Koch, G. G. (1995). *Categorical data analysis using the sas system*. Cary, NC: SAS Institute. (Maura E. Stokes, Charles S. Davis, Gary G. Koch. 28 cm. "55320" on spine. Includes index.)

Stone, M. (1974). Cross-validatory choice and assessment of statistical predictions. *Journal of the Royal Statistical Society, Series B, Methodological, 36*, 111-147.

Stuart, A., & Ord, J. K. (1991). *Kendall's advanced theory of statistics* (5th ed., Vol. 2). New York: The Clarendon Press Oxford University Press.

Subbaiah, P., & Mudholkar, G. S. (1978). A comparison of two tests of significance of mean vectors. *Journal of the American Statistical Association, 73*, 414-418.

Sugiura, N. (1972). Locally best invariant test for sphericity and the limiting distribution. *Annals of Mathematical Statistics, 43*, 1312-1316.

Sugiura, N. (1978). Further analysis of the data by akaike's information criterion and the finite correction. *Communications in Statistics, Part A - Theory and Methods, 7*, 13-26.

Swamy, P. (1971). *Statistical inference in random coefficient regression models*. New York: Springer-Verlag.

Takeuchi, K., Yanai, H., & Mukherjee, B. N. (1982). *The foundations of multivariate analysis: a unified approach by means of projection onto linear subspaces*. New York: Wiley.

Tanaka, J. S. (1993). Multifaceted conceptions of fit in structural equation models. In K. A. Bollen & J. S. Long (Eds.), *Testing structural equation models* (p. 136-162). Newbury Park, CA: Sage.

Tatsuoka, M. M., & Lohnes, P. R. (1988). *Multivariate analysis: techniques for educational and psychological research* (2nd ed.). New York: MacMillan.

Taylor, D. J., & Muller, K. E. (1995). Computing confidence bounds for power and sample size of the general linear univariate model. *The American Statistician, 49,* 43-47.

Taylor, D. J., & Muller, K. E. (1996). Bias in linear model power and sample size calculations due to estimating noncentrality. *Communications in Statistics, Part A - Theory and Methods, 25,* 1595-1610.

Telser, L. G. (1964). Iterative estimation of a set of linear regression equations. *Journal of the American Statistical Association, 59,* 845-862.

Theil, H. (1971). *Principles of econometrics.* New York: Wiley.

Thomas, D. R. (1983). Univariate repeated measures techniques applied to multi-variate data. *Psychometrika, 48,* 451-464.

Thompson, R. (1973). The estimation of variance and covariance components with an application when records are subject to culling. *Biometrics, 29,* 527-550.

Thum, Y. M. (1997). Hierarchical linear models for multivariate outcomes. *Journal of Educational and Behavioral Statistics, 22,* 77-108.

Tibshirani, R. (1996). Regression shrinkage and selection via the lasso. *Journal of the Royal Statistical Society, Series B, Methodological, 58,* 267-288.

Timm, N. H. (1970). The estimation of variance-covariance and correlation matrices from incomplete data. *Psychometrika, 35*(417-437).

Timm, N. H. (1975). *Multivariate analysis with applications in education and psychology.* Monterey, CA: Brooks/Cole Publication Co.

Timm, N. H. (1980a). The analysis of nonorthogonal manova designs employing a restricted full rank multivariate linear model. *Multivariate statistical analysis,* 257-273.

Timm, N. H. (1980b). Multivariate analysis of variance of repeated measurements. In P. R. Krishnaiah (Ed.), *Handbook of statistics* (Vol. 1 - Analysis of Variance, p. 41-87). New York: North-Holland.

Timm, N. H. (1993a). Manova and mancova: an overview. In G. Keren & C. Lewis (Eds.), *A handbook for data analysis in the behavioral sciences: statistical issues.* Mahwah, NJ: Lawrence Erlbaum Associates, Inc.

Timm, N. H. (1993b). *Multivariate analysis with applications in education and psychology.* Oberlin, OH: The Digital Printshop.

Timm, N. H. (1995). Simultaneous inference using finite intersection tests: a better mousetrap. *Multivariate Behavioral Research, 30,* 461-512.

Timm, N. H. (1996). A note on the manova model with double linear restrictions. *Communications in Statistics, Part A - Theory and Methods, 25,* 1391-1395.

Timm, N. H. (2002). *Applied multivariate analysis.* New York: Springer.

Timm, N. H., & Al-Subaihi, A. A. (2001). Testing model specification in seemingly unrelated regression models. *Communications in Statistics, Part A - Theory and Methods, 30,* 579-590.

Timm, N. H., & Carlson, J. E. (1975). Analysis of variance through full rank models. *Multivariate Behavioral Research Monograph, 75*(1), 120.

Timm, N. H., & Lin, J. (2005). *Computer program for krishnaiah's finite intersection test using visual fortran* (Unpublished manuscript). University of Pittsburgh.

Timm, N. H., & Mieczkowski, T. (1997). *Univariate & multivariate general linear models: theory and applications using sas.* Cary, NC: SAS Institute.

Ting, N., Burdick, R. K., Graybill, F. A., Jeyaratnam, S., & Lu, T.-F. C. (1990). Confidence intervals on linear combinations of variance components that are unrestricted in sign. *Journal of Statistical Computation and Simulation, 35,* 135-143.

Toothaker, L. E. (1991). *Multiple comparisons for researchers.* Newbury Park, CA: Sage Publications.

Troendle, J. F. (1995). A stepwise resampling method of multiple hypothesis testing. *Journal of the American Statistical Association, 90,* 370-378.

Tubbs, J. D., Lewis, T. O., & Duran, B. S. (1975). A note on the analysis of the manova model and its application to growth curves. *Communications in Statistics, Part A - Theory and Methods, 4*(7), 643–653.

Tukey, J. W. (1949). Comparing individual means in the analysis of variance. *Biometrics, 5,* 99-114.

Tukey, J. W. (1953). The problem of multiple comparisons. *Unpublished Manuscript.*

Šidák, Z. (1967). Rectangular confidence regions for the means of multivariate normal distributions. *Journal of the American Statistical Association, 62,* 626-633.

Velilla, S., & Barrio, J. A. (1994). A discriminant rule under transformation. *Technometrics, 36,* 348-353.

Verbyla, A. P. (1986). Conditioning in the growth curve model. *Biometrika, 73,* 475-483.

Verbyla, A. P. (1988). Analysis of repeated measures designs with changing covariates. *Biometrika, 75,* 172-174.

Verbyla, A. P., & Venables, W. N. (1988). An extension of the growth curve model. *Biometrika, 75,* 129-138.

Vonesh, E. F., & Carter, R. L. (1987). Efficient inference for random-coefficient growth curve models with unbalanced data. *Biometrics, 43,* 617-628.

Vonesh, E. F., & Chinchilli, V. M. (1997). *Linear and nonlinear models for the analysis of repeated measurements.* New York: M. Dekker.

Wald, A. (1943). Tests of statistical hypothesis concerning serveral parameters when the number of variables is large. *Transactions of the American Mathematical Society, 54,* 426-482.

Ware, J. H. (1985). Linear models for the analysis of longitudinal studies. *The American Statistician, 39,* 95-101.

Wegge, L. L. (1971). The finite sampling distribution of least squares estimators with stochastic regressors. *Econometrica, 39,* 241-251.

Weng, L.-J., & Cheng, C. P. (1997). Why might relative fit indices differ between estimators? *Structural Equation Modeling, 4,* 121-128.

Werts, C. E., & Linn, R. L. (1970). Path analysis: Psychological examples. *Psychological Bulletin, 74*(3), 193-212.

Westfall, P. H., & Young, S. S. (1993). On adjusting p-values for multiplicity. *Biometrics, 49*, 941-944.

Wheaton, B., Muthén, B. O., Alwin, D. F., & Summers, G. F. (1977). Assessing reliability and stability in panel models. *Sociological Methodology, 8*, 84-136.

Wiley, D. E. (1973). The identification problem for structural equation models with unmeasured variables. In A. S. Goldberger & O. D. Duncan (Eds.), *Structural equation models in the social sciences* (p. 69-83). New York: Academic Press.

Wilks, S. S. (1932). Certain generalizations in the analysis of variance. *Biometrika, 24*, 471-494.

Williams, T. O. J., Eaves, R. C., & Cox, C. (2002). Confirmatory factor analysis of an instrument designed to measure affective and cognitive arousal. *Educational and Psychological Measurement, 62*, 264-283.

Winer, B. J. (1971). *Statistical principles in experimental design* (2d ed.). New York: McGraw-Hill.

Wolfinger, R., Tobias, R., & Sall, J. (1994). Computing gaussian likelihoods and their derivatives for general linear mixed models. *SIAM Journal on Scientific and Statistical Computing, 15*, 1294-1310.

Woolson, R. F., & Clarke, W. R. (1987). Estimation of growth norms from incomplete longitudinal data. *Biometrical Journal, 29*, 937-952.

Wright, S. (1918). On the nature of size factors. *Genetics, 3*, 367-374.

Wright, S. (1921). Correlation and causation. *Journal of Agricultural Research, 20*, 557-585.

Wright, S. (1934). The method of path coefficients. *Annals of mathematical statistics, 5*, 161-215.

Wright, S. (1960). Path coefficients and path regressions: alternative or complementary concepts? *Biometrics, 16*, 189-202.

Yanai, H. (1974). Unification of various techniques of multivariate analysis by means of generalized coefficient of determination. *Japanese Journal of Behaviormetrics, 1*, 46-54.

Yuan, K. H. (2005). Fit indices versus test statistics. *Multivariate Behavioral Research, 40*, 115-148.

Yuan, K. H., & Bentler, P. M. (1997). Finite sample distribution-free test statistics for nested structural models. *Behaviormetrika, 24*(1), 19-26.

Yuan, K. H., & Bentler, P. M. (1998). Normal theory based test statistics in structural equation modeling. *British Journal of Mathematical and Statistical Psychology, 51*, 289-309.

Yuan, K. H., & Bentler, P. M. (2000). Three likelihood-based methods for mean and covariance structure analysis with nonnormal missing data. *Sociological Methodology*, 165-200.

Zellner, A. (1962). An efficient method of estimating seemingly unrelated regressions and tests for aggregation bias. *Journal of the American Statistical Association, 57*, 348-368.

Zellner, A. (1963). Estimators for seemingly unrelated regression equations: some exact finite sample results. *Journal of the American Statistical Association*, *58*, 977-992.

Author Index

Subject Index

QA279/.T56/2007
Univariate and multivariate general linear models :
theory and applications with SAS.